An Overview of Science & The Scientific Method

Edited by Paul F. Kisak

Contents

1 Science   1
    1.1 History 1
        1.1.1 Antiquity 2
        1.1.2 Medieval science 3
        1.1.3 Renaissance and early modern science 3
        1.1.4 Age of Enlightenment 5
        1.1.5 19th century 5
        1.1.6 20th century and beyond 5
    1.2 Scientific method 6
        1.2.1 Mathematics and formal sciences 7
    1.3 Scientific community 7
        1.3.1 Branches and fields 8
        1.3.2 Institutions 8
        1.3.3 Literature 8
    1.4 Science and society 9
        1.4.1 Women in science 9
        1.4.2 Science policy 9
        1.4.3 Media perspectives 10
        1.4.4 Political usage 10
        1.4.5 Science and the public 11
    1.5 Philosophy of science 11
        1.5.1 Certainty and science 12
        1.5.2 Fringe science, pseudoscience, and junk science 13
    1.6 Scientific practice 13
        1.6.1 Basic and applied research 14
        1.6.2 Research in practice 14
        1.6.3 Practical impacts of scientific research 14
    1.7 See also 14
    1.8 Notes 14

	1.9	References	17
	1.10	Sources	20
	1.11	Further reading	20
	1.12	External links	21

2 Formal science — 22

	2.1	History	22
	2.2	Differences from other forms of science	22
	2.3	See also	23
	2.4	References	23
	2.5	Further reading	23
	2.6	External links	23
	2.7	References	23

3 Logic — 24

	3.1	Concepts	24
		3.1.1 Logical form	24
		3.1.2 Semantics	25
		3.1.3 Inference	25
		3.1.4 Logical systems	26
		3.1.5 Logic and rationality	26
		3.1.6 Rival conceptions	27
	3.2	History	27
	3.3	Types	28
		3.3.1 Syllogistic logic	28
		3.3.2 Propositional logic	29
		3.3.3 Predicate logic	29
		3.3.4 Modal logic	29
		3.3.5 Informal reasoning and dialectic	30
		3.3.6 Mathematical logic	30
		3.3.7 Philosophical logic	30
		3.3.8 Computational logic	31
		3.3.9 Non-classical logic	31
	3.4	Controversies	32
		3.4.1 "Is Logic Empirical?"	32
		3.4.2 Implication: Strict or material	32
		3.4.3 Tolerating the impossible	32
		3.4.4 Rejection of logical truth	33
	3.5	See also	33

3.6	Notes and references	33
3.7	Bibliography	35
3.8	External links	36

4 Mathematics   37

4.1	History	38
	4.1.1 Etymology	39
4.2	Definitions of mathematics	39
	4.2.1 Mathematics as science	40
4.3	Inspiration, pure and applied mathematics, and aesthetics	41
4.4	Notation, language, and rigor	42
4.5	Fields of mathematics	42
	4.5.1 Foundations and philosophy	43
	4.5.2 Pure mathematics	43
	4.5.3 Applied mathematics	45
4.6	Mathematical awards	45
4.7	See also	45
4.8	Notes	46
4.9	Footnotes	46
4.10	References	47
4.11	Further reading	48
4.12	External links	48

5 Mathematical logic   50

5.1	Subfields and scope	50
5.2	History	50
	5.2.1 Early history	51
	5.2.2 19th century	51
	5.2.3 20th century	52
5.3	Formal logical systems	53
	5.3.1 First-order logic	53
	5.3.2 Other classical logics	54
	5.3.3 Nonclassical and modal logic	54
	5.3.4 Algebraic logic	54
5.4	Set theory	54
5.5	Model theory	55
5.6	Recursion theory	56
	5.6.1 Algorithmically unsolvable problems	56
5.7	Proof theory and constructive mathematics	56

		5.8	Applications	56
		5.9	Connections with computer science	57
		5.10	Foundations of mathematics	57
		5.11	See also	58
		5.12	Notes	58
		5.13	References	58
			5.13.1 Undergraduate texts	58
			5.13.2 Graduate texts	59
			5.13.3 Research papers, monographs, texts, and surveys	59
			5.13.4 Classical papers, texts, and collections	60
		5.14	External links	61
6	**Mathematical statistics**			**63**
	6.1	Introduction		63
	6.2	Topics		63
		6.2.1	Probability distributions	63
		6.2.2	Statistical inferences	64
		6.2.3	Regression	64
		6.2.4	Nonparametric statistics	65
	6.3	Statistics, mathematics, and mathematical statistics		65
	6.4	See also		65
	6.5	References		65
	6.6	Additional reading		66
7	**Theoretical computer science**			**67**
	7.1	History		67
	7.2	Topics		68
		7.2.1	Algorithms	68
		7.2.2	Data structures	68
		7.2.3	Computational complexity theory	68
		7.2.4	Distributed computation	68
		7.2.5	Parallel computation	69
		7.2.6	Very-large-scale integration	69
		7.2.7	Machine learning	69
		7.2.8	Computational biology	69
		7.2.9	Computational geometry	69
		7.2.10	Information theory	70
		7.2.11	Cryptography	70
		7.2.12	Quantum computation	70

		7.2.13	Information-based complexity	71
		7.2.14	Computational number theory	71
		7.2.15	Symbolic computation	71
		7.2.16	Program semantics	71
		7.2.17	Formal methods	71
		7.2.18	Automata theory	71
		7.2.19	Coding theory	72
		7.2.20	Computational learning theory	72
	7.3	Organizations		72
	7.4	Journals and newsletters		72
	7.5	Conferences		72
	7.6	See also		73
	7.7	Notes		73
	7.8	Further reading		75
	7.9	External links		75
8	**Outline of physical science**			**76**
	8.1	Definition		76
	8.2	Branches of physical science		76
	8.3	History of physical science		77
	8.4	General principles of the physical sciences		83
		8.4.1	Basic principles of physics	83
		8.4.2	Basic principles of astronomy	83
		8.4.3	Basic principles of chemistry	83
		8.4.4	Basic principles of earth science	84
	8.5	Notable physical scientists		85
		8.5.1	Earth scientists	85
	8.6	See also		85
	8.7	Notes		85
	8.8	References		85
		8.8.1	Works cited	86
	8.9	External links		86
9	**List of life sciences**			**87**
	9.1	Biology and its branches		87
	9.2	Medicine and its branches		88
	9.3	New and other life science types		89
	9.4	References		92
	9.5	Further reading		94

10 Social science — 95
10.1 History — 95
10.2 Branches — 96
10.2.1 Anthropology — 96
10.2.2 Communication studies — 97
10.2.3 Economics — 97
10.2.4 Education — 97
10.2.5 Geography — 98
10.2.6 History — 98
10.2.7 Law — 99
10.2.8 Linguistics — 99
10.2.9 Political science — 100
10.2.10 Psychology — 100
10.2.11 Sociology — 101
10.3 Additional fields of study — 102
10.4 Methodology — 103
10.4.1 Social research — 103
10.4.2 Theory — 104
10.5 Education and degrees — 104
10.6 See also — 105
10.6.1 General — 105
10.6.2 Methods — 105
10.6.3 Areas — 105
10.6.4 History — 105
10.6.5 Lists — 105
10.6.6 People — 105
10.6.7 Other — 105
10.7 Notes and references — 105
10.8 Bibliography — 106
10.8.1 20th and 21st centuries sources — 106
10.8.2 19th century sources — 106
10.8.3 General sources — 107
10.8.4 Academic resources — 107
10.8.5 Opponents and critics — 107
10.9 External links — 107

11 Applied science — 109
11.1 Branches of applied science — 109
11.2 In education — 109

11.3 See also	109
11.4 References	109

12 Interdisciplinarity — 111

- 12.1 Development . . . 111
- 12.2 Barriers . . . 112
- 12.3 Interdisciplinary studies and studies of interdisciplinarity . . . 112
  - 12.3.1 Politics of interdisciplinary studies . . . 113
- 12.4 Historical examples . . . 113
- 12.5 Efforts to simplify and defend the concept . . . 114
- 12.6 Quotations . . . 114
- 12.7 See also . . . 115
- 12.8 References . . . 115
- 12.9 Further reading . . . 116
- 12.10 External links . . . 117

13 Philosophy of science — 119

- 13.1 Introduction . . . 119
  - 13.1.1 Defining science . . . 119
  - 13.1.2 Scientific explanation . . . 120
  - 13.1.3 Justifying science . . . 120
  - 13.1.4 Observation inseparable from theory . . . 121
  - 13.1.5 The purpose of science . . . 121
  - 13.1.6 Values and science . . . 122
- 13.2 History . . . 122
  - 13.2.1 Pre-modern . . . 122
  - 13.2.2 Modern . . . 122
  - 13.2.3 Logical positivism . . . 123
  - 13.2.4 Thomas Kuhn . . . 123
- 13.3 Current approaches . . . 124
  - 13.3.1 Axiomatic assumptions . . . 124
  - 13.3.2 Coherentism . . . 124
  - 13.3.3 Anything goes . . . 125
  - 13.3.4 Sociology of scientific knowledge . . . 125
  - 13.3.5 Continental philosophy . . . 126
- 13.4 Other topics . . . 126
  - 13.4.1 Reductionism . . . 126
  - 13.4.2 Social accountability . . . 126
- 13.5 Philosophy of particular sciences . . . 127

- 13.5.1 Philosophy of statistics 127
- 13.5.2 Philosophy of mathematics 127
- 13.5.3 Philosophy of physics 127
- 13.5.4 Philosophy of chemistry 127
- 13.5.5 Philosophy of biology 128
- 13.5.6 Philosophy of medicine 128
- 13.5.7 Philosophy of psychology 128
- 13.5.8 Philosophy of psychiatry 129
- 13.5.9 Philosophy of economics 129
- 13.5.10 Philosophy of social science 130
- 13.6 See also 130
- 13.7 References 130
- 13.8 Cited texts 134
- 13.9 Further reading 134
- 13.10 External links 135

14 History of science — 136

- 14.1 Early cultures 136
  - 14.1.1 Africa 136
  - 14.1.2 Ancient Near East 137
  - 14.1.3 Greco-Roman world 137
  - 14.1.4 India 139
  - 14.1.5 China 141
- 14.2 Science in the Middle Ages 142
  - 14.2.1 Islamic world 143
  - 14.2.2 Europe 144
- 14.3 Impact of science in Europe 145
  - 14.3.1 Age of Enlightenment 146
  - 14.3.2 Romanticism in science 146
- 14.4 Modern science 147
  - 14.4.1 Natural sciences 147
  - 14.4.2 Social sciences 152
  - 14.4.3 Emerging disciplines 156
- 14.5 Academic study 156
  - 14.5.1 Theories and sociology of the history of science 156
  - 14.5.2 The Plight of Many Scientific Innovators 157
- 14.6 See also 157
- 14.7 Notes and references 158
- 14.8 Further reading 163

15 Outline of science — 165

- 14.9 External links — 164
- 15.1 Essence of science — 165
- 15.2 Scientific method — 165
- 15.3 Branches of science — 166
  - 15.3.1 Natural science — 166
  - 15.3.2 Formal science — 166
  - 15.3.3 Social science — 175
  - 15.3.4 Applied science — 175
- 15.4 How scientific fields differ — 175
- 15.5 Politics of science — 175
- 15.6 History of science — 176
  - 15.6.1 By period — 176
  - 15.6.2 By field — 177
  - 15.6.3 By region — 177
- 15.7 Philosophy of science — 177
- 15.8 Scientific community — 178
  - 15.8.1 Scientific organizations — 178
  - 15.8.2 Scientists — 178
- 15.9 Science education — 181
- 15.10 See also — 182
- 15.11 References — 182

16 Objectivity (philosophy) — 183

- 16.1 Objectivism — 183
- 16.2 Objectivity in ethics — 183
  - 16.2.1 Ethical subjectivism — 184
  - 16.2.2 Ethical objectivism — 184
- 16.3 See also — 184
- 16.4 Further reading — 184
- 16.5 External links — 185

17 Logical reasoning — 186

- 17.1 See also — 186
- 17.2 References — 186

18 Scientific method — 187

- 18.1 Overview — 188
  - 18.1.1 Process — 188

	18.1.2 DNA example	189
	18.1.3 Other components	189
18.2	Scientific inquiry	189
	18.2.1 Properties of scientific inquiry	190
	18.2.2 Beliefs and biases	190
18.3	Elements of the scientific method	191
	18.3.1 Characterizations	191
	18.3.2 Hypothesis development	193
	18.3.3 Predictions from the hypothesis	194
	18.3.4 Experiments	195
	18.3.5 Evaluation and improvement	195
	18.3.6 Confirmation	196
18.4	Models of scientific inquiry	196
	18.4.1 Classical model	196
	18.4.2 Pragmatic model	196
18.5	Communication and community	198
	18.5.1 Peer review evaluation	198
	18.5.2 Documentation and replication	198
	18.5.3 Dimensions of practice	199
18.6	Philosophy and sociology of science	199
	18.6.1 Role of chance in discovery	200
18.7	History	200
18.8	Relationship with mathematics	201
18.9	Relationship with statistics	201
18.10	See also	202
	18.10.1 Problems and issues	202
	18.10.2 History, philosophy, sociology	202
18.11	Notes	202
18.12	References	208
18.13	Further reading	210
18.14	External links	211

19 Outline of scientific method — **212**

19.1	Nature of scientific method	212
19.2	Elements of scientific method	212
	19.2.1 Observation	212
	19.2.2 Hypothesis	212
	19.2.3 Experiment	212
	19.2.4 Theory	212

	19.2.5 Evaluation by scientific community	213
19.3	Scientific method concepts	213
	19.3.1 Empirical methods	213
	19.3.2 Use of statistics	213
	19.3.3 Paradigm change	213
	19.3.4 Problem of induction	213
	19.3.5 Scientific creativity	213
	19.3.6 Deviations from the scientific method	213
	19.3.7 Critique of scientific method	213
	19.3.8 Relationship of scientific method to technology	213
	19.3.9 Aesthetics in the scientific method	213
19.4	History of scientific method	214
	19.4.1 Publications	214
	19.4.2 Persons influential in the development of scientific method	214
	19.4.3 Why didn't the scientific method arise elsewhere?	214
19.5	See also	214
19.6	External links	214

20 Empirical research   215

20.1	Terminology	215
20.2	Usage	215
	20.2.1 Scientific research	215
20.3	Empirical cycle	216
20.4	See also	217
20.5	References	217
20.6	External links	217

21 Deductive-nomological model   218

21.1	Form	218
21.2	Roots	218
21.3	Growth	219
21.4	Decline	219
21.5	Strengths	220
21.6	Weaknesses	220
21.7	Covering action	221
21.8	See also	222
21.9	Notes	222
21.10	Sources	230
21.11	Further reading	233

22 Scientific modelling — 234

- 22.1 Overview — 234
- 22.2 Basics of scientific modelling — 235
  - 22.2.1 Modelling as a substitute for direct measurement and experimentation — 235
  - 22.2.2 Simulation — 235
  - 22.2.3 Structure — 235
  - 22.2.4 Systems — 235
  - 22.2.5 Generating a model — 235
  - 22.2.6 Evaluating a model — 236
  - 22.2.7 Visualization — 236
  - 22.2.8 Space mapping — 236
- 22.3 Types of scientific modelling — 236
- 22.4 Applications — 236
  - 22.4.1 Modelling and simulation — 236
  - 22.4.2 Model-based learning in education — 237
- 22.5 See also — 237
- 22.6 References — 238
- 22.7 Further reading — 238
- 22.8 External links — 239

23 Hypothetico-deductive model — 240

- 23.1 Example — 240
- 23.2 Discussion — 240
- 23.3 See also — 241
  - 23.3.1 Types of inference — 241
- 23.4 Citations — 241
- 23.5 References — 241

24 Branches of science — 243

- 24.1 Natural/Pure Science — 243
  - 24.1.1 Physical science — 243
  - 24.1.2 Life science — 245
- 24.2 Social sciences — 245
- 24.3 Formal sciences — 246
  - 24.3.1 Decision theory — 246
  - 24.3.2 Logic — 246
  - 24.3.3 Mathematics — 246
  - 24.3.4 Statistics — 246
  - 24.3.5 Systems theory — 246

	24.3.6 Theoretical computer science	247
24.4	Applied sciences	247
24.5	See also	247
24.6	Notes	247
24.7	References	248

25 Exact sciences — 249
- 25.1 See also — 249
- 25.2 References — 249

26 History of scientific method — 251
- 26.1 Early methodology — 251
  - 26.1.1 Aristotle — 252
  - 26.1.2 Epicurus — 253
- 26.2 Emergence of inductive experimental method — 254
  - 26.2.1 Ibn al-Haytham — 254
  - 26.2.2 Al-Biruni — 255
  - 26.2.3 Ibn Sina (Avicenna) — 255
  - 26.2.4 Robert Grosseteste — 256
  - 26.2.5 Roger Bacon — 256
  - 26.2.6 Renaissance humanism and medicine — 256
  - 26.2.7 Skepticism as a basis for understanding — 257
  - 26.2.8 Francis Bacon's eliminative induction — 258
  - 26.2.9 Descartes — 258
  - 26.2.10 Galileo Galilei — 259
  - 26.2.11 Isaac Newton — 260
- 26.3 Integrating deductive and inductive method — 261
  - 26.3.1 Charles Sanders Peirce — 262
  - 26.3.2 Popper and Kuhn — 263
- 26.4 Mention of the topic — 263
- 26.5 Current issues — 264
- 26.6 Science and pseudoscience — 264
- 26.7 See also — 264
- 26.8 Notes and references — 264
- 26.9 Sources — 269
- 26.10 Text and image sources, contributors, and licenses — 270
  - 26.10.1 Text — 270
  - 26.10.2 Images — 282

Chapter 1

Science

Science (from Latin *scientia*, meaning "knowledge")[1][2]:58 is a systematic enterprise that builds and organizes knowledge in the form of testable explanations and predictions about the universe.[lower-alpha 1]

Contemporary science is typically subdivided into the natural sciences, which study the material universe; the social sciences, which study people and societies; and the formal sciences, which study logic and mathematics. The formal sciences are often excluded as they do not depend on empirical observations.[3] Disciplines which use science, like engineering and medicine, may also be considered to be applied sciences.[4]

From classical antiquity through the 19th century, science as a type of knowledge was more closely linked to philosophy than it is now, and in the Western world the term "natural philosophy" once encompassed fields of study that are today associated with science, such as astronomy, medicine, and physics.[5][lower-alpha 2] However, during the Islamic Golden Age foundations for the scientific method were laid by Ibn al-Haytham in his *Book of Optics*.[6][7][8][9][10] While the classification of the material world by the ancient Indians and Greeks into air, earth, fire and water was more philosophical, medieval Middle Easterns used practical and experimental observation to classify materials.[11]

In the 17th and 18th centuries, scientists increasingly sought to formulate knowledge in terms of physical laws. Over the course of the 19th century, the word "science" became increasingly associated with the scientific method itself as a disciplined way to study the natural world. It was during this time that scientific disciplines such as biology, chemistry, and physics reached their modern shapes. That same time period also included the origin of the terms "scientist" and "scientific community", the founding of scientific institutions, and the increasing significance of their interactions with society and other aspects of culture.[12][13]

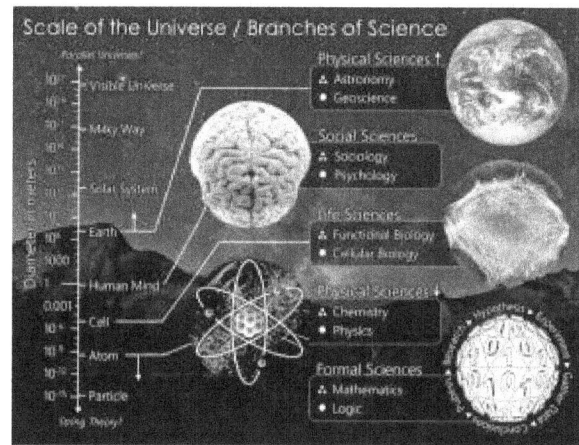

The scale of the universe mapped to the branches of science, with formal sciences as the foundation.[14]: Vol.1, Chaps.1,2,&3.

1.1 History

Main article: History of science

Science in a broad sense existed before the modern era and in many historical civilizations.[lower-alpha 3] Modern science is distinct in its approach and successful in its results, so it now defines what science is in the strictest sense of the term.[15]

Science in its original sense was a word for a type of knowledge rather than a specialized word for the pursuit of such knowledge. In particular, it was the type of knowledge which people can communicate to each other and share. For example, knowledge about the working of natural things was gathered long before recorded history and led to the development of complex abstract thought. This is shown by the construction of complex calendars, techniques for making poisonous plants edible, and buildings such as the Pyramids. However, no consistent conscientious distinction was made between knowledge of such things, which are true in every community, and other types of communal knowledge, such as mythologies and legal systems.

1.1.1 Antiquity

See also: Nature (philosophy)
Before the invention or discovery of the concept of "nature"

Maize, known in some English-speaking countries as corn, is a large grain plant domesticated by indigenous peoples in Mesoamerica in prehistoric times

(ancient Greek *phusis*) by the Pre-Socratic philosophers, the same words tend to be used to describe the *natural* "way" in which a plant grows,[16] and the "way" in which, for example, one tribe worships a particular god. For this reason, it is claimed these men were the first philosophers in the strict sense, and also the first people to clearly distinguish "nature" and "convention."[17]:209 Science was therefore distinguished as the knowledge of nature and things which are true for every community, and the name of the specialized pursuit of such knowledge was *philosophy* — the realm of the first philosopher-physicists. They were mainly speculators or theorists, particularly interested in astronomy. In contrast, trying to use knowledge of nature to imitate nature (artifice or technology, Greek *technē*) was seen by classical scientists as a more appropriate interest for lower class artisans.[18]

A major turning point in the history of early philosophical science was the controversial but successful attempt by Socrates to apply philosophy to the study of human things, including human nature, the nature of political communities, and human knowledge itself. He criticized the older type of study of physics as too purely speculative and lacking in self-criticism. He was particularly concerned that some of the early physicists treated nature as if it could be assumed that it had no intelligent order, explaining things merely in terms of motion and matter. The study of human things had been the realm of mythology and tradition, however, so Socrates was executed as a heretic.[20]: 30e Aristotle later created a less controversial systematic programme of Socratic philosophy which was teleological and human-centred. He rejected many of the conclusions of earlier scientists. For example, in his physics, the sun goes

Aristotle, 384–322 BCE, one of the early figures in the development of the scientific method[19]

around the earth, and many things have it as part of their nature that they are for humans. Each thing has a formal cause and final cause and a role in the rational cosmic order. Motion and change is described as the actualization of potentials already in things, according to what types of things they are. While the Socratics insisted that philosophy should be used to consider the practical question of the best way to live for a human being (a study Aristotle divided into ethics and political philosophy), they did not argue for any other types of applied science.

Aristotle maintained the sharp distinction between science and the practical knowledge of artisans, treating theoretical speculation as the highest type of human activity, practical thinking about good living as something less lofty, and the knowledge of artisans as something only suitable for the lower classes. In contrast to modern science, Aristotle's influential emphasis was upon the "theoretical" steps of deducing universal rules from raw data and did not treat the gathering of experience and raw data as part of science itself.[lower-alpha 4]

1.1. HISTORY

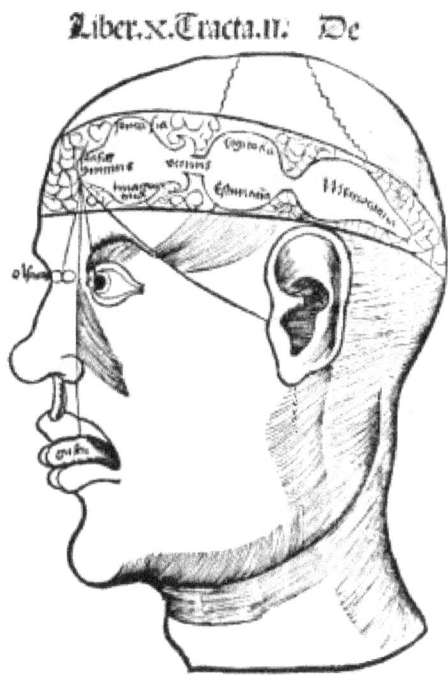

De potentiis anime sensitive, Gregor Reisch (1504) Margarita philosophica. Medieval science postulated a ventricle of the brain as the location for our common sense,[21] where the forms from our sensory systems commingled.

1.1.2 Medieval science

During late antiquity and the early Middle Ages, the Aristotelian approach to inquiries on natural phenomena was used. Some ancient knowledge was lost, or in some cases kept in obscurity, during the fall of the Roman Empire and periodic political struggles. However, the general fields of science (or "natural philosophy" as it was called) and much of the general knowledge from the ancient world remained preserved through the works of the early Latin encyclopedists like Isidore of Seville. In the Byzantine empire, many Greek science texts were preserved in Syriac translations done by groups such as the Nestorians and Monophysites.[23] Many of these were later on translated into Arabic under the Caliphate, during which many types of classical learning were preserved and in some cases improved upon.[23][lower-alpha 6]

The House of Wisdom was established in Abbasid-era Baghdad, Iraq.[24] It is considered to have been a major intellectual center during the Islamic Golden Age, where Muslim scholars such as al-Kindi and Ibn Sahl in Baghdad and Ibn al-Haytham in Cairo flourished from the ninth to the thirteenth centuries until the Mongol sack of Baghdad. Ibn

Ibn al-Haytham (Alhazen), 965–1039 Basra, Buyid Emirate. The Muslim scholar who is considered by some to be the father of modern scientific methodology due to his emphasis on experimental data and reproducibility of its results.[22][lower-alpha 5]

al-Haytham, known later to the West as Alhazen, furthered the Aristotelian viewpoint[25] by emphasizing experimental data.[lower-alpha 7][26]

In the later medieval period, as demand for translations grew (for example, from the Toledo School of Translators), western Europeans began collecting texts written not only in Latin, but also Latin translations from Greek, Arabic, and Hebrew. In particular, the texts of Aristotle, Ptolemy,[lower-alpha 8] and Euclid, preserved in the Houses of Wisdom, were sought amongst Catholic scholars. In Europe, the Latin translation of Alhazen's *Book of Optics* directly influenced Roger Bacon (13th century) in England, who argued for more experimental science as demonstrated by Alhazen. By the late Middle Ages, a synthesis of Catholicism and Aristotelianism known as Scholasticism was flourishing in western Europe, which had become a new geographic center of science, but all aspects of scholasticism were criticized in the 15th and 16th centuries.

1.1.3 Renaissance and early modern science

Main article: Scientific revolution

Medieval science carried on the views of the Hellenist civilization of Socrates, Plato, and Aristotle, as shown by Alhazen's lost work *A Book in which I have Summarized*

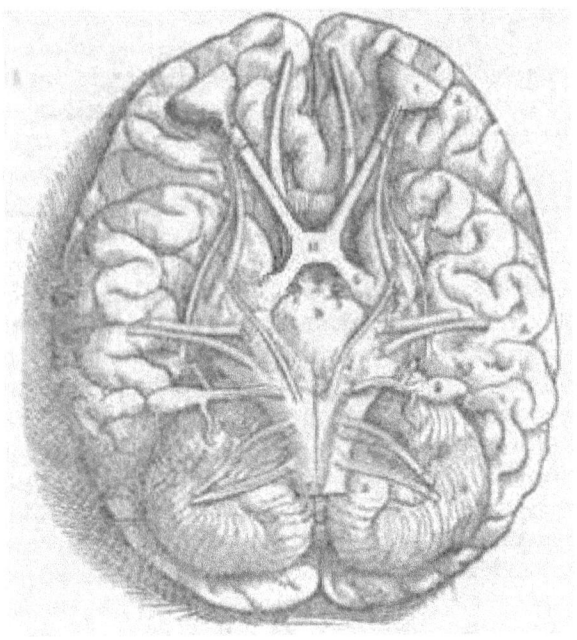

Galen (129–c. 216) noted the optic chiasm is X-shaped. (Engraving from Vesalius, 1543)

Galileo Galilei, father of modern science.[28]: Vol. 24, No. 1, p. 36

the Science of Optics from the Two Books of Euclid and Ptolemy, to which I have added the Notions of the First Discourse which is Missing from Ptolemy's Book from Ibn Abi Usaibia's catalog, as cited in (Smith 2001).[91(vol.1), p. xv] Alhazen conclusively disproved Ptolemy's theory of vision, but he retained Aristotle's ontology; Roger Bacon, Vitello, and John Peckham each built up a scholastic ontology upon Alhazen's *Book of Optics*, a causal chain beginning with sensation, perception, and finally apperception of the individual and universal forms of Aristotle.[27] This model of vision became known as Perspectivism, which was exploited and studied by the artists of the Renaissance.

A. Mark Smith points out the perspectivist theory of vision, which pivots on three of Aristotle's four causes, formal, material, and final, "is remarkably economical, reasonable, and coherent."[29] Although Alhacen knew that a scene imaged through an aperture is inverted, he argued that vision is about perception. This was overturned by Kepler,[30]:102 who modelled the eye as a water-filled glass sphere with an aperture in front of it to model the entrance pupil. He found that all the light from a single point of the scene was imaged at a single point at the back of the glass sphere. The optical chain ends on the retina at the back of the eye and the image is inverted.[lower-alpha 9]

Copernicus formulated a heliocentric model of the solar system unlike the geocentric model of Ptolemy's *Almagest*.

Galileo made innovative use of experiment and mathematics. However, he became persecuted after Pope Urban VIII blessed Galileo to write about the Copernican system. Galileo had used arguments from the Pope and put them in the voice of the simpleton in the work "Dialogue Concerning the Two Chief World Systems," which greatly offended him.[31]

In Northern Europe, the new technology of the printing press was widely used to publish many arguments, including some that disagreed widely with contemporary ideas of nature. René Descartes and Francis Bacon published philosophical arguments in favor of a new type of non-Aristotelian science. Descartes argued that mathematics could be used in order to study nature, as Galileo had done, and Bacon emphasized the importance of experiment over contemplation. Bacon questioned the Aristotelian concepts of formal cause and final cause, and promoted the idea that science should study the laws of "simple" natures, such as heat, rather than assuming that there is any specific nature, or "formal cause," of each complex type of thing. This new modern science began to see itself as describing "laws of nature". This updated approach to studies in nature was seen as mechanistic. Bacon also argued that science should aim for the first time at practical inventions for the improvement of all human life.

1.1. HISTORY

Isaac Newton, shown here in a 1689 portrait, made seminal contributions to classical mechanics, gravity, and optics. Newton shares credit with Gottfried Leibniz for the development of calculus.

1.1.4 Age of Enlightenment

In the 17th and 18th centuries, the project of modernity, as had been promoted by Bacon and Descartes, led to rapid scientific advance and the successful development of a new type of natural science, mathematical, methodically experimental, and deliberately innovative. Newton and Leibniz succeeded in developing a new physics, now referred to as classical mechanics, which could be confirmed by experiment and explained using mathematics. Leibniz also incorporated terms from Aristotelian physics, but now being used in a new non-teleological way, for example, "energy" and "potential" (modern versions of Aristotelian "*energeia* and *potentia*"). In the style of Bacon, he assumed that different types of things all work according to the same general laws of nature, with no special formal or final causes for each type of thing. It is during this period that the word "science" gradually became more commonly used to refer to a *type of pursuit* of a type of knowledge, especially knowledge of nature — coming close in meaning to the old term "natural philosophy."

1.1.5 19th century

Charles Darwin in 1854, by then working towards publication of On the Origin of Species

Both John Herschel and William Whewell systematized methodology: the latter coined the term scientist.[32] When Charles Darwin published *On the Origin of Species* he established evolution as the prevailing explanation of biological complexity. His theory of natural selection provided a natural explanation of how species originated, but this only gained wide acceptance a century later. John Dalton developed the idea of atoms. The laws of thermodynamics and the electromagnetic theory were also established in the 19th century, which raised new questions which could not easily be answered using Newton's framework. The phenomena that would allow the deconstruction of the atom were discovered in the last decade of the 19th century: the discovery of X-rays inspired the discovery of radioactivity. In the next year came the discovery of the first subatomic particle, the electron.

1.1.6 20th century and beyond

Einstein's theory of relativity and the development of quantum mechanics led to the replacement of classical mechanics with a new physics which contains two parts that describe different types of events in nature.

Combustion and chemical reactions were studied by Michael Faraday and reported in his lectures before the Royal Institution: The Chemical History of a Candle, *1861*

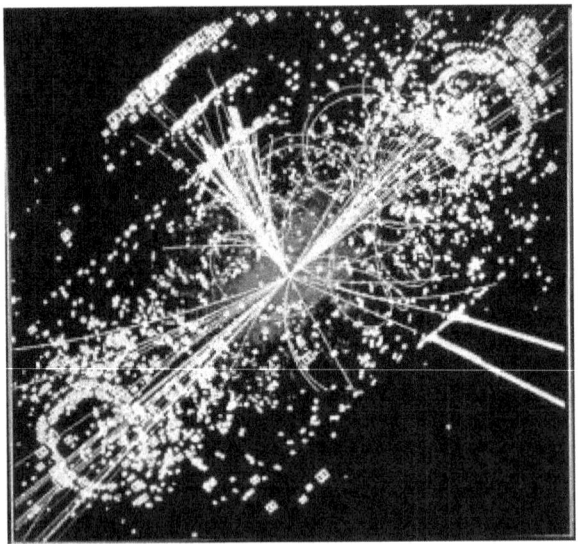

A simulated event in the CMS detector of the Large Hadron Collider, featuring a possible appearance of the Higgs boson

In the first half of the century, the development of artificial fertilizer made global human population growth possible. At the same time, the structure of the atom and its nucleus was discovered, leading to the release of "atomic energy" (nuclear power). In addition, the extensive use of scientific innovation stimulated by the wars of this century led to antibiotics and increased life expectancy, revolutions in transportation (automobiles and aircraft), the development of ICBMs, a space race, and a nuclear arms race, all giving a widespread public appreciation of the importance of modern science.

Widespread use of integrated circuits in the last quarter of the 20th century combined with communications satellites led to a revolution in information technology and the rise of the global internet and mobile computing, including smartphones.

More recently, it has been argued that the ultimate purpose of science is to make sense of human beings and our nature. For example, in his book *Consilience*, E. O. Wilson said: "The human condition is the most important frontier of the natural sciences".[2]:334

1.2 Scientific method

Main article: Scientific method

The scientific method seeks to explain the events of nature in a reproducible way.[lower-alpha 10] An explanatory thought experiment or hypothesis is put forward as explanation using principles such as parsimony (also known as "Occam's Razor") and are generally expected to seek consilience—fitting well with other accepted facts related to the phenomena.[2] This new explanation is used to make falsifiable predictions that are testable by experiment or observation. The predictions are to be posted before a confirming experiment or observation is sought, as proof that no tampering has occurred. Disproof of a prediction is evidence of progress.[lower-alpha 11][lower-alpha 12] This is done partly through observation of natural phenomena, but also through experimentation that tries to simulate natural events under controlled conditions as appropriate to the discipline (in the observational sciences, such as astronomy or geology, a predicted observation might take the place of a controlled experiment). Experimentation is especially important in science to help establish causal relationships (to avoid the correlation fallacy).

When a hypothesis proves unsatisfactory, it is either modified or discarded.[33] If the hypothesis survived testing, it may become adopted into the framework of a scientific theory, a logically reasoned, self-consistent model or framework for describing the behavior of certain natural phenomena. A theory typically describes the behavior of much broader sets of phenomena than a hypothesis; commonly, a large number of hypotheses can be logically bound together by a single theory. Thus a theory is a hypothesis explaining various other hypotheses. In that vein, theories are formulated according to most of the same scientific principles as hypotheses. In addition to testing hypotheses, scientists may also generate a model, an attempt to describe or depict the phenomenon in terms of a logical, physical or mathematical representation and to generate new hypotheses that can be tested, based on observable phenomena.[34]

While performing experiments to test hypotheses, scientists may have a preference for one outcome over another, and so it is important to ensure that science as a whole can eliminate this bias.[35][36] This can be achieved by careful experimental design, transparency, and a thorough peer review process of the experimental results as well as any conclusions.[37][38] After the results of an experiment are announced or published, it is normal practice for independent researchers to double-check how the research was performed, and to follow up by performing similar experiments to determine how dependable the results might be.[39] Taken in its entirety, the scientific method allows for highly creative problem solving while minimizing any effects of subjective bias on the part of its users (especially the confirmation bias).[40]

1.2.1 Mathematics and formal sciences

Main article: Mathematics

Mathematics is essential to the sciences. One important

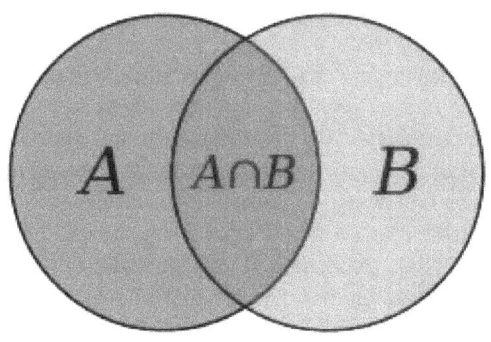

A Venn diagram illustrating the intersection of two sets.

function of mathematics in science is the role it plays in the expression of scientific models. Observing and collecting measurements, as well as hypothesizing and predicting, often require extensive use of mathematics. For example, arithmetic, algebra, geometry, trigonometry, and calculus are all essential to physics. Virtually every branch of mathematics has applications in science, including "pure" areas such as number theory and topology.

Statistical methods, which are mathematical techniques for summarizing and analyzing data, allow scientists to assess the level of reliability and the range of variation in experimental results. Statistical analysis plays a fundamental role in many areas of both the natural sciences and social sciences.

Computational science applies computing power to simulate real-world situations, enabling a better understanding of scientific problems than formal mathematics alone can achieve. According to the Society for Industrial and Applied Mathematics, computation is now as important as theory and experiment in advancing scientific knowledge.[41]

A great amount of interest was taken in the study of formal logic in the early 20th century among mathematicians and philosophers with the rise of set theory and its use for the foundations of mathematics. Notable mathematicians and philosophers who contributed to this field include: Gottlob Frege, Giuseppe Peano, George Boole, Ernst Zermelo, Abraham Fraenkel, David Hilbert, Bertrand Russell, and Alfred Whitehead among many others. Various axiomatic systems such as Peano arithmetic, the Zermelo–Fraenkel system of set theory, as well as the system in Principia Mathematica, were thought by many to prove the foundations of math. However, in 1931, with the publication of Kurt Gödel's incompleteness theorem, much of their effort was undermined.[42] Formal logic is still studied today at universities by students of mathematics, philosophy, and computer science. For example, Boolean algebra is employed by all modern computers to function, and thus is an extremely useful branch of knowledge for programmers.

Whether mathematics itself is properly classified as science has been a matter of some debate. Some thinkers see mathematicians as scientists, regarding physical experiments as inessential or mathematical proofs as equivalent to experiments. Others do not see mathematics as a science because it does not require an experimental test of its theories and hypotheses. Mathematical theorems and formulas are obtained by logical derivations which presume axiomatic systems, rather than the combination of empirical observation and logical reasoning that has come to be known as the scientific method. In general, mathematics is classified as formal science, while natural and social sciences are classified as empirical sciences.[43]

1.3 Scientific community

Main article: Scientific community

The scientific community is the group of all interacting scientists. It includes many sub-communities working on particular scientific fields, and within particular institutions; interdisciplinary and cross-institutional activities are also significant.

1.3.1 Branches and fields

Main article: Branches of science

Scientific fields are commonly divided into two major

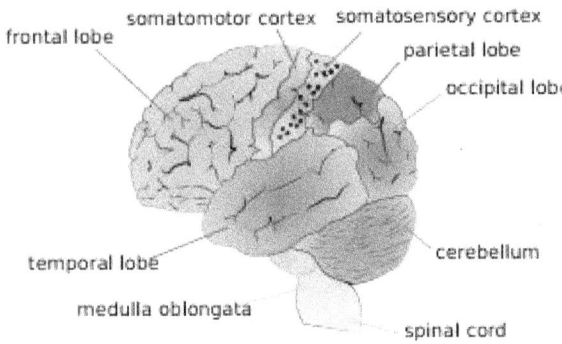

The somatosensory system is located throughout our bodies but is integrated in the brain.

groups: natural sciences, which study natural phenomena (including biological life), and social sciences, which study human behavior and societies. These are both empirical sciences, which means their knowledge must be based on observable phenomena and capable of being tested for its validity by other researchers working under the same conditions.[44] There are also related disciplines that are grouped into interdisciplinary applied sciences, such as engineering and medicine. Within these categories are specialized scientific fields that can include parts of other scientific disciplines but often possess their own nomenclature and expertise.[45]

Mathematics, which is classified as a formal science,[46][47] has both similarities and differences with the empirical sciences (the natural and social sciences). It is similar to empirical sciences in that it involves an objective, careful and systematic study of an area of knowledge; it is different because of its method of verifying its knowledge, using *a priori* rather than empirical methods.[48] The formal sciences, which also include statistics and logic, are vital to the empirical sciences. Major advances in formal science have often led to major advances in the empirical sciences. The formal sciences are essential in the formation of hypotheses, theories, and laws,[49] both in discovering and describing how things work (natural sciences) and how people think and act (social sciences).

Apart from its broad meaning, the word "science" sometimes may specifically refer to fundamental sciences (maths and natural sciences) alone. Science schools or faculties within many institutions are separate from those for medicine or engineering, each of which is an applied science.

1.3.2 Institutions

Learned societies for the communication and promotion of scientific thought and experimentation have existed since the Renaissance period.[50] The oldest surviving institution is the Italian *Accademia dei Lincei* which was established in 1603.[51] The respective National Academies of Science are distinguished institutions that exist in a number of countries, beginning with the British Royal Society in 1660[52] and the French *Académie des Sciences* in 1666.[53]

International scientific organizations, such as the International Council for Science, have since been formed to promote cooperation between the scientific communities of different nations. Many governments have dedicated agencies to support scientific research. Prominent scientific organizations include the National Science Foundation in the U.S., the National Scientific and Technical Research Council in Argentina, CSIRO in Australia, Centre national de la recherche scientifique in France, the Max Planck Society and Deutsche Forschungsgemeinschaft in Germany, and CSIC in Spain.

1.3.3 Literature

Main article: Scientific literature

An enormous range of scientific literature is published.[54] Scientific journals communicate and document the results of research carried out in universities and various other research institutions, serving as an archival record of science. The first scientific journals, *Journal des Sçavans* followed by the *Philosophical Transactions*, began publication in 1665. Since that time the total number of active periodicals has steadily increased. In 1981, one estimate for the number of scientific and technical journals in publication was 11,500.[55] The United States National Library of Medicine currently indexes 5,516 journals that contain articles on topics related to the life sciences. Although the journals are in 39 languages, 91 percent of the indexed articles are published in English.[56]

Most scientific journals cover a single scientific field and publish the research within that field; the research is normally expressed in the form of a scientific paper. Science has become so pervasive in modern societies that it is generally considered necessary to communicate the achievements, news, and ambitions of scientists to a wider populace.

Science magazines such as *New Scientist*, *Science & Vie*, and *Scientific American* cater to the needs of a much wider readership and provide a non-technical summary of popular areas of research, including notable discoveries and advances in certain fields of research. Science books engage the interest of many more people. Tangentially, the science fiction genre, primarily fantastic in nature, engages the public imagination and transmits the ideas, if not the methods, of science.

Recent efforts to intensify or develop links between science and non-scientific disciplines such as literature or more specifically, poetry, include the *Creative Writing Science* resource developed through the Royal Literary Fund.[57]

1.4 Science and society

"Science and society" redirects here. For the academic journal, see Science & Society.

1.4.1 Women in science

Main article: Women in science
Science has historically been a male-dominated field, with some notable exceptions.[lower-alpha 13] Women faced considerable discrimination in science, much as they did in other areas of male-dominated societies, such as frequently being passed over for job opportunities and denied credit for their work.[lower-alpha 14] For example, Christine Ladd (1847–1930) was able to enter a PhD program as "C. Ladd"; Christine "Kitty" Ladd completed the requirements in 1882, but was awarded her degree only in 1926, after a career which spanned the algebra of logic (see truth table), color vision, and psychology. Her work preceded notable researchers like Ludwig Wittgenstein and Charles Sanders Peirce. The achievements of women in science have been attributed to their defiance of their traditional role as laborers within the domestic sphere.[59]

In the late 20th century, active recruitment of women and elimination of institutional discrimination on the basis of sex greatly increased the number of women scientists, but large gender disparities remain in some fields; over half of new biologists are female, while 80% of PhDs in physics are given to men. Feminists claim this is the result of culture rather than an innate difference between the sexes, and some experiments have shown that parents challenge and explain more to boys than girls, asking them to reflect more deeply and logically.[60]: 258–61. In the early part of the 21st century, in America, women earned 50.3% bachelor's degrees, 45.6% master's degrees, and 40.7% of PhDs in science and engineering fields with women earning more

Marie Curie was the first person to be awarded two Nobel Prizes: Physics in 1903 and Chemistry in 1911[58]

than half of the degrees in three fields: Psychology (about 70%), Social Sciences (about 50%), and Biology (about 50-60%). However, when it comes to the Physical Sciences, Geosciences, Math, Engineering, and Computer Science, women earned less than half the degrees.[61] However, lifestyle choice also plays a major role in female engagement in science; women with young children are 28% less likely to take tenure-track positions due to work-life balance issues,[62] and female graduate students' interest in careers in research declines dramatically over the course of graduate school, whereas that of their male colleagues remains unchanged.[63]

1.4.2 Science policy

Main articles: Science policy, History of science policy, Funding of science, and Economics of science
Science policy is an area of public policy concerned with the policies that affect the conduct of the scientific enterprise, including research funding, often in pursuance of other national policy goals such as technological innovation to promote commercial product development, weapons development, health care and environmental monitoring. Science policy also refers to the act of applying scientific knowledge and consensus to the development of public

President Clinton meets the 1998 U.S. Nobel Prize winners in the White House

policies. Science policy thus deals with the entire domain of issues that involve the natural sciences. In accordance with public policy being concerned about the well-being of its citizens, science policy's goal is to consider how science and technology can best serve the public.

State policy has influenced the funding of public works and science for thousands of years, dating at least from the time of the Mohists, who inspired the study of logic during the period of the Hundred Schools of Thought, and the study of defensive fortifications during the Warring States period in China. In Great Britain, governmental approval of the Royal Society in the 17th century recognized a scientific community which exists to this day. The professionalization of science, begun in the 19th century, was partly enabled by the creation of scientific organizations such as the National Academy of Sciences, the Kaiser Wilhelm Institute, and state funding of universities of their respective nations. Public policy can directly affect the funding of capital equipment and intellectual infrastructure for industrial research by providing tax incentives to those organizations that fund research. Vannevar Bush, director of the Office of Scientific Research and Development for the United States government, the forerunner of the National Science Foundation, wrote in July 1945 that "Science is a proper concern of government."[64]

Science and technology research is often funded through a competitive process in which potential research projects are evaluated and only the most promising receive funding. Such processes, which are run by government, corporations, or foundations, allocate scarce funds. Total research funding in most developed countries is between 1.5% and 3% of GDP.[65] In the OECD, around two-thirds of research and development in scientific and technical fields is carried out by industry, and 20% and 10% respectively by universities and government. The government funding proportion in certain industries is higher, and it dominates research in social science and humanities. Similarly, with some excep-

tions (e.g. biotechnology) government provides the bulk of the funds for basic scientific research. In commercial research and development, all but the most research-oriented corporations focus more heavily on near-term commercialisation possibilities rather than "blue-sky" ideas or technologies (such as nuclear fusion).

1.4.3 Media perspectives

The mass media face a number of pressures that can prevent them from accurately depicting competing scientific claims in terms of their credibility within the scientific community as a whole. Determining how much weight to give different sides in a scientific debate may require considerable expertise regarding the matter.[66] Few journalists have real scientific knowledge, and even beat reporters who know a great deal about certain scientific issues may be ignorant about other scientific issues that they are suddenly asked to cover.[67][68]

1.4.4 Political usage

See also: Politicization of science

Many issues damage the relationship of science to the media and the use of science and scientific arguments by politicians. As a very broad generalisation, many politicians seek certainties and *facts* whilst scientists typically offer probabilities and caveats. However, politicians' ability to be heard in the mass media frequently distorts the scientific understanding by the public. Examples in the United Kingdom include the controversy over the MMR inoculation, and the 1988 forced resignation of a Government Minister, Edwina Currie, for revealing the high probability that battery farmed eggs were contaminated with *Salmonella*.[69]

John Horgan, Chris Mooney, and researchers from the US and Canada have described Scientific Certainty Argumentation Methods (SCAMs), where an organization or think tank makes it their only goal to cast doubt on supported science because it conflicts with political agendas.[70][71][72][73] Hank Campbell and microbiologist Alex Berezow have described "feel-good fallacies" used in politics, especially on the left, where politicians frame their positions in a way that makes people feel good about supporting certain policies even when scientific evidence shows there is no need to worry or there is no need for dramatic change on current programs.[74]: Vol. 78, No. 1, 2–38

1.4.5 Science and the public

Various activities are developed to facilitate communication between the general public and science/scientists, such as science outreach, public awareness of science, science communication, science festivals, citizen science, science journalism, public science, and popular science. See Science and the public for related concepts.

Science is represented by the 'S' in STEM fields.

1.5 Philosophy of science

See also: Philosophy of science

Working scientists usually take for granted a set of basic assumptions that are needed to justify the scientific method: (1) that there is an objective reality shared by all rational observers; (2) that this objective reality is governed by natural laws; (3) that these laws can be discovered by means of systematic observation and experimentation.[15] Philosophy of science seeks a deep understanding of what these underlying assumptions mean and whether they are valid.

The belief that scientific theories should and do represent metaphysical reality is known as realism. It can be contrasted with anti-realism, the view that the success of science does not depend on it being accurate about unobservable entities such as electrons. One form of anti-realism is idealism, the belief that the mind or consciousness is the most basic essence, and that each mind generates its own reality.[lower-alpha 15] In an idealistic world view, what is true for one mind need not be true for other minds.

The Sand Reckoner is a work by Archimedes in which he sets out to determine an upper bound for the number of grains of sand that fit into the universe. In order to do this, he had to estimate the size of the universe according to the contemporary model, and invent a way to analyze extremely large numbers.

There are different schools of thought in philosophy of science. The most popular position is empiricism,[lower-alpha 16] which holds that knowledge is created by a process involving observation and that scientific theories are the result of generalizations from such observations.[75] Empiricism generally encompasses inductivism, a position that tries to explain the way general theories can be justified by the finite number of observations humans can make and hence the finite amount of empirical evidence available to confirm scientific theories. This is necessary because the number of predictions those theories make is infinite, which means that they cannot be known from the finite amount of evidence using deductive logic only. Many versions of empiricism exist, with the predominant ones being Bayesianism[76] and the hypothetico-deductive method.[77]:236

Empiricism has stood in contrast to rationalism, the position originally associated with Descartes, which holds that knowledge is created by the human intellect, not by observation.[77]:20 Critical rationalism is a contrasting 20th-century approach to science, first defined by Austrian-British philosopher Karl Popper. Popper rejected the way that empiricism describes the connection between theory and observation. He claimed that theories are not generated by observation, but that observation is made in the light of theories and that the only way a theory can be affected by observation is when it comes in conflict with it.[77]:63–67 Popper proposed replacing verifiability with falsifiability as the landmark of scientific theories and replacing induction with falsification as the empirical method.[77]:68 Popper further claimed that there is actually only one universal method, not specific to science: the negative method of criticism, trial and error.[78] It covers all products of the human mind, including science, mathematics, philosophy, and art.[79]

Another approach, instrumentalism, colloquially termed "shut up and multiply,"[80] emphasizes the utility of theories as instruments for explaining and predicting phenomena.[81] It views scientific theories as black boxes with only their input (initial conditions) and output (predictions) being relevant. Consequences, theoretical entities, and logical structure are claimed to be something that should simply be ignored and that scientists shouldn't make a fuss about (see interpretations of quantum mechanics). Close to instrumentalism is constructive empiricism, according to which the main criterion for the success of a scientific theory is whether what it says about observable entities is true.

Paul Feyerabend advanced the idea of epistemological anarchism, which holds that there are no useful and exception-free methodological rules governing the progress of science or the growth of knowledge and that the idea that science can or should operate according to universal and fixed rules are unrealistic, pernicious and detrimental to science itself.[82] Feyerabend advocates treating science as an ideology alongside others such as religion, magic, and mythology, and considers the dominance of science in society authoritarian and unjustified. He also contended (along with Imre Lakatos) that the demarcation problem of distinguishing science from pseudoscience on objective grounds is not possible and thus fatal to the notion of science running according to fixed, universal rules.[82] Feyerabend also stated that science does not have evidence for its philosophical precepts, particularly the notion of uniformity of law and process across time and space.[83]

Finally, another approach often cited in debates of scientific skepticism against controversial movements like "creation science" is methodological naturalism. Its main point is that a difference between natural and supernatural explanations should be made and that science should be restricted methodologically to natural explanations.[lower-alpha 17] That the restriction is merely methodological (rather than ontological) means that science should not consider supernatural explanations itself, but should not claim them to be wrong either. Instead, supernatural explanations should be left a matter of personal belief outside the scope of science. Methodological naturalism maintains that proper science requires strict adherence to empirical study and independent verification as a process for properly developing and evaluating explanations for observable phenomena.[84] The absence of these standards, arguments from authority, biased observational studies and other common fallacies are frequently cited by supporters of methodological naturalism as characteristic of the non-science they criticize.

1.5.1 Certainty and science

A scientific theory is empirical[lower-alpha 16][85] and is always open to falsification if new evidence is presented. That is, no theory is ever considered strictly certain as science accepts the concept of fallibilism.[lower-alpha 18] The philosopher of science Karl Popper sharply distinguished truth from certainty. He wrote that scientific knowledge "consists in the search for truth," but it "is not the search for certainty ... All human knowledge is fallible and therefore uncertain."[86]:4

New scientific knowledge rarely results in vast changes in our understanding. According to psychologist Keith Stanovich, it may be the media's overuse of words like "breakthrough" that leads the public to imagine that science is constantly proving everything it thought was true to be false.[87]:119-38 While there are such famous cases as the theory of relativity that required a complete reconceptualization, these are extreme exceptions. Knowledge in science is gained by a gradual synthesis of information from different experiments by various researchers across different branches of science: it is more like a climb than a leap.[87]:123 Theories vary in the extent to which they have been tested and verified, as well as their acceptance in the scientific community.[lower-alpha 19] For exam-

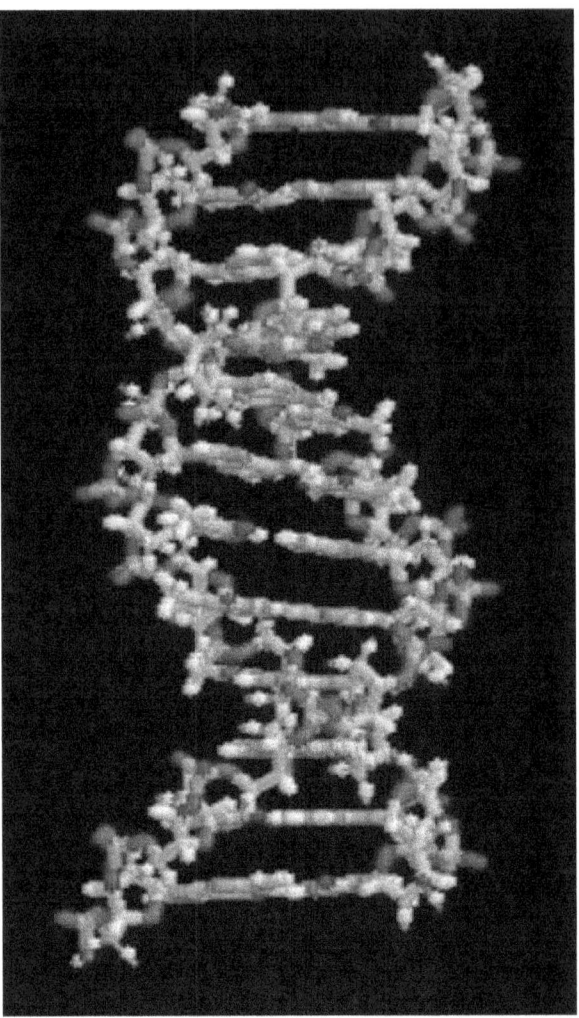

The DNA double helix is a molecule that encodes the genetic instructions used in the development and functioning of all known living organisms and many viruses.

ple, heliocentric theory, the theory of evolution, relativity theory, and germ theory still bear the name "theory" even though, in practice, they are considered factual.[88] Philosopher Barry Stroud adds that, although the best definition for "knowledge" is contested, being skeptical and entertaining the *possibility* that one is incorrect is compatible with being correct. Ironically, then, the scientist adhering to proper scientific approaches will doubt themselves even once they possess the truth.[89] The fallibilist C. S. Peirce argued that inquiry is the struggle to resolve actual doubt and that merely quarrelsome, verbal, or hyperbolic doubt is fruitless[90]—but also that the inquirer should try to attain genuine doubt rather than resting uncritically on common sense.[91] He held that the successful sciences trust not to any single chain of inference (no stronger than its weakest link) but to the cable of multiple and various arguments intimately connected.[92]

Stanovich also asserts that science avoids searching for a "magic bullet"; it avoids the single-cause fallacy. This means a scientist would not ask merely "What is *the* cause of ...", but rather "What *are* the most significant *causes* of ...". This is especially the case in the more macroscopic fields of science (e.g. psychology, physical cosmology).[87]:141-47 Of course, research often analyzes few factors at once, but these are always added to the long list of factors that are most important to consider.[87]:141-47 For example, knowing the details of only a person's genetics, or their history and upbringing, or the current situation may not explain a behavior, but a deep understanding of all these variables combined can be very predictive.

1.5.2 Fringe science, pseudoscience, and junk science

An area of study or speculation that masquerades as science in an attempt to claim a legitimacy that it would not otherwise be able to achieve is sometimes referred to as pseudoscience, fringe science, or junk science.[lower-alpha 20] Physicist Richard Feynman coined the term "cargo cult science" for cases in which researchers believe they are doing science because their activities have the outward appearance of science but actually lack the "kind of utter honesty" that allows their results to be rigorously evaluated.[93] Various types of commercial advertising, ranging from hype to fraud, may fall into these categories.

There can also be an element of political or ideological bias on all sides of scientific debates. Sometimes, research may be characterized as "bad science," research that may be well-intended but is actually incorrect, obsolete, incomplete, or over-simplified expositions of scientific ideas. The term "scientific misconduct" refers to situations such as where researchers have intentionally misrepresented their published data or have purposely given credit for a discovery to the wrong person.[94]

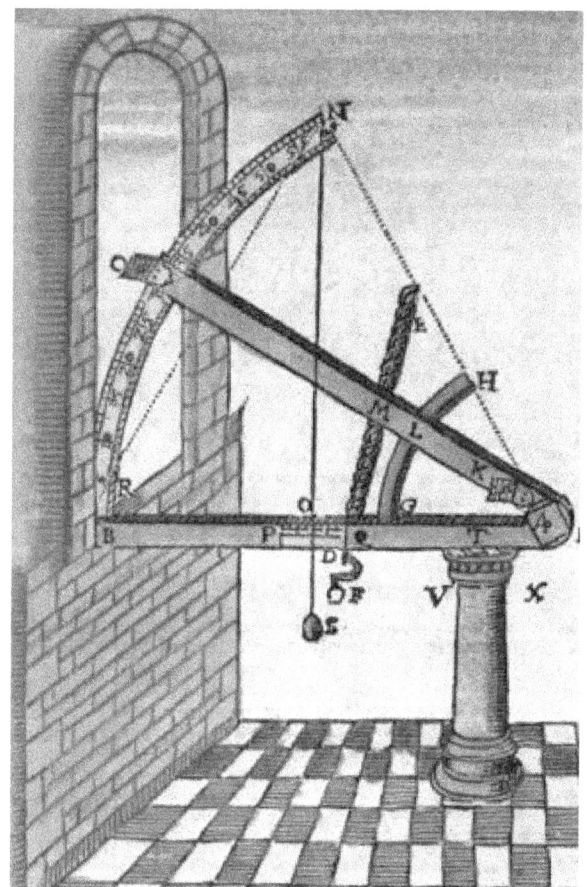

Astronomy became much more accurate after Tycho Brahe devised his scientific instruments for measuring angles between two celestial bodies, before the invention of the telescope. Brahe's observations were the basis for Kepler's laws.

"If a man will begin with certainties, he shall end in doubts; but if he will be content to begin with doubts, he shall end in certainties."
— Francis Bacon, "The Advancement of Learning", Book 1, v. 8

1.6 Scientific practice

Although encyclopedias such as Pliny's (fl. 77 AD) *Natural History* offered purported fact, they proved unreliable. A skeptical point of view, demanding a method of proof, was the practical position taken to deal with unreliable knowledge. As early as 1000 years ago, scholars such as Alhazen (*Doubts Concerning Ptolemy*), Roger Bacon, Witelo, John Pecham, Francis Bacon (1605), and C. S. Peirce (1839–1914) provided the community to address these points of uncertainty. In particular, fallacious reasoning can be exposed, such as "affirming the consequent."

The methods of inquiry into a problem have been known for thousands of years,[95] and extend beyond theory to practice. The use of measurements, for example, is a practical approach to settle disputes in the community.

John Ziman points out that intersubjective pattern recognition is fundamental to the creation of all scientific knowledge.[96]:44 Ziman shows how scientists can identify patterns to each other across centuries; he refers to this ability as "perceptual consensibility."[97]:46 He then makes consensibility, leading to consensus, the touchstone of reliable knowledge.[97]:104

1.6.1 Basic and applied research

Anthropogenic pollution has an effect on the Earth's environment and climate

Although some scientific research is applied research into specific problems, a great deal of our understanding comes from the curiosity-driven undertaking of basic research. This leads to options for technological advance that were not planned or sometimes even imaginable. This point was made by Michael Faraday when allegedly in response to the question "what is the *use* of basic research?" he responded: "Sir, what is the use of a new-born child?".[98] For example, research into the effects of red light on the human eye's rod cells did not seem to have any practical purpose; eventually, the discovery that our night vision is not troubled by red light would lead search and rescue teams (among others) to adopt red light in the cockpits of jets and helicopters.[87]:106–10 In a nutshell, basic research is the search for knowledge and applied research is the search for solutions to practical problems using this knowledge. Finally, even basic research can take unexpected turns, and there is some sense in which the scientific method is built to harness luck.

1.6.2 Research in practice

Due to the increasing complexity of information and specialization of scientists, most of the cutting-edge research today is done by well-funded groups of scientists, rather than individuals.[99] D.K. Simonton notes that due to the breadth of very precise and far reaching tools already used by researchers today and the amount of research generated so far, creation of new disciplines or revolutions within a discipline may no longer be possible as it is unlikely that some phenomenon that merits its own discipline has been overlooked. Hybridizing of disciplines and finessing knowledge is, in his view, the future of science.[99]

1.6.3 Practical impacts of scientific research

Discoveries in fundamental science can be world-changing. For example:

1.7 See also

- Antiquarian science books
- Criticism of science
- Human timeline
- Index of branches of science
- Life timeline
- Normative science
- Outline of science
- Pathological science
- Protoscience
-
- Science wars
- Scientific dissent
- Sociology of scientific knowledge

1.8 Notes

[1] "... modern science is a discovery as well as an invention. It was a discovery that nature generally acts regularly enough to be described by laws and even by mathematics; and required invention to devise the techniques, abstractions, apparatus, and organization for exhibiting the regularities and securing their law-like descriptions."— Heilbron 2003, p. vii
"science". *Merriam-Webster Online Dictionary*. Merriam-Webster, Inc. Retrieved October 16, 2011. **3 a:** knowledge or a system of knowledge covering general truths or the operation of general laws especially as obtained and tested through scientific method **b:** such knowledge or such a system of knowledge concerned with the physical world and its phenomena.

[2] Isaac Newton's Philosophiae Naturalis Principia Mathematica (1687), for example, is translated "Mathematical Principles of Natural Philosophy", and reflects the then-current use of the words "natural philosophy", akin to "systematic study of nature"

[3] "The historian ... requires a very broad definition of "science" — one that ... will help us to understand the modern scientific enterprise. We need to be broad and inclusive, rather than narrow and exclusive ... and we should expect that the farther back we go [in time] the broader we will need to be." — David Pingree (1992). "Hellenophilia versus the History of Science" *Isis* **83** 554–63, as cited in (Lindberg 2007, p. 3), *The beginnings of Western science: the European Scientific tradition in philosophical, religious, and institutional context*, Second ed. Chicago: Univ. of Chicago Press ISBN 978-0-226-48205-7

- See Grant, Edward (1 January 1997). "History of Science: When Did Modern Science Begin?". *The American Scholar*. **66** (1): 105–13. JSTOR 41212592.
  - *History of science#Early cultures*
  - *History of science#Ancient Near East*, Mesopotamia
  - *History of science#Ancient Near East*, Egypt
  - *History of Science in China*
  - *History of science#India*

[4] "... [A] man knows a thing scientifically when he possesses a conviction arrived at in a certain way, and when the first principles on which that conviction rests are known to him with certainty—for unless he is more certain of his first principles than of the conclusion drawn from them he will only possess the knowledge in question accidentally." — Aristotle. *Nicomachean Ethics* (H. Rackham, ed. ed.).

[5] Tracey Tokuhama-Espinosa (2010). *Mind, Brain, and Education Science: A Comprehensive Guide to the New Brain-Based Teaching*. W. W. Norton & Company. p. 39. ISBN 978-0-393-70607-9. Alhazen (or Al-Haytham; 965–1039 C.E.) was perhaps one of the greatest physicists of all times and a product of the Islamic Golden Age or Islamic Renaissance (7th–13th centuries). He made significant contributions to anatomy, astronomy, engineering, mathematics, medicine, ophthalmology, philosophy, physics, psychology, and visual perception and is primarily attributed as the inventor of the scientific method, for which author Bradley Steffens (2006) describes him as the "first scientist".

[6] Alhacen had access to the optics books of Euclid and Ptolemy, as is shown by the title of his lost work *A Book in which I have Summarized the Science of Optics from the Two Books of Euclid and Ptolemy, to which I have added the Notions of the First Discourse which is Missing from Ptolemy's Book* From Ibn Abi Usaibia's catalog, as cited in (Smith 2001).⁹¹(vol.1), p. xv

[7] "[Ibn al-Haytham] followed Ptolemy's bridge building ... into a grand synthesis of light and vision. Part of his effort consisted in devising ranges of experiments, of a kind probed before but now undertaken on larger scale."— Cohen 2010, p. 59

[8] The translator, Gerard of Cremona (c. 1114–87), inspired by his love of the Almagest, came to Toledo, where he knew he could find the Almagest in Arabic. There he found Arabic books of every description, and learned Arabic in order to translate these books into Latin, being aware of 'the poverty of the Latins'. —As cited by Charles Burnett (2001) "The Coherence of the Arabic-Latin Translation Program in Toledo in the Twelfth Century", pp. 250, 255, 257. *Science in Context* **14**(1/2), 249–88 (2001). doi:10.1017/0269889701000096

[9] Kepler, Johannes (1604) *Ad Vitellionem paralipomena, quibus astronomiae pars opticae traditur* (Supplements to Witelo, in which the optical part of astronomy is treated) as cited in Smith, A. Mark (1 January 2004). "What Is the History of Medieval Optics Really about?". *Proceedings of the American Philosophical Society*. **148** (2): 180–94.

- The full title translation is from p. 60 of James R. Voelkel (2001) *Johannes Kepler and the New Astronomy* Oxford University Press. Kepler was driven to this experiment after observing the partial solar eclipse at Graz, July 10, 1600. He used Tycho Brahe's method of observation, which was to project the image of the sun on a piece of paper through a pinhole aperture, instead of looking directly at the sun. He disagreed with Brahe's conclusion that total eclipses of the sun were impossible, because there were historical accounts of total eclipses. Instead he deduced that the size of the aperture controls the sharpness of the projected image (the larger the aperture, the more accurate the image — this fact is now fundamental for optical system design). Voelkel, p. 61, notes that Kepler's experiments produced the first correct account of vision and the eye, because he realized he could not accurately write about astronomical observation by ignoring the eye.

[10] di Francia 1976, p. 13: "The amazing point is that for the first time since the discovery of mathematics, a method has been introduced, the results of which have an intersubjective value!" *(Author's punctuation)*

[11] di Francia 1976, pp. 4–5: "One learns in a laboratory; one learns how to make experiments only by experimenting, and one learns how to work with his hands only by using them. The first and fundamental form of experimentation in physics is to teach young people to work with their hands. Then they should be taken into a laboratory and taught to work with measuring instruments — each student carrying out real experiments in physics. This form of teaching is indispensable and cannot be read in a book."

[12] Fara 2009, p. 204: "Whatever their discipline, scientists claimed to share a common scientific method that ... distinguished them from non-scientists."

[13] Women in science have included:

- Hypatia (c. 350–415 CE), of the Library of Alexandria.
- Trotula of Salerno, a physician c. 1060 CE.
- Caroline Herschel, one of the first professional astronomers of the 18th and 19th centuries.

- Christine Ladd-Franklin, a doctoral student of C. S. Peirce, who published Wittgenstein's proposition 5.101 in her dissertation, 40 years before Wittgenstein's publication of Tractatus Logico-Philosophicus.
- Henrietta Leavitt, a professional human computer and astronomer, who first published the significant relationship between the luminosity of Cepheid variable stars and their distance from Earth. This allowed Hubble to make the discovery of the expanding universe, which led to the Big Bang theory.
- Emmy Noether, who proved the conservation of energy and other constants of motion in 1915.
- Marie Curie, who made discoveries relating to radioactivity along with her husband, and for whom Curium is named.
- Rosalind Franklin, who worked with X-ray diffraction.

[14] Nina Byers, Contributions of 20th Century Women to Physics which provides details on 83 female physicists of the 20th century. By 1976, more women were physicists, and the 83 who were detailed were joined by other women in noticeably larger numbers.

[15] This realization is the topic of intersubjective verifiability, as recounted, for example, by Max Born (1949, 1965) *Natural Philosophy of Cause and Chance*, who points out that all knowledge, including natural or social science, is also subjective. p. 162: "Thus it dawned upon me that fundamentally everything is subjective, everything without exception. That was a shock."

[16] In his investigation of the law of falling bodies, Galileo (1638) serves as example for scientific investigation: *Two New Sciences* "A piece of wooden moulding or scantling, about 12 cubits long, half a cubit wide, and three fingerbreadths thick, was taken; on its edge was cut a channel a little more than one finger in breadth; having made this groove very straight, smooth, and polished, and having lined it with parchment, also as smooth and polished as possible, we rolled along it a hard, smooth, and very round bronze ball. Having placed this board in a sloping position, by lifting one end some one or two cubits above the other, we rolled the ball, as I was just saying, along the channel, noting, in a manner presently to be described, the time required to make the descent. We . . . now rolled the ball only one-quarter the length of the channel; and having measured the time of its descent, we found it precisely one-half of the former. Next we tried other distances, comparing the time for the whole length with that for the half, or with that for two-thirds, or three-fourths, or indeed for any fraction; in such experiments, repeated many, many, times." Galileo solved the problem of time measurement by weighing a jet of water collected during the descent of the bronze ball, as stated in his *Two New Sciences*.

[17] Godfrey-Smith 2003, p. 151 credits Willard Van Orman Quine (1969) "Epistemology Naturalized" *Ontological Relativity and Other Essays* New York: Columbia University Press, as well as John Dewey, with the basic ideas of naturalism — Naturalized Epistemology, but Godfrey-Smith diverges from Quine's position: according to Godfrey-Smith, "A naturalist can think that science can contribute to *answers* to philosophical questions, without thinking that philosophical questions can be replaced by science questions.".

[18] "No amount of experimentation can ever prove me right; a single experiment can prove me wrong." —Albert Einstein, noted by Alice Calaprice (ed. 2005) *The New Quotable Einstein* Princeton University Press and Hebrew University of Jerusalem, ISBN 0-691-12074-9 p. 291. Calaprice denotes this not as an exact quotation, but as a paraphrase of a translation of A. Einstein's "Induction and Deduction". *Collected Papers of Albert Einstein* 7 Document 28. Volume 7 is *The Berlin Years: Writings, 1918–1921*. A. Einstein; M. Janssen, R. Schulmann, et al., eds.

[19] Fleck, Ludwik (1979). Trenn, Thaddeus J.; Merton, Robert K., eds. *Genesis and Development of a Scientific Fact*. Chicago: University of Chicago Press. ISBN 0-226-25325-2. Claims that before a specific fact "existed", it had to be created as part of a social agreement within a community. Steven Shapin (1980) "A view of scientific thought" *Science* ccvii (Mar 7, 1980) 1065–66 states "[To Fleck,] facts are invented, not discovered. Moreover, the appearance of scientific facts as discovered things is itself a social construction: a *made* thing. "

[20] "*Pseudoscientific – pretending to be scientific, falsely represented as being scientific*", from the *Oxford American Dictionary*, published by the Oxford English Dictionary; Hansson, Sven Ove (1996)."Defining Pseudoscience", Philosophia Naturalis, 33: 169–176, as cited in "Science and Pseudoscience" (2008) in Stanford Encyclopedia of Philosophy. The Stanford article states: "Many writers on pseudoscience have emphasized that pseudoscience is non-science posing as science. The foremost modern classic on the subject (Gardner 1957) bears the title Fads and Fallacies in the Name of Science. According to Brian Baigrie (1988, 438), "[w]hat is objectionable about these beliefs is that they masquerade as genuinely scientific ones." These and many other authors assume that to be pseudoscientific, an activity or a teaching has to satisfy the following two criteria (Hansson 1996): (1) it is not scientific, and (2) its major proponents try to create the impression that it is scientific".

- For example, Hewitt et al. *Conceptual Physical Science* Addison Wesley; 3 edition (July 18, 2003) ISBN 0-321-05173-4, Bennett et al. *The Cosmic Perspective* 3e Addison Wesley; 3 edition (July 25, 2003) ISBN 0-8053-8738-2; *See also*, e.g., Gauch HG Jr. *Scientific Method in Practice* (2003).
- A 2006 National Science Foundation report on Science and engineering indicators quoted Michael Shermer's (1997) definition of pseudoscience: "'claims presented so that they appear [to be] scientific even though they lack supporting evidence and plausibility"(p. 33). In contrast, science is "a set of methods

designed to describe and interpret observed and inferred phenomena, past or present, and aimed at building a testable body of knowledge open to rejection or confirmation"(p. 17)'.Shermer M. (1997). *Why People Believe Weird Things: Pseudoscience, Superstition, and Other Confusions of Our Time*. New York: W. H. Freeman and Company. ISBN 0-7167-3090-1. as cited by National Science Board. National Science Foundation, Division of Science Resources Statistics (2006). "Science and Technology: Public Attitudes and Understanding". *Science and engineering indicators 2006*. Archived from the original on February 1, 2013.

- "A pretended or spurious science; a collection of related beliefs about the world mistakenly regarded as being based on scientific method or as having the status that scientific truths now have." from the *Oxford English Dictionary*, second edition 1989.

[21] Evicting Einstein, March 26, 2004, NASA. *"Both [relativity and quantum mechanics] are extremely successful. The Global Positioning System (GPS), for instance, wouldn't be possible without the theory of relativity. Computers, telecommunications, and the Internet, meanwhile, are spin-offs of quantum mechanics."*

1.9 References

[1] Harper, Douglas. "science". *Online Etymology Dictionary*. Retrieved September 20, 2014.

[2] Wilson, Edward (1999). *Consilience: The Unity of Knowledge*. New York: Vintage. ISBN 0-679-76867-X.

[3] Editorial Staff (March 7, 2008). "The Branches of Science". South Carolina State University. Retrieved October 28, 2014.

[4] Editorial Staff (March 7, 2008). "Scientific Method: Relationships among Scientific Paradigms". Seed Magazine. Retrieved September 12, 2007.

[5] Lindberg 2007, p. 3.

[6] Haq, Syed (2009). "Science in Islam". Oxford Dictionary of the Middle Ages. ISSN 1703-7603. Retrieved 2014-10-22.

[7] G. J. Toomer. Review on JSTOR, Toomer's 1964 review of Matthias Schramm (1963) *Ibn Al-Haythams Weg Zur Physik* Toomer p. 464: "Schramm sums up [Ibn Al-Haytham's] achievement in the development of scientific method."

[8] "International Year of Light – Ibn Al-Haytham and the Legacy of Arabic Optics".

[9] Al-Khalili, Jim (4 January 2009). "The 'first true scientist'". BBC News. Retrieved 24 September 2013.

[10] Gorini, Rosanna (October 2003). "Al-Haytham the man of experience. First steps in the science of vision" (PDF). *Journal of the International Society for the History of Islamic Medicine*. **2** (4): 53–55. Retrieved 2008-09-25.

[11] *Science and Islam*, Jim Al-Khalili. BBC, 2009

[12] Cahan, David, ed. (2003). *From Natural Philosophy to the Sciences: Writing the History of Nineteenth-Century Science*. Chicago: University of Chicago Press. ISBN 0-226-08928-2.

[13] The *Oxford English Dictionary* dates the origin of the word "scientist" to 1834.

[14] Feynman, Richard. *The Feynman Lectures on Physics*. **1**.

[15] Heilbron 2003, p. vii

[16] See the quotation in Homer (8th century BCE) *Odyssey* 10.302–3

[17] "Progress or Return" in *An Introduction to Political Philosophy: Ten Essays by Leo Strauss* (Expanded version of *Political Philosophy: Six Essays by Leo Strauss*, 1975.) Ed. Hilail Gilden. Detroit: Wayne State UP, 1989.

[18] Cropsey; Strauss (eds.). *History of Political Philosophy* (3rd ed.). p. 209.

[19] Mitchell, Jacqueline S. (February 18, 2003). "The Origins of Science". *Scientific American Frontiers*. PBS. Archived from the original on March 3, 2003. Retrieved November 3, 2016.

[20] "Plato, Apology, section 30". *Perseus Digital Library*. Tufts University. 1966. Retrieved November 1, 2016.

[21]
- Smith, A. Mark (June 2004). "What is the History of Medieval Optics Really About?". *Proceedings of the American Philosophical Society*, **148** (2): 180–94. JSTOR 1558283[189]

[22] Jim Al-Khalili (January 4, 2009). "The 'first true scientist'". BBC News.

[23] Grant, Edward (2007). *A History of Natural Philosophy: From the Ancient World to the Nineteenth Century*. Cambridge University Press. pp. 62–67. ISBN 978-0-521-68957-1.

[24] "Bayt al-Hikmah". Encyclopædia Britannica. Retrieved November 3, 2016.

[25] Smith, A. Mark (December 1981). "Getting the Big Picture in Perspectivist Optics". *Isis*. **72** (4): 568–89. JSTOR 231249. doi:10.1086/352843.

[26] "Science in Islam". *Oxford Dictionary of the Middle Ages*. 2009.

[27] Smith 2001 p. lxxii. via JSTOR

[28] "Galileo and the Birth of Modern Science". *American Heritage of Invention and Technology*. **24**.

[29] Smith, A. Mark (1981). "Getting the Big Picture in Perspectivist Optics" *Isis* **72**(#4 — Dec. 1981), pp. 568–89 p. 588 via JSTOR

[30] Cohen, H. Floris (2010). *How modern science came into the world. Four civilizations, one 17th-century breakthrough.* (Second ed.). Amsterdam: Amsterdam University Press. ISBN 9789089642394.

[31] van Helden, Al (1995). "Pope Urban VIII". *The Galileo Project*. Retrieved November 3, 2016.

[32] Ross, Sydney (1962). "Scientist: The story of a word" (PDF). *Annals of Science*. **18** (2): 65–85. doi:10.1080/00033796200202722. Retrieved 2011-03-08. To be exact, the person coined the term *scientist* was referred to in Whewell 1834 only as "some ingenious gentleman." Ross added a comment that this "some ingenious gentleman" was Whewell himself, without giving the reason for the identification. Ross 1962, p. 72.

[33] Nola & Irzik 2005, p. 208.

[34] Nola & Irzik 2005, pp. 199–201.

[35] van Gelder, Tim (1999). ""Heads I win, tails you lose": A Foray Into the Psychology of Philosophy" (PDF). University of Melbourne. Archived from the original (PDF) on April 9, 2008. Retrieved March 28, 2008.

[36] Pease, Craig (September 6, 2006). "Chapter 23. Deliberate bias: Conflict creates bad science". *Science for Business, Law and Journalism*. Vermont Law School. Archived from the original on June 19, 2010.

[37] Shatz, David (2004). *Peer Review: A Critical Inquiry*. Rowman & Littlefield. ISBN 0-7425-1434-X. OCLC 54989960.

[38] Krimsky, Sheldon (2003). *Science in the Private Interest: Has the Lure of Profits Corrupted the Virtue of Biomedical Research*. Rowman & Littlefield. ISBN 0-7425-1479-X. OCLC 185926306.

[39] Bulger, Ruth Ellen; Heitman, Elizabeth; Reiser, Stanley Joel (2002). *The Ethical Dimensions of the Biological and Health Sciences* (2nd ed.). Cambridge University Press. ISBN 0-521-00886-7. OCLC 47791316.

[40] Backer, Patricia Ryaby (October 29, 2004). "What is the scientific method?". San Jose State University. Archived from the original on April 8, 2008. Retrieved March 28, 2008.

[41] "SIAM: Graduate Education for Computational Science and Engineering". Society for Industrial and Applied Mathematics. Retrieved November 4, 2016.

[42] "Incompleteness theorem". Encyclopædia Britannica.

[43] Bunge, Mario Augusto (1998). *Philosophy of Science: From Problem to Theory*. Transaction Publishers. p. 24. ISBN 0-7658-0413-1.

[44] Popper 2002, p. 20.

[45] "Scientific Method: Relationships Among Scientific Paradigms". Seed Magazine. March 7, 2007. Retrieved November 4, 2016.

[46] Tomalin, Marcus (2006). *Linguistics and the Formal Sciences*. Cambridge.org. doi:10.2277/0521854814. Retrieved February 5, 2012.

[47] "The Formal Sciences: Their Scope, Their Foundations, and Their Unity". *Synthese*. **133**.

[48] Popper 2002, pp. 10–11.

[49] Popper 2002, pp. 79–82.

[50] Parrott, Jim (August 9, 2007). "Chronicle for Societies Founded from 1323 to 1599". Scholarly Societies Project. Retrieved September 11, 2007.

[51] "Accademia Nazionale dei Lincei" (in Italian). 2006. Retrieved September 11, 2007.

[52] "History of the Royal Society". The Royal Society. Retrieved October 16, 2011.

[53] Meynell, G.G. "The French Academy of Sciences, 1666–91: A reassessment of the French Académie royale des sciences under Colbert (1666–83) and Louvois (1683–91)". Archived from the original on January 18, 2012. Retrieved October 13, 2011.

[54] Ziman, J.M. (1980). "The proliferation of scientific literature: a natural process". *Science*. **208** (4442): 369–71. Bibcode:1980Sci...208..369Z. PMID 7367863. doi:10.1126/science.7367863.

[55] Subramanyam, Krishna; Subramanyam, Bhadriraju (1981). *Scientific and Technical Information Resources*. CRC Press. ISBN 0-8247-8297-6. OCLC 232950234.

[56] "MEDLINE Fact Sheet". Washington DC: United States National Library of Medicine. Retrieved October 15, 2011.

[57] Petrucci, Mario. "Creative Writing – Science". Retrieved April 27, 2008.

[58] "Nobel Prize Facts". Nobel Foundation. Retrieved 2015-10-11.

[59] Spanier, Bonnie (1995). "From Molecules to Brains, Normal Science Supports Sexist Beliefs about Difference". *Im/partial Science: Gender Identity in Molecular Biology*. Indiana University Press. ISBN 9780253209689.

[60] "Parents explain more often to boys than to girls during shared scientific thinking". *Psychol. Sci.* **12**.

[61] Rosser, Sue V. *Breaking into the Lab: Engineering Progress for Women in Science*. New York: New York University Press. p. 7. ISBN 978-0-8147-7645-2.

[62] Goulden, Mark; Frasch, Karie; Mason, Mary Ann (2009). *Staying Competitive: Patching America's Leaky Pipeline in the Sciences*. University of Berkeley Law.

[63] *Change of Heart: Career intentions and the chemistry PhD*. Royal Society of Chemistry. 2008.

[64] Bush, Vannevar (July 1945). "Science the Endless Frontier". National Science Foundation. Retrieved November 4, 2016.

[65] "Main Science and Technology Indicators – 2008-1" (PDF). OECD. Archived from the original (PDF) on October 19, 2010.

[66] Dickson, David (October 11, 2004). "Science journalism must keep a critical edge". Science and Development Network. Archived from the original on June 21, 2010.

[67] Mooney, Chris (Nov–Dec 2004). "Blinded By Science, How 'Balanced' Coverage Lets the Scientific Fringe Hijack Reality". **43** (4). Columbia Journalism Review. Retrieved February 20, 2008.

[68] McIlwaine, S.; Nguyen, D. A. (2005). "Are Journalism Students Equipped to Write About Science?". *Australian Studies in Journalism*. **14**: 41–60. Retrieved February 20, 2008.

[69] "1988: Egg industry fury over salmonella claim". *BBC News*. December 3, 1988. Retrieved November 4, 2016.

[70] "– No Title –". *Truth Tobacco Industry Documents*. UC San Francisco. August 21, 1969. Retrieved November 4, 2016. Doubt is our product since it is the best means of competing with the 'body of fact' that exists in the mind of the general public. It is also the means of establishing a controversy.

[71] Horgan, John (December 18, 2005). "Political Science". *The New York Times*. Retrieved November 4, 2016.

[72] Mooney, Chris (2005). *The Republican War on Science*. Basic Books. ISBN 0-465-04676-2.

[73] "Scientific Certainty Argumentation Methods (SCAMs): Science and the Politics of Doubt". *Sociological Inquiry*. **78**.

[74] Berezow, Alex; Campbell, Alex. *Science Left Behind: Feel-good Fallacies and the Rise of the Anti-Scientific Left* (1st ed.). New York: PublicAffairs. ISBN 978-1-61039-164-1.

[75] "... [T]he logical empiricists thought that the great aim of science was to discover and establish *generalizations*." — Godfrey-Smith 2003, p. 41

[76] Godfrey-Smith 2003, p. 203.

[77] Godfrey-Smith 2003

[78] Popper called this *Conjecture and Refutation*. Godfrey-Smith 2003, pp. 117–8

[79] Popper, Karl (1972). *Objective Knowledge*.

[80] "Shut up and multiply". *LessWrong Wiki*. September 13, 2015. Retrieved November 4, 2016.

[81] Newton-Smith, W. H. (1994). *The Rationality of Science*. London: Routledge. p. 30. ISBN 0-7100-0913-5.

[82] Feyerabend 1993.

[83] Feyerabend, Paul (1987). *Farewell To Reason*. Verso. p. 100. ISBN 0-86091-184-5.

[84] Brugger, E. Christian (2004). "Casebeer, William D. Natural Ethical Facts: Evolution, Connectionism, and Moral Cognition". *The Review of Metaphysics*. **58** (2).

[85] Winther, Rasmus Grønfeldt (2015). "The Structure of Scientific Theories". *Stanford Encyclopedia of Philosophy*. Retrieved November 4, 2016.

[86] Popper 1996.

[87] Stanovich 2007

[88] Dawkins, Richard; Coyne, Jerry (September 2, 2005). "One side can be wrong". *The Guardian*. London.

[89] "Barry Stroud on Scepticism". philosophy bites. December 16, 2007. Retrieved February 5, 2012.

[90] Peirce (1877). "The Fixation of Belief", Popular Science Monthly, v. 12, pp. 1–15, see §IV on pp. 6–7. Reprinted *Collected Papers* v. 5, paragraphs 358–87 (see 374–6), *Writings* v. 3, pp. 242–57 (see 247–8), *Essential Peirce* v. 1, pp. 109–23 (see 114–15), and elsewhere.

[91] Peirce (1905), "Issues of Pragmaticism", *The Monist*, v. XV, n. 4, pp. 481–99, see "Character V" on p. 491. Reprinted in *Collected Papers* v. 5, paragraphs 438–63 (see 451), *Essential Peirce* v. 2, pp. 346–59 (see 353), and elsewhere.

[92] Peirce (1868), "Some Consequences of Four Incapacities", *Journal of Speculative Philosophy* v. 2, n. 3, pp. 140–57, see p. 141. Reprinted in *Collected Papers*, v. 5, paragraphs 264–317, *Writings* v. 2, pp. 211–42, *Essential Peirce* v. 1, pp. 28–55, and elsewhere.

[93] Feynman, Richard (1974). "Cargo Cult Science". *Center for Theoretical Neuroscience*. Columbia University. Archived from the original on March 4, 2005. Retrieved November 4, 2016.

[94] "Coping with fraud" (PDF). *The COPE Report 1999*: 11–18. Archived from the original (PDF) on September 28, 2007. Retrieved July 21, 2011. It is 10 years, to the month, since Stephen Lock ... Reproduced with kind permission of the Editor, The Lancet.

[95] In mathematics, Plato's *Meno* demonstrates that it is possible to know logical propositions, such as the Pythagorean theorem, and even to prove them, as cited by Crease 2009, pp. 35–41

[96] Ziman cites Polanyi 1958 chapter 12, as referenced in Ziman 1978

[97] Ziman 1978

[98] "To Live at All Is Miracle Enough — Richard Dawkins". RichardDawkins.net. May 10, 2006. Archived from the original on January 19, 2012. Retrieved February 5, 2012.

[99] Simonton, Dean Keith (2013). "After Einstein: Scientific genius is extinct". *Nature*. **493** (7434): 602–02. Bibcode:2013Natur.493..602S. PMID 23364725. doi:10.1038/493602a.

1.10 Sources

- Crease, Robert P. (2009). *The Great Equations*. New York: W.W. Norton. ISBN 978-0-393-06204-5.

- di Francia, Giuliano Toraldo (1976). *The Investigation of the Physical World*. Originally published in Italian as *L'Indagine del Mondo Fisico* by Giulio Einaudi editore 1976; first published in English by Cambridge University Press 1981. Cambridge: Cambridge University Press. ISBN 0-521-29925-X.

- Fara, Patricia (2009). *Science : a four thousand year history*. Oxford: Oxford University Press. p. 408. ISBN 978-0-19-922689-4.

- Feyerabend, Paul (1993). *Against Method* (3rd ed.). London: Verso. ISBN 0-86091-646-4.

- Godfrey-Smith, Peter (2003). *Theory and Reality*. Chicago 60637: University of Chicago. p. 272. ISBN 0-226-30062-5.

- Heilbron, J. L. (editor-in-chief) (2003). *The Oxford Companion to the History of Modern Science*. New York: Oxford University Press. ISBN 0-19-511229-6.

- Lindberg, David C. (2007). *The beginnings of Western science: the European Scientific tradition in philosophical, religious, and institutional context* (Second ed.). Chicago: Univ. of Chicago Press. ISBN 978-0-226-48205-7.

- Nola, Robert; Irzik, Gürol (2005). *Philosophy, science, education and culture*. Science & technology education library. **28**. Springer. ISBN 1-4020-3769-4.

- Polanyi, Michael (1958). *Personal Knowledge: Towards a Post-Critical Philosophy*. University of Chicago Press. ISBN 0-226-67288-3

- Popper, Karl Raimund (1996) [1984]. *In search of a better world: lectures and essays from thirty years*. New York, NY: Routledge. ISBN 0-415-13548-6.

- Popper, Karl R. (2002) [1959]. *The Logic of Scientific Discovery*. New York, NY: Routledge Classics. ISBN 0-415-27844-9. OCLC 59377149.

- Stanovich, Keith E. (2007). *How to Think Straight About Psychology*. Boston: Pearson Education. ISBN 978-0-205-68590-5.

- Ziman, John (1978). *Reliable knowledge: An exploration of the grounds for belief in science*. Cambridge: Cambridge University Press. p. 197. ISBN 0-521-22087-4

1.11 Further reading

- Augros, Robert M., Stanciu, George N., *The New Story of Science: mind and the universe*, Lake Bluff, Ill.: Regnery Gateway, c1984. ISBN 0-89526-833-7

- Becker, Ernest (1968). *The structure of evil; an essay on the unification of the science of man*. New York: G. Braziller.

- Cole, K. C., *Things your teacher never told you about science: Nine shocking revelations* Newsday, Long Island, New York, March 23, 1986, pp. 21+

- Crease, Robert P. (2011). *World in the Balance: the historic quest for an absolute system of measurement*. New York: W.W. Norton. p. 317. ISBN 978-0-393-07298-3.

- Feyerabend, Paul (2005). *Science, history of the philosophy*, as cited in Honderich, Ted (2005). *The Oxford companion to philosophy*. Oxford Oxfordshire: Oxford University Press. ISBN 0-19-926479-1. OCLC 173262485.

- Feynman, Richard P. (1999). Robbins, Jeffrey, ed. *The pleasure of finding things out the best short works of Richard P. Feynman*. Cambridge, Mass.: Perseus Books. ISBN 0465013120.

- Feynman, R.P. (1999). *The Pleasure of Finding Things Out: The Best Short Works of Richard P. Feynman*. Perseus Books Group. ISBN 0-465-02395-9. OCLC 181597764.

- Feynman, Richard "Cargo Cult Science"

- Gaukroger, Stephen (2006). *The Emergence of a Scientific Culture: Science and the Shaping of Modernity 1210–1685*. Oxford: Oxford University Press. ISBN 0-19-929644-8.

- Gopnik, Alison, "Finding Our Inner Scientist", Daedalus, Winter 2004.

- Krige, John, and Dominique Pestre, eds., *Science in the Twentieth Century*, Routledge 2003, ISBN 0-415-28606-9

- Levin, Yuval (2008). *Imagining the Future: Science and American Democracy*. New York, Encounter Books. ISBN 1-59403-209-2

- Lindberg, D. C. (1976). *Theories of Vision from al-Kindi to Kepler*. Chicago: Univ. of Chicago Pr.

- Kuhn, Thomas, *The Structure of Scientific Revolutions*, 1962.

- William F., McComas (1998). "The principal elements of the nature of science: Dispelling the myths". In McComas, William F. *The nature of science in science education: rationales and strategies* (PDF). Springer. ISBN 978-0-7923-6168-8.

- Needham, Joseph (1954). "Science and Civilisation in China: Introductory Orientations". **1**. Cambridge University Press.

- Obler, Paul C.; Estrin, Herman A. (1962). *The New Scientist: Essays on the Methods and Values of Modern Science*. Anchor Books, Doubleday.

- Papineau, David. (2005). *Science, problems of the philosophy of.*, as cited in Honderich, Ted (2005). *The Oxford companion to philosophy*. Oxford Oxfordshire: Oxford University Press. ISBN 0-19-926479-1. OCLC 173262485.

- Parkin, D. (1991). "Simultaneity and Sequencing in the Oracular Speech of Kenyan Diviners". In Philip M. Peek. *African Divination Systems: Ways of Knowing*. Indianapolis, IN: Indiana University Press.

- Russell, Bertrand (1985) [1952]. *The Impact of Science on Society*. London: Unwin. ISBN 0-04-300090-8.

- Rutherford, F. James; Ahlgren, Andrew (1990). *Science for all Americans*. New York, NY: American Association for the Advancement of Science, Oxford University Press. ISBN 0-19-506771-1.

- Smith, A. Mark (2001). Written at Philadelphia. *Alhacen's Theory of Visual Perception: A Critical Edition, with English Translation and Commentary, of the First Three Books of Alhacen's* De Aspectibus, *the Medieval Latin Version of Ibn al-Haytham's* Kitāb al-Manāẓir, *2 vols*. Transactions of the American Philosophical Society. **91**. Philadelphia: American Philosophical Society. ISBN 0-87169-914-1. OCLC 47168716. Books I-III (2001 — **91**(4)) Vol 1 Commentary and Latin text via JSTOR; — **91**(5) Vol 2 English translation, Book I:TOC pp. 339–41, Book II:TOC pp. 415–16, Book III:TOC pp. 559–60, Notes 681ff, Bibl. via JSTOR

- Thurs, Daniel Patrick (2007). *Science Talk: Changing Notions of Science in American Popular Culture*. New Brunswick, NJ: Rutgers University Press. pp. 22–52. ISBN 978-0-8135-4073-3.

1.12 External links

Publications

- "*GCSE Science textbook*". Wikibooks.org

Resources

- Euroscience:
    - "ESOF: Euroscience Open Forum". Archived from the original on June 10, 2010.

- Science Development in the *Latin American docta*

- Classification of the Sciences in *Dictionary of the History of Ideas*. (Dictionary's new electronic format is badly botched, entries after "Design" are inaccessible. *Internet Archive* old version).

- "Nature of Science" *University of California Museum of Paleontology*

- United States Science Initiative Selected science information provided by US Government agencies, including research & development results

- How science works *University of California Museum of Paleontology*

Chapter 2

Formal science

Formal sciences are language disciplines concerned with formal systems, such as logic, mathematics, statistics, theoretical computer science, information theory, game theory, systems theory, decision theory, and theoretical linguistics. Whereas the natural sciences and social sciences seek to characterize physical systems and social systems respectively using empirical methods, the formal sciences are language tools concerned with characterizing abstract structures described by sign systems. The formal sciences aid the natural sciences by providing information about the structures the latter use to describe the world, and what inferences may be made about them.

2.1 History

Formal sciences began before the formulation of the scientific method, with the most ancient mathematical texts dating back to 1800 BC (Babylonian mathematics), 1600 BC (Egyptian mathematics) and 1000 BC (Indian mathematics). From then on different cultures such as the Indian, Greek and Islamic mathematicians made major contributions to mathematics, while the Chinese and Japanese, independently of more distant cultures, developed their own mathematical tradition.

Besides mathematics, logic is another example of one of oldest subjects in the field of the formal sciences. As an explicit analysis of the methods of reasoning, logic received sustained development originally in three places: India from the 6th century BC, China in the 5th century BC, and Greece between the 4th century BC and the 1st century BC. The formally sophisticated treatment of modern logic descends from the Greek tradition, being informed from the transmission of Aristotelian logic, which was then further developed by Islamic logicians. The Indian tradition also continued into the early modern period. The native Chinese tradition did not survive beyond antiquity, though Indian logic was later adopted in medieval China.

As a number of other disciplines of formal science rely heavily on mathematics, they did not exist until mathematics had developed into a relatively advanced level. Pierre de Fermat and Blaise Pascal (1654), and Christiaan Huygens (1657) started the earliest study of probability theory. In the early 1800s, Gauss and Laplace developed the mathematical theory of statistics, which also explained the use of statistics in insurance and governmental accounting. Mathematical statistics was recognized as a mathematical discipline in the early 20th century.

In the mid-20th century, mathematics was broadened and enriched by the rise of new mathematical sciences and engineering disciplines such as operations research and systems engineering. These sciences benefited from basic research in electrical engineering and then by the development of electrical computing, which also stimulated information theory, numerical analysis (scientific computing), and theoretical computer science. Theoretical computer science also benefits from the discipline of mathematical logic, which included the theory of computation.

2.2 Differences from other forms of science

> One reason why mathematics enjoys special esteem, above all other sciences, is that its laws are absolutely certain and indisputable, while those of other sciences are to some extent debatable and in constant danger of being overthrown by newly discovered facts.
> — Albert Einstein[1]

As opposed to empirical sciences (natural and social), the formal sciences do not involve empirical procedures. They also do not presuppose knowledge of contingent facts, or describe the real world. In this sense, formal sciences are both logically and methodologically a priori, for their content and validity are independent of any empirical procedures.

Although formal sciences are conceptual systems, lacking empirical content, this does not mean that they have no relation to the real world. But this relation is such that their formal statements hold in all possible conceivable worlds (see valid formula) – whereas, statements based on empirical theories, such as, say, general relativity or evolutionary biology, do not hold in all possible worlds, and may eventually turn out not to hold in this world as well. That is why formal sciences are applicable in all domains and useful in all empirical sciences.

Because of their non-empirical nature, formal sciences are construed by outlining a set of axioms and definitions from which other statements (theorems) are deduced. In other words, theories in formal sciences contain no synthetic statements; all their statements are analytic.[2][3]

2.3 See also

- Rationalism
- Abstract structure
- Abstraction in mathematics
- Abstraction in computer science
- Formal grammar
- Formal language
- Formal method
- Formal system
- Mathematical model

2.4 References

[1] Albert Einstein (1923). "Geometry and Experience". *Sidelights on relativity*. Courier Dover Publications. p. 27. Reprinted by Dover (2010). ISBN 978-0-486-24511-9.

[2] Carnap, Rudolf (1938). "Logical Foundations of the Unity of Science". *International Encyclopaedia of Unified Science*. I. Chicago: University of Chicago Press.

[3] Bill, Thompson (2007). "2.4 Formal Science and Applied Mathematics", *The Nature of Statistical Evidence*, Lecture Notes in Statistics, **189** (1st ed.), Springer, p. 15

2.5 Further reading

- Mario Bunge (1985). *Philosophy of Science and Technology*. Springer.
- Mario Bunge (1998). *Philosophy of Science*. Rev. ed. of: *Scientific research*. Berlin, New York: Springer-Verlag, 1967.
- C. West Churchman (1940). *Elements of Logic and Formal Science*, J.B. Lippincott Co., New York.
- James Franklin (1994). The formal sciences discover the philosophers' stone. In: *Studies in History and Philosophy of Science*. Vol. 25, No. 4, pp. 513–533, 1994
- Stephen Leacock (1906). *Elements of Political Science*. Houghton, Mifflin Co, 417 pp.
- Bernt P. Stigum (1990). *Toward a Formal Science of Economics*. MIT Press
- Marcus Tomalin (2006), *Linguistics and the Formal Sciences*. Cambridge University Press
- William L. Twining (1997). *Law in Context: Enlarging a Discipline*. 365 pp.

2.6 External links

- Interdisciplinary conferences — *Foundations of the Formal Sciences*

2.7 References

- Popper, Karl R. (2002) [1959]. *The Logic of Scientific Discovery*. New York, NY: Routledge Classics. ISBN 0-415-27844-9. OCLC 59377149.

Chapter 3

Logic

This article is about the systematic study of the form of arguments. For other uses, see Logic (disambiguation).

Logic (from the Ancient Greek: λογική, *logikē*[1]), originally meaning "the word" or "what is spoken" (but coming to mean "thought" or "reason"), is generally held to consist of the systematic study of the form of arguments. A valid argument is one where there is a specific relation of logical support between the assumptions of the argument and its conclusion. (In ordinary discourse, the conclusion of such an argument may be signified by words like *therefore*, *hence*, *ergo* and so on.)

There is no universal agreement as to the exact scope and subject matter of logic (see § Rival conceptions, below), but it has traditionally included the classification of arguments, the systematic exposition of the 'logical form' common to all valid arguments, the study of inference, including fallacies, and the study of semantics, including paradoxes. Historically, logic has been studied in philosophy (since ancient times) and mathematics (since the mid-1800s), and recently logic has been studied in computer science, linguistics, psychology, and other fields.

3.1 Concepts

The concept of logical form is central to logic. The validity of an argument is determined by its logical form, not by its content. Traditional Aristotelian syllogistic logic and modern symbolic logic are examples of formal logic.

- **Informal logic** is the study of natural language arguments. The study of fallacies is an important branch of informal logic. Since much informal argument is not strictly speaking deductive, on some conceptions of logic, informal logic is not logic at all. See 'Rival conceptions', below.

- **Formal logic** is the study of inference with purely formal content. An inference possesses a *purely formal content* if it can be expressed as a particular application of a wholly abstract rule, that is, a rule that is not about any particular thing or property. The works of Aristotle contain the earliest known formal study of logic. Modern formal logic follows and expands on Aristotle.[2] In many definitions of logic, logical inference and inference with purely formal content are the same. This does not render the notion of informal logic vacuous, because no formal logic captures all of the nuances of natural language.

- **Symbolic logic** is the study of symbolic abstractions that capture the formal features of logical inference.[3][4] Symbolic logic is often divided into two main branches: propositional logic and predicate logic.

- **Mathematical logic** is an extension of symbolic logic into other areas, in particular to the study of model theory, proof theory, set theory, and recursion theory.

However, agreement on what logic is has remained elusive, and although the field of universal logic has studied the common structure of logics, in 2007 Mossakowski et al. commented that "it is embarrassing that there is no widely acceptable formal definition of 'a logic'".[5]

3.1.1 Logical form

Main article: Logical form

Logic is generally considered **formal** when it analyzes and represents the *form* of any valid argument type. The form of an argument is displayed by representing its sentences in the formal grammar and symbolism of a logical language to make its content usable in formal inference. Simply put, formalising simply means translating English sentences into the language of logic.

This is called showing the *logical form* of the argument. It is necessary because indicative sentences of ordinary language show a considerable variety of form and complexity that makes their use in inference impractical. It requires, first, ignoring those grammatical features irrelevant to logic (such as gender and declension, if the argument is in Latin), replacing conjunctions irrelevant to logic (such as "but") with logical conjunctions like "and" and replacing ambiguous, or alternative logical expressions ("any", "every", etc.) with expressions of a standard type (such as "all", or the universal quantifier ∀).

Second, certain parts of the sentence must be replaced with schematic letters. Thus, for example, the expression "all Ps are Qs" shows the logical form common to the sentences "all men are mortals", "all cats are carnivores", "all Greeks are philosophers", and so on. The schema can further be condensed into the formula $A(P,Q)$, where the letter A indicates the judgement 'all - are -'.

The importance of form was recognised from ancient times. Aristotle uses variable letters to represent valid inferences in *Prior Analytics*, leading Jan Łukasiewicz to say that the introduction of variables was "one of Aristotle's greatest inventions".[6] According to the followers of Aristotle (such as Ammonius), only the logical principles stated in schematic terms belong to logic, not those given in concrete terms. The concrete terms "man", "mortal", etc., are analogous to the substitution values of the schematic placeholders P, Q, R, which were called the "matter" (Greek *hyle*) of the inference.

There is a big difference between the kinds of formulas seen in traditional term logic and the predicate calculus that is the fundamental advance of modern logic. The formula $A(P,Q)$ (all Ps are Qs) of traditional logic corresponds to the more complex formula $\forall x.(P(x) \to Q(x))$ in predicate logic, involving the logical connectives for universal quantification and implication rather than just the predicate letter A and using variable arguments $P(x)$ where traditional logic uses just the term letter P. With the complexity comes power, and the advent of the predicate calculus inaugurated revolutionary growth of the subject.

3.1.2 Semantics

Main article: Semantics of logic

The validity of an argument depends upon the meaning or *semantics* of the sentences that make it up.

Aristotle's Organon, especially On Interpretation, gives a cursory outline of semantics which the scholastic logicians, particularly in the thirteenth and fourteenth century, developed into a complex and sophisticated theory, called Supposition Theory. This showed how the truth of simple sentences, expressed schematically, depend on how the terms 'supposit' or *stand for* certain extra-linguistic items. For example, in part II of his Summa Logicae, William of Ockham presents a comprehensive account of the necessary and sufficient conditions for the truth of simple sentences, in order to show which arguments are valid and which are not. Thus "every A is B' is true if and only if there is something for which 'A' stands, and there is nothing for which 'A' stands, for which 'B' does not also stand." [7]

Early modern logic defined semantics purely as a relation between ideas. Antoine Arnauld in the Port Royal Logic, says that 'after conceiving things by our ideas, we compare these ideas, and, finding that some belong together and some do not, we unite or separate them. This is called *affirming* or *denying*, and in general *judging*.[8] Thus truth and falsity are no more than the agreement or disagreement of ideas. This suggests obvious difficulties, leading Locke to distinguish between 'real' truth, when our ideas have 'real existence' and 'imaginary' or 'verbal' truth, where ideas like harpies or centaurs exist only in the mind.[9] This view (psychologism) was taken to the extreme in the nineteenth century, and is generally held by modern logicians to signify a low point in the decline of logic before the twentieth century.

Modern semantics is in some ways closer to the medieval view, in rejecting such psychological truth-conditions. However, the introduction of quantification, needed to solve the problem of multiple generality, rendered impossible the kind of subject-predicate analysis that underlies medieval semantics. The main modern approach is *model-theoretic semantics*, based on Alfred Tarski's semantic theory of truth. The approach assumes that the meaning of the various parts of the propositions are given by the possible ways we can give a recursively specified group of interpretation functions from them to some predefined domain of discourse: an interpretation of first-order predicate logic is given by a mapping from terms to a universe of individuals, and a mapping from propositions to the truth values "true" and "false". Model-theoretic semantics is one of the fundamental concepts of model theory. Modern semantics also admits rival approaches, such as the proof-theoretic semantics that associates the meaning of propositions with the roles that they can play in inferences, an approach that ultimately derives from the work of Gerhard Gentzen on structural proof theory and is heavily influenced by Ludwig Wittgenstein's later philosophy, especially his aphorism "meaning is use".

3.1.3 Inference

Inference is not to be confused with *implication*. An implication is a sentence of the form 'If p then q', and can be true or false. The Stoic logician Philo of Megara was the first to define the truth conditions of such an implication: false only when the antecedent p is true and the consequent q is false, in all other cases true. An inference, on the other hand, consists of two separately asserted propositions of the form 'p therefore q'. An inference is not true or false, but valid or invalid. However, there is a connection between implication and inference, as follows: if the implication 'if p then q' is *true*, the inference 'p therefore q' is *valid*. This was given an apparently paradoxical formulation by Philo, who said that the implication 'if it is day, it is night' is true only at night, so the inference 'it is day, therefore it is night' is valid in the night, but not in the day.

The theory of inference (or 'consequences') was systematically developed in medieval times by logicians such as William of Ockham and Walter Burley. It is uniquely medieval, though it has its origins in Aristotle's Topics and Boethius' *De Syllogismis hypotheticis*. This is why many terms in logic are Latin. For example, the rule that licenses the move from the implication 'if p then q' plus the assertion of its antecedent p, to the assertion of the consequent q is known as modus ponens (or 'mode of positing'). Its Latin formulation is 'Posito antecedente ponitur consequens'. The Latin formulations of many other rules such as 'ex falso quodlibet' (anything follows from a falsehood), 'reductio ad absurdum' (disproof by showing the consequence is absurd) also date from this period.

However, the theory of consequences, or of the so-called 'hypothetical syllogism' was never fully integrated into the theory of the 'categorical syllogism'. This was partly because of the resistance to reducing the categorical judgment 'Every S is P' to the so-called hypothetical judgment 'if anything is S, it is P'. The first was thought to imply 'some S is P', the second was not, and as late as 1911 in the Encyclopædia Britannica article on Logic, we find the Oxford logician T.H. Case arguing against Sigwart's and Brentano's modern analysis of the universal proposition.

3.1.4 Logical systems

Main article: Formal system

A formal system is an organization of terms used for the analysis of deduction. It consists of an alphabet, a language over the alphabet to construct sentences, and a rule for deriving sentences. Among the important properties that logical systems can have are:

- **Consistency**, which means that no theorem of the system contradicts another.[10]

- **Validity**, which means that the system's rules of proof never allow a false inference from true premises.

- **Completeness**, which means that if a formula is true, it can be proven, i.e. is a *theorem* of the system.

- **Soundness**, meaning that if any formula is a theorem of the system, it is true. This is the converse of completeness. (Note that in a distinct philosophical use of the term, an argument is sound when it is both valid and its premises are true).[11]

Some logical systems do not have all four properties. As an example, Kurt Gödel's incompleteness theorems show that sufficiently complex formal systems of arithmetic cannot be consistent and complete;[4] however, first-order predicate logics not extended by specific axioms to be arithmetic formal systems with equality can be complete and consistent.[12]

3.1.5 Logic and rationality

Main article: Logic and rationality

As the study of argument is of clear importance to the reasons that we hold things to be true, logic is of essential importance to rationality. Here we have defined logic to be "the systematic study of the form of arguments"; the reasoning behind argument is of several sorts, but only some of these arguments fall under the aegis of logic proper.

Deductive reasoning concerns the logical consequence of given premises and is the form of reasoning most closely connected to logic. On a narrow conception of logic (see below) logic concerns just deductive reasoning, although such a narrow conception controversially excludes most of what is called informal logic from the discipline.

There are other forms of reasoning that are rational but that are generally not taken to be part of logic. These include inductive reasoning, which covers forms of inference that move from collections of particular judgements to universal judgements, and abductive reasoning,[13] which is a form of inference that goes from observation to a hypothesis that accounts for the reliable data (observation) and seeks to explain relevant evidence. The American philosopher Charles Sanders Peirce (1839–1914) first introduced the term as "guessing".[14] Peirce said that to *abduce* a hypothetical explanation a from an observed surprising circumstance b is to surmise that a may be true because then b would be a matter of course.[15] Thus, to abduce a from b involves determining that a is sufficient (or nearly sufficient), but not necessary, for b.

While inductive and abductive inference are not part of logic proper, the methodology of logic has been applied to them with some degree of success. For example, the notion of deductive validity (where an inference is deductively valid if and only if there is no possible situation in which all the premises are true but the conclusion false) exists in an analogy to the notion of inductive validity, or "strength", where an inference is inductively strong if and only if its premises give some degree of probability to its conclusion. Whereas the notion of deductive validity can be rigorously stated for systems of formal logic in terms of the well-understood notions of semantics, inductive validity requires us to define a reliable generalization of some set of observations. The task of providing this definition may be approached in various ways, some less formal than others; some of these definitions may use logical association rule induction, while others may use mathematical models of probability such as decision trees.

3.1.6 Rival conceptions

Main article: Rival conceptions of logic

Logic arose (see below) from a concern with correctness of argumentation. Modern logicians usually wish to ensure that logic studies just those arguments that arise from appropriately general forms of inference. For example, Thomas Hofweber writes in the *Stanford Encyclopedia of Philosophy* that logic "does not, however, cover good reasoning as a whole. That is the job of the theory of rationality. Rather it deals with inferences whose validity can be traced back to the formal features of the representations that are involved in that inference, be they linguistic, mental, or other representations."[16]

Logic has been defined as "the study of arguments correct in virtue of their form". This has not been the definition taken in this article, but the idea that logic treats special forms of argument, deductive argument, rather than argument in general, has a history in logic that dates back at least to logicism in mathematics (19th and 20th centuries) and the advent of the influence of mathematical logic on philosophy. A consequence of taking logic to treat special kinds of argument is that it leads to identification of special kinds of truth, the logical truths (with logic equivalently being the study of logical truth), and excludes many of the original objects of study of logic that are treated as informal logic. Robert Brandom has argued against the idea that logic is the study of a special kind of logical truth, arguing that instead one can talk of the logic of material inference (in the terminology of Wilfred Sellars), with logic making explicit the commitments that were originally implicit in informal inference.[17]

3.2 History

Main article: History of logic

In Europe, logic was first developed by Aristotle.[18]

Aristotle, 384–322 BCE.

Aristotelian logic became widely accepted in science and mathematics and remained in wide use in the West until the early 19th century.[19] Aristotle's system of logic was responsible for the introduction of hypothetical syllogism,[20] temporal modal logic,[21][22] and inductive logic,[23] as well as influential terms such as terms, predicables, syllogisms and propositions. In Europe during the later medieval period, major efforts were made to show that Aristotle's ideas were compatible with Christian faith. During the High Middle Ages, logic became a main focus of philosophers, who would engage in critical logical analyses of philosophical arguments, often using variations of the methodology of scholasticism. In 1323, William of Ockham's influential *Summa Logicae* was released. By the 18th century, the structured approach to arguments had degenerated and fallen out of favour, as depicted in Holberg's satirical play *Erasmus Montanus*.

The Chinese logical philosopher Gongsun Long (c. 325–250 BCE) proposed the paradox "One and one cannot become two, since neither becomes two."[24] In China, the tradition of scholarly investigation into logic, however, was repressed by the Qin dynasty following the legalist philosophy of Han Feizi.

In India, innovations in the scholastic school, called Nyaya, continued from ancient times into the early 18th century with the Navya-Nyaya school. By the 16th century, it developed theories resembling modern logic, such as Gottlob Frege's "distinction between sense and reference of proper names" and his "definition of number", as well as the theory of "restrictive conditions for universals" anticipating some of the developments in modern set theory.[25] Since 1824, Indian logic attracted the attention of many Western scholars, and has had an influence on important 19th-century logicians such as Charles Babbage, Augustus De Morgan, and George Boole.[26] In the 20th century, Western philosophers like Stanislaw Schayer and Klaus Glashoff have explored Indian logic more extensively.

The syllogistic logic developed by Aristotle predominated in the West until the mid-19th century, when interest in the foundations of mathematics stimulated the development of symbolic logic (now called mathematical logic). In 1854, George Boole published *An Investigation of the Laws of Thought on Which are Founded the Mathematical Theories of Logic and Probabilities*, introducing symbolic logic and the principles of what is now known as Boolean logic. In 1879, Gottlob Frege published *Begriffsschrift*, which inaugurated modern logic with the invention of quantifier notation. From 1910 to 1913, Alfred North Whitehead and Bertrand Russell published *Principia Mathematica*[3] on the foundations of mathematics, attempting to derive mathematical truths from axioms and inference rules in symbolic logic. In 1931, Gödel raised serious problems with the foundationalist program and logic ceased to focus on such issues.

The development of logic since Frege, Russell, and Wittgenstein had a profound influence on the practice of philosophy and the perceived nature of philosophical problems (see analytic philosophy) and philosophy of mathematics. Logic, especially sentential logic, is implemented in computer logic circuits and is fundamental to computer science. Logic is commonly taught by university philosophy departments, often as a compulsory discipline.

3.3 Types

3.3.1 Syllogistic logic

Main article: Aristotelian logic

The *Organon* was Aristotle's body of work on logic, with the *Prior Analytics* constituting the first explicit work in formal logic, introducing the syllogistic.[27] The parts of syllogistic

A depiction from the 15th century of the square of opposition, which expresses the fundamental dualities of syllogistic.

logic, also known by the name term logic, are the analysis of the judgements into propositions consisting of two terms that are related by one of a fixed number of relations, and the expression of inferences by means of syllogisms that consist of two propositions sharing a common term as premise, and a conclusion that is a proposition involving the two unrelated terms from the premises.

Aristotle's work was regarded in classical times and from medieval times in Europe and the Middle East as the very picture of a fully worked out system. However, it was not alone: the Stoics proposed a system of propositional logic that was studied by medieval logicians. Also, the problem of multiple generality was recognized in medieval times. Nonetheless, problems with syllogistic logic were not seen as being in need of revolutionary solutions.

Today, some academics claim that Aristotle's system is generally seen as having little more than historical value (though there is some current interest in extending term logics), regarded as made obsolete by the advent of propositional logic and the predicate calculus. Others use Aristotle in argumentation theory to help develop and critically question argumentation schemes that are used in artificial intelligence and legal arguments.

I was upset. I had always believed logic was a universal weapon, and now I realized how its validity depended on the way it was employed.[28]

3.3.2 Propositional logic

Main article: Propositional calculus

A propositional calculus or logic (also a sentential calculus) is a formal system in which formulae representing propositions can be formed by combining atomic propositions using logical connectives, and in which a system of formal proof rules establishes certain formulae as "theorems". An example of a theorem of propositional logic is $A \to B \to A$, which says that if A holds, then B implies A.

3.3.3 Predicate logic

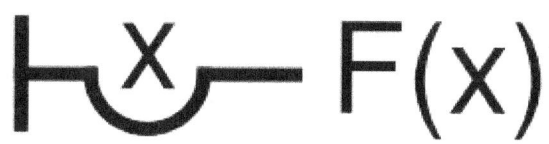

Gottlob Frege's Begriffschrift *introduced the notion of quantifier in a graphical notation, which here represents the judgement that* $\forall x. F(x)$ *is true.*

Main article: Predicate logic

Predicate logic is the generic term for symbolic formal systems such as first-order logic, second-order logic, many-sorted logic, and infinitary logic. It provides an account of quantifiers general enough to express a wide set of arguments occurring in natural language. For example, Bertrand Russell's famous barber paradox, "there is a man who shaves all and only men who do not shave themselves" can be formalised by the sentence $(\exists x)(\text{man}(x) \land (\forall y)(\text{man}(y) \to (\text{shaves}(x,y) \leftrightarrow \neg\text{shaves}(y,y))))$, using the non-logical predicate $\text{man}(x)$ to indicate that x is a man, and the non-logical relation $\text{shaves}(x,y)$ to indicate that x shaves y; all other symbols of the formulae are logical, expressing the universal and existential quantifiers, conjunction, implication, negation and biconditional.

Whilst Aristotelian syllogistic logic specifies a small number of forms that the relevant part of the involved judgements may take, predicate logic allows sentences to be analysed into subject and argument in several additional ways—allowing predicate logic to solve the problem of multiple generality that had perplexed medieval logicians.

The development of predicate logic is usually attributed to Gottlob Frege, who is also credited as one of the founders of analytical philosophy, but the formulation of predicate logic most often used today is the first-order logic presented in Principles of Mathematical Logic by David Hilbert and Wilhelm Ackermann in 1928. The analytical generality of predicate logic allowed the formalization of mathematics, drove the investigation of set theory, and allowed the development of Alfred Tarski's approach to model theory. It provides the foundation of modern mathematical logic.

Frege's original system of predicate logic was second-order, rather than first-order. Second-order logic is most prominently defended (against the criticism of Willard Van Orman Quine and others) by George Boolos and Stewart Shapiro.

3.3.4 Modal logic

Main article: Modal logic

In languages, modality deals with the phenomenon that subparts of a sentence may have their semantics modified by special verbs or modal particles. For example, "*We go to the games*" can be modified to give "*We should go to the games*", and "*We can go to the games*" and perhaps "*We will go to the games*". More abstractly, we might say that modality affects the circumstances in which we take an assertion to be satisfied. Confusing modality is known as the modal fallacy.

Aristotle's logic is in large parts concerned with the theory of non-modalized logic. Although, there are passages in his work, such as the famous sea-battle argument in *De Interpretatione* § 9, that are now seen as anticipations of modal logic and its connection with potentiality and time, the earliest formal system of modal logic was developed by Avicenna, whom ultimately developed a theory of "temporally modalized" syllogistic.[29]

While the study of necessity and possibility remained important to philosophers, little logical innovation happened until the landmark investigations of Clarence Irving Lewis in 1918, who formulated a family of rival axiomatizations of the alethic modalities. His work unleashed a torrent of new work on the topic, expanding the kinds of modality treated to include deontic logic and epistemic logic. The seminal work of Arthur Prior applied the same formal language to treat temporal logic and paved the way for the marriage of the two subjects. Saul Kripke discovered (contemporaneously with rivals) his theory of frame semantics, which revolutionized the formal technology available to modal logicians and gave a new graph-theoretic way of looking at modality that has driven many applications in computational linguistics and computer science, such as dynamic logic.

3.3.5 Informal reasoning and dialectic

Main articles: Informal logic and Logic and dialectic

The motivation for the study of logic in ancient times was clear: it is so that one may learn to distinguish good arguments from bad arguments, and so become more effective in argument and oratory, and perhaps also to become a better person. Half of the works of Aristotle's Organon treat inference as it occurs in an informal setting, side by side with the development of the syllogistic, and in the Aristotelian school, these informal works on logic were seen as complementary to Aristotle's treatment of rhetoric.

This ancient motivation is still alive, although it no longer takes centre stage in the picture of logic; typically dialectical logic forms the heart of a course in critical thinking, a compulsory course at many universities. Dialectic has been linked to logic since ancient times, but it has not been until recent decades that European and American logicians have attempted to provide mathematical foundations for logic and dialectic by formalising dialectical logic. Dialectical logic is also the name given to the special treatment of dialectic in Hegelian and Marxist thought. There have been pre-formal treatises on argument and dialectic, from authors such as Stephen Toulmin (*The Uses of Argument*), Nicholas Rescher (*Dialectics*),[30][31][32] and van Eemeren and Grootendorst (Pragma-dialectics). Theories of defeasible reasoning can provide a foundation for the formalisation of dialectical logic and dialectic itself can be formalised as moves in a game, where an advocate for the truth of a proposition and an opponent argue. Such games can provide a formal game semantics for many logics.

Argumentation theory is the study and research of informal logic, fallacies, and critical questions as they relate to every day and practical situations. Specific types of dialogue can be analyzed and questioned to reveal premises, conclusions, and fallacies. Argumentation theory is now applied in artificial intelligence and law.

3.3.6 Mathematical logic

Main article: Mathematical logic

Mathematical logic comprises two distinct areas of research: the first is the application of the techniques of formal logic to mathematics and mathematical reasoning, and the second, in the other direction, the application of mathematical techniques to the representation and analysis of formal logic.[33]

The earliest use of mathematics and geometry in relation to logic and philosophy goes back to the ancient Greeks such as Euclid, Plato, and Aristotle.[34] Many other ancient and medieval philosophers applied mathematical ideas and methods to their philosophical claims.[35]

One of the boldest attempts to apply logic to mathematics was the logicism pioneered by philosopher-logicians such as Gottlob Frege and Bertrand Russell. Mathematical theories were supposed to be logical tautologies, and the programme was to show this by means of a reduction of mathematics to logic.[3] The various attempts to carry this out met with failure, from the crippling of Frege's project in his *Grundgesetze* by Russell's paradox, to the defeat of Hilbert's program by Gödel's incompleteness theorems.

Both the statement of Hilbert's program and its refutation by Gödel depended upon their work establishing the second area of mathematical logic, the application of mathematics to logic in the form of proof theory.[36] Despite the negative nature of the incompleteness theorems, Gödel's completeness theorem, a result in model theory and another application of mathematics to logic, can be understood as showing how close logicism came to being true: every rigorously defined mathematical theory can be exactly captured by a first-order logical theory; Frege's proof calculus is enough to *describe* the whole of mathematics, though not *equivalent* to it.

If proof theory and model theory have been the foundation of mathematical logic, they have been but two of the four pillars of the subject.[37] Set theory originated in the study of the infinite by Georg Cantor, and it has been the source of many of the most challenging and important issues in mathematical logic, from Cantor's theorem, through the status of the Axiom of Choice and the question of the independence of the continuum hypothesis, to the modern debate on large cardinal axioms.

Recursion theory captures the idea of computation in logical and arithmetic terms; its most classical achievements are the undecidability of the Entscheidungsproblem by Alan Turing, and his presentation of the Church–Turing thesis.[38] Today recursion theory is mostly concerned with the more refined problem of complexity classes—when is a problem efficiently solvable?—and the classification of degrees of unsolvability.[39]

3.3.7 Philosophical logic

Main article: Philosophical logic

Philosophical logic deals with formal descriptions of ordinary, non-specialist ("natural") language, that is strictly only about the arguments within philosophy's other branches. Most philosophers assume that the bulk of everyday reasoning can be captured in logic if a method or methods to translate ordinary language into that logic can be found. Philosophical logic is essentially a continuation of the traditional discipline called "logic" before the invention of mathematical logic. Philosophical logic has a much greater concern with the connection between natural language and logic. As a result, philosophical logicians have contributed a great deal to the development of non-standard logics (e.g. free logics, tense logics) as well as various extensions of classical logic (e.g. modal logics) and non-standard semantics for such logics (e.g. Kripke's supervaluationism in the semantics of logic).

Logic and the philosophy of language are closely related. Philosophy of language has to do with the study of how our language engages and interacts with our thinking. Logic has an immediate impact on other areas of study. Studying logic and the relationship between logic and ordinary speech can help a person better structure his own arguments and critique the arguments of others. Many popular arguments are filled with errors because so many people are untrained in logic and unaware of how to formulate an argument correctly.[40][41]

3.3.8 Computational logic

Main article: Logic in computer science

Logic cut to the heart of computer science as it

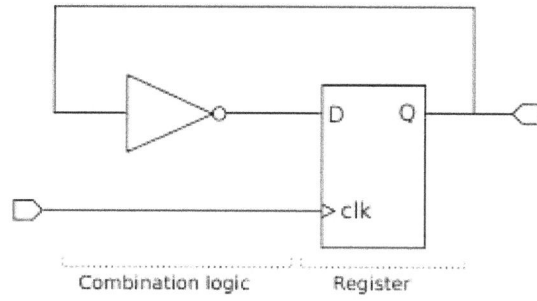

A simple toggling circuit is expressed using a logic gate and a synchronous register.

emerged as a discipline: Alan Turing's work on the *Entscheidungsproblem* followed from Kurt Gödel's work on the incompleteness theorems. The notion of the general purpose computer that came from this work was of fundamental importance to the designers of the computer machinery in the 1940s.

In the 1950s and 1960s, researchers predicted that when human knowledge could be expressed using logic with mathematical notation, it would be possible to create a machine that reasons, or artificial intelligence. This was more difficult than expected because of the complexity of human reasoning. In logic programming, a program consists of a set of axioms and rules. Logic programming systems such as Prolog compute the consequences of the axioms and rules in order to answer a query.

Today, logic is extensively applied in the fields of Artificial Intelligence and Computer Science, and these fields provide a rich source of problems in formal and informal logic. Argumentation theory is one good example of how logic is being applied to artificial intelligence. The ACM Computing Classification System in particular regards:

- Section F.3 on Logics and meanings of programs and F.4 on Mathematical logic and formal languages as part of the theory of computer science: this work covers formal semantics of programming languages, as well as work of formal methods such as Hoare logic;

- Boolean logic as fundamental to computer hardware: particularly, the system's section B.2 on Arithmetic and logic structures, relating to operatives AND, [[negation|NOT]], and OR;

- Many fundamental logical formalisms are essential to section I.2 on artificial intelligence, for example modal logic and default logic in Knowledge representation formalisms and methods, Horn clauses in logic programming, and description logic.

Furthermore, computers can be used as tools for logicians. For example, in symbolic logic and mathematical logic, proofs by humans can be computer-assisted. Using automated theorem proving, the machines can find and check proofs, as well as work with proofs too lengthy to write out by hand.

3.3.9 Non-classical logic

Main article: Non-classical logic

The logics discussed above are all "bivalent" or "two-valued"; that is, they are most naturally understood as dividing propositions into true and false propositions. Non-classical logics are those systems that reject various rules of Classical logic.

Hegel developed his own dialectic logic that extended Kant's transcendental logic but also brought it back to ground by assuring us that "neither in heaven nor in earth, neither in the world of mind nor of nature, is there anywhere such an abstract 'either–or' as the understanding maintains. Whatever exists is concrete, with difference and opposition in itself".[42]

In 1910, Nicolai A. Vasiliev extended the law of excluded middle and the law of contradiction and proposed the law of excluded fourth and logic tolerant to contradiction.[43] In the early 20th century Jan Łukasiewicz investigated the extension of the traditional true/false values to include a third value, "possible", so inventing ternary logic, the first multi-valued logic in the Western tradition.[44]

Logics such as fuzzy logic have since been devised with an infinite number of "degrees of truth", represented by a real number between 0 and 1.[45]

Intuitionistic logic was proposed by L.E.J. Brouwer as the correct logic for reasoning about mathematics, based upon his rejection of the law of the excluded middle as part of his intuitionism. Brouwer rejected formalization in mathematics, but his student Arend Heyting studied intuitionistic logic formally, as did Gerhard Gentzen. Intuitionistic logic is of great interest to computer scientists, as it is a constructive logic and can be applied for extracting verified programs from proofs.

Modal logic is not truth conditional, and so it has often been proposed as a non-classical logic. However, modal logic is normally formalized with the principle of the excluded middle, and its relational semantics is bivalent, so this inclusion is disputable.

3.4 Controversies

3.4.1 "Is Logic Empirical?"

For more details on this topic, see Is Logic Empirical?.

What is the epistemological status of the laws of logic? What sort of argument is appropriate for criticizing purported principles of logic? In an influential paper entitled "Is Logic Empirical?"[46] Hilary Putnam, building on a suggestion of W. V. Quine, argued that in general the facts of propositional logic have a similar epistemological status as facts about the physical universe, for example as the laws of mechanics or of general relativity, and in particular that what physicists have learned about quantum mechanics provides a compelling case for abandoning certain familiar principles of classical logic: if we want to be realists about the physical phenomena described by quantum theory, then we should abandon the principle of distributivity, substituting for classical logic the quantum logic proposed by Garrett Birkhoff and John von Neumann.[47]

Another paper of the same name by Michael Dummett argues that Putnam's desire for realism mandates the law of distributivity.[48] Distributivity of logic is essential for the realist's understanding of how propositions are true of the world in just the same way as he has argued the principle of bivalence is. In this way, the question, "Is Logic Empirical?" can be seen to lead naturally into the fundamental controversy in metaphysics on realism versus anti-realism.

3.4.2 Implication: Strict or material

Main article: Paradoxes of material implication

The notion of implication formalized in classical logic does not comfortably translate into natural language by means of "if ... then ...", due to a number of problems called the paradoxes of material implication.

The first class of paradoxes involves counterfactuals, such as *If the moon is made of green cheese, then 2+2=5*, which are puzzling because natural language does not support the principle of explosion. Eliminating this class of paradoxes was the reason for C. I. Lewis's formulation of strict implication, which eventually led to more radically revisionist logics such as relevance logic.

The second class of paradoxes involves redundant premises, falsely suggesting that we know the succedent because of the antecedent: thus "if that man gets elected, granny will die" is materially true since granny is mortal, regardless of the man's election prospects. Such sentences violate the Gricean maxim of relevance, and can be modelled by logics that reject the principle of monotonicity of entailment, such as relevance logic.

3.4.3 Tolerating the impossible

Main article: Paraconsistent logic

Hegel was deeply critical of any simplified notion of the Law of Non-Contradiction. It was based on Leibniz's idea that this law of logic also requires a sufficient ground to specify from what point of view (or time) one says that something cannot contradict itself. A building, for example, both moves and does not move; the ground for the first is our solar system and for the second the earth. In Hegelian dialectic, the law of non-contradiction, of identity, itself relies upon difference and so is not independently assertable.

Closely related to questions arising from the paradoxes of implication comes the suggestion that logic ought to tolerate inconsistency. Relevance logic and paraconsistent logic are the most important approaches here, though the concerns are different: a key consequence of classical logic and some of its rivals, such as intuitionistic logic, is that they respect the principle of explosion, which means that the logic collapses if it is capable of deriving a contradiction. Graham

Priest, the main proponent of dialetheism, has argued for paraconsistency on the grounds that there are in fact, true contradictions.[49]

3.4.4 Rejection of logical truth

The philosophical vein of various kinds of skepticism contains many kinds of doubt and rejection of the various bases on which logic rests, such as the idea of logical form, correct inference, or meaning, typically leading to the conclusion that there are no logical truths. Observe that this is opposite to the usual views in philosophical skepticism, where logic directs skeptical enquiry to doubt received wisdoms, as in the work of Sextus Empiricus.

Friedrich Nietzsche provides a strong example of the rejection of the usual basis of logic: his radical rejection of idealization led him to reject truth as a "... mobile army of metaphors, metonyms, and anthropomorphisms—in short ... metaphors which are worn out and without sensuous power; coins which have lost their pictures and now matter only as metal, no longer as coins."[50] His rejection of truth did not lead him to reject the idea of either inference or logic completely, but rather suggested that "logic [came] into existence in man's head [out] of illogic, whose realm originally must have been immense. Innumerable beings who made inferences in a way different from ours perished".[51] Thus there is the idea that logical inference has a use as a tool for human survival, but that its existence does not support the existence of truth, nor does it have a reality beyond the instrumental: "Logic, too, also rests on assumptions that do not correspond to anything in the real world".[52]

This position held by Nietzsche however, has come under extreme scrutiny for several reasons. Some philosophers, such as Jürgen Habermas, claim his position is self-refuting—and accuse Nietzsche of not even having a coherent perspective, let alone a theory of knowledge.[53] Georg Lukács, in his book *The Destruction of Reason*, asserts that, "Were we to study Nietzsche's statements in this area from a logico-philosophical angle, we would be confronted by a dizzy chaos of the most lurid assertions, arbitrary and violently incompatible."[54] Bertrand Russell described Nietzsche's irrational claims with "He is fond of expressing himself paradoxically and with a view to shocking conventional readers" in his book *A History of Western Philosophy*.[55]

3.5 See also

- Digital electronics (also known as *digital logic* or logic gates)
- Fallacies
- List of logicians
- List of logic journals
- List of logic symbols
- Logic puzzle
- Mathematics
    - List of mathematics articles
    - Outline of mathematics
- Metalogic
- Outline of logic
- Philosophy
    - List of philosophy topics
    - Outline of philosophy
- Reason
- Truth
- Vector logic

3.6 Notes and references

[1] "possessed of reason, intellectual, dialectical, argumentative", also related to λόγος (*logos*), "word, thought, idea, argument, account, reason, or principle" (Liddell & Scott 1999; Online Etymology Dictionary 2001).

[2] Aristotle (2001). "Posterior Analytics". In Mckeon, Richard. *The Basic Works*. Modern Library. ISBN 0-375-75799-6.

[3] Whitehead, Alfred North; Russell, Bertrand (1967). *Principia Mathematica to *56*. Cambridge University Press. ISBN 0-521-62606-4.

[4] For a more modern treatment, see Hamilton, A. G. (1980). *Logic for Mathematicians*. Cambridge University Press. ISBN 0-521-29291-3.

[5] T. Mossakowski, J. A. Goguen, R. Diaconescu, A. Tarlecki, "What is a Logic?", Logica Universalis 2007 Birkhauser, pp. 113–133.

[6] Łukasiewicz, Jan (1957). *Aristotle's syllogistic from the standpoint of modern formal logic* (2nd ed.). Oxford University Press. p. 7. ISBN 978-0-19-824144-7.

[7] *Summa Logicae* Part II c.4 transl. as *Ockam's Theory of Propositions*, A. Freddoso and H. Schuurman, St Augustine's Press 1998, p.96

[8] Arnauld, *Logic or the Art of Thinking* Part 2 Chapter 3.

[9] Locke, 1690. An Essay Concerning Human Understanding. IV. v. 1-8)

[10] Bergmann, Merrie; Moor, James; Nelson, Jack (2009). *The Logic Book* (Fifth ed.). New York, NY: McGraw-Hill. ISBN 978-0-07-353563-0.

[11] *Internet Encyclopedia of Philosophy*. Validity and Soundness

[12] Mendelson, Elliott (1964). "Quantification Theory: Completeness Theorems". *Introduction to Mathematical Logic*. Van Nostrand. ISBN 0-412-80830-7.

[13] On abductive reasoning, see:

- Magnani, L. "Abduction, Reason, and Science: Processes of Discovery and Explanation". *Kluwer Academic Plenum Publishers, New York, 2001.* xvii. 205 pages. Hard cover, ISBN 0-306-46514-0.
- R. Josephson, J. & G. Josephson, S. "Abductive Inference: Computation, Philosophy, Technology" *Cambridge University Press, New York & Cambridge (U.K.).* viii. 306 pages. Hard cover (1994). ISBN 0-521-43461-0. Paperback (1996), ISBN 0-521-57545-1.
- Bunt, H. & Black, W. "Abduction, Belief and Context in Dialogue: Studies in Computational Pragmatics" *(Natural Language Processing, 1.) John Benjamins, Amsterdam & Philadelphia, 2000.* vi. 471 pages. Hard cover, ISBN 90-272-4983-0 (Europe).

1-58619-794-2 (U.S.)

[14] See Abduction and Retroduction at *Commens Dictionary of Peirce's Terms*, and see Peirce's papers:

- "On the Logic of drawing History from Ancient Documents especially from Testimonies" (1901), *Collected Papers* v. 7, paragraph 219.
- "PAP" ["Prolegomena to an Apology for Pragmatism"], MS 293 c. 1906, *New Elements of Mathematics* v. 4, pp. 319-320.
- A Letter to F. A. Woods (1913), *Collected Papers* v. 8, paragraphs 385-388.

[15] Peirce, C. S. (1903). Harvard lectures on pragmatism. *Collected Papers* v. 5, paragraphs 188–189.

[16] Hofweber, T. (2004). "Logic and Ontology". In Zalta, Edward N. *Stanford Encyclopedia of Philosophy*.

[17] Brandom, Robert (2000). *Articulating Reasons*. Cambridge, MA: Harvard University Press. ISBN 0-674-00158-3.

[18] E.g., Kline (1972, p.53) wrote "A major achievement of Aristotle was the founding of the science of logic".

[19] "Aristotle", MTU Department of Chemistry.

[20] Jonathan Lear (1986). "*Aristotle and Logical Theory*". Cambridge University Press. p.34. ISBN 0-521-31178-0

[21] Simo Knuuttila (1981). "*Reforging the great chain of being: studies of the history of modal theories*". Springer Science & Business. p.71. ISBN 90-277-1125-9

[22] Michael Fisher, Dov M. Gabbay, Lluís Vila (2005). "*Handbook of temporal reasoning in artificial intelligence*". Elsevier. p.119. ISBN 0-444-51493-7

[23] Harold Joseph Berman (1983). "*Law and revolution: the formation of the Western legal tradition*". Harvard University Press. p.133. ISBN 0-674-51776-8

[24] The four Catuṣkoṭi logical divisions are formally very close to the four opposed propositions of the Greek tetralemma, which in turn are analogous to the four truth values of modern relevance logic Cf. Belnap (1977); Jayatilleke, K. N., (1967, The logic of four alternatives, in *Philosophy East and West*, University of Hawaii Press).

[25] Kisor Kumar Chakrabarti (June 1976). "Some Comparisons Between Frege's Logic and Navya-Nyaya Logic". *Philosophy and Phenomenological Research*. International Phenomenological Society. **36** (4): 554–563. JSTOR 2106873. doi:10.2307/2106873. This paper consists of three parts. The first part deals with Frege's distinction between sense and reference of proper names and a similar distinction in Navya-Nyaya logic. In the second part we have compared Frege's definition of number to the Navya-Nyaya definition of number. In the third part we have shown how the study of the so-called 'restrictive conditions for universals' in Navya-Nyaya logic anticipated some of the developments of modern set theory.

[26] Jonardon Ganeri (2001). *Indian logic: a reader*. Routledge. pp. vii, 5, 7. ISBN 0-7007-1306-9.

[27] "Aristotle", *Encyclopædia Britannica*.

[28] Eco, Umberto (1980). *The Name of the Rose*. London: Vintage. p. 253. ISBN 9780099466031.

[29] "History of logic: Arabic logic". Encyclopædia Britannica.

[30] Rescher, Nicholas (1978). "Dialectics: A Controversy-Oriented Approach to the Theory of Knowledge". *Informal Logic*. **1** (#3).

[31] Hetherington, Stephen (2006). "Nicholas Rescher: Philosophical Dialectics". *Notre Dame Philosophical Reviews* (2006.07.16).

[32] Rescher, Nicholas (2009). Jacquette,Dale. ed. *Reason, Method, and Value: A Reader on the Philosophy of Nicholas Rescher*. Ontos Verlag.

[33] Stolyar, Abram A. (1983). *Introduction to Elementary Mathematical Logic*. Dover Publications. p. 3. ISBN 0-486-64561-4.

[34] Barnes, Jonathan (1995). *The Cambridge Companion to Aristotle*. Cambridge University Press. p. 27. ISBN 0-521-42294-9.

[35] Aristotle (1989). *Prior Analytics*. Hackett Publishing Co. p. 115. ISBN 978-0-87220-064-7.

[36] Mendelson, Elliott (1964). "Formal Number Theory: Gödel's Incompleteness Theorem". *Introduction to Mathematical Logic*. Monterey, Calif.: Wadsworth & Brooks/Cole Advanced Books & Software. OCLC 13580200.

[37] Barwise (1982) divides the subject of mathematical logic into model theory, proof theory, set theory and recursion theory.

[38] Brookshear, J. Glenn (1989). "Computability: Foundations of Recursive Function Theory". *Theory of computation: formal languages, automata, and complexity*. Redwood City, Calif.: Benjamin/Cummings Pub. Co. ISBN 0-8053-0143-7.

[39] Brookshear, J. Glenn (1989). "Complexity". *Theory of computation: formal languages, automata, and complexity*. Redwood City, Calif.: Benjamin/Cummings Pub. Co. ISBN 0-8053-0143-7.

[40] Goldman, Alvin I. (1986). *Epistemology and Cognition*. Harvard University Press. p. 293. ISBN 9780674258969, untrained subjects are prone to commit various sorts of fallacies and mistakes.

[41] Demetriou, A.; Efklides, A., eds. (1994), *Intelligence, Mind, and Reasoning: Structure and Development*. Advances in Psychology, **106**. Elsevier. p. 194. ISBN 9780080867601.

[42] Hegel, G. W. F (1971) [1817]. *Philosophy of Mind*. Encyclopedia of the Philosophical Sciences. trans. William Wallace. Oxford: Clarendon Press. p. 174. ISBN 0-19-875014-5.

[43] Joseph E. Brenner (3 August 2008). *Logic in Reality*. Springer. pp. 28–30. ISBN 978-1-4020-8374-7. Retrieved 9 April 2012.

[44] Zegarelli, Mark (2010), *Logic For Dummies*, John Wiley & Sons, p. 30, ISBN 9781118053072.

[45] Hájek, Petr (2006). "Fuzzy Logic". In Zalta, Edward N. *Stanford Encyclopedia of Philosophy*.

[46] Putnam, H. (1969). "Is Logic Empirical?". *Boston Studies in the Philosophy of Science*. **5**.

[47] Birkhoff, G.; von Neumann, J. (1936). "The Logic of Quantum Mechanics". *Annals of Mathematics*. Annals of Mathematics. **37** (4): 823–843. JSTOR 1968621. doi:10.2307/1968621.

[48] Dummett, M. (1978). "Is Logic Empirical?". *Truth and Other Enigmas*. ISBN 0-674-91076-1.

[49] Priest, Graham (2008). "Dialetheism". In Zalta, Edward N. *Stanford Encyclopedia of Philosophy*.

[50] Nietzsche, 1873, On Truth and Lies in a Nonmoral Sense.

[51] Nietzsche, 1882, *The Gay Science*.

[52] Nietzsche, 1878, *Human, All Too Human*

[53] Babette Babich, Habermas, Nietzsche, and Critical Theory

[54] Georg Lukács. "The Destruction of Reason by Georg Lukács 1952". Marxists.org. Retrieved 2013-06-16.

[55] Russell, Bertrand (1945), *A History of Western Philosophy And Its Connection with Political and Social Circumstances from the Earliest Times to the Present Day* (PDF), Simon and Schuster, p. 762, archived from the original (PDF) on 28 May 2014

3.7 Bibliography

- Barwise, J. (1982). *Handbook of Mathematical Logic*. Elsevier. ISBN 9780080933641.

- Belnap, N. (1977). "A useful four-valued logic". In Dunn & Eppstein, *Modern uses of multiple-valued logic*. Reidel: Boston.

- Bocheński, J. M. (1959). *A précis of mathematical logic*. Translated from the French and German editions by Otto Bird. D. Reidel, Dordrecht, South Holland.

- Bocheński, J. M. (1970). *A history of formal logic*. 2nd Edition. Translated and edited from the German edition by Ivo Thomas. Chelsea Publishing, New York.

- Brookshear, J. Glenn (1989). *Theory of computation: formal languages, automata, and complexity*. Redwood City, Calif.: Benjamin/Cummings Pub. Co. ISBN 0-8053-0143-7.

- Cohen, R.S. and Wartofsky, M.W. (1974). *Logical and Epistemological Studies in Contemporary Physics*. Boston Studies in the Philosophy of Science. D. Reidel Publishing Company: Dordrecht, Netherlands. ISBN 90-277-0377-9.

- Finkelstein, D. (1969). "Matter, Space, and Logic". in R.S. Cohen and M.W. Wartofsky (eds. 1974).

- Gabbay, D.M., and Guenthner, F. (eds., 2001–2005). *Handbook of Philosophical Logic*. 13 vols., 2nd edition. Kluwer Publishers: Dordrecht.

- Haack, Susan (1996). *Deviant Logic, Fuzzy Logic: Beyond the Formalism*, University of Chicago Press.

- Harper, Robert (2001). "Logic". *Online Etymology Dictionary*. Retrieved 8 May 2009.

- Hilbert, D., and Ackermann, W. (1928). *Grundzüge der theoretischen Logik (Principles of Mathematical Logic)*. Springer-Verlag. OCLC 2085765* Hodges, W. (2001). *Logic. An introduction to Elementary Logic*, Penguin Books.

- Hofweber, T. (2004), Logic and Ontology. *Stanford Encyclopedia of Philosophy*. Edward N. Zalta (ed.).

- Hughes, R.I.G. (1993, ed.). *A Philosophical Companion to First-Order Logic*. Hackett Publishing.

- Kline, Morris (1972). *Mathematical Thought From Ancient to Modern Times*. Oxford University Press. ISBN 0-19-506135-7.

- Kneale, William, and Kneale, Martha. (1962). *The Development of Logic*. Oxford University Press, London, UK.

- Liddell, Henry George; Scott, Robert. "Logikos". *A Greek-English Lexicon*. Perseus Project. Retrieved 8 May 2009.

- Mendelson, Elliott, (1964). *Introduction to Mathematical Logic*. Wadsworth & Brooks/Cole Advanced Books & Software: Monterey, Calif. OCLC 13580200

- Smith, B. (1989). "Logic and the Sachverhalt". *The Monist* 72(1):52–69.

- Whitehead, Alfred North and Bertrand Russell (1910). *Principia Mathematica*. Cambridge University Press: Cambridge, England. OCLC 1041146

3.8 External links

- Logic at PhilPapers

- Logic at the Indiana Philosophy Ontology Project

- "Logic". *Internet Encyclopedia of Philosophy*.

- Hazewinkel, Michiel, ed. (2001) [1994], "Logical calculus", *Encyclopedia of Mathematics*, Springer Science+Business Media B.V. / Kluwer Academic Publishers, ISBN 978-1-55608-010-4

- An Outline for Verbal Logic

- Introductions and tutorials

    - An Introduction to Philosophical Logic, by Paul Newall, aimed at beginners.
    - forall x: an introduction to formal logic, by P.D. Magnus, covers sentential and quantified logic.

- Logic Self-Taught: A Workbook (originally prepared for on-line logic instruction).

    - Nicholas Rescher. (1964). *Introduction to Logic*, St. Martin's Press.

- Essays

    - "Symbolic Logic" and "The Game of Logic", Lewis Carroll, 1896.
    - Math & Logic: The history of formal mathematical, logical, linguistic and methodological ideas. In *The Dictionary of the History of Ideas*.

- Online Tools

    - Interactive Syllogistic Machine A web based syllogistic machine for exploring fallacies, figures, terms, and modes of syllogisms.

- Reference material

    - Translation Tips, by Peter Suber, for translating from English into logical notation.
    - Ontology and History of Logic. An Introduction with an annotated bibliography.

- Reading lists

    - The London Philosophy Study Guide offers many suggestions on what to read, depending on the student's familiarity with the subject:
        - Logic & Metaphysics
        - Set Theory and Further Logic
        - Mathematical Logic

Chapter 4

Mathematics

This article is about the study of topics such as quantity and structure. For other uses, see Mathematics (disambiguation). Math redirects here. For other uses, see Math (disambiguation).

Euclid (holding calipers), Greek mathematician, 3rd century BC, as imagined by Raphael in this detail from The School of Athens.[lower-alpha 1]

Mathematics (from Greek μάθημα *máthēma*, "knowledge, study, learning"; often shortened to maths or math) is the study of topics such as quantity (numbers),[1] structure,[2] space,[1] and change.[3][4][5] There is a range of views among mathematicians and philosophers as to the exact scope and definition of mathematics.[6][7]

Mathematicians seek out patterns[8][9] and use them to formulate new conjectures. Mathematicians resolve the truth or falsity of conjectures by mathematical proof. When mathematical structures are good models of real phenomena, then mathematical reasoning can provide insight or predictions about nature. Through the use of abstraction and logic, mathematics developed from counting, calculation, measurement, and the systematic study of the shapes and motions of physical objects. Practical mathematics has been a human activity from as far back as written records exist. The research required to solve mathematical problems can take years or even centuries of sustained inquiry.

Rigorous arguments first appeared in Greek mathematics, most notably in Euclid's *Elements*. Since the pioneering work of Giuseppe Peano (1858–1932), David Hilbert (1862–1943), and others on axiomatic systems in the late 19th century, it has become customary to view mathematical research as establishing truth by rigorous deduction from appropriately chosen axioms and definitions. Mathematics developed at a relatively slow pace until the Renaissance, when mathematical innovations interacting with new scientific discoveries led to a rapid increase in the rate of mathematical discovery that has continued to the present day.[10]

Galileo Galilei (1564–1642) said, "The universe cannot be read until we have learned the language and become familiar with the characters in which it is written. It is written in mathematical language, and the letters are triangles, circles and other geometrical figures, without which means it is humanly impossible to comprehend a single word. Without these, one is wandering about in a dark labyrinth."[11] Carl Friedrich Gauss (1777–1855) referred to mathematics as "the Queen of the Sciences".[12] Benjamin Peirce (1809–1880) called mathematics "the science that draws necessary conclusions".[13] David Hilbert said of mathematics: "We are not speaking here of arbitrariness in any sense. Mathematics is not like a game whose tasks are determined by arbitrarily stipulated rules. Rather, it is a conceptual system possessing internal necessity that can only be so and by no means otherwise."[14] Albert Einstein (1879–1955) stated that "as far as the laws of mathematics refer to reality, they are not certain; and as far as they are certain, they do not refer to reality."[15]

Mathematics is essential in many fields, including natural science, engineering, medicine, finance and the social sciences. Applied mathematics has led to entirely new mathematical disciplines, such as statistics and game theory. Mathematicians also engage in pure mathematics, or mathematics for its own sake, without having any application in mind. There is no clear line separating pure and applied mathematics, and practical applications for what began as pure mathematics are often discovered.[16]

4.1 History

Main article: History of mathematics

The history of mathematics can be seen as an ever-increasing series of abstractions. The first abstraction, which is shared by many animals,[17] was probably that of numbers: the realization that a collection of two apples and a collection of two oranges (for example) have something in common, namely quantity of their members.

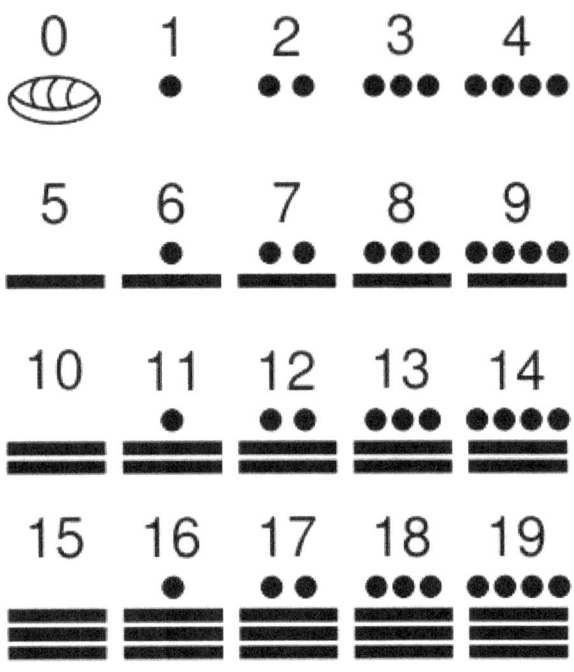

Mayan numerals

Greek mathematician Pythagoras (c. 570 BC – c. 495 BC), commonly credited with discovering the Pythagorean theorem

As evidenced by tallies found on bone, in addition to recognizing how to count physical objects, prehistoric peoples may have also recognized how to count abstract quantities, like time – days, seasons, years.[18]

Evidence for more complex mathematics does not appear until around 3000 BC, when the Babylonians and Egyptians began using arithmetic, algebra and geometry for taxation and other financial calculations, for building and construction, and for astronomy.[19] The earliest uses of mathematics were in trading, land measurement, painting and weaving patterns and the recording of time.

In Babylonian mathematics elementary arithmetic (addition, subtraction, multiplication and division) first appears in the archaeological record. Numeracy pre-dated writing and numeral systems have been many and diverse, with the first known written numerals created by Egyptians in Middle Kingdom texts such as the Rhind Mathematical Papyrus.

Between 600 and 300 BC the Ancient Greeks began a systematic study of mathematics in its own right with Greek mathematics.[20]

During the Golden Age of Islam, especially during the 9th and 10th centuries, mathematics saw many important innovations building on Greek mathematics: most of them include the contributions from Persian mathematicians such as Al-Khwarismi, Omar Khayyam and Sharaf al-Dīn al-Tūsī.

Mathematics has since been greatly extended, and there has been a fruitful interaction between mathematics and science, to the benefit of both. Mathematical discoveries continue to be made today. According to Mikhail B. Sevryuk, in the January 2006 issue of the *Bulletin of the American Mathematical Society*, "The number of papers and books included in the *Mathematical Reviews* database since 1940 (the first year of operation of MR) is now more than 1.9 million, and more than 75 thousand items are

Persian mathematician Al-Khwarizmi (c. 780 – c. 850), the inventor of Algebra.

added to the database each year. The overwhelming majority of works in this ocean contain new mathematical theorems and their proofs."[21]

4.1.1 Etymology

The word *mathematics* comes from the Greek μάθημα (*máthēma*), which, in the ancient Greek language, means "that which is learnt",[22] "what one gets to know", hence also "study" and "science", and in modern Greek just "lesson". The word *máthēma* is derived from μανθάνω (*manthano*), while the modern Greek equivalent is μαθαίνω (*mathaino*), both of which mean "to learn". In Greece, the word for "mathematics" came to have the narrower and more technical meaning "mathematical study" even in Classical times.[23] Its adjective is μαθηματικός (*mathēmatikós*), meaning "related to learning" or "studious", which likewise further came to mean "mathematical". In particular, μαθηματικὴ τέχνη (*mathēmatikḗ tékhnē*), Latin: *ars mathematica*, meant "the mathematical art".

Similarly, one of the two main schools of thought in Pythagoreanism was known as the *mathēmatikoi* (μαθηματικοί) – which at the time meant "teachers" rather than "mathematicians" in the modern sense.

In Latin, and in English until around 1700, the term *mathematics* more commonly meant "astrology" (or sometimes "astronomy") rather than "mathematics"; the meaning gradually changed to its present one from about 1500 to 1800. This has resulted in several mistranslations: a particularly notorious one is Saint Augustine's warning that Christians should beware of *mathematici* meaning astrologers, which is sometimes mistranslated as a condemnation of mathematicians.[24]

The apparent plural form in English, like the French plural form *les mathématiques* (and the less commonly used singular derivative *la mathématique*), goes back to the Latin neuter plural *mathematica* (Cicero), based on the Greek plural τα μαθηματικά (*ta mathēmatiká*), used by Aristotle (384–322 BC), and meaning roughly "all things mathematical"; although it is plausible that English borrowed only the adjective *mathematic(al)* and formed the noun *mathematics* anew, after the pattern of physics and metaphysics, which were inherited from the Greek.[25] In English, the noun *mathematics* takes singular verb forms. It is often shortened to *maths* or, in English-speaking North America, *math*.[26]

4.2 Definitions of mathematics

Main article: Definitions of mathematics

Aristotle defined mathematics as "the science of quantity", and this definition prevailed until the 18th century.[27] Starting in the 19th century, when the study of mathematics increased in rigor and began to address abstract topics such as group theory and projective geometry, which have no clear-cut relation to quantity and measurement, mathematicians and philosophers began to propose a variety of new definitions.[28] Some of these definitions emphasize the deductive character of much of mathematics, some emphasize its abstractness, some emphasize certain topics within mathematics. Today, no consensus on the definition of mathematics prevails, even among professionals.[6] There is not even consensus on whether mathematics is an art or a science.[7] A great many professional mathematicians take no interest in a definition of mathematics, or consider it undefinable.[6] Some just say, "Mathematics is what mathematicians do."[6]

Three leading types of definition of mathematics are called logicist, intuitionist, and formalist, each reflecting a different philosophical school of thought.[29] All have severe problems, none has widespread acceptance, and no reconciliation seems possible.[29]

An early definition of mathematics in terms of logic was Benjamin Peirce's "the science that draws necessary conclusions" (1870).[30] In the *Principia Mathematica*, Bertrand Russell and Alfred North Whitehead advanced the philo-

truth". In formal systems, an axiom is a combination of tokens that is included in a given formal system without needing to be derived using the rules of the system.

4.2.1 Mathematics as science

Leonardo Fibonacci, the Italian mathematician who introduced the Hindu–Arabic numeral system invented between the 1st and 4th centuries by Indian mathematicians, to the Western World

sophical program known as logicism, and attempted to prove that all mathematical concepts, statements, and principles can be defined and proved entirely in terms of symbolic logic. A logicist definition of mathematics is Russell's "All Mathematics is Symbolic Logic" (1903).[31]

Intuitionist definitions, developing from the philosophy of mathematician L.E.J. Brouwer, identify mathematics with certain mental phenomena. An example of an intuitionist definition is "Mathematics is the mental activity which consists in carrying out constructs one after the other."[29] A peculiarity of intuitionism is that it rejects some mathematical ideas considered valid according to other definitions. In particular, while other philosophies of mathematics allow objects that can be proved to exist even though they cannot be constructed, intuitionism allows only mathematical objects that one can actually construct.

Formalist definitions identify mathematics with its symbols and the rules for operating on them. Haskell Curry defined mathematics simply as "the science of formal systems".[32] A formal system is a set of symbols, or *tokens*, and some *rules* telling how the tokens may be combined into *formulas*. In formal systems, the word *axiom* has a special meaning, different from the ordinary meaning of "a self-evident

Carl Friedrich Gauss, known as the prince of mathematicians

Gauss referred to mathematics as "the Queen of the Sciences".[112] In the original Latin *Regina Scientiarum*, as well as in German *Königin der Wissenschaften*, the word corresponding to *science* means a "field of knowledge", and this was the original meaning of "science" in English, also: mathematics is in this sense a field of knowledge. The specialization restricting the meaning of "science" to *natural science* follows the rise of Baconian science, which contrasted "natural science" to scholasticism, the Aristotelean method of inquiring from first principles. The role of empirical experimentation and observation is negligible in mathematics, compared to natural sciences such as biology, chemistry, or physics. Albert Einstein stated that "as far as the laws of mathematics refer to reality, they are not certain; and as far as they are certain, they do not refer to reality."[15] More recently, Marcus du Sautoy has called mathematics "the Queen of Science ... the main driving force behind scientific discovery".[33]

Many philosophers believe that mathematics is not experimentally falsifiable, and thus not a science according to the definition of Karl Popper.[34] However, in the 1930s Gödel's incompleteness theorems convinced many mathematicians that mathematics cannot be reduced to logic alone, and Karl Popper concluded that "most mathematical theories are, like those of physics and biology, hypothetico-deductive: pure mathematics therefore turns out to be much closer to the natural sciences whose hypotheses are conjectures, than it seemed even recently."[35] Other thinkers, notably Imre Lakatos, have applied a version of falsificationism to mathematics itself.

An alternative view is that certain scientific fields (such as theoretical physics) are mathematics with axioms that are intended to correspond to reality. The theoretical physicist J.M. Ziman proposed that science is "public knowledge", and thus includes mathematics.[36] Mathematics shares much in common with many fields in the physical sciences, notably the exploration of the logical consequences of assumptions. Intuition and experimentation also play a role in the formulation of conjectures in both mathematics and the (other) sciences. Experimental mathematics continues to grow in importance within mathematics, and computation and simulation are playing an increasing role in both the sciences and mathematics.

The opinions of mathematicians on this matter are varied. Many mathematicians feel that to call their area a science is to downplay the importance of its aesthetic side, and its history in the traditional seven liberal arts; others feel that to ignore its connection to the sciences is to turn a blind eye to the fact that the interface between mathematics and its applications in science and engineering has driven much development in mathematics. One way this difference of viewpoint plays out is in the philosophical debate as to whether mathematics is *created* (as in art) or *discovered* (as in science). It is common to see universities divided into sections that include a division of *Science and Mathematics*, indicating that the fields are seen as being allied but that they do not coincide. In practice, mathematicians are typically grouped with scientists at the gross level but separated at finer levels. This is one of many issues considered in the philosophy of mathematics.

4.3 Inspiration, pure and applied mathematics, and aesthetics

Main article: Mathematical beauty

Isaac Newton (left) and Gottfried Wilhelm Leibniz (right), developers of infinitesimal calculus

Mathematics arises from many different kinds of problems. At first these were found in commerce, land measurement, architecture and later astronomy; today, all sciences suggest problems studied by mathematicians, and many problems arise within mathematics itself. For example, the physicist Richard Feynman invented the path integral formulation of quantum mechanics using a combination of mathematical reasoning and physical insight, and today's string theory, a still-developing scientific theory which attempts to unify the four fundamental forces of nature, continues to inspire new mathematics.[37]

Some mathematics is relevant only in the area that inspired it, and is applied to solve further problems in that area. But often mathematics inspired by one area proves useful in many areas, and joins the general stock of mathematical concepts. A distinction is often made between pure mathematics and applied mathematics. However pure mathematics topics often turn out to have applications, e.g. number theory in cryptography. This remarkable fact, that even the "purest" mathematics often turns out to have practical applications, is what Eugene Wigner has called "the unreasonable effectiveness of mathematics".[38] As in most areas of study, the explosion of knowledge in the scientific age has led to specialization: there are now hundreds of specialized areas in mathematics and the latest Mathematics Subject Classification runs to 46 pages.[39] Several areas of applied mathematics have merged with related traditions outside of mathematics and become disciplines in their own right, including statistics, operations research, and computer science.

For those who are mathematically inclined, there is often a definite aesthetic aspect to much of mathematics. Many mathematicians talk about the *elegance* of mathematics, its intrinsic aesthetics and inner beauty. Simplicity and generality are valued. There is beauty in a simple and elegant proof, such as Euclid's proof that there are infinitely many prime numbers, and in an elegant numerical method that speeds calculation, such as the fast Fourier transform. G.H. Hardy in *A Mathematician's Apology* expressed the belief that these aesthetic considerations are, in themselves, sufficient to justify the study of pure mathematics. He identified criteria such as significance, unexpectedness, inevitability, and economy as factors that contribute to a mathematical aesthetic.[40] Mathematicians often strive to find proofs that are particularly elegant, proofs from "The Book" of God according to Paul Erdős.[41][42] The popularity of recreational mathematics is another sign of the pleasure many find in solving mathematical questions.

4.4 Notation, language, and rigor

Main article: Mathematical notation

Most of the mathematical notation in use today was not

Leonhard Euler, who created and popularized much of the mathematical notation used today

invented until the 16th century.[43] Before that, mathematics was written out in words, limiting mathematical discovery.[44] Euler (1707–1783) was responsible for many of the notations in use today. Modern notation makes mathematics much easier for the professional, but beginners often find it daunting. It is compressed: a few symbols contain a great deal of information. Like musical notation, modern mathematical notation has a strict syntax and encodes information that would be difficult to write in any other way.

Mathematical language can be difficult to understand for beginners. Common words such as *or* and *only* have more precise meanings than in everyday speech. Moreover, words such as *open* and *field* have specialized mathematical meanings. Technical terms such as *homeomorphism* and *integrable* have precise meanings in mathematics. Additionally, shorthand phrases such as *iff* for "if and only if" belong to mathematical jargon. There is a reason for special notation and technical vocabulary: mathematics requires more precision than everyday speech. Mathematicians refer to this precision of language and logic as "rigor".

Mathematical proof is fundamentally a matter of rigor. Mathematicians want their theorems to follow from axioms by means of systematic reasoning. This is to avoid mistaken "theorems", based on fallible intuitions, of which many instances have occurred in the history of the subject.[45] The level of rigor expected in mathematics has varied over time: the Greeks expected detailed arguments, but at the time of Isaac Newton the methods employed were less rigorous. Problems inherent in the definitions used by Newton would lead to a resurgence of careful analysis and formal proof in the 19th century. Misunderstanding the rigor is a cause for some of the common misconceptions of mathematics. Today, mathematicians continue to argue among themselves about computer-assisted proofs. Since large computations are hard to verify, such proofs may not be sufficiently rigorous.[46]

Axioms in traditional thought were "self-evident truths", but that conception is problematic.[47] At a formal level, an axiom is just a string of symbols, which has an intrinsic meaning only in the context of all derivable formulas of an axiomatic system. It was the goal of Hilbert's program to put all of mathematics on a firm axiomatic basis, but according to Gödel's incompleteness theorem every (sufficiently powerful) axiomatic system has undecidable formulas; and so a final axiomatization of mathematics is impossible. Nonetheless mathematics is often imagined to be (as far as its formal content) nothing but set theory in some axiomatization, in the sense that every mathematical statement or proof could be cast into formulas within set theory.[48]

4.5 Fields of mathematics

See also: Areas of mathematics and Glossary of areas of mathematics

Mathematics can, broadly speaking, be subdivided into

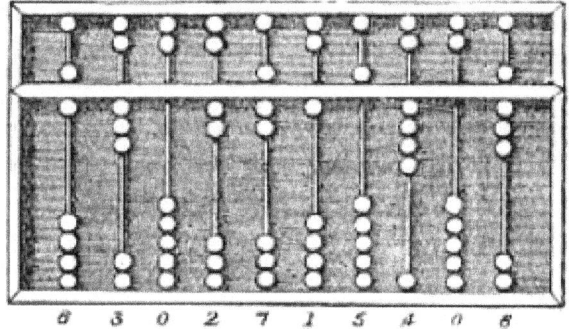

An abacus, a simple calculating tool used since ancient times

the study of quantity, structure, space, and change (i.e. arithmetic, algebra, geometry, and analysis). In addition to these main concerns, there are also subdivisions dedicated to exploring links from the heart of mathematics to other fields: to logic, to set theory (foundations), to the empirical mathematics of the various sciences (applied mathematics), and more recently to the rigorous study of uncertainty. While some areas might seem unrelated, the Langlands program has found connections between areas previously thought unconnected, such as Galois groups, Riemann surfaces and number theory.

4.5.1 Foundations and philosophy

In order to clarify the foundations of mathematics, the fields of mathematical logic and set theory were developed. Mathematical logic includes the mathematical study of logic and the applications of formal logic to other areas of mathematics; set theory is the branch of mathematics that studies sets or collections of objects. Category theory, which deals in an abstract way with mathematical structures and relationships between them, is still in development. The phrase "crisis of foundations" describes the search for a rigorous foundation for mathematics that took place from approximately 1900 to 1930.[49] Some disagreement about the foundations of mathematics continues to the present day. The crisis of foundations was stimulated by a number of controversies at the time, including the controversy over Cantor's set theory and the Brouwer–Hilbert controversy.

Mathematical logic is concerned with setting mathematics within a rigorous axiomatic framework, and studying the implications of such a framework. As such, it is home to Gödel's incompleteness theorems which (informally) imply that any effective formal system that contains basic arithmetic, if *sound* (meaning that all theorems that can be proved are true), is necessarily *incomplete* (meaning that there are true theorems which cannot be proved *in that sys-*

tem). Whatever finite collection of number-theoretical axioms is taken as a foundation, Gödel showed how to construct a formal statement that is a true number-theoretical fact, but which does not follow from those axioms. Therefore, no formal system is a complete axiomatization of full number theory. Modern logic is divided into recursion theory, model theory, and proof theory, and is closely linked to theoretical computer science, as well as to category theory. In the context of recursion theory, the impossibility of a full axiomatization of number theory can also be formally demonstrated as a consequence of the MRDP theorem.

Theoretical computer science includes computability theory, computational complexity theory, and information theory. Computability theory examines the limitations of various theoretical models of the computer, including the most well-known model – the Turing machine. Complexity theory is the study of tractability by computer; some problems, although theoretically solvable by computer, are so expensive in terms of time or space that solving them is likely to remain practically unfeasible, even with the rapid advancement of computer hardware. A famous problem is the "**P = NP?**" problem, one of the Millennium Prize Problems.[50] Finally, information theory is concerned with the amount of data that can be stored on a given medium, and hence deals with concepts such as compression and entropy.

4.5.2 Pure mathematics

Quantity

Main article: Arithmetic

The study of quantity starts with numbers, first the familiar natural numbers and integers ("whole numbers") and arithmetical operations on them, which are characterized in arithmetic. The deeper properties of integers are studied in number theory, from which come such popular results as Fermat's Last Theorem. The twin prime conjecture and Goldbach's conjecture are two unsolved problems in number theory.

As the number system is further developed, the integers are recognized as a subset of the rational numbers ("fractions"). These, in turn, are contained within the real numbers, which are used to represent continuous quantities. Real numbers are generalized to complex numbers. These are the first steps of a hierarchy of numbers that goes on to include quaternions and octonions. Consideration of the natural numbers also leads to the transfinite numbers, which formalize the concept of "infinity". According to the fundamental theorem of algebra all solutions of equations in one unknown with complex coefficients are complex numbers, regardless of degree. Another area of study is the size of sets, which is described with the cardinal numbers. These include the aleph numbers, which allow meaningful comparison of the size of infinitely large sets.

Structure

Main article: Algebra

Many mathematical objects, such as sets of numbers and functions, exhibit internal structure as a consequence of operations or relations that are defined on the set. Mathematics then studies properties of those sets that can be expressed in terms of that structure; for instance number theory studies properties of the set of integers that can be expressed in terms of arithmetic operations. Moreover, it frequently happens that different such structured sets (or structures) exhibit similar properties, which makes it possible, by a further step of abstraction, to state axioms for a class of structures, and then study at once the whole class of structures satisfying these axioms. Thus one can study groups, rings, fields and other abstract systems; together such studies (for structures defined by algebraic operations) constitute the domain of abstract algebra.

By its great generality, abstract algebra can often be applied to seemingly unrelated problems; for instance a number of ancient problems concerning compass and straightedge constructions were finally solved using Galois theory, which involves field theory and group theory. Another example of an algebraic theory is linear algebra, which is the general study of vector spaces, whose elements called vectors have both quantity and direction, and can be used to model (relations between) points in space. This is one example of the phenomenon that the originally unrelated areas of geometry and algebra have very strong interactions in modern mathematics. Combinatorics studies ways of enumerating the number of objects that fit a given structure.

Space

Main article: Geometry

The study of space originates with geometry – in particular, Euclidean geometry, which combines space and numbers, and encompasses the well-known Pythagorean theorem. Trigonometry is the branch of mathematics that deals with relationships between the sides and the angles of triangles and with the trigonometric functions. The modern study of space generalizes these ideas to include higher-dimensional geometry, non-Euclidean geometries (which play a central role in general relativity) and topology. Quantity and space both play a role in analytic geometry, differential geometry, and algebraic geometry. Convex and discrete geometry were developed to solve problems in number theory and functional analysis but now are pursued with an eye on applications in optimization and computer science. Within differential geometry are the concepts of fiber bundles and calculus on manifolds, in particular, vector and tensor calculus. Within algebraic geometry is the description of geometric objects as solution sets of polynomial equations, combining the concepts of quantity and space, and also the study of topological groups, which combine structure and space. Lie groups are used to study space, structure, and change. Topology in all its many ramifications may have been the greatest growth area in 20th-century mathematics; it includes point-set topology, set-theoretic topology, algebraic topology and differential topology. In particular, instances of modern-day topology are metrizability theory, axiomatic set theory, homotopy theory, and Morse theory. Topology also includes the now solved Poincaré conjecture, and the still unsolved areas of the Hodge conjecture. Other results in geometry and topology, including the four color theorem and Kepler conjecture, have been proved only with the help of computers.

Change

Main article: Calculus

Understanding and describing change is a common theme in the natural sciences, and calculus was developed as a powerful tool to investigate it. Functions arise here, as a central concept describing a changing quantity. The rigorous study of real numbers and functions of a real variable is known as real analysis, with complex analysis the equivalent field for the complex numbers. Functional analysis focuses attention on (typically infinite-dimensional) spaces of functions. One of many applications of functional analysis is quantum mechanics. Many problems lead naturally to relationships between a quantity and its rate of change, and these are studied as differential equations. Many phenomena in nature can be described by dynamical systems; chaos theory makes precise the ways in which many of these systems exhibit unpredictable yet still deterministic behavior.

4.5.3 Applied mathematics

Main article: Applied mathematics

Applied mathematics concerns itself with mathematical methods that are typically used in science, engineering, business, and industry. Thus, "applied mathematics" is a mathematical science with specialized knowledge. The term *applied mathematics* also describes the professional specialty in which mathematicians work on practical problems; as a profession focused on practical problems, *applied mathematics* focuses on the "formulation, study, and use of mathematical models" in science, engineering, and other areas of mathematical practice.

In the past, practical applications have motivated the development of mathematical theories, which then became the subject of study in pure mathematics, where mathematics is developed primarily for its own sake. Thus, the activity of applied mathematics is vitally connected with research in pure mathematics.

Statistics and other decision sciences

Main article: Statistics

Applied mathematics has significant overlap with the discipline of statistics, whose theory is formulated mathematically, especially with probability theory. Statisticians (working as part of a research project) "create data that makes sense" with random sampling and with randomized experiments;[51] the design of a statistical sample or experiment specifies the analysis of the data (before the data be available). When reconsidering data from experiments and samples or when analyzing data from observational studies, statisticians "make sense of the data" using the art of modelling and the theory of inference – with model selection and estimation; the estimated models and consequential predictions should be tested on new data.[lower-alpha 2]

Statistical theory studies decision problems such as minimizing the risk (expected loss) of a statistical action, such as using a procedure in, for example, parameter estimation, hypothesis testing, and selecting the best. In these traditional areas of mathematical statistics, a statistical-decision problem is formulated by minimizing an objective function, like expected loss or cost, under specific constraints: For example, designing a survey often involves minimizing the cost of estimating a population mean with a given level of confidence.[52] Because of its use of optimization, the mathematical theory of statistics shares concerns with other decision sciences, such as operations research, control theory, and mathematical economics.[53]

Computational mathematics

Computational mathematics proposes and studies methods for solving mathematical problems that are typically too large for human numerical capacity. Numerical analysis studies methods for problems in analysis using functional analysis and approximation theory; numerical analysis includes the study of approximation and discretization broadly with special concern for rounding errors. Numerical analysis and, more broadly, scientific computing also study non-analytic topics of mathematical science, especially algorithmic matrix and graph theory. Other areas of computational mathematics include computer algebra and symbolic computation.

4.6 Mathematical awards

Arguably the most prestigious award in mathematics is the Fields Medal,[54][55] established in 1936 and awarded every four years (except around World War II) to as many as four individuals. The Fields Medal is often considered a mathematical equivalent to the Nobel Prize.

The Wolf Prize in Mathematics, instituted in 1978, recognizes lifetime achievement, and another major international award, the Abel Prize, was instituted in 2003. The Chern Medal was introduced in 2010 to recognize lifetime achievement. These accolades are awarded in recognition of a particular body of work, which may be innovational, or provide a solution to an outstanding problem in an established field.

A famous list of 23 open problems, called "Hilbert's problems", was compiled in 1900 by German mathematician David Hilbert. This list achieved great celebrity among mathematicians, and at least nine of the problems have now been solved. A new list of seven important problems, titled the "Millennium Prize Problems", was published in 2000. A solution to each of these problems carries a $1 million reward, and only one (the Riemann hypothesis) is duplicated in Hilbert's problems.

4.7 See also

- Philosophy of mathematics
- Lists of mathematics topics
- Mathematics and art
- Mathematics education
- National Museum of Mathematics

- Relationship between mathematics and physics
- Science, Technology, Engineering, and Mathematics

4.8 Notes

[1] No likeness or description of Euclid's physical appearance made during his lifetime survived antiquity. Therefore, Euclid's depiction in works of art depends on the artist's imagination (see *Euclid*).

[2] Like other mathematical sciences such as physics and computer science, statistics is an autonomous discipline rather than a branch of applied mathematics. Like research physicists and computer scientists, research statisticians are mathematical scientists. Many statisticians have a degree in mathematics, and some statisticians are also mathematicians.

4.9 Footnotes

[1] "mathematics, n.". *Oxford English Dictionary*. Oxford University Press. 2012. Retrieved June 16, 2012. The science of space, number, quantity, and arrangement, whose methods involve logical reasoning and usually the use of symbolic notation, and which includes geometry, arithmetic, algebra, and analysis.

[2] Kneebone, G.T. (1963). *Mathematical Logic and the Foundations of Mathematics: An Introductory Survey*. Dover. pp. 4. ISBN 0-486-41712-3. Mathematics ... is simply the study of abstract structures, or formal patterns of connectedness.

[3] LaTorre, Donald R.; Kenelly, John W.; Biggers, Sherry S.; Carpenter, Laurel R.; Reed, Iris B.; Harris, Cynthia R. (2011). *Calculus Concepts: An Informal Approach to the Mathematics of Change*. Cengage Learning. p. 2. ISBN 1-4390-4957-2. Calculus is the study of change—how things change, and how quickly they change.

[4] Ramana (2007). *Applied Mathematics*. Tata McGraw–Hill Education. p. 2.10. ISBN 0-07-066753-5. The mathematical study of change, motion, growth or decay is calculus.

[5] Ziegler, Günter M. (2011). "What Is Mathematics?". *An Invitation to Mathematics: From Competitions to Research*. Springer. p. 7. ISBN 3-642-19532-6.

[6] Mura, Roberta (Dec 1993). "Images of Mathematics Held by University Teachers of Mathematical Sciences". *Educational Studies in Mathematics*. 25 (4): 375–85.

[7] Tobies, Renate & Helmut Neunzert (2012). *Iris Runge: A Life at the Crossroads of Mathematics, Science, and Industry*. Springer. p. 9. ISBN 3-0348-0229-3. It is first necessary to ask what is meant by *mathematics* in general. Illustrious scholars have debated this matter until they were blue in the face, and yet no consensus has been reached about whether mathematics is a natural science, a branch of the humanities, or an art form.

[8] Steen, L.A. (April 29, 1988). *The Science of Patterns* Science, 240: 611–16. And summarized at Association for Supervision and Curriculum Development, www.ascd.org.

[9] Devlin, Keith, *Mathematics: The Science of Patterns: The Search for Order in Life, Mind and the Universe* (Scientific American Paperback Library) 1996, ISBN 978-0-7167-5047-5

[10] Eves

[11] Marcus du Sautoy, *A Brief History of Mathematics: 1. Newton and Leibniz*, BBC Radio 4, September 27, 2010.

[12] Waltershausen

[13] Peirce, p. 97.

[14] Hilbert, D. (1919–20), Natur und Mathematisches Erkennen: Vorlesungen, gehalten 1919–1920 in Göttingen. Nach der Ausarbeitung von Paul Bernays (Edited and with an English introduction by David E. Rowe), Basel, Birkhäuser (1992).

[15] Einstein, p. 28. The quote is Einstein's answer to the question: "how can it be that mathematics, being after all a product of human thought which is independent of experience, is so admirably appropriate to the objects of reality?" He, too, is concerned with *The Unreasonable Effectiveness of Mathematics in the Natural Sciences*.

[16] Peterson

[17] Dehaene, Stanislas; Dehaene-Lambertz, Ghislaine; Cohen, Laurent (Aug 1998). "Abstract representations of numbers in the animal and human brain". *Trends in Neuroscience*. 21 (8): 355–61. PMID 9720604. doi:10.1016/S0166-2236(98)01263-6.

[18] See, for example, Raymond L. Wilder, *Evolution of Mathematical Concepts; an Elementary Study*, passim

[19] Kline 1990, Chapter 1.

[20] "*A History of Greek Mathematics: From Thales to Euclid*". Thomas Little Heath (1981). ISBN 0-486-24073-8

[21] Sevryuk 2006, pp. 101–09.

[22] "mathematic". Online Etymology Dictionary.

[23] Both senses can be found in Plato. μαθηματική. Liddell, Henry George; Scott, Robert; *A Greek–English Lexicon* at the Perseus Project

[24] Cipra, Barry A. (1982). "St. Augustine v. The Mathematicians". *osu.edu*. Ohio State University Mathematics department. Archived from the original on July 16, 2014. Retrieved July 14, 2014.

4.10. REFERENCES

[25] *The Oxford Dictionary of English Etymology*, Oxford English Dictionary, sub "mathematics", "mathematic", "mathematics"

[26] "maths, n." and "math, n.3", *Oxford English Dictionary*, on-line version (2012).

[27] James Franklin, "Aristotelian Realism" in *Philosophy of Mathematics*, ed. A.D. Irvine, p. 104, Elsevier (2009).

[28] Cajori, Florian (1893). *A History of Mathematics*. American Mathematical Society (1991 reprint). pp. 285–86. ISBN 0-8218-2102-4.

[29] Snapper, Ernst (September 1979). "The Three Crises in Mathematics: Logicism, Intuitionism, and Formalism". *Mathematics Magazine*. **52** (4): 207–16. JSTOR 2689412. doi:10.2307/2689412.

[30] Peirce, Benjamin (1882). *Linear Associative Algebra*. p. 1.

[31] Bertrand Russell, *The Principles of Mathematics*, p. 5. University Press, Cambridge (1903)

[32] Curry, Haskell (1951). *Outlines of a Formalist Philosophy of Mathematics*. Elsevier. p. 56. ISBN 0-444-53368-0.

[33] Marcus du Sautoy, *A Brief History of Mathematics: 10. Nicolas Bourbaki*, BBC Radio 4, October 1, 2010.

[34] Shasha, Dennis Elliot; Lazere, Cathy A. (1998). *Out of Their Minds: The Lives and Discoveries of 15 Great Computer Scientists*. Springer. p. 228.

[35] Popper 1995, p. 56

[36] Ziman

[37] Johnson, Gerald W.; Lapidus, Michel L. (2002). *The Feynman Integral and Feynman's Operational Calculus*. Oxford University Press. ISBN 0-8218-2413-9.

[38] Wigner, Eugene (1960). "The Unreasonable Effectiveness of Mathematics in the Natural Sciences". *Communications on Pure and Applied Mathematics*. **13** (1): 1–14. doi:10.1002/cpa.3160130102.

[39] "Mathematics Subject Classification 2010" (PDF). Retrieved November 9, 2010.

[40] Hardy, G.H. (1940). *A Mathematician's Apology*. Cambridge University Press. ISBN 0-521-42706-1.

[41] Gold, Bonnie; Simons, Rogers A. (2008). *Proof and Other Dilemmas: Mathematics and Philosophy*. MAA.

[42] Aigner, Martin; Ziegler, Günter M. (2001). *Proofs from The Book*. Springer. ISBN 3-540-40460-0.

[43] "Earliest Uses of Various Mathematical Symbols". Retrieved September 14, 2014.

[44] Kline, p. 140, on Diophantus; p. 261, on Vieta.

[45] See *false proof* for simple examples of what can go wrong in a formal proof.

[46] Ivars Peterson, *The Mathematical Tourist*, Freeman, 1988. ISBN 0-7167-1953-3. p. 4 "A few complain that the computer program can't be verified properly", (in reference to the Haken–Apple proof of the Four Color Theorem).

[47] "The method of 'postulating' what we want has many advantages; they are the same as the advantages of theft over honest toil." Bertrand Russell (1919). *Introduction to Mathematical Philosophy*, New York and London, p. 71.

[48] Patrick Suppes, *Axiomatic Set Theory*, Dover, 1972. ISBN 0-486-61630-4. p. 1. "Among the many branches of modern mathematics set theory occupies a unique place: with a few rare exceptions the entities which are studied and analyzed in mathematics may be regarded as certain particular sets or classes of objects."

[49] Luke Howard Hodgkin & Luke Hodgkin, *A History of Mathematics*, Oxford University Press, 2005.

[50] Clay Mathematics Institute, P=NP, claymath.org

[51] Rao, C.R. (1997) *Statistics and Truth: Putting Chance to Work*, World Scientific. ISBN 981-02-3111-3

[52] Rao, C.R. (1981). "Foreword". In Arthanari, T.S.; Dodge, Yadolah. *Mathematical programming in statistics*. Wiley Series in Probability and Mathematical Statistics. New York: Wiley. pp. vii–viii. ISBN 0-471-08073-X. MR 607328.

[53] Whittle (1994, pp. 10–11 and 14–18): Whittle, Peter (1994). "Almost home". In Kelly, F.P. *Probability, statistics and optimisation: A Tribute to Peter Whittle* (previously "A realised path: The Cambridge Statistical Laboratory upto 1993 (revised 2002)" ed.). Chichester: John Wiley. pp. 1–28. ISBN 0-471-94829-2.

[54] Monastyrsky 2001: "*The Fields Medal is now indisputably the best known and most influential award in mathematics.*"

[55] Riehm 2002, pp. 778–82.

4.10 References

- Courant, Richard and H. Robbins, *What Is Mathematics? : An Elementary Approach to Ideas and Methods*, Oxford University Press, USA; 2 edition (July 18, 1996). ISBN 0-19-510519-2.

- Einstein, Albert (1923). *Sidelights on Relativity: I. Ether and relativity. II. Geometry and experience (translated by G.B. Jeffery, D.Sc., and W. Perrett, Ph.D)*. E.P. Dutton & Co., New York.

- du Sautoy, Marcus, *A Brief History of Mathematics*, BBC Radio 4 (2010).

- Eves, Howard, *An Introduction to the History of Mathematics*, Sixth Edition, Saunders, 1990, ISBN 0-03-029558-0.

- Kline, Morris, *Mathematical Thought from Ancient to Modern Times*, Oxford University Press, USA; Paperback edition (March 1, 1990). ISBN 0-19-506135-7.

- Monastyrsky, Michael (2001). "Some Trends in Modern Mathematics and the Fields Medal" (PDF). Canadian Mathematical Society. Retrieved July 28, 2006.

- Oxford English Dictionary, second edition, ed. John Simpson and Edmund Weiner, Clarendon Press, 1989, ISBN 0-19-861186-2.

- *The Oxford Dictionary of English Etymology*, 1983 reprint. ISBN 0-19-861112-9.

- Pappas, Theoni, *The Joy Of Mathematics*, Wide World Publishing; Revised edition (June 1989). ISBN 0-933174-65-9.

- Peirce, Benjamin (1881). Peirce, Charles Sanders, ed. "Linear associative algebra". *American Journal of Mathematics* (Corrected, expanded, and annotated revision with an 1875 paper by B. Peirce and annotations by his son, C.S. Peirce, of the 1872 lithograph ed.). Johns Hopkins University. **4** (1–4): 97–229. JSTOR 2369153. doi:10.2307/2369153. Corrected, expanded, and annotated revision with an 1875 paper by B. Peirce and annotations by his son, C. S. Peirce, of the 1872 lithograph ed. *Google* Eprint and as an extract, D. Van Nostrand, 1882, *Google* Eprint..

- Peterson, Ivars, *Mathematical Tourist, New and Updated Snapshots of Modern Mathematics*, Owl Books, 2001, ISBN 0-8050-7159-8.

- Popper, Karl R. (1995). "On knowledge". *In Search of a Better World: Lectures and Essays from Thirty Years*. Routledge. ISBN 0-415-13548-6.

- Riehm, Carl (August 2002). "The Early History of the Fields Medal" (PDF). *Notices of the AMS*. AMS. **49** (7): 778–72.

- Sevryuk, Mikhail B. (January 2006). "Book Reviews" (PDF). *Bulletin of the American Mathematical Society*. **43** (1): 101–09. doi:10.1090/S0273-0979-05-01069-4. Retrieved June 24, 2006.

- Waltershausen, Wolfgang Sartorius von (1965) [first published 1856]. *Gauss zum Gedächtniss*. Sändig Reprint Verlag H. R. Wohlwend. ASIN B0000BN5SQ. ISBN 3-253-01702-8. ASIN 3253017028.

4.11 Further reading

- Benson, Donald C., *The Moment of Proof: Mathematical Epiphanies*, Oxford University Press, USA; New Ed edition (December 14, 2000). ISBN 0-19-513919-4.

- Boyer, Carl B., *A History of Mathematics*, Wiley; 2nd edition, revised by Uta C. Merzbach, (March 6, 1991). ISBN 0-471-54397-7. – A concise history of mathematics from the Concept of Number to contemporary Mathematics.

- Davis, Philip J. and Hersh, Reuben, *The Mathematical Experience*. Mariner Books; Reprint edition (January 14, 1999). ISBN 0-395-92968-7.

- Gullberg, Jan, *Mathematics – From the Birth of Numbers*. W. W. Norton & Company; 1st edition (October 1997). ISBN 0-393-04002-X.

- Hazewinkel, Michiel (ed.), *Encyclopaedia of Mathematics*. Kluwer Academic Publishers 2000. – A translated and expanded version of a Soviet mathematics encyclopedia, in ten (expensive) volumes, the most complete and authoritative work available. Also in paperback and on CD-ROM, and online.

- Jourdain, Philip E. B., *The Nature of Mathematics*, in *The World of Mathematics*, James R. Newman, editor, Dover Publications, 2003, ISBN 0-486-43268-8.

- Maier, Annaliese, *At the Threshold of Exact Science: Selected Writings of Annaliese Maier on Late Medieval Natural Philosophy*, edited by Steven Sargent, Philadelphia: University of Pennsylvania Press, 1982.

4.12 External links

- Mathematics at *Encyclopædia Britannica*

- Mathematics on *In Our Time* at the BBC.

- Free Mathematics books Free Mathematics books collection.

- Encyclopaedia of Mathematics online encyclopaedia from Springer, Graduate-level reference work with over 8,000 entries, illuminating nearly 50,000 notions in mathematics.

- HyperMath site at Georgia State University

- FreeScience Library The mathematics section of FreeScience library

4.12. EXTERNAL LINKS

- Rusin, Dave: *The Mathematical Atlas*. A guided tour through the various branches of modern mathematics. (Can also be found at NIU.edu.)

- Cain, George: Online Mathematics Textbooks available free online.

- Tricki, Wiki-style site that is intended to develop into a large store of useful mathematical problem-solving techniques.

- Mathematical Structures, list information about classes of mathematical structures.

- Mathematician Biographies. The MacTutor History of Mathematics archive Extensive history and quotes from all famous mathematicians.

- *Metamath*. A site and a language, that formalize mathematics from its foundations.

- Nrich, a prize-winning site for students from age five from Cambridge University

- Open Problem Garden, a wiki of open problems in mathematics

- *Planet Math*. An online mathematics encyclopedia under construction, focusing on modern mathematics. Uses the Attribution-ShareAlike license, allowing article exchange with Wikipedia. Uses TeX markup.

- Some mathematics applets, at MIT

- Weisstein, Eric et al.: *Wolfram MathWorld: World of Mathematics*. An online encyclopedia of mathematics.

- Patrick Jones' Video Tutorials on Mathematics

- Citizendium: Theory (mathematics).

- du Sautoy, Marcus, *A Brief History of Mathematics*, BBC Radio 4 (2010).

- Maths.SE A Q&A site for mathematics

- MathOverflow A Q&A site for research-level mathematics

Chapter 5

Mathematical logic

For Quine's theory sometimes called "Mathematical Logic", see New Foundations.
"Mathematical formalism" redirects here. For the philosophical view, see Formalism (philosophy of mathematics).

Mathematical logic is a subfield of mathematics exploring the applications of formal logic to mathematics. It bears close connections to metamathematics, the foundations of mathematics, and theoretical computer science.[1] The unifying themes in mathematical logic include the study of the expressive power of formal systems and the deductive power of formal proof systems.

Mathematical logic is often divided into the fields of set theory, model theory, recursion theory, and proof theory. These areas share basic results on logic, particularly first-order logic, and definability. In computer science (particularly in the ACM Classification) mathematical logic encompasses additional topics not detailed in this article; see Logic in computer science for those.

Since its inception, mathematical logic has both contributed to, and has been motivated by, the study of foundations of mathematics. This study began in the late 19th century with the development of axiomatic frameworks for geometry, arithmetic, and analysis. In the early 20th century it was shaped by David Hilbert's program to prove the consistency of foundational theories. Results of Kurt Gödel, Gerhard Gentzen, and others provided partial resolution to the program, and clarified the issues involved in proving consistency. Work in set theory showed that almost all ordinary mathematics can be formalized in terms of sets, although there are some theorems that cannot be proven in common axiom systems for set theory. Contemporary work in the foundations of mathematics often focuses on establishing which parts of mathematics can be formalized in particular formal systems (as in reverse mathematics) rather than trying to find theories in which all of mathematics can be developed.

5.1 Subfields and scope

The *Handbook of Mathematical Logic* (Barwise 1989) makes a rough division of contemporary mathematical logic into four areas:

1. set theory
2. model theory
3. recursion theory, and
4. proof theory and constructive mathematics (considered as parts of a single area).

Each area has a distinct focus, although many techniques and results are shared among multiple areas. The borderlines amongst these fields, and the lines separating mathematical logic and other fields of mathematics, are not always sharp. Gödel's incompleteness theorem marks not only a milestone in recursion theory and proof theory, but has also led to Löb's theorem in modal logic. The method of forcing is employed in set theory, model theory, and recursion theory, as well as in the study of intuitionistic mathematics.

The mathematical field of category theory uses many formal axiomatic methods, and includes the study of categorical logic, but category theory is not ordinarily considered a subfield of mathematical logic. Because of its applicability in diverse fields of mathematics, mathematicians including Saunders Mac Lane have proposed category theory as a foundational system for mathematics, independent of set theory. These foundations use toposes, which resemble generalized models of set theory that may employ classical or nonclassical logic.

5.2 History

Mathematical logic emerged in the mid-19th century as a subfield of mathematics independent of the traditional study of logic (Ferreirós 2001, p. 443). "Mathematical logic, also called 'logistic', 'symbolic logic', the 'algebra of logic', and, more recently, simply 'formal logic', is the set of logical theories elaborated in the course of the last [nineteenth] century with the aid of an artificial notation and a rigorously deductive method."[2] Before this emergence, logic was studied with rhetoric, with *calculationes*,[3] through the syllogism, and with philosophy. The first half of the 20th century saw an explosion of fundamental results, accompanied by vigorous debate over the foundations of mathematics.

5.2.1 Early history

Further information: History of logic

Theories of logic were developed in many cultures in history, including China, India, Greece and the Islamic world. In 18th-century Europe, attempts to treat the operations of formal logic in a symbolic or algebraic way had been made by philosophical mathematicians including Leibniz and Lambert, but their labors remained isolated and little known.

5.2.2 19th century

In the middle of the nineteenth century, George Boole and then Augustus De Morgan presented systematic mathematical treatments of logic. Their work, building on work by algebraists such as George Peacock, extended the traditional Aristotelian doctrine of logic into a sufficient framework for the study of foundations of mathematics (Katz 1998, p. 686).

Charles Sanders Peirce built upon the work of Boole to develop a logical system for relations and quantifiers, which he published in several papers from 1870 to 1885. Gottlob Frege presented an independent development of logic with quantifiers in his *Begriffsschrift*, published in 1879, a work generally considered as marking a turning point in the history of logic. Frege's work remained obscure, however, until Bertrand Russell began to promote it near the turn of the century. The two-dimensional notation Frege developed was never widely adopted and is unused in contemporary texts.

From 1890 to 1905, Ernst Schröder published *Vorlesungen über die Algebra der Logik* in three volumes. This work summarized and extended the work of Boole, De Morgan, and Peirce, and was a comprehensive reference to symbolic logic as it was understood at the end of the 19th century.

Foundational theories

Concerns that mathematics had not been built on a proper foundation led to the development of axiomatic systems for fundamental areas of mathematics such as arithmetic, analysis, and geometry.

In logic, the term *arithmetic* refers to the theory of the natural numbers. Giuseppe Peano (1889) published a set of axioms for arithmetic that came to bear his name (Peano axioms), using a variation of the logical system of Boole and Schröder but adding quantifiers. Peano was unaware of Frege's work at the time. Around the same time Richard Dedekind showed that the natural numbers are uniquely characterized by their induction properties. Dedekind (1888) proposed a different characterization, which lacked the formal logical character of Peano's axioms. Dedekind's work, however, proved theorems inaccessible in Peano's system, including the uniqueness of the set of natural numbers (up to isomorphism) and the recursive definitions of addition and multiplication from the successor function and mathematical induction.

In the mid-19th century, flaws in Euclid's axioms for geometry became known (Katz 1998, p. 774). In addition to the independence of the parallel postulate, established by Nikolai Lobachevsky in 1826 (Lobachevsky 1840), mathematicians discovered that certain theorems taken for granted by Euclid were not in fact provable from his axioms. Among these is the theorem that a line contains at least two points, or that circles of the same radius whose centers are separated by that radius must intersect. Hilbert (1899) developed a complete set of axioms for geometry, building on previous work by Pasch (1882). The success in axiomatizing geometry motivated Hilbert to seek complete axiomatizations of other areas of mathematics, such as the natural numbers and the real line. This would prove to be a major area of research in the first half of the 20th century.

The 19th century saw great advances in the theory of real analysis, including theories of convergence of functions and Fourier series. Mathematicians such as Karl Weierstrass began to construct functions that stretched intuition, such as nowhere-differentiable continuous functions. Previous conceptions of a function as a rule for computation, or a smooth graph, were no longer adequate. Weierstrass began to advocate the arithmetization of analysis, which sought to axiomatize analysis using properties of the natural numbers. The modern (ε, δ)-definition of limit and continuous functions was already developed by Bolzano in 1817 (Felscher 2000), but remained relatively unknown. Cauchy in 1821 defined continuity in terms of infinitesimals (see Cours d'Analyse, page 34). In 1858, Dedekind proposed a definition of the real numbers in terms of Dedekind cuts of rational numbers (Dedekind 1872), a definition still employed in contemporary texts.

Georg Cantor developed the fundamental concepts of infinite set theory. His early results developed the theory of cardinality and proved that the reals and the natural numbers have different cardinalities (Cantor 1874). Over the next twenty years, Cantor developed a theory of transfinite numbers in a series of publications. In 1891, he published a new proof of the uncountability of the real numbers that introduced the diagonal argument, and used this method to prove Cantor's theorem that no set can have the same cardinality as its powerset. Cantor believed that every set could be well-ordered, but was unable to produce a proof for this result, leaving it as an open problem in 1895 (Katz 1998, p. 807).

5.2.3 20th century

In the early decades of the 20th century, the main areas of study were set theory and formal logic. The discovery of paradoxes in informal set theory caused some to wonder whether mathematics itself is inconsistent, and to look for proofs of consistency.

In 1900, Hilbert posed a famous list of 23 problems for the next century. The first two of these were to resolve the continuum hypothesis and prove the consistency of elementary arithmetic, respectively; the tenth was to produce a method that could decide whether a multivariate polynomial equation over the integers has a solution. Subsequent work to resolve these problems shaped the direction of mathematical logic, as did the effort to resolve Hilbert's *Entscheidungsproblem*, posed in 1928. This problem asked for a procedure that would decide, given a formalized mathematical statement, whether the statement is true or false.

Set theory and paradoxes

Ernst Zermelo (1904) gave a proof that every set could be well-ordered, a result Georg Cantor had been unable to obtain. To achieve the proof, Zermelo introduced the axiom of choice, which drew heated debate and research among mathematicians and the pioneers of set theory. The immediate criticism of the method led Zermelo to publish a second exposition of his result, directly addressing criticisms of his proof (Zermelo 1908a). This paper led to the general acceptance of the axiom of choice in the mathematics community.

Skepticism about the axiom of choice was reinforced by recently discovered paradoxes in naive set theory. Cesare Burali-Forti (1897) was the first to state a paradox: the Burali-Forti paradox shows that the collection of all ordinal numbers cannot form a set. Very soon thereafter, Bertrand Russell discovered Russell's paradox in 1901, and Jules Richard (1905) discovered Richard's paradox.

Zermelo (1908b) provided the first set of axioms for set theory. These axioms, together with the additional axiom of replacement proposed by Abraham Fraenkel, are now called Zermelo–Fraenkel set theory (ZF). Zermelo's axioms incorporated the principle of limitation of size to avoid Russell's paradox.

In 1910, the first volume of *Principia Mathematica* by Russell and Alfred North Whitehead was published. This seminal work developed the theory of functions and cardinality in a completely formal framework of type theory, which Russell and Whitehead developed in an effort to avoid the paradoxes. *Principia Mathematica* is considered one of the most influential works of the 20th century, although the framework of type theory did not prove popular as a foundational theory for mathematics (Ferreirós 2001, p. 445).

Fraenkel (1922) proved that the axiom of choice cannot be proved from the axioms of Zermelo's set theory with urelements. Later work by Paul Cohen (1966) showed that the addition of urelements is not needed, and the axiom of choice is unprovable in ZF. Cohen's proof developed the method of forcing, which is now an important tool for establishing independence results in set theory.[4]

Symbolic logic

Leopold Löwenheim (1915) and Thoralf Skolem (1920) obtained the Löwenheim–Skolem theorem, which says that first-order logic cannot control the cardinalities of infinite structures. Skolem realized that this theorem would apply to first-order formalizations of set theory, and that it implies any such formalization has a countable model. This counterintuitive fact became known as Skolem's paradox.

In his doctoral thesis, Kurt Gödel (1929) proved the completeness theorem, which establishes a correspondence between syntax and semantics in first-order logic. Gödel used the completeness theorem to prove the compactness theorem, demonstrating the finitary nature of first-order logical consequence. These results helped establish first-order logic as the dominant logic used by mathematicians.

In 1931, Gödel published *On Formally Undecidable Propositions of Principia Mathematica and Related Systems*, which proved the incompleteness (in a different meaning of the word) of all sufficiently strong, effective first-order theories. This result, known as Gödel's incompleteness theorem, establishes severe limitations on axiomatic foundations for mathematics, striking a strong blow to Hilbert's program. It showed the impossibility of providing a consistency proof of arithmetic within any formal theory of arithmetic. Hilbert, however, did not acknowledge the importance of the incompleteness theorem for some time.[5]

Gödel's theorem shows that a consistency proof of any sufficiently strong, effective axiom system cannot be obtained in the system itself, if the system is consistent, nor in any weaker system. This leaves open the possibility of consistency proofs that cannot be formalized within the system they consider. Gentzen (1936) proved the consistency of arithmetic using a finitistic system together with a principle of transfinite induction. Gentzen's result introduced the ideas of cut elimination and proof-theoretic ordinals, which became key tools in proof theory. Gödel (1958) gave a different consistency proof, which reduces the consistency of classical arithmetic to that of intuitionistic arithmetic in higher types.

Beginnings of the other branches

Alfred Tarski developed the basics of model theory.

Beginning in 1935, a group of prominent mathematicians collaborated under the pseudonym Nicolas Bourbaki to publish a series of encyclopedic mathematics texts. These texts, written in an austere and axiomatic style, emphasized rigorous presentation and set-theoretic foundations. Terminology coined by these texts, such as the words *bijection*, *injection*, and *surjection*, and the set-theoretic foundations the texts employed, were widely adopted throughout mathematics.

The study of computability came to be known as recursion theory, because early formalizations by Gödel and Kleene relied on recursive definitions of functions.[6] When these definitions were shown equivalent to Turing's formalization involving Turing machines, it became clear that a new concept – the computable function – had been discovered, and that this definition was robust enough to admit numerous independent characterizations. In his work on the incompleteness theorems in 1931, Gödel lacked a rigorous concept of an effective formal system; he immediately realized that the new definitions of computability could be used for this purpose, allowing him to state the incompleteness theorems in generality that could only be implied in the original paper.

Numerous results in recursion theory were obtained in the 1940s by Stephen Cole Kleene and Emil Leon Post. Kleene (1943) introduced the concepts of relative computability, foreshadowed by Turing (1939), and the arithmetical hierarchy. Kleene later generalized recursion theory to higher-order functionals. Kleene and Kreisel studied formal versions of intuitionistic mathematics, particularly in the context of proof theory.

5.3 Formal logical systems

At its core, mathematical logic deals with mathematical concepts expressed using formal logical systems. These systems, though they differ in many details, share the common property of considering only expressions in a fixed formal language. The systems of propositional logic and first-order logic are the most widely studied today, because of their applicability to foundations of mathematics and because of their desirable proof-theoretic properties.[7] Stronger classical logics such as second-order logic or infinitary logic are also studied, along with nonclassical logics such as intuitionistic logic.

5.3.1 First-order logic

Main article: First-order logic

First-order logic is a particular formal system of logic. Its syntax involves only finite expressions as well-formed formulas, while its semantics are characterized by the limitation of all quantifiers to a fixed domain of discourse.

Early results from formal logic established limitations of first-order logic. The Löwenheim–Skolem theorem (1919) showed that if a set of sentences in a countable first-order language has an infinite model then it has at least one model of each infinite cardinality. This shows that it is impossible for a set of first-order axioms to characterize the natural numbers, the real numbers, or any other infinite structure up to isomorphism. As the goal of early foundational studies was to produce axiomatic theories for all parts of mathematics, this limitation was particularly stark.

Gödel's completeness theorem (Gödel 1929) established the equivalence between semantic and syntactic definitions of logical consequence in first-order logic. It shows that if a particular sentence is true in every model that satisfies a particular set of axioms, then there must be a finite deduction of the sentence from the axioms. The compactness theorem first appeared as a lemma in Gödel's proof of the completeness theorem, and it took many years before logicians grasped its significance and began to apply it routinely. It says that a set of sentences has a model if and only if every finite subset has a model, or in other words that an inconsistent set of formulas must have a finite inconsistent subset. The completeness and compactness theorems allow for sophisticated analysis of logical consequence in first-order logic and the development of model theory, and they are a key reason for the prominence of first-order logic in mathematics.

Gödel's incompleteness theorems (Gödel 1931) establish additional limits on first-order axiomatizations. The **first**

incompleteness theorem states that for any consistent, effectively given (defined below) logical system that is capable of interpreting arithmetic, there exists a statement that is true (in the sense that it holds for the natural numbers) but not provable within that logical system (and which indeed may fail in some non-standard models of arithmetic which may be consistent with the logical system). For example, in every logical system capable of expressing the Peano axioms, the Gödel sentence holds for the natural numbers but cannot be proved.

Here a logical system is said to be effectively given if it is possible to decide, given any formula in the language of the system, whether the formula is an axiom, and one which can express the Peano axioms is called "sufficiently strong." When applied to first-order logic, the first incompleteness theorem implies that any sufficiently strong, consistent, effective first-order theory has models that are not elementarily equivalent, a stronger limitation than the one established by the Löwenheim–Skolem theorem. The **second incompleteness theorem** states that no sufficiently strong, consistent, effective axiom system for arithmetic can prove its own consistency, which has been interpreted to show that Hilbert's program cannot be completed.

5.3.2 Other classical logics

Many logics besides first-order logic are studied. These include infinitary logics, which allow for formulas to provide an infinite amount of information, and higher-order logics, which include a portion of set theory directly in their semantics.

The most well studied infinitary logic is $L_{\omega_1,\omega}$. In this logic, quantifiers may only be nested to finite depths, as in first-order logic, but formulas may have finite or countably infinite conjunctions and disjunctions within them. Thus, for example, it is possible to say that an object is a whole number using a formula of $L_{\omega_1,\omega}$ such as

$$(x = 0) \vee (x = 1) \vee (x = 2) \vee \cdots.$$

Higher-order logics allow for quantification not only of elements of the domain of discourse, but subsets of the domain of discourse, sets of such subsets, and other objects of higher type. The semantics are defined so that, rather than having a separate domain for each higher-type quantifier to range over, the quantifiers instead range over all objects of the appropriate type. The logics studied before the development of first-order logic, for example Frege's logic, had similar set-theoretic aspects. Although higher-order logics are more expressive, allowing complete axiomatizations of structures such as the natural numbers, they do not satisfy analogues of the completeness and compactness theorems from first-order logic, and are thus less amenable to proof-theoretic analysis.

Another type of logics are fixed-point logics that allow inductive definitions, like one writes for primitive recursive functions.

One can formally define an extension of first-order logic — a notion which encompasses all logics in this section because they behave like first-order logic in certain fundamental ways, but does not encompass all logics in general, e.g. it does not encompass intuitionistic, modal or fuzzy logic. Lindström's theorem implies that the only extension of first-order logic satisfying both the compactness theorem and the Downward Löwenheim–Skolem theorem is first-order logic.

5.3.3 Nonclassical and modal logic

Modal logics include additional modal operators, such as an operator which states that a particular formula is not only true, but necessarily true. Although modal logic is not often used to axiomatize mathematics, it has been used to study the properties of first-order provability (Solovay 1976) and set-theoretic forcing (Hamkins and Löwe 2007).

Intuitionistic logic was developed by Heyting to study Brouwer's program of intuitionism, in which Brouwer himself avoided formalization. Intuitionistic logic specifically does not include the law of the excluded middle, which states that each sentence is either true or its negation is true. Kleene's work with the proof theory of intuitionistic logic showed that constructive information can be recovered from intuitionistic proofs. For example, any provably total function in intuitionistic arithmetic is computable; this is not true in classical theories of arithmetic such as Peano arithmetic.

5.3.4 Algebraic logic

Algebraic logic uses the methods of abstract algebra to study the semantics of formal logics. A fundamental example is the use of Boolean algebras to represent truth values in classical propositional logic, and the use of Heyting algebras to represent truth values in intuitionistic propositional logic. Stronger logics, such as first-order logic and higher-order logic, are studied using more complicated algebraic structures such as cylindric algebras.

5.4 Set theory

Main article: Set theory

Set theory is the study of sets, which are abstract collections of objects. Many of the basic notions, such as ordinal and cardinal numbers, were developed informally by Cantor before formal axiomatizations of set theory were developed. The first such axiomatization, due to Zermelo (1908b), was extended slightly to become Zermelo–Fraenkel set theory (ZF), which is now the most widely used foundational theory for mathematics.

Other formalizations of set theory have been proposed, including von Neumann–Bernays–Gödel set theory (NBG), Morse–Kelley set theory (MK), and New Foundations (NF). Of these, ZF, NBG, and MK are similar in describing a cumulative hierarchy of sets. New Foundations takes a different approach; it allows objects such as the set of all sets at the cost of restrictions on its set-existence axioms. The system of Kripke–Platek set theory is closely related to generalized recursion theory.

Two famous statements in set theory are the axiom of choice and the continuum hypothesis. The axiom of choice, first stated by Zermelo (1904), was proved independent of ZF by Fraenkel (1922), but has come to be widely accepted by mathematicians. It states that given a collection of nonempty sets there is a single set C that contains exactly one element from each set in the collection. The set C is said to "choose" one element from each set in the collection. While the ability to make such a choice is considered obvious by some, since each set in the collection is nonempty, the lack of a general, concrete rule by which the choice can be made renders the axiom nonconstructive. Stefan Banach and Alfred Tarski (1924) showed that the axiom of choice can be used to decompose a solid ball into a finite number of pieces which can then be rearranged, with no scaling, to make two solid balls of the original size. This theorem, known as the Banach–Tarski paradox, is one of many counterintuitive results of the axiom of choice.

The continuum hypothesis, first proposed as a conjecture by Cantor, was listed by David Hilbert as one of his 23 problems in 1900. Gödel showed that the continuum hypothesis cannot be disproven from the axioms of Zermelo–Fraenkel set theory (with or without the axiom of choice), by developing the constructible universe of set theory in which the continuum hypothesis must hold. In 1963, Paul Cohen showed that the continuum hypothesis cannot be proven from the axioms of Zermelo–Fraenkel set theory (Cohen 1966). This independence result did not completely settle Hilbert's question, however, as it is possible that new axioms for set theory could resolve the hypothesis. Recent work along these lines has been conducted by W. Hugh Woodin, although its importance is not yet clear (Woodin 2001).

Contemporary research in set theory includes the study of large cardinals and determinacy. Large cardinals are cardinal numbers with particular properties so strong that the existence of such cardinals cannot be proved in ZFC. The existence of the smallest large cardinal typically studied, an inaccessible cardinal, already implies the consistency of ZFC. Despite the fact that large cardinals have extremely high cardinality, their existence has many ramifications for the structure of the real line. *Determinacy* refers to the possible existence of winning strategies for certain two-player games (the games are said to be *determined*). The existence of these strategies implies structural properties of the real line and other Polish spaces.

5.5 Model theory

Main article: Model theory

Model theory studies the models of various formal theories. Here a theory is a set of formulas in a particular formal logic and signature, while a model is a structure that gives a concrete interpretation of the theory. Model theory is closely related to universal algebra and algebraic geometry, although the methods of model theory focus more on logical considerations than those fields.

The set of all models of a particular theory is called an elementary class; classical model theory seeks to determine the properties of models in a particular elementary class, or determine whether certain classes of structures form elementary classes.

The method of quantifier elimination can be used to show that definable sets in particular theories cannot be too complicated. Tarski (1948) established quantifier elimination for real-closed fields, a result which also shows the theory of the field of real numbers is decidable. (He also noted that his methods were equally applicable to algebraically closed fields of arbitrary characteristic.) A modern subfield developing from this is concerned with o-minimal structures.

Morley's categoricity theorem, proved by Michael D. Morley (1965), states that if a first-order theory in a countable language is categorical in some uncountable cardinality, i.e. all models of this cardinality are isomorphic, then it is categorical in all uncountable cardinalities.

A trivial consequence of the continuum hypothesis is that a complete theory with less than continuum many nonisomorphic countable models can have only countably many. Vaught's conjecture, named after Robert Lawson Vaught, says that this is true even independently of the continuum hypothesis. Many special cases of this conjecture have been established.

5.6 Recursion theory

Main article: Recursion theory

Recursion theory, also called **computability theory**, studies the properties of computable functions and the Turing degrees, which divide the uncomputable functions into sets that have the same level of uncomputability. Recursion theory also includes the study of generalized computability and definability. Recursion theory grew from the work of Rózsa Péter, Alonzo Church and Alan Turing in the 1930s, which was greatly extended by Kleene and Post in the 1940s.[8]

Classical recursion theory focuses on the computability of functions from the natural numbers to the natural numbers. The fundamental results establish a robust, canonical class of computable functions with numerous independent, equivalent characterizations using Turing machines, λ calculus, and other systems. More advanced results concern the structure of the Turing degrees and the lattice of recursively enumerable sets.

Generalized recursion theory extends the ideas of recursion theory to computations that are no longer necessarily finite. It includes the study of computability in higher types as well as areas such as hyperarithmetical theory and α-recursion theory.

Contemporary research in recursion theory includes the study of applications such as algorithmic randomness, computable model theory, and reverse mathematics, as well as new results in pure recursion theory.

5.6.1 Algorithmically unsolvable problems

An important subfield of recursion theory studies algorithmic unsolvability; a decision problem or function problem is **algorithmically unsolvable** if there is no possible computable algorithm that returns the correct answer for all legal inputs to the problem. The first results about unsolvability, obtained independently by Church and Turing in 1936, showed that the Entscheidungsproblem is algorithmically unsolvable. Turing proved this by establishing the unsolvability of the halting problem, a result with far-ranging implications in both recursion theory and computer science.

There are many known examples of undecidable problems from ordinary mathematics. The word problem for groups was proved algorithmically unsolvable by Pyotr Novikov in 1955 and independently by W. Boone in 1959. The busy beaver problem, developed by Tibor Radó in 1962, is another well-known example.

Hilbert's tenth problem asked for an algorithm to determine whether a multivariate polynomial equation with integer coefficients has a solution in the integers. Partial progress was made by Julia Robinson, Martin Davis and Hilary Putnam. The algorithmic unsolvability of the problem was proved by Yuri Matiyasevich in 1970 (Davis 1973).

5.7 Proof theory and constructive mathematics

Main article: Proof theory

Proof theory is the study of formal proofs in various logical deduction systems. These proofs are represented as formal mathematical objects, facilitating their analysis by mathematical techniques. Several deduction systems are commonly considered, including Hilbert-style deduction systems, systems of natural deduction, and the sequent calculus developed by Gentzen.

The study of **constructive mathematics**, in the context of mathematical logic, includes the study of systems in non-classical logic such as intuitionistic logic, as well as the study of predicative systems. An early proponent of predicativism was Hermann Weyl, who showed it is possible to develop a large part of real analysis using only predicative methods (Weyl 1918).

Because proofs are entirely finitary, whereas truth in a structure is not, it is common for work in constructive mathematics to emphasize provability. The relationship between provability in classical (or nonconstructive) systems and provability in intuitionistic (or constructive, respectively) systems is of particular interest. Results such as the Gödel–Gentzen negative translation show that it is possible to embed (or *translate*) classical logic into intuitionistic logic, allowing some properties about intuitionistic proofs to be transferred back to classical proofs.

Recent developments in proof theory include the study of proof mining by Ulrich Kohlenbach and the study of proof-theoretic ordinals by Michael Rathjen.

5.8 Applications

"Mathematical logic has been successfully applied not only to mathematics and its foundations (G. Frege, B. Russell, D. Hilbert, P. Bernays, H. Scholz, R. Carnap, S. Lesniewski, T. Skolem), but also to physics (R. Carnap, A. Dittrich, B. Russell, C. E. Shannon, A. N. Whitehead, H. Reichenbach, P. Fevrier), to biology (J. H. Woodger, A. Tarski), to psychology (F. B. Fitch, C. G. Hempel), to law and morals (K. Menger, U. Klug, P. Oppenheim), to economics (J. Neumann, O. Morgenstern), to practical questions (E. C.

Berkeley, E. Stamm), and even to metaphysics (J. [Jan] Salamucha,[9] H. Scholz, J. M. Bochenski). Its applications to the history of logic have proven extremely fruitful (J. Lukasiewicz, H. Scholz, B. Mates, A. Becker, E. Moody, J. Salamucha, K. Duerr, Z. Jordan, P. Boehner, J. M. Bochenski, S. [Stanislaw] T. Schayer,[10] D. Ingalls)."[11] "Applications have also been made to theology (F. Drewnowski, J. Salamucha, I. Thomas)."[12]

5.9 Connections with computer science

Main article: Logic in computer science

The study of computability theory in computer science is closely related to the study of computability in mathematical logic. There is a difference of emphasis, however. Computer scientists often focus on concrete programming languages and feasible computability, while researchers in mathematical logic often focus on computability as a theoretical concept and on noncomputability.

The theory of semantics of programming languages is related to model theory, as is program verification (in particular, model checking). The Curry–Howard isomorphism between proofs and programs relates to proof theory, especially intuitionistic logic. Formal calculi such as the lambda calculus and combinatory logic are now studied as idealized programming languages.

Computer science also contributes to mathematics by developing techniques for the automatic checking or even finding of proofs, such as automated theorem proving and logic programming.

Descriptive complexity theory relates logics to computational complexity. The first significant result in this area, Fagin's theorem (1974) established that NP is precisely the set of languages expressible by sentences of existential second-order logic.

5.10 Foundations of mathematics

Main article: Foundations of mathematics

In the 19th century, mathematicians became aware of logical gaps and inconsistencies in their field. It was shown that Euclid's axioms for geometry, which had been taught for centuries as an example of the axiomatic method, were incomplete. The use of infinitesimals, and the very definition of function, came into question in analysis, as pathological examples such as Weierstrass' nowhere-differentiable continuous function were discovered.

Cantor's study of arbitrary infinite sets also drew criticism. Leopold Kronecker famously stated "God made the integers; all else is the work of man," endorsing a return to the study of finite, concrete objects in mathematics. Although Kronecker's argument was carried forward by constructivists in the 20th century, the mathematical community as a whole rejected them. David Hilbert argued in favor of the study of the infinite, saying "No one shall expel us from the Paradise that Cantor has created."

Mathematicians began to search for axiom systems that could be used to formalize large parts of mathematics. In addition to removing ambiguity from previously naive terms such as function, it was hoped that this axiomatization would allow for consistency proofs. In the 19th century, the main method of proving the consistency of a set of axioms was to provide a model for it. Thus, for example, non-Euclidean geometry can be proved consistent by defining *point* to mean a point on a fixed sphere and *line* to mean a great circle on the sphere. The resulting structure, a model of elliptic geometry, satisfies the axioms of plane geometry except the parallel postulate.

With the development of formal logic, Hilbert asked whether it would be possible to prove that an axiom system is consistent by analyzing the structure of possible proofs in the system, and showing through this analysis that it is impossible to prove a contradiction. This idea led to the study of proof theory. Moreover, Hilbert proposed that the analysis should be entirely concrete, using the term *finitary* to refer to the methods he would allow but not precisely defining them. This project, known as Hilbert's program, was seriously affected by Gödel's incompleteness theorems, which show that the consistency of formal theories of arithmetic cannot be established using methods formalizable in those theories. Gentzen showed that it is possible to produce a proof of the consistency of arithmetic in a finitary system augmented with axioms of transfinite induction, and the techniques he developed to do so were seminal in proof theory.

A second thread in the history of foundations of mathematics involves nonclassical logics and constructive mathematics. The study of constructive mathematics includes many different programs with various definitions of *constructive*. At the most accommodating end, proofs in ZF set theory that do not use the axiom of choice are called constructive by many mathematicians. More limited versions of constructivism limit themselves to natural numbers, number-theoretic functions, and sets of natural numbers (which can be used to represent real numbers, facilitating the study of mathematical analysis). A common idea is that a concrete means of computing the values of the function must be known before the function itself can be said to exist.

In the early 20th century, Luitzen Egbertus Jan Brouwer founded intuitionism as a philosophy of mathematics. This philosophy, poorly understood at first, stated that in order for a mathematical statement to be true to a mathematician, that person must be able to *intuit* the statement, to not only believe its truth but understand the reason for its truth. A consequence of this definition of truth was the rejection of the law of the excluded middle, for there are statements that, according to Brouwer, could not be claimed to be true while their negations also could not be claimed true. Brouwer's philosophy was influential, and the cause of bitter disputes among prominent mathematicians. Later, Kleene and Kreisel would study formalized versions of intuitionistic logic (Brouwer rejected formalization, and presented his work in unformalized natural language). With the advent of the BHK interpretation and Kripke models, intuitionism became easier to reconcile with classical mathematics.

5.11 See also

- Knowledge representation and reasoning
- List of computability and complexity topics
- List of first-order theories
- List of logic symbols
- List of mathematical logic topics
- List of set theory topics
- Metalogic

5.12 Notes

[1] Undergraduate texts include Boolos, Burgess, and Jeffrey (2002), Enderton (2001), and Mendelson (1997). A classic graduate text by Shoenfield (2001) first appeared in 1967.

[2] Jozef Maria Bochenski, *A Precis of Mathematical Logic* (1959), rev. and trans., Albert Menne, ed. and trans., Otto Bird, Dordrecht, South Holland: Reidel, Sec. 0.1, p. 1.

[3] Richard Swineshead (1498), *Calculationes Suiseth Anglici*, Papie: Per Franciscum Gyrardengum.

[4] See also Cohen 2008.

[5] In the foreword to the 1934 first edition of "Grundlagen der Mathematik" (Hilbert & Bernays 1934), Bernays wrote the following, which is reminiscent of the famous note by Frege when informed of Russell's paradox.

"Die Ausführung dieses Vorhabens hat eine wesentliche Verzögerung dadurch erfahren, daß in einem Stadium, in dem die Darstellung schon ihrem Abschuß nahe war, durch das Erscheinen der Arbeiten von Herbrand und von Gödel eine veränderte Situation im Gebiet der Beweistheorie entstand, welche die Berücksichtigung neuer Einsichten zur Aufgabe machte. Dabei ist der Umfang des Buches angewachsen, so daß eine Teilung in zwei Bände angezeigt erschien."

Translation:

"Carrying out this plan [by Hilbert for an exposition on proof theory for mathematical logic] has experienced an essential delay because, at the stage at which the exposition was already near to its conclusion, there occurred an altered situation in the area of proof theory due to the appearance of works by Herbrand and Gödel, which necessitated the consideration of new insights. Thus the scope of this book has grown, so that a division into two volumes seemed advisable."

So certainly Hilbert was aware of the importance of Gödel's work by 1934. The second volume in 1939 included a form of Gentzen's consistency proof for arithmetic.

[6] A detailed study of this terminology is given by Soare (1996).

[7] Ferreirós (2001) surveys the rise of first-order logic over other formal logics in the early 20th century.

[8] Soare, Robert Irving (22 December 2011). "Computability Theory and Applications: The Art of Classical Computability" (PDF). *Department of Mathematics*. University of Chicago. Retrieved 23 August 2017.

[9] "Jan Salamucha", http://pl.wikipedia.org/wiki/Jan_Salamucha .

[10] "Stanislaw Schayer", http://pl.wikipedia.org/wiki/Stanislaw_Schayer .

[11] Jozef Maria Bochenski, *A Precis of Mathematical Logic*, rev. and trans., Albert Menne, ed. and trans., Otto Bird, Dordrecht, South Holland: Reidel, Sec. 0.3, p. 2.

[12] Jozef Maria Bochenski, *A Precis of Mathematical Logic*, rev. and trans., Albert Menne, ed. and trans., Otto Bird, Dordrecht, South Holland: Reidel, Sec. 0.3, p. 2.

5.13 References

5.13.1 Undergraduate texts

- Walicki, Michał (2011), *Introduction to Mathematical Logic*, Singapore: World Scientific Publishing, ISBN 978-981-4343-87-9.

- Boolos, George; Burgess, John; Jeffrey, Richard (2002), *Computability and Logic* (4th ed.), Cambridge: Cambridge University Press, ISBN 978-0-521-00758-0.

- Crossley, J.N.; Ash, C.J.; Brickhill, C.J.; Stillwell, J.C.; Williams, N.H. (1972), *What is mathematical logic?*, London-Oxford-New York: Oxford University Press, ISBN 0-19-888087-1, Zbl 0251.02001.

- Enderton, Herbert (2001), *A mathematical introduction to logic* (2nd ed.), Boston, MA: Academic Press, ISBN 978-0-12-238452-3.

- Hamilton, A.G. (1988), *Logic for Mathematicians* (2nd ed.), Cambridge: Cambridge University Press, ISBN 978-0-521-36865-0.

- Ebbinghaus, H.-D.; Flum, J.; Thomas, W. (1994), *Mathematical Logic* (2nd ed.), New York: Springer, ISBN 0-387-94258-0.

- Katz, Robert (1964), *Axiomatic Analysis*, Boston, MA: D. C. Heath and Company.

- Mendelson, Elliott (1997), *Introduction to Mathematical Logic* (4th ed.), London: Chapman & Hall, ISBN 978-0-412-80830-2.

- Rautenberg, Wolfgang (2010), *A Concise Introduction to Mathematical Logic* (3rd ed.), New York: Springer Science+Business Media, ISBN 978-1-4419-1220-6, doi:10.1007/978-1-4419-1221-3.

- Schwichtenberg, Helmut (2003–2004), *Mathematical Logic* (PDF), Munich, Germany: Mathematisches Institut der Universität München, retrieved 2016-02-24.

- Shawn Hedman, *A first course in logic: an introduction to model theory, proof theory, computability, and complexity*, Oxford University Press, 2004, ISBN 0-19-852981-3. Covers logics in close relation with computability theory and complexity theory

- van Dalen, Dirk (2013), *Logic and Structure*, Berlin: Springer-Verlag, ISBN 978-1-4471-4557-8.

5.13.2 Graduate texts

- Andrews, Peter B. (2002), *An Introduction to Mathematical Logic and Type Theory: To Truth Through Proof* (2nd ed.), Boston: Kluwer Academic Publishers, ISBN 978-1-4020-0763-7.

- Barwise, Jon, ed. (1989). *Handbook of Mathematical Logic*. Studies in Logic and the Foundations of Mathematics. North Holland. ISBN 978-0-444-86388-1..

- Hodges, Wilfrid (1997), *A shorter model theory*, Cambridge: Cambridge University Press, ISBN 978-0-521-58713-6.

- Jech, Thomas (2003), *Set Theory: Millennium Edition*. Springer Monographs in Mathematics, Berlin, New York: Springer-Verlag, ISBN 978-3-540-44085-7.

- Kleene, Stephen Cole.(1952), *Introduction to Metamathematics*. New York: Van Nostrand. (Ishi Press: 2009 reprint).

- Kleene, Stephen Cole. (1967), *Mathematical Logic*. John Wiley. Dover reprint, 2002. ISBN 0-486-42533-9.

- Shoenfield, Joseph R. (2001) [1967], *Mathematical Logic* (2nd ed.), A K Peters, ISBN 978-1-56881-135-2.

- Troelstra, Anne Sjerp; Schwichtenberg, Helmut (2000), *Basic Proof Theory*. Cambridge Tracts in Theoretical Computer Science (2nd ed.), Cambridge: Cambridge University Press, ISBN 978-0-521-77911-1.

5.13.3 Research papers, monographs, texts, and surveys

- Augusto, Luis M. (2017). *Logical consequences. Theory and applications: An introduction*. London: College Publications. ISBN 978-1-84890-236-7.

- Cohen, P. J. (1966), *Set Theory and the Continuum Hypothesis*, Menlo Park, CA: W. A. Benjamin.

- Cohen, Paul Joseph (2008) [1966]. *Set theory and the continuum hypothesis*. Mineola, New York: Dover Publications. ISBN 978-0-486-46921-8..

- J.D. Sneed, *The Logical Structure of Mathematical Physics*. Reidel, Dordrecht, 1971 (revised edition 1979).

- Davis, Martin (1973), "Hilbert's tenth problem is unsolvable", *The American Mathematical Monthly*, The American Mathematical Monthly, Vol. 80, No. 3, **80** (3): 233–269, JSTOR 2318447, doi:10.2307/2318447, reprinted as an appendix in Martin Davis, Computability and Unsolvability, Dover reprint 1982. JStor

- Felscher, Walter (2000), "Bolzano, Cauchy, Epsilon, Delta", *The American Mathematical Monthly*, The American Mathematical Monthly, Vol. 107, No. 9, **107** (9): 844–862, JSTOR 2695743, doi:10.2307/2695743, JSTOR(subscription required)

- Ferreirós, José (2001), "The Road to Modern Logic- An Interpretation", *Bulletin of Symbolic Logic*, The Bulletin of Symbolic Logic, Vol. 7, No. 4, **7** (4): 441–484, JSTOR 2687794, doi:10.2307/2687794.

- Hamkins, Joel David; Benedikt Löwe, "The modal logic of forcing", *Transactions of the American Mathematical Society*, arXiv:math/0509616, doi:10.1090/s0002-9947-07-04297-3

- Katz, Victor J. (1998), *A History of Mathematics*, Addison–Wesley, ISBN 0-321-01618-1.

- Morley, Michael (1965), "Categoricity in Power", *Transactions of the American Mathematical Society*, Transactions of the American Mathematical Society, Vol. 114, No. 2, **114** (2): 514–538, JSTOR 1994188, doi:10.2307/1994188.

- Soare, Robert I. (1996), "Computability and recursion", *Bulletin of Symbolic Logic*, The Bulletin of Symbolic Logic, Vol. 2, No. 3, **2** (3): 284–321, CiteSeerX 10.1.1.35.5803, JSTOR 420992, doi:10.2307/420992.

- Solovay, Robert M. (1976), "Provability Interpretations of Modal Logic", *Israel Journal of Mathematics*, **25** (3–4): 287–304, doi:10.1007/BF02757006.

- Woodin, W. Hugh (2001), "The Continuum Hypothesis, Part I", *Notices of the American Mathematical Society*, **48** (6). PDF

5.13.4 Classical papers, texts, and collections

- Burali-Forti, Cesare (1897), *A question on transfinite numbers*, reprinted in van Heijenoort 1976, pp. 104–111.

- Dedekind, Richard (1872), *Stetigkeit und irrationale Zahlen*. English translation of title: "Consistency and irrational numbers".

- Dedekind, Richard (1888), *Was sind und was sollen die Zahlen?* Two English translations:
  - 1963 (1901). *Essays on the Theory of Numbers*. Beman, W. W., ed. and trans. Dover.
  - 1996. In *From Kant to Hilbert: A Source Book in the Foundations of Mathematics*, 2 vols. Ewald, William B., ed., Oxford University Press: 787–832.

- Fraenkel, Abraham A. (1922), "Der Begriff 'definit' und die Unabhängigkeit des Auswahlsaxioms", *Sitzungsberichte der Preussischen Akademie der Wissenschaften, Physikalisch-mathematische Klasse*, pp. 253–257 (German), reprinted in English translation as "The notion of 'definite' and the independence of the axiom of choice", van Heijenoort 1976, pp. 284–289.

- Frege, Gottlob (1879), *Begriffsschrift, eine der arithmetischen nachgebildete Formelsprache des reinen Denkens*. Halle a. S.: Louis Nebert. Translation: *Concept Script, a formal language of pure thought modelled upon that of arithmetic*, by S. Bauer-Mengelberg in Jean Van Heijenoort, ed., 1967. *From Frege to Gödel: A Source Book in Mathematical Logic, 1879–1931*. Harvard University Press.

- Frege, Gottlob (1884), *Die Grundlagen der Arithmetik: eine logisch-mathematische Untersuchung über den Begriff der Zahl*. Breslau: W. Koebner. Translation: J. L. Austin, 1974. *The Foundations of Arithmetic: A logico-mathematical enquiry into the concept of number*, 2nd ed. Blackwell.

- Gentzen, Gerhard (1936), "Die Widerspruchsfreiheit der reinen Zahlentheorie", *Mathematische Annalen*, **112**: 132–213, doi:10.1007/BF01565428, reprinted in English translation in Gentzen's *Collected works*, M. E. Szabo, ed., North-Holland, Amsterdam, 1969.

- Gödel, Kurt (1929), *Über die Vollständigkeit des Logikkalküls*, doctoral dissertation, University Of Vienna. English translation of title: "Completeness of the logical calculus".

- Gödel, Kurt (1930), "Die Vollständigkeit der Axiome des logischen Funktionen-kalküls", *Monatshefte für Mathematik und Physik*, **37**: 349–360, doi:10.1007/BF01696781. English translation of title: "The completeness of the axioms of the calculus of logical functions".

- Gödel, Kurt (1931), "Über formal unentscheidbare Sätze der Principia Mathematica und verwandter Systeme I", *Monatshefte für Mathematik und Physik*, **38** (1): 173–198, doi:10.1007/BF01700692, see On Formally Undecidable Propositions of Principia Mathematica and Related Systems for details on English translations.

- Gödel, Kurt (1958), "Über eine bisher noch nicht benützte Erweiterung des finiten Standpunktes", *Dialectica. International Journal of Philosophy*, **12** (3–4): 280–287, doi:10.1111/j.1746-8361.1958.tb01464.x, reprinted in English translation

- in Gödel's *Collected Works*, vol II, Solomon Feferman et al., eds. Oxford University Press, 1990.
- van Heijenoort, Jean, ed. (1976) [1967], *From Frege to Gödel: A Source Book in Mathematical Logic, 1879–1931* (3rd ed.), Cambridge, Mass: Harvard University Press, ISBN 0-674-32449-8, (pbk.)
- Hilbert, David (1899), *Grundlagen der Geometrie*, Leipzig: Teubner, English 1902 edition (*The Foundations of Geometry*) republished 1980, Open Court, Chicago.
- David, Hilbert (1929), "Probleme der Grundlegung der Mathematik", *Mathematische Annalen*, **102**: 1–9, doi:10.1007/BF01782335. Lecture given at the International Congress of Mathematicians, 3 September 1928. Published in English translation as "The Grounding of Elementary Number Theory", in Mancosu 1998, pp. 266–273.
- Hilbert, David; Bernays, Paul (1934). *Grundlagen der Mathematik. I. Die Grundlehren der mathematischen Wissenschaften*. **40**. Berlin, New York: Springer-Verlag. ISBN 978-3-540-04134-4. JFM 60.0017.02. MR 0237246.
- Kleene, Stephen Cole (1943), "Recursive Predicates and Quantifiers", *American Mathematical Society Transactions*, Transactions of the American Mathematical Society, Vol. 53, No. 1, **54** (1): 41–73, JSTOR 1990131, doi:10.2307/1990131.
- Lobachevsky, Nikolai (1840), *Geometrishe Untersuchungen zur Theorie der Parellellinien* (German). Reprinted in English translation as "Geometric Investigations on the Theory of Parallel Lines" in *Non-Euclidean Geometry*, Robert Bonola (ed.), Dover, 1955. ISBN 0-486-60027-0
- Löwenheim, Leopold (1915), "Über Möglichkeiten im Relativkalkül", *Mathematische Annalen*, **76** (4): 447–470, ISSN 0025-5831, doi:10.1007/BF01458217 (German). Translated as "On possibilities in the calculus of relatives" in Jean van Heijenoort, 1967, *A Source Book in Mathematical Logic, 1879–1931*. Harvard Univ. Press: 228–251.
- Mancosu, Paolo, ed. (1998), *From Brouwer to Hilbert. The Debate on the Foundations of Mathematics in the 1920s*. Oxford: Oxford University Press.
- Pasch, Moritz (1882), *Vorlesungen über neuere Geometrie*.
- Peano, Giuseppe (1889), *Arithmetices principia, nova methodo exposita* (Latin), excerpt reprinted in English translation as "The principles of arithmetic, presented by a new method", van Heijenoort 1976, pp. 83–97.
- Richard, Jules (1905), "Les principes des mathématiques et le problème des ensembles", *Revue générale des sciences pures et appliquées*, **16**: 541 (French), reprinted in English translation as "The principles of mathematics and the problems of sets", van Heijenoort 1976, pp. 142–144.
- Skolem, Thoralf (1920), "Logisch-kombinatorische Untersuchungen über die Erfüllbarkeit oder Beweisbarkeit mathematischer Sätze nebst einem Theoreme über dichte Mengen", *Videnskapsselskapet Skrifter, I. Matematisk-naturvidenskabelig Klasse*, **6**: 1–36.
- Tarski, Alfred (1948), *A decision method for elementary algebra and geometry*, Santa Monica, California: RAND Corporation
- Turing, Alan M. (1939), "Systems of Logic Based on Ordinals", *Proceedings of the London Mathematical Society*, **45** (2): 161–228, doi:10.1112/plms/s2-45.1.161
- Zermelo, Ernst (1904), "Beweis, daß jede Menge wohlgeordnet werden kann", *Mathematische Annalen*, **59** (4): 514–516, doi:10.1007/BF01445300 (German), reprinted in English translation as "Proof that every set can be well-ordered", van Heijenoort 1976, pp. 139–141.
- Zermelo, Ernst (1908a), "Neuer Beweis für die Möglichkeit einer Wohlordnung", *Mathematische Annalen*, **65**: 107–128, ISSN 0025-5831, doi:10.1007/BF01450054 (German), reprinted in English translation as "A new proof of the possibility of a well-ordering", van Heijenoort 1976, pp. 183–198.
- Zermelo, Ernst (1908b), "Untersuchungen über die Grundlagen der Mengenlehre", *Mathematische Annalen*. **65** (2): 261–281, doi:10.1007/BF01449999.

5.14 External links

- Hazewinkel, Michiel, ed. (2001) [1994], "Mathematical logic", *Encyclopedia of Mathematics*, Springer Science+Business Media B.V. / Kluwer Academic Publishers, ISBN 978-1-55608-010-4
- Polyvalued logic and Quantity Relation Logic
- *forall x: an introduction to formal logic*, a free textbook by P. D. Magnus.
- *A Problem Course in Mathematical Logic*, a free textbook by Stefan Bilaniuk.

- Detlovs, Vilnis, and Podnieks, Karlis (University of Latvia), *Introduction to Mathematical Logic*. (hypertextbook).

- In the Stanford Encyclopedia of Philosophy:

    Classical Logic by Stewart Shapiro.
    First-order Model Theory by Wilfrid Hodges.

- In the London Philosophy Study Guide:

    Mathematical Logic
    Set Theory & Further Logic
    Philosophy of Mathematics

Chapter 6

Mathematical statistics

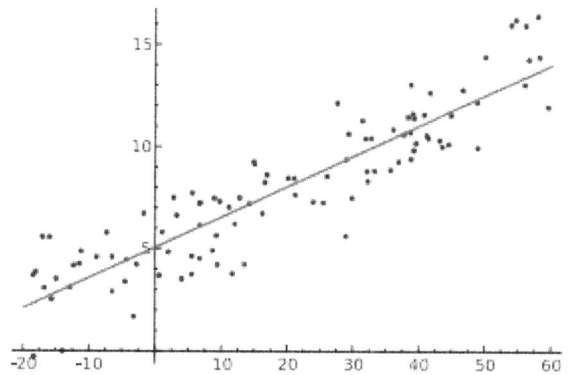

Illustration of linear regression on a data set. Regression analysis is an important part of mathematical statistics.

Mathematical statistics is the application of mathematics to statistics, which was originally conceived as the science of the state — the collection and analysis of facts about a country: its economy, land, military, population, and so on. Mathematical techniques which are used for this include mathematical analysis, linear algebra, stochastic analysis, differential equations, and measure-theoretic probability theory.[1][2]

6.1 Introduction

Statistical science is concerned with the planning of studies, especially with the design of randomized experiments and with the planning of surveys using random sampling. The initial analysis of the data from properly randomized studies often follows the study protocol. The data from a randomized study can be analyzed to consider secondary hypotheses or to suggest new ideas. A secondary analysis of the data from a planned study uses tools from data analysis.

Data analysis is divided into:

- descriptive statistics - the part of statistics that describes data, i.e. summarises the data and their typical properties.

- inferential statistics - the part of statistics that draws conclusions from data (using some model for the data): For example, inferential statistics involves selecting a model for the data, checking whether the data fulfill the conditions of a particular model, and with quantifying the involved uncertainty (e.g. using confidence intervals).

While the tools of data analysis work best on data from randomized studies, they are also applied to other kinds of data --- for example, from natural experiments and observational studies, in which case the inference is dependent on the model chosen by the statistician, and so subjective.[3]

Mathematical statistics has been inspired by and has extended many options in applied statistics.

6.2 Topics

The following are some of the important topics in mathematical statistics:[4][5]

6.2.1 Probability distributions

Main article: Probability distribution

A probability distribution assigns a probability to each measurable subset of the possible outcomes of a random experiment, survey, or procedure of statistical inference. Examples are found in experiments whose sample space is non-numerical, where the distribution would be a categorical distribution; experiments whose sample space is encoded by discrete random variables, where the distribution can be specified by a probability mass function; and experiments with sample spaces encoded by continuous random variables, where the distribution can be specified by a probability density function. More complex experiments, such as those involving stochastic processes defined in continuous time, may demand the use of more general probability measures.

A probability distribution can either be univariate or multivariate. A univariate distribution gives the probabilities of a single random variable taking on various alternative values; a multivariate distribution (a joint probability distribution) gives the probabilities of a random vector—a set of two or more random variables—taking on various combinations of values. Important and commonly encountered univariate probability distributions include the binomial distribution, the hypergeometric distribution, and the normal distribution. The multivariate normal distribution is a commonly encountered multivariate distribution.

Special distributions

- Normal distribution (Gaussian distribution), the most common continuous distribution

- Bernoulli distribution, for the outcome of a single Bernoulli trial (e.g. success/failure, yes/no)

- Binomial distribution, for the number of "positive occurrences" (e.g. successes, yes votes, etc.) given a fixed total number of independent occurrences

- Negative binomial distribution, for binomial-type observations but where the quantity of interest is the number of failures before a given number of successes occurs

- Geometric distribution, for binomial-type observations but where the quantity of interest is the number of failures before the first success; a special c*Discrete uniform distribution, for a finite set of values (e.g. the outcome of a fair die)

- Continuous uniform distribution, for continuously distributed values

- Poisson distribution, for the number of occurrences of a Poisson-type event in a given period of time

- Exponential distribution, for the time before the next Poisson-type event occurs

- Gamma distribution, for the time before the next k Poisson-type events occur

- Chi-squared distribution, the distribution of a sum of squared standard normal variables; useful e.g. for inference regarding the sample variance of normally distributed samples (see chi-squared test)

- Student's t distribution, the distribution of the ratio of a standard normal variable and the square root of a scaled chi squared variable; useful for inference regarding the mean of normally distributed samples with unknown variance (see Student's t-test)

- Beta distribution, for a single probability (real number between 0 and 1); conjugate to the Bernoulli distribution and binomial distribution

6.2.2 Statistical inferences

Main article: Statistical inference

Statistical inference is the process of drawing conclusions from data that are subject to random variation, for example, observational errors or sampling variation.[6] Initial requirements of such a system of procedures for inference and induction are that the system should produce reasonable answers when applied to well-defined situations and that it should be general enough to be applied across a range of situations. Inferential statistics are used to test hypotheses and make estimations using sample data. Whereas descriptive statistics describe a sample, inferential statistics infer predictions about a larger population that the sample represents.

The outcome of statistical inference may be an answer to the question "what should be done next?", where this might be a decision about making further experiments or surveys, or about drawing a conclusion before implementing some organizational or governmental policy. For the most part, statistical inference makes propositions about populations, using data drawn from the population of interest via some form of random sampling. More generally, data about a random process is obtained from its observed behavior during a finite period of time. Given a parameter or hypothesis about which one wishes to make inference, statistical inference most often uses:

- a statistical model of the random process that is supposed to generate the data, which is known when randomization has been used, and

- a particular realization of the random process; i.e., a set of data.

6.2.3 Regression

Main article: Regression analysis

In statistics, **regression analysis** is a statistical process for estimating the relationships among variables. It includes many techniques for modeling and analyzing several variables, when the focus is on the relationship between a dependent variable and one or more independent variables. More specifically, regression analysis helps one understand how the typical value of the dependent variable (or 'criterion variable') changes when any one of the independent variables is varied, while the other independent variables are held fixed. Most commonly, regression analysis estimates the conditional expectation of the dependent variable given the independent variables – that is, the average value of the dependent variable when the independent variables are fixed. Less commonly, the focus is on a quantile, or other location parameter of the conditional distribution of the dependent variable given the independent variables. In all cases, the estimation target is a function of the independent variables called the **regression function**. In regression analysis, it is also of interest to characterize the variation of the dependent variable around the regression function which can be described by a probability distribution.

Many techniques for carrying out regression analysis have been developed. Familiar methods such as linear regression and ordinary least squares regression are parametric, in that the regression function is defined in terms of a finite number of unknown parameters that are estimated from the data. Nonparametric regression refers to techniques that allow the regression function to lie in a specified set of functions, which may be infinite-dimensional.

6.2.4 Nonparametric statistics

Main article: Nonparametric statistics

Nonparametric statistics are statistics not based on parameterized families of probability distributions. They include both descriptive and inferential statistics. The typical parameters are the mean, variance, etc. Unlike parametric statistics, nonparametric statistics make no assumptions about the probability distributions of the variables being assessed.

Non-parametric methods are widely used for studying populations that take on a ranked order (such as movie reviews receiving one to four stars). The use of non-parametric methods may be necessary when data have a ranking but no clear numerical interpretation, such as when assessing preferences. In terms of levels of measurement, non-parametric methods result in "ordinal" data.

As non-parametric methods make fewer assumptions, their applicability is much wider than the corresponding parametric methods. In particular, they may be applied in situations where less is known about the application in question. Also, due to the reliance on fewer assumptions, non-parametric methods are more robust.

Another justification for the use of non-parametric methods is simplicity. In certain cases, even when the use of parametric methods is justified, non-parametric methods may be easier to use. Due both to this simplicity and to their greater robustness, non-parametric methods are seen by some statisticians as leaving less room for improper use and misunderstanding.

6.3 Statistics, mathematics, and mathematical statistics

Mathematical statistics has substantial overlap with the discipline of statistics. Statistical theorists study and improve statistical procedures with mathematics, and statistical research often raises mathematical questions. Statistical theory relies on probability and decision theory.

Mathematicians and statisticians like Gauss, Laplace, and C. S. Peirce used decision theory with probability distributions and loss functions (or utility functions). The decision-theoretic approach to statistical inference was reinvigorated by Abraham Wald and his successors,[7][8][9][10][11][12][13] and makes extensive use of scientific computing, analysis, and optimization; for the design of experiments, statisticians use algebra and combinatorics.

6.4 See also

- Asymptotic theory (statistics)

6.5 References

[1] Lakshmikantham,, ed. by D. Kannan,... V. (2002). *Handbook of stochastic analysis and applications*. New York: M. Dekker. ISBN 0824706609.

[2] Schervish, Mark J. (1995). *Theory of statistics* (Corr. 2nd print. ed.). New York: Springer. ISBN 0387945466.

[3] Freedman, D.A. (2005) *Statistical Models: Theory and Practice*, Cambridge University Press. ISBN 978-0-521-67105-7

[4] Hogg, R. V., A. Craig, and J. W. McKean. "Intro to Mathematical Statistics." (2005).

[5] Larsen, Richard J. and Marx, Morris L. "An Introduction to Mathematical Statistics and Its Applications" (2012). Prentice Hall.

[6] Upton, G., Cook, I. (2008) *Oxford Dictionary of Statistics*, OUP. ISBN 978-0-19-954145-4

[7] Wald, Abraham (1947). *Sequential analysis*. New York: John Wiley and Sons. ISBN 0-471-91806-7. See Dover reprint, 2004: ISBN 0-486-43912-7

[8] Wald, Abraham (1950). *Statistical Decision Functions*. John Wiley and Sons, New York.

[9] Lehmann, Erich (1997). *Testing Statistical Hypotheses* (2nd ed.). ISBN 0-387-94919-4.

[10] Lehmann, Erich; Cassella, George (1998). *Theory of Point Estimation* (2nd ed.). ISBN 0-387-98502-6.

[11] Bickel, Peter J.; Doksum, Kjell A. (2001). *Mathematical Statistics: Basic and Selected Topics*. 1 (Second (updated printing 2007) ed.). Pearson Prentice-Hall.

[12] Le Cam, Lucien (1986). *Asymptotic Methods in Statistical Decision Theory*. Springer-Verlag. ISBN 0-387-96307-3.

[13] Liese, Friedrich & Miescke, Klaus-J. (2008). *Statistical Decision Theory: Estimation, Testing, and Selection*. Springer.

6.6 Additional reading

- Borovkov, A. A. (1999). *Mathematical Statistics*. CRC Press. ISBN 90-5699-018-7

- Virtual Laboratories in Probability and Statistics (Univ. of Ala.-Huntsville)

- StatiBot, interactive online expert system on statistical tests.

Chapter 7

Theoretical computer science

This article is about the branch of computer science and mathematics. For the journal, see Theoretical Computer Science (journal).

Theoretical computer science, or TCS, is a subset of

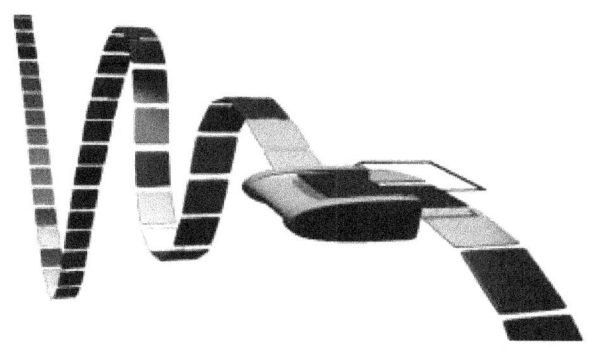

An artistic representation of a Turing machine. Turing machines are used to model general computing devices.

general computer science and mathematics that focuses on more mathematical topics of computing and includes the theory of computation.

It is not easy to circumscribe the theoretical areas precisely. The ACM's Special Interest Group on Algorithms and Computation Theory (SIGACT) provides the following description:[1]

> TCS covers a wide variety of topics including algorithms, data structures, computational complexity, parallel and distributed computation, probabilistic computation, quantum computation, automata theory, information theory, cryptography, program semantics and verification, machine learning, computational biology, computational economics, computational geometry, and computational number theory and algebra. Work in this field is often distinguished by its emphasis on mathematical technique and rigor.

In this list, the ACM's journal Transactions on Computation Theory includes coding theory and computational learning theory, as well as theoretical computer science aspects of areas such as databases, information retrieval, economic models, and networks.[2] Despite this broad scope, the "theory people" in computer science self-identify as different from the "applied people". Some characterize themselves as doing the "(more fundamental) 'science(s)' underlying the field of computing."[3] Other "theory-applied people" suggest that it is impossible to separate theory and application. This means that the so-called "theory people" regularly use experimental science(s) done in less-theoretical areas such as software system research. It also means that there is more cooperation than mutually exclusive competition between theory and application.

7.1 History

Main article: History of computer science

While logical inference and mathematical proof had existed previously, in 1931 Kurt Gödel proved with his incompleteness theorem that there are fundamental limitations on what statements could be proved or disproved.

These developments have led to the modern study of logic and computability, and indeed the field of theoretical computer science as a whole. Information theory was added to the field with a 1948 mathematical theory of communication by Claude Shannon. In the same decade, Donald Hebb introduced a mathematical model of learning in the brain. With mounting biological data supporting this hypothesis with some modification, the fields of neural networks and parallel distributed processing were established. In 1971, Stephen Cook and, working independently, Leonid Levin, proved that there exist practically relevant problems that are NP-complete – a landmark result in computational complexity theory.

With the development of quantum mechanics in the beginning of the 20th century came the concept that mathematical operations could be performed on an entire particle wavefunction. In other words, one could compute functions on multiple states simultaneously. This led to the concept of a quantum computer in the latter half of the 20th century that took off in the 1990s when Peter Shor showed that such methods could be used to factor large numbers in polynomial time, which, if implemented, would render most modern public key cryptography systems uselessly insecure.

Modern theoretical computer science research is based on these basic developments, but includes many other mathematical and interdisciplinary problems that have been posed.

7.2 Topics

7.2.1 Algorithms

Main article: Algorithm

An algorithm is a step-by-step procedure for calculations. Algorithms are used for calculation, data processing, and automated reasoning.

An algorithm is an effective method expressed as a finite list[4] of well-defined instructions[5] for calculating a function.[6] Starting from an initial state and initial input (perhaps empty),[7] the instructions describe a computation that, when executed, proceeds through a finite[8] number of well-defined successive states, eventually producing "output"[9] and terminating at a final ending state. The transition from one state to the next is not necessarily deterministic; some algorithms, known as randomized algorithms, incorporate random input.[10]

7.2.2 Data structures

Main article: Data structure

A data structure is a particular way of organizing data in a computer so that it can be used efficiently.[11][12]

Different kinds of data structures are suited to different kinds of applications, and some are highly specialized to specific tasks. For example, databases use B-tree indexes for small percentages of data retrieval and compilers and databases use dynamic hash tables as look up tables.

Data structures provide a means to manage large amounts of data efficiently for uses such as large databases and internet indexing services. Usually, efficient data structures are key to designing efficient algorithms. Some formal design methods and programming languages emphasize data structures, rather than algorithms, as the key organizing factor in software design. Storing and retrieving can be carried out on data stored in both main memory and in secondary memory.

7.2.3 Computational complexity theory

Main article: Computational complexity theory

Computational complexity theory is a branch of the theory of computation that focuses on classifying computational problems according to their inherent difficulty, and relating those classes to each other. A computational problem is understood to be a task that is in principle amenable to being solved by a computer, which is equivalent to stating that the problem may be solved by mechanical application of mathematical steps, such as an algorithm.

A problem is regarded as inherently difficult if its solution requires significant resources, whatever the algorithm used. The theory formalizes this intuition, by introducing mathematical models of computation to study these problems and quantifying the amount of resources needed to solve them, such as time and storage. Other complexity measures are also used, such as the amount of communication (used in communication complexity), the number of gates in a circuit (used in circuit complexity) and the number of processors (used in parallel computing). One of the roles of computational complexity theory is to determine the practical limits on what computers can and cannot do.

7.2.4 Distributed computation

Main article: Distributed computation

Distributed computing studies distributed systems. A distributed system is a software system in which components located on networked computers communicate and coordinate their actions by passing messages.[13] The components interact with each other in order to achieve a common goal. Three significant characteristics of distributed systems are: concurrency of components, lack of a global clock, and independent failure of components.[13] Examples of distributed systems vary from SOA-based systems to massively multiplayer online games to peer-to-peer applications.

A computer program that runs in a distributed system is called a **distributed program**, and distributed programming is the process of writing such programs.[14] There are many alternatives for the message passing mechanism, including RPC-like connectors and message queues. An important goal and challenge of distributed systems is location transparency.

7.2.5 Parallel computation

Main article: Parallel computation

Parallel computing is a form of computation in which many calculations are carried out simultaneously,[15] operating on the principle that large problems can often be divided into smaller ones, which are then solved "in parallel". There are several different forms of parallel computing: bit-level, instruction level, data, and task parallelism. Parallelism has been employed for many years, mainly in high-performance computing, but interest in it has grown lately due to the physical constraints preventing frequency scaling.[16] As power consumption (and consequently heat generation) by computers has become a concern in recent years,[17] parallel computing has become the dominant paradigm in computer architecture, mainly in the form of multi-core processors.[18]

Parallel computer programs are more difficult to write than sequential ones,[19] because concurrency introduces several new classes of potential software bugs, of which race conditions are the most common. Communication and synchronization between the different subtasks are typically some of the greatest obstacles to getting good parallel program performance.

The maximum possible speed-up of a single program as a result of parallelization is known as Amdahl's law.

7.2.6 Very-large-scale integration

Main article: VLSI

Very-large-scale integration (**VLSI**) is the process of creating an integrated circuit (IC) by combining thousands of transistors into a single chip. VLSI began in the 1970s when complex semiconductor and communication technologies were being developed. The microprocessor is a VLSI device. Before the introduction of VLSI technology most ICs had a limited set of functions they could perform. An electronic circuit might consist of a CPU, ROM, RAM and other glue logic. VLSI allows IC makers to add all of these circuits into one chip.

7.2.7 Machine learning

Main article: Machine learning

Machine learning is a scientific discipline that deals with the construction and study of algorithms that can learn from data.[20] Such algorithms operate by building a model based on inputs[21]:2 and using that to make predictions or decisions, rather than following only explicitly programmed instructions.

Machine learning can be considered a subfield of computer science and statistics. It has strong ties to artificial intelligence and optimization, which deliver methods, theory and application domains to the field. Machine learning is employed in a range of computing tasks where designing and programming explicit, rule-based algorithms is infeasible. Example applications include spam filtering, optical character recognition (OCR),[22] search engines and computer vision. Machine learning is sometimes conflated with data mining,[23] although that focuses more on exploratory data analysis.[24] Machine learning and pattern recognition "can be viewed as two facets of the same field."[21]:vii

7.2.8 Computational biology

Main article: Computational biology

Computational biology involves the development and application of data-analytical and theoretical methods, mathematical modeling and computational simulation techniques to the study of biological, behavioral, and social systems.[25] The field is broadly defined and includes foundations in computer science, applied mathematics, animation, statistics, biochemistry, chemistry, biophysics, molecular biology, genetics, genomics, ecology, evolution, anatomy, neuroscience, and visualization.[26]

Computational biology is different from biological computation, which is a subfield of computer science and computer engineering using bioengineering and biology to build computers, but is similar to bioinformatics, which is an interdisciplinary science using computers to store and process biological data.

7.2.9 Computational geometry

Main article: Computational geometry

Computational geometry is a branch of computer science devoted to the study of algorithms that can be stated in terms of geometry. Some purely geometrical problems arise out of the study of computational geometric algorithms, and such problems are also considered to be part of computational geometry. While modern computational geometry is a recent development, it is one of the oldest fields of computing with history stretching back to antiquity. An ancient precursor is the Sanskrit treatise Shulba Sutras, or "Rules of the Chord", that is a book of algorithms written in 800 BCE. The book prescribes step-by-step procedures for constructing geometric objects like altars using a peg and chord.

The main impetus for the development of computational geometry as a discipline was progress in computer graphics and computer-aided design and manufacturing (CAD/CAM), but many problems in computational geometry are classical in nature, and may come from mathematical visualization.

Other important applications of computational geometry include robotics (motion planning and visibility problems), geographic information systems (GIS) (geometrical location and search, route planning), integrated circuit design (IC geometry design and verification), computer-aided engineering (CAE) (mesh generation), computer vision (3D reconstruction).

7.2.10 Information theory

Main article: Information theory

Information theory is a branch of applied mathematics, electrical engineering, and computer science involving the quantification of information. Information theory was developed by Claude E. Shannon to find fundamental limits on signal processing operations such as compressing data and on reliably storing and communicating data. Since its inception it has broadened to find applications in many other areas, including statistical inference, natural language processing, cryptography, neurobiology,[27] the evolution[28] and function[29] of molecular codes, model selection in statistics,[30] thermal physics,[31] quantum computing, linguistics, plagiarism detection,[32] pattern recognition, anomaly detection and other forms of data analysis.[33]

Applications of fundamental topics of information theory include lossless data compression (e.g. ZIP files), lossy data compression (e.g. MP3s and JPEGs), and channel coding (e.g. for Digital Subscriber Line (DSL)). The field is at the intersection of mathematics, statistics, computer science, physics, neurobiology, and electrical engineering. Its impact has been crucial to the success of the Voyager missions to deep space, the invention of the compact disc, the feasibility of mobile phones, the development of the Internet, the study of linguistics and of human perception, the understanding of black holes, and numerous other fields. Important sub-fields of information theory are source coding, channel coding, algorithmic complexity theory, algorithmic information theory, information-theoretic security, and measures of information.

7.2.11 Cryptography

Main article: Cryptography

Cryptography is the practice and study of techniques for secure communication in the presence of third parties (called adversaries).[34] More generally, it is about constructing and analyzing protocols that overcome the influence of adversaries[35] and that are related to various aspects in information security such as data confidentiality, data integrity, authentication, and non-repudiation.[36] Modern cryptography intersects the disciplines of mathematics, computer science, and electrical engineering. Applications of cryptography include ATM cards, computer passwords, and electronic commerce.

Modern cryptography is heavily based on mathematical theory and computer science practice; cryptographic algorithms are designed around computational hardness assumptions, making such algorithms hard to break in practice by any adversary. It is theoretically possible to break such a system, but it is infeasible to do so by any known practical means. These schemes are therefore termed computationally secure; theoretical advances, e.g., improvements in integer factorization algorithms, and faster computing technology require these solutions to be continually adapted. There exist information-theoretically secure schemes that provably cannot be broken even with unlimited computing power—an example is the one-time pad—but these schemes are more difficult to implement than the best theoretically breakable but computationally secure mechanisms.

7.2.12 Quantum computation

Main article: Quantum computation

A quantum computer is a computation system that makes direct use of quantum-mechanical phenomena, such as superposition and entanglement, to perform operations on data.[37] Quantum computers are different from digital computers based on transistors. Whereas digital computers require data to be encoded into binary digits (bits), each of which is always in one of two definite states (0 or 1), quantum computation uses qubits (quantum bits), which can be in superpositions of states. A theoretical model is the quantum Turing machine, also known as the universal quantum computer. Quantum computers share theoretical similarities with non-deterministic and probabilistic computers; one example is the ability to be in more than one state simultaneously. The field of quantum computing was first introduced by Yuri Manin in 1980[38] and Richard Feynman in 1982.[39][40] A quantum computer with spins as quantum bits was also formulated for use as a quantum space–time in 1968.[41]

As of 2014, quantum computing is still in its infancy but experiments have been carried out in which quantum computational operations were executed on a very small number of qubits.[42] Both practical and theoretical research continues, and many national governments and military funding agencies support quantum computing research to develop quantum computers for both civilian and national security purposes, such as cryptanalysis.[43]

7.2.13 Information-based complexity

Main article: Information-based complexity

Information-based complexity (IBC) studies optimal algorithms and computational complexity for continuous problems. IBC has studied continuous problems as path integration, partial differential equations, systems of ordinary differential equations, nonlinear equations, integral equations, fixed points, and very-high-dimensional integration.

7.2.14 Computational number theory

Main article: Computational number theory

Computational number theory, also known as **algorithmic number theory**, is the study of algorithms for performing number theoretic computations. The best known problem in the field is integer factorization.

7.2.15 Symbolic computation

Main article: Symbolic computation

Computer algebra, also called symbolic computation or algebraic computation is a scientific area that refers to the study and development of algorithms and software for manipulating mathematical expressions and other mathematical objects. Although, properly speaking, computer algebra should be a subfield of scientific computing, they are generally considered as distinct fields because scientific computing is usually based on numerical computation with approximate floating point numbers, while symbolic computation emphasizes *exact* computation with expressions containing variables that have not any given value and are thus manipulated as symbols (therefore the name of *symbolic computation*).

Software applications that perform symbolic calculations are called *computer algebra systems*, with the term *system* alluding to the complexity of the main applications that include, at least, a method to represent mathematical data in a computer, a user programming language (usually different from the language used for the implementation), a dedicated memory manager, a user interface for the input/output of mathematical expressions, a large set of routines to perform usual operations, like simplification of expressions, differentiation using chain rule, polynomial factorization, indefinite integration, etc.

7.2.16 Program semantics

Main article: Program semantics

In programming language theory, **semantics** is the field concerned with the rigorous mathematical study of the meaning of programming languages. It does so by evaluating the meaning of syntactically legal strings defined by a specific programming language, showing the computation involved. In such a case that the evaluation would be of syntactically illegal strings, the result would be non-computation. Semantics describes the processes a computer follows when executing a program in that specific language. This can be shown by describing the relationship between the input and output of a program, or an explanation of how the program will execute on a certain platform, hence creating a model of computation.

7.2.17 Formal methods

Main article: Formal methods

Formal methods are a particular kind of mathematics based techniques for the specification, development and verification of software and hardware systems.[44] The use of formal methods for software and hardware design is motivated by the expectation that, as in other engineering disciplines, performing appropriate mathematical analysis can contribute to the reliability and robustness of a design.[45]

Formal methods are best described as the application of a fairly broad variety of theoretical computer science fundamentals, in particular logic calculi, formal languages, automata theory, and program semantics, but also type systems and algebraic data types to problems in software and hardware specification and verification.[46]

7.2.18 Automata theory

Main article: Automata theory

Automata theory is the study of *abstract machines* and *automata*, as well as the computational problems that can be solved using them. It is a theory in theoretical computer science, under Discrete mathematics (a section of Mathematics and also of Computer Science). *Automata* comes from the Greek word αὐτόματα meaning "self-acting".

Automata Theory is the study of self-operating virtual machines to help in logical understanding of input and output process, without or with intermediate stage(s) of computation (or any function / process).

7.2.19 Coding theory

Main article: Coding theory

Coding theory is the study of the properties of codes and their fitness for a specific application. Codes are used for data compression, cryptography, error-correction and more recently also for network coding. Codes are studied by various scientific disciplines—such as information theory, electrical engineering, mathematics, and computer science—for the purpose of designing efficient and reliable data transmission methods. This typically involves the removal of redundancy and the correction (or detection) of errors in the transmitted data.

7.2.20 Computational learning theory

Main article: Computational learning theory

Theoretical results in machine learning mainly deal with a type of inductive learning called supervised learning. In supervised learning, an algorithm is given samples that are labeled in some useful way. For example, the samples might be descriptions of mushrooms, and the labels could be whether or not the mushrooms are edible. The algorithm takes these previously labeled samples and uses them to induce a classifier. This classifier is a function that assigns labels to samples including the samples that have never been previously seen by the algorithm. The goal of the supervised learning algorithm is to optimize some measure of performance such as minimizing the number of mistakes made on new samples.

7.3 Organizations

- European Association for Theoretical Computer Science

- SIGACT

7.4 Journals and newsletters

- *Information and Computation*
- *Theory of Computing* (open access journal)
- *Formal Aspects of Computing*
- *Journal of the ACM*
- *SIAM Journal on Computing* (SICOMP)
- *SIGACT News*
- *Theoretical Computer Science*
- *Theory of Computing Systems*
- *International Journal of Foundations of Computer Science*
- *Chicago Journal of Theoretical Computer Science* (open access journal)
- *Foundations and Trends in Theoretical Computer Science*
- *Journal of Automata, Languages and Combinatorics*
- *Acta Informatica*
- *Fundamenta Informaticae*
- *ACM Transactions on Computation Theory*
- *Computational Complexity*
- *Journal of Complexity*
- ACM Transactions on Algorithms
- Information Processing Letters

7.5 Conferences

- Annual ACM Symposium on Theory of Computing (STOC)[47]
- Annual IEEE Symposium on Foundations of Computer Science (FOCS)[47]
- ACM–SIAM Symposium on Discrete Algorithms (SODA)[47]
- IEEE Symposium on Logic in Computer Science (LICS)[47]

- Computational Complexity Conference (CCC)[48]
- International Colloquium on Automata, Languages and Programming (ICALP)[48]
- Annual Symposium on Computational Geometry (SoCG)[48]
- ACM Symposium on Principles of Distributed Computing (PODC)[47]
- ACM Symposium on Parallelism in Algorithms and Architectures (SPAA)[48]
- Annual Conference on Learning Theory (COLT)[48]
- Symposium on Theoretical Aspects of Computer Science (STACS)[48]
- European Symposium on Algorithms (ESA)[48]
- Workshop on Approximation Algorithms for Combinatorial Optimization Problems (APPROX)[48]
- Workshop on Randomization and Computation (RANDOM)[48]
- International Symposium on Algorithms and Computation (ISAAC)[48]
- International Symposium on Fundamentals of Computation Theory (FCT)[49]
- International Workshop on Graph-Theoretic Concepts in Computer Science (WG)

7.6 See also

- Formal science
- Unsolved problems in computer science
- List of important publications in theoretical computer science

7.7 Notes

[1] "SIGACT". Retrieved 2017-01-19.

[2] "ToCT". Retrieved 2010-06-09.

[3] "Challenges for Theoretical Computer Science: Theory as the Scientific Foundation of Computing". Retrieved 2009-03-29.

[4] "Any classical mathematical algorithm, for example, can be described in a finite number of English words" (Rogers 1987:2).

[5] Well defined with respect to the agent that executes the algorithm: "There is a computing agent, usually human, which can react to the instructions and carry out the computations" (Rogers 1987:2).

[6] "an algorithm is a procedure for computing a *function* (with respect to some chosen notation for integers) ... this limitation (to numerical functions) results in no loss of generality". (Rogers 1987:1).

[7] "An algorithm has zero or more inputs, i.e., quantities which are given to it initially before the algorithm begins" (Knuth 1973:5).

[8] "A procedure which has all the characteristics of an algorithm except that it possibly lacks finiteness may be called a 'computational method'" (Knuth 1973:5).

[9] "An algorithm has one or more outputs, i.e. quantities which have a specified relation to the inputs" (Knuth 1973:5).

[10] Whether or not a process with random interior processes (not including the input) is an algorithm is debatable. Rogers opines that: "a computation is carried out in a discrete stepwise fashion, without use of continuous methods or analogue devices . . . carried forward deterministically, without resort to random methods or devices. e.g.. dice" Rogers 1987:2.

[11] Paul E. Black (ed.), entry for *data structure* in *Dictionary of Algorithms and Data Structures*. U.S. National Institute of Standards and Technology. 15 December 2004. Online version Accessed May 21, 2009.

[12] Entry *data structure* in the Encyclopædia Britannica (2009) Online entry accessed on May 21, 2009.

[13] Coulouris, George; Jean Dollimore; Tim Kindberg; Gordon Blair (2011). *Distributed Systems: Concepts and Design (5th Edition)*. Boston: Addison-Wesley. ISBN 0-132-14301-1.

[14] Andrews (2000), Dolev (2000), Ghosh (2007), p. 10.

[15] Gottlieb, Allan; Almasi, George S. (1989). *Highly parallel computing*. Redwood City, Calif.: Benjamin/Cummings. ISBN 0-8053-0177-1.

[16] S.V. Adve et al. (November 2008). "Parallel Computing Research at Illinois: The UPCRC Agenda" (PDF). Parallel@Illinois, University of Illinois at Urbana-Champaign. "The main techniques for these performance benefits – increased clock frequency and smarter but increasingly complex architectures – are now hitting the so-called power wall. The computer industry has accepted that future performance increases must largely come from increasing the number of processors (or cores) on a die, rather than making a single core go faster."

[17] Asanovic et al. Old [conventional wisdom]: Power is free, but transistors are expensive. New [conventional wisdom] is [that] power is expensive, but transistors are "free".

[18] Asanovic, Krste et al. (December 18, 2006). "The Landscape of Parallel Computing Research: A View from Berkeley" (PDF). University of California, Berkeley. Technical Report No. UCB/EECS-2006-183. "Old [conventional wisdom]: Increasing clock frequency is the primary method of improving processor performance. New [conventional wisdom]: Increasing parallelism is the primary method of improving processor performance ... Even representatives from Intel, a company generally associated with the 'higher clock-speed is better' position, warned that traditional approaches to maximizing performance through maximizing clock speed have been pushed to their limit."

[19] Hennessy, John L.; Patterson, David A.; Larus, James R. (1999). *Computer organization and design : the hardware/software interface* (2. ed., 3rd print. ed.). San Francisco: Kaufmann. ISBN 1-55860-428-6.

[20] Ron Kovahi; Foster Provost (1998). "Glossary of terms". *Machine Learning*. **30**: 271–274.

[21] C. M. Bishop (2006). *Pattern Recognition and Machine Learning*. Springer. ISBN 0-387-31073-8.

[22] Wernick, Yang, Brankov, Yourganov and Strother, Machine Learning in Medical Imaging, *IEEE Signal Processing Magazine*, vol. 27, no. 4, July 2010, pp. 25-38

[23] Mannila, Heikki (1996). *Data mining: machine learning, statistics, and databases*. Int'l Conf. Scientific and Statistical Database Management. IEEE Computer Society.

[24] Friedman, Jerome H. (1998). "Data Mining and Statistics: What's the connection?". *Computing Science and Statistics*. **29** (1): 3–9.

[25] "NIH working definition of bioinformatics and computational biology" (PDF). Biomedical Information Science and Technology Initiative. 17 July 2000. Retrieved 18 August 2012.

[26] "About the CCMB". Center for Computational Molecular Biology. Retrieved 18 August 2012.

[27] F. Rieke; D. Warland; R Ruyter van Steveninck; W Bialek (1997). *Spikes: Exploring the Neural Code*. The MIT press. ISBN 978-0262681087.

[28] cf. Huelsenbeck, J. P., F. Ronquist, R. Nielsen and J. P. Bollback (2001) Bayesian inference of phylogeny and its impact on evolutionary biology, *Science* **294**:2310-2314

[29] Rando Allikmets, Wyeth W. Wasserman, Amy Hutchinson, Philip Smallwood, Jeremy Nathans, Peter K. Rogan, Thomas D. Schneider, Michael Dean (1998) Organization of the ABCR gene: analysis of promoter and splice junction sequences, *Gene* **215**:1, 111-122

[30] Burnham, K. P. and Anderson D. R. (2002) *Model Selection and Multimodel Inference: A Practical Information-Theoretic Approach, Second Edition* (Springer Science, New York) ISBN 978-0-387-95364-9.

[31] Jaynes, E. T. (1957) Information Theory and Statistical Mechanics, *Phys. Rev.* **106**:620

[32] Charles H. Bennett, Ming Li, and Bin Ma (2003) Chain Letters and Evolutionary Histories, *Scientific American* **288**:6, 76-81

[33] David R. Anderson (November 1, 2003). "Some background on why people in the empirical sciences may want to better understand the information-theoretic methods" (pdf). Retrieved 2010-06-23.

[34] Rivest, Ronald L. (1990). "Cryptology". In J. Van Leeuwen. *Handbook of Theoretical Computer Science*. **1**. Elsevier.

[35] Bellare, Mihir; Rogaway, Phillip (21 September 2005). "Introduction". *Introduction to Modern Cryptography*. p. 10.

[36] Menezes, A. J.; van Oorschot, P. C.; Vanstone, S. A. *Handbook of Applied Cryptography*. ISBN 0-8493-8523-7.

[37] "Quantum Computing with Molecules" article in *Scientific American* by Neil Gershenfeld and Isaac L. Chuang

[38] Manin, Yu. I. (1980). *Vychislimoe i nevychislimoe [Computable and Noncomputable]* (in Russian). Sov.Radio. pp. 13–15. Retrieved 4 March 2013.

[39] Feynman, R. P. (1982). "Simulating physics with computers". *International Journal of Theoretical Physics*. **21** (6): 467–488. doi:10.1007/BF02650179.

[40] Deutsch, David (1992-01-06). "Quantum computation". *Physics World*.

[41] Finkelstein, David (1968). "Space-Time Structure in High Energy Interactions". In Gudehus, T.; Kaiser, G. *Fundamental Interactions at High Energy*. New York: Gordon & Breach.

[42] "New qubit control bodes well for future of quantum computing". Retrieved 26 October 2014.

[43] Quantum Information Science and Technology Roadmap for a sense of where the research is heading.

[44] R. W. Butler (2001-08-06). "What is Formal Methods?". Retrieved 2006-11-16.

[45] C. Michael Holloway. "Why Engineers Should Consider Formal Methods" (PDF). 16th Digital Avionics Systems Conference (27–30 October 1997). Retrieved 2006-11-16.

[46] Monin, pp.3-4

[47] The 2007 Australian Ranking of ICT Conferences: tier A+.

[48] The 2007 Australian Ranking of ICT Conferences: tier A.

[49] FCT 2011 (retrieved 2013-06-03)

7.8 Further reading

- Martin Davis, Ron Sigal, Elaine J. Weyuker, *Computability, complexity, and languages: fundamentals of theoretical computer science*, 2nd ed., Academic Press, 1994, ISBN 0-12-206382-1. Covers theory of computation, but also program semantics and quantification theory. Aimed at graduate students.

7.9 External links

- SIGACT directory of additional theory links
- Theory Matters Wiki Theoretical Computer Science (TCS) Advocacy Wiki
- Usenet comp.theory
- List of academic conferences in the area of theoretical computer science at confsearch
- Theoretical Computer Science - StackExchange, a Question and Answer site for researchers in theoretical computer science
- Computer Science Animated
- http://theory.csail.mit.edu/ @ Massachusetts Institute of Technology

Chapter 8

Outline of physical science

"Physical Science" redirects here. It is not to be confused with Physics.

Physical science is a branch of natural science that studies non-living systems, in contrast to life science. It in turn has many branches, each referred to as a "physical science", together called the "physical sciences". However, the term "physical" creates an unintended, somewhat arbitrary distinction, since many branches of physical science also study biological phenomena and branches of chemistry such as organic chemistry.

8.1 Definition

Physical science can be described as all of the following:

- A branch of science (a systematic enterprise that builds and organizes knowledge in the form of testable explanations and predictions about the universe).[1][2][3]

    - A branch of natural science – natural science is a major branch of science that tries to explain and predict nature's phenomena, based on empirical evidence. In natural science, hypotheses must be verified scientifically to be regarded as scientific theory. Validity, accuracy, and social mechanisms ensuring quality control, such as peer review and repeatability of findings, are amongst the criteria and methods used for this purpose. Natural science can be broken into two main branches: life science (for example biology) and physical science. Each of these branches, and all of their sub-branches, are referred to as natural sciences.

8.2 Branches of physical science

- Physics – natural and physical science that involves the study of matter[4] and its motion through space and time, along with related concepts such as energy and force.[5] More broadly, it is the general analysis of nature, conducted in order to understand how the universe behaves.[lower-alpha 1][6][7]

    - Branches of physics

- Astronomy – study of celestial objects (such as stars, galaxies, planets, moons, asteroids, comets and nebulae), the physics, chemistry, and evolution of such objects, and phenomena that originate outside the atmosphere of Earth, including supernovae explosions, gamma ray bursts, and cosmic microwave background radiation.

    - Branches of astronomy

- Chemistry – studies the composition, structure, properties and change of matter.[8][9] In this realm, chemistry deals with such topics as the properties of individual atoms, the manner in which atoms form chemical bonds in the formation of compounds, the interactions of substances through intermolecular forces to give matter its general properties, and the interactions between substances through chemical reactions to form different substances.

    - Branches of chemistry

- Earth science – all-embracing term referring to the fields of science dealing with planet Earth. Earth science is the study of how the natural environment (ecosphere or Earth system) works and how it evolved to its current state. It includes the study of the atmosphere, hydrosphere, lithosphere, and biosphere.

    - Branches of Earth science

8.3 History of physical science

History of physical science – history of the branch of natural science that studies non-living systems, in contrast to the biological sciences. It in turn has many branches, each referred to as a "physical science", together called the "physical sciences". However, the term "physical" creates an unintended, somewhat arbitrary distinction, since many branches of physical science also study biological phenomena (organic chemistry, for example).

- History of physics – history of the physical science that studies matter and its motion through space-time, and related concepts such as energy and force
    - History of acoustics – history of the study of mechanical waves in solids, liquids, and gases (such as vibration and sound)
    - History of agrophysics – history of the study of physics applied to agroecosystems
        - History of soil physics – history of the study of soil physical properties and processes.
    - History of astrophysics – history of the study of the physical aspects of celestial objects
    - History of astronomy – history of the studies the universe beyond Earth, including its formation and development, and the evolution, physics, chemistry, meteorology, and motion of celestial objects (such as galaxies, planets, etc.) and phenomena that originate outside the atmosphere of Earth (such as the cosmic background radiation).
        - History of astrodynamics – history of the application of ballistics and celestial mechanics to the practical problems concerning the motion of rockets and other spacecraft.
        - History of astrometry – history of the branch of astronomy that involves precise measurements of the positions and movements of stars and other celestial bodies.
        - History of cosmology – history of the discipline that deals with the nature of the Universe as a whole.
        - History of extragalactic astronomy – history of the branch of astronomy concerned with objects outside our own Milky Way Galaxy
        - History of galactic astronomy – history of the study of our own Milky Way galaxy and all its contents.
        - History of physical cosmology – history of the study of the largest-scale structures and dynamics of the universe and is concerned with fundamental questions about its formation and evolution.
    - History of planetary science – history of the scientific study of planets (including Earth), moons, and planetary systems, in particular those of the Solar System and the processes that form them.
    - History of stellar astronomy – history of the natural science that deals with the study of celestial objects (such as stars, planets, comets, nebulae, star clusters and galaxies) and phenomena that originate outside the atmosphere of Earth (such as cosmic background radiation)
- History of atmospheric physics – history of the study of the application of physics to the atmosphere
- History of atomic, molecular, and optical physics – history of the study of how matter and light interact
- History of biophysics – history of the study of physical processes relating to biology
    - History of medical physics – history of the application of physics concepts, theories and methods to medicine.
    - History of neurophysics – history of the branch of biophysics dealing with the nervous system.
- History of chemical physics – history of the branch of physics that studies chemical processes from the point of view of physics.
- History of computational physics – history of the study and implementation of numerical algorithms to solve problems in physics for which a quantitative theory already exists.
- History of condensed matter physics – history of the study of the physical properties of condensed phases of matter.
- History of cryogenics – history of the cryogenics is the study of the production of very low temperature (below −150 °C, −238 °F or 123K) and the behavior of materials at those temperatures.
- Dynamics – history of the study of the causes of motion and changes in motion
- History of econophysics – history of the interdisciplinary research field, applying theories and methods originally developed by physicists in order to solve problems in economics
- History of electromagnetism – history of the branch of science concerned with the forces that occur between electrically charged particles.

- History of geophysics – history of the physics of the Earth and its environment in space; also the study of the Earth using quantitative physical methods
- History of materials physics – history of the use of physics to describe materials in many different ways such as force, heat, light and mechanics.
- History of mathematical physics – history of the application of mathematics to problems in physics and the development of mathematical methods for such applications and for the formulation of physical theories.
- History of mechanics – history of the branch of physics concerned with the behavior of physical bodies when subjected to forces or displacements, and the subsequent effects of the bodies on their environment.
    - History of biomechanics – history of the study of the structure and function of biological systems such as humans, animals, plants, organs, and cells by means of the methods of mechanics.
    - History of classical mechanics – history of the one of the two major sub-fields of mechanics, which is concerned with the set of physical laws describing the motion of bodies under the action of a system of forces.
    - History of continuum mechanics – history of the branch of mechanics that deals with the analysis of the kinematics and the mechanical behavior of materials modeled as a continuous mass rather than as discrete particles.
    - History of fluid mechanics – history of the study of fluids and the forces on them.
    - History of quantum mechanics – history of the branch of physics dealing with physical phenomena where the action is on the order of the Planck constant.
    - History of thermodynamics – history of the branch of physical science concerned with heat and its relation to other forms of energy and work.
- History of nuclear physics – history of the field of physics that studies the building blocks and interactions of atomic nuclei.
- History of optics – history of the branch of physics which involves the behavior and properties of light, including its interactions with matter and the construction of instruments that use or detect it.
- History of particle physics – history of the branch of physics that studies the existence and interactions of particles that are the constituents of what is usually referred to as matter or radiation.
- History of psychophysics – history of the quantitatively investigates the relationship between physical stimuli and the sensations and perceptions they affect.
- History of plasma physics – history of the state of matter similar to gas in which a certain portion of the particles are ionized.
- History of polymer physics – history of the field of physics that studies polymers, their fluctuations, mechanical properties, as well as the kinetics of reactions involving degradation and polymerisation of polymers and monomers respectively.
- History of quantum physics – history of the branch of physics dealing with physical phenomena where the action is on the order of the Planck constant.
- Relativity –
- History of statics – history of the branch of mechanics concerned with the analysis of loads (force, torque/moment) on physical systems in static equilibrium, that is, in a state where the relative positions of subsystems do not vary over time, or where components and structures are at a constant velocity.
- History of solid state physics – history of the study of rigid matter, or solids, through methods such as quantum mechanics, crystallography, electromagnetism, and metallurgy.
- History of vehicle dynamics – history of the dynamics of vehicles, here assumed to be ground vehicles.

- History of chemistry – history of the physical science of atomic matter (matter that is composed of chemical elements), especially its chemical reactions, but also including its properties, structure, composition, behavior, and changes as they relate the chemical reactions
    - History of analytical chemistry – history of the study of the separation, identification, and quantification of the chemical components of natural and artificial materials.
    - History of astrochemistry – history of the study of the abundance and reactions of chemical elements and molecules in the universe, and their interaction with radiation.

- History of cosmochemistry – history of the study of the chemical composition of matter in the universe and the processes that led to those compositions
- History of atmospheric chemistry – history of the branch of atmospheric science in which the chemistry of the Earth's atmosphere and that of other planets is studied. It is a multidisciplinary field of research and draws on environmental chemistry, physics, meteorology, computer modeling, oceanography, geology and volcanology and other disciplines
- History of biochemistry – history of the study of chemical processes in living organisms, including, but not limited to, living matter. Biochemistry governs all living organisms and living processes.
    - History of agrochemistry – history of the study of both chemistry and biochemistry which are important in agricultural production, the processing of raw products into foods and beverages, and in environmental monitoring and remediation.
    - History of bioinorganic chemistry – history of the examines the role of metals in biology.
    - History of bioorganic chemistry – history of the rapidly growing scientific discipline that combines organic chemistry and biochemistry.
    - History of biophysical chemistry – history of the new branch of chemistry that covers a broad spectrum of research activities involving biological systems.
    - History of environmental chemistry – history of the scientific study of the chemical and biochemical phenomena that occur in natural places.
    - History of immunochemistry – history of the branch of chemistry that involves the study of the reactions and components on the immune system.
    - History of medicinal chemistry – history of the discipline at the intersection of chemistry, especially synthetic organic chemistry, and pharmacology and various other biological specialties, where they are involved with design, chemical synthesis and development for market of pharmaceutical agents (drugs).
    - History of pharmacology – history of the branch of medicine and biology concerned with the study of drug action.
- History of natural product chemistry – history of the chemical compound or substance produced by a living organism – history of the found in nature that usually has a pharmacological or biological activity for use in pharmaceutical drug discovery and drug design.
- History of neurochemistry – history of the specific study of neurochemicals, which include neurotransmitters and other molecules such as neuro-active drugs that influence neuron function.
- History of computational chemistry – history of the branch of chemistry that uses principles of computer science to assist in solving chemical problems.
    - History of chemo-informatics – history of the use of computer and informational techniques, applied to a range of problems in the field of chemistry.
    - History of molecular mechanics – history of the uses Newtonian mechanics to model molecular systems.
- History of Flavor chemistry – history of the someone who uses chemistry to engineer artificial and natural flavors.
- History of Flow chemistry – history of the chemical reaction is run in a continuously flowing stream rather than in batch production.
- History of geochemistry – history of the study of the mechanisms behind major geological systems using chemistry
    - History of aqueous geochemistry – history of the study of the role of various elements in watersheds, including copper, sulfur, mercury, and how elemental fluxes are exchanged through atmospheric-terrestrial-aquatic interactions
    - History of isotope geochemistry – history of the study of the relative and absolute concentrations of the elements and their isotopes using chemistry and geology
    - History of ocean chemistry – history of the studies the chemistry of marine environments including the influences of different variables.
    - History of organic geochemistry – history of the study of the impacts and processes that organisms have had on Earth
    - History of regional, environmental and exploration geochemistry – history of the

study of the spatial variation in the chemical composition of materials at the surface of the Earth
- History of inorganic chemistry – history of the branch of chemistry concerned with the properties and behavior of inorganic compounds.
- History of nuclear chemistry – history of the subfield of chemistry dealing with radioactivity, nuclear processes and nuclear properties.
    - History of radiochemistry – history of the chemistry of radioactive materials, where radioactive isotopes of elements are used to study the properties and chemical reactions of non-radioactive isotopes (often within radiochemistry the absence of radioactivity leads to a substance being described as being inactive as the isotopes are stable).
- History of organic chemistry – history of the study of the structure, properties, composition, reactions, and preparation (by synthesis or by other means) of carbon-based compounds, hydrocarbons, and their derivatives.
    - History of petrochemistry – history of the branch of chemistry that studies the transformation of crude oil (petroleum) and natural gas into useful products or raw materials.
- History of organometallic chemistry – history of the study of chemical compounds containing bonds between carbon and a metal.
- History of photochemistry – history of the study of chemical reactions that proceed with the absorption of light by atoms or molecules..
- History of physical chemistry – history of the study of macroscopic, atomic, subatomic, and particulate phenomena in chemical systems in terms of physical laws and concepts.
    - History of chemical kinetics – history of the study of rates of chemical processes.
    - History of chemical thermodynamics – history of the study of the interrelation of heat and work with chemical reactions or with physical changes of state within the confines of the laws of thermodynamics.
    - History of electrochemistry – history of the branch of chemistry that studies chemical reactions which take place in a solution at the interface of an electron conductor (a metal or a semiconductor) and an ionic conductor (the electrolyte), and which involve electron transfer between the electrode and the electrolyte or species in solution.
- History of Femtochemistry – history of the Femtochemistry is the science that studies chemical reactions on extremely short timescales, approximately 10^{-15} seconds (one femtosecond, hence the name).
- History of mathematical chemistry – history of the area of research engaged in novel applications of mathematics to chemistry; it concerns itself principally with the mathematical modeling of chemical phenomena.
- History of mechanochemistry – history of the coupling of the mechanical and the chemical phenomena on a molecular scale and includes mechanical breakage, chemical behaviour of mechanically stressed solids (e.g., stress-corrosion cracking), tribology, polymer degradation under shear, cavitation-related phenomena (e.g., sonochemistry and sonoluminescence), shock wave chemistry and physics, and even the burgeoning field of molecular machines.
- History of physical organic chemistry – history of the study of the interrelationships between structure and reactivity in organic molecules.
- History of quantum chemistry – history of the branch of chemistry whose primary focus is the application of quantum mechanics in physical models and experiments of chemical systems.
- History of sonochemistry – history of the study of the effect of sonic waves and wave properties on chemical systems.
- History of stereochemistry – history of the study of the relative spatial arrangement of atoms within molecules.
- History of supramolecular chemistry – history of the area of chemistry beyond the molecules and focuses on the chemical systems made up of a discrete number of assembled molecular subunits or components.
- History of thermochemistry – history of the study of the energy and heat associated with chemical reactions and/or physical transformations.
- History of phytochemistry – history of the strict sense of the word the study of phytochemicals.
- History of polymer chemistry – history of the multidisciplinary science that deals with the chemical synthesis and chemical properties of polymers or macromolecules.
- History of solid-state chemistry – history of the study of the synthesis, structure, and properties

of solid phase materials, particularly, but not necessarily exclusively of, non-molecular solids
- Multidisciplinary fields involving chemistry
  - History of chemical biology – history of the scientific discipline spanning the fields of chemistry and biology that involves the application of chemical techniques and tools, often compounds produced through synthetic chemistry, to the study and manipulation of biological systems.
  - History of chemical engineering – history of the branch of engineering that deals with physical science (e.g., chemistry and physics), and life sciences (e.g., biology, microbiology and biochemistry) with mathematics and economics, to the process of converting raw materials or chemicals into more useful or valuable forms.
  - History of chemical oceanography – history of the study of the behavior of the chemical elements within the Earth's oceans.
  - History of chemical physics – history of the branch of physics that studies chemical processes from the point of view of physics.
  - History of materials science – history of the interdisciplinary field applying the properties of matter to various areas of science and engineering.
  - History of nanotechnology – history of the study of manipulating matter on an atomic and molecular scale
  - History of oenology – history of the science and study of all aspects of wine and winemaking except vine-growing and grape-harvesting, which is a subfield called viticulture.
  - History of spectroscopy – history of the study of the interaction between matter and radiated energy
  - History of surface science – history of the Surface science is the study of physical and chemical phenomena that occur at the interface of two phases, including solid–liquid interfaces, solid–gas interfaces, solid–vacuum interfaces, and liquid–gas interfaces.
- History of earth science – history of the all-embracing term for the sciences related to the planet Earth. Earth science, and all of its branches, are branches of physical science.
- History of atmospheric sciences – history of the umbrella term for the study of the atmosphere, its processes, the effects other systems have on the atmosphere, and the effects of the atmosphere on these other systems.
  - History of climatology
  - History of meteorology
    - History of atmospheric chemistry
- History of biogeography – history of the study of the distribution of species (biology), organisms, and ecosystems in geographic space and through geological time.
- History of cartography – history of the study and practice of making maps or globes.
- History of climatology – history of the study of climate, scientifically defined as weather conditions averaged over a period of time
- History of coastal geography – history of the study of the dynamic interface between the ocean and the land, incorporating both the physical geography (i.e. coastal geomorphology, geology and oceanography) and the human geography (sociology and history) of the coast.
- History of environmental science – history of an integrated, quantitative, and interdisciplinary approach to the study of environmental systems.
  - History of ecology – history of the scientific study of the distribution and abundance of living organisms and how the distribution and abundance are affected by interactions between the organisms and their environment.
    - History of Freshwater biology – history of the scientific biological study of freshwater ecosystems and is a branch of limnology
    - History of marine biology – history of the scientific study of organisms in the ocean or other marine or brackish bodies of water
    - History of parasitology – history of the Parasitology is the study of parasites, their hosts, and the relationship between them.
    - History of population dynamics – history of the Population dynamics is the branch of life sciences that studies short-term and long-term changes in the size and age composition of populations, and the biological and environmental processes influencing those changes.

- History of environmental chemistry – history of the Environmental chemistry is the scientific study of the chemical and biochemical phenomena that occur in natural places.
- History of environmental soil science – history of the Environmental soil science is the study of the interaction of humans with the pedosphere as well as critical aspects of the biosphere, the lithosphere, the hydrosphere, and the atmosphere.
- History of environmental geology – history of the Environmental geology, like hydrogeology, is an applied science concerned with the practical application of the principles of geology in the solving of environmental problems.
- History of toxicology – history of the branch of biology, chemistry, and medicine concerned with the study of the adverse effects of chemicals on living organisms.

- History of geodesy – history of the scientific discipline that deals with the measurement and representation of the Earth, including its gravitational field, in a three-dimensional time-varying space
- History of geography – history of the science that studies the lands, features, inhabitants, and phenomena of Earth
- History of geoinformatics – history of the science and the technology which develops and uses information science infrastructure to address the problems of geography, geosciences and related branches of engineering.
- History of geology – history of the study of the Earth, with the general exclusion of present-day life, flow within the ocean, and the atmosphere.
  - History of planetary geology – history of the planetary science discipline concerned with the geology of the celestial bodies such as the planets and their moons, asteroids, comets, and meteorites.
- History of geomorphology – history of the scientific study of landforms and the processes that shape them
- History of geostatistics – history of the branch of statistics focusing on spatial or spatiotemporal datasets
- History of geophysics – history of the physics of the Earth and its environment in space; also the study of the Earth using quantitative physical methods.
- History of glaciology – history of the study of glaciers, or more generally ice and natural phenomena that involve ice.
- History of hydrology – history of the study of the movement, distribution, and quality of water on Earth and other planets, including the hydrologic cycle, water resources and environmental watershed sustainability.
- History of hydrogeology – history of the area of geology that deals with the distribution and movement of groundwater in the soil and rocks of the Earth's crust (commonly in aquifers).
- History of mineralogy – history of the study of chemistry, crystal structure, and physical (including optical) properties of minerals.
- History of meteorology – history of the interdisciplinary scientific study of the atmosphere which explains and forecasts weather events.
- History of oceanography – history of the branch of Earth science that studies the ocean
- History of paleoclimatology – history of the study of changes in climate taken on the scale of the entire history of Earth
- History of paleontology – history of the study of prehistoric life
- History of petrology – history of the branch of geology that studies the origin, composition, distribution and structure of rocks.
- History of limnology – history of the study of inland waters
- History of seismology – history of the scientific study of earthquakes and the propagation of elastic waves through the Earth or through other planet-like bodies
- History of soil science – history of the study of soil as a natural resource on the surface of the earth including soil formation, classification and mapping; physical, chemical, biological, and fertility properties of soils; and these properties in relation to the use and management of soils.
- History of topography – history of the study of surface shape and features of the Earth and other observable astronomical objects including planets, moons, and asteroids.
- History of volcanology – history of the study of volcanoes, lava, magma, and related geological, geophysical and geochemical phenomena.

8.4 General principles of the physical sciences

- Principle – law or rule that has to be, or usually is to be followed, or can be desirably followed, or is an inevitable consequence of something, such as the laws observed in nature or the way that a system is constructed. The principles of such a system are understood by its users as the essential characteristics of the system, or reflecting system's designed purpose, and the effective operation or use of which would be impossible if any one of the principles was to be ignored.

8.4.1 Basic principles of physics

Physics – branch of science that studies matter[4] and its motion through space and time, along with related concepts such as energy and force.[5] Physics is one of the "fundamental sciences" because the other natural sciences (like biology, geology etc.) deal with systems that seem to obey the laws of physics. According to physics, the physical laws of matter, energy and the fundamental forces of nature govern the interactions between particles and physical entities (such as planets, molecules, atoms or the subatomic particles). Some of the basic pursuits of physics, which include some of the most prominent developments in modern science in the last millennium, include:

- Describing the nature, measuring and quantifying of bodies and their motion, dynamics etc.
  - Newton's laws of motion
  - Mass, force and weight
  - Momentum and conservation of energy
  - Gravity, theories of gravity
  - Energy, work, and their relationship
  - Motion, position, and energy
  - Different forms of Energy, their interconversion and the inevitable loss of energy in the form of heat (Thermodynamics)
  - Energy conservation, conversion, and transfer.
  - Energy source the transfer of energy from one source to work in another.
- Kinetic molecular theory
  - Phases of matter and phase transitions
  - Temperature and thermometers
  - Energy and heat
  - Heat flow: conduction, convection, and radiation
- The three laws of thermodynamics
- The principles of waves and sound
- The principles of electricity, magnetism, and electromagnetism
- The principles, sources, and properties of light

8.4.2 Basic principles of astronomy

Astronomy – science of celestial bodies and their interactions in space. Its studies includes the following:

- The life and characteristics of stars and galaxies
- Origins of the universe. Physical science uses the Big Bang theory as the commonly accepted scientific theory of the origin of the universe.
- A heliocentric Solar System. Ancient cultures saw the Earth as the centre of the Solar System or universe (geocentrism). In the 16th century, Nicolaus Copernicus advanced the ideas of heliocentrism, recognizing the Sun as the centre of the Solar System.
- The structure of solar systems, planets, comets, asteroids, and meteors
- The shape and structure of Earth (roughly spherical, see also Spherical Earth)
- Earth in the Solar System
- Time measurement
- The composition and features of the Moon
- Interactions of the Earth and Moon

(Note: Astronomy should not be confused with astrology, which assumes[10] that people's destiny and human affairs in general correlate to the apparent positions of astronomical objects in the sky - although the two fields share a common origin, they are quite different; astronomers embrace the scientific method, while astrologers do not.)

8.4.3 Basic principles of chemistry

Chemistry – branch of science that studies the composition, structure, properties and change of matter.[8][9] Chemistry is chiefly concerned with atoms and molecules and their interactions and transformations, for example, the properties of the chemical bonds formed between atoms to create chemical compounds. As such, chemistry studies the involvement of electrons and various forms of energy in photochemical reactions, oxidation-reduction reactions, changes in phases of matter, and separation of mixtures. Preparation and properties of complex substances, such as alloys, polymers, biological molecules, and pharmaceutical agents are considered in specialized fields of chemistry.

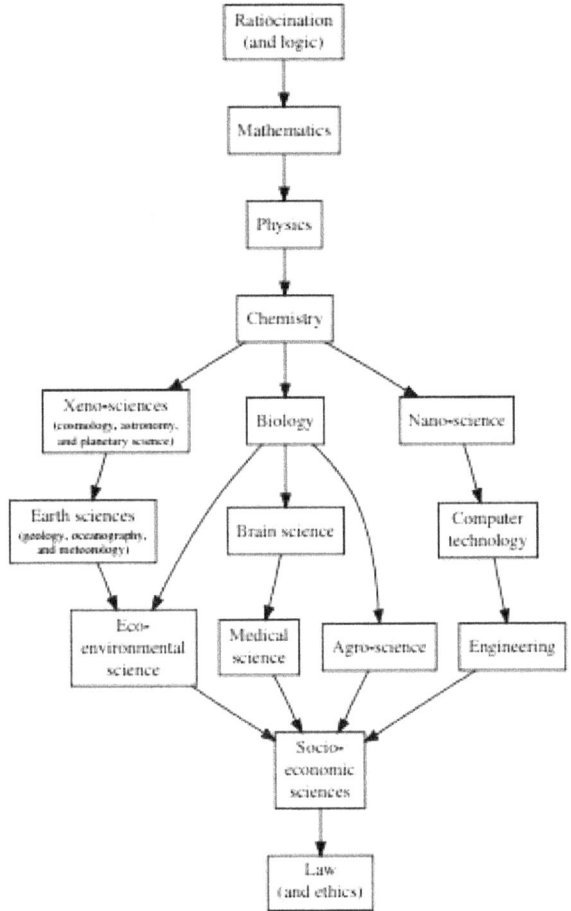

Chemistry, the central science, partial ordering of the sciences proposed by Balaban and Klein.

- Physical chemistry
  - Chemical thermodynamics
  - Reaction kinetics
  - Molecular structure
  - Quantum chemistry
  - Spectroscopy
- Theoretical chemistry
  - Electron configuration
  - Molecular modelling
  - Molecular dynamics
  - Statistical mechanics
- Computational chemistry
  - Mathematical chemistry
  - Cheminformatics
- Nuclear chemistry
  - The nature of the atomic nucleus
  - Characterization of radioactive decay
  - Nuclear reactions
- Organic chemistry
  - Organic compounds
  - Organic reaction
  - Functional groups
  - Organic synthesis
- Inorganic chemistry
  - Inorganic compounds
  - Crystal structure
  - Coordination chemistry
  - Solid-state chemistry
- Biochemistry
- Analytical chemistry
  - Instrumental analysis
  - Electroanalytical method
  - Wet chemistry
- Electrochemistry
  - Redox reaction
- Materials chemistry

8.4.4 Basic principles of earth science

Earth science – the science of the planet Earth, as of 2014 the only identified life-bearing planet. Its studies include the following:

- The water cycle and the process of transpiration
- Freshwater
- Oceanography
  - Weathering and erosion
  - Rocks
- Agrophysics

- Soil science
  - Pedogenesis
  - Soil fertility
- Earth's tectonic structure
- Geomorphology and geophysics
  - Physical geography
  - Seismology: stress, strain, and earthquakes
  - Characteristics of mountains and volcanoes
- Characteristics and formation of fossils
- Atmospheric sciences – the branches of science that study the atmosphere, its processes, the effects other systems have on the atmosphere, and the effects of the atmosphere on these other systems.
  - Atmosphere of Earth
  - Atmospheric pressure and winds
  - Evaporation, condensation, and humidity
  - Fog and clouds
- Meteorology, weather, climatology, and climate
  - Hydrology, clouds and precipitation
  - Air masses and weather fronts
  - Major storms: thunderstorms, tornadoes, and hurricanes
  - Major climate groups
- Speleology
  - Cave

8.5 Notable physical scientists

- List of physicists
- List of astronomers
- List of chemists

8.5.1 Earth scientists

- List of Russian earth scientists

8.6 See also

- Outline of science
  - Outline of natural science
    - Outline of physical science
      - Outline of earth science
  - Outline of formal science
  - Outline of social science
  - Outline of applied science

8.7 Notes

[1] The term 'universe' is defined as everything that physically exists: the entirety of space and time, all forms of matter, energy and momentum, and the physical laws and constants that govern them. However, the term 'universe' may also be used in slightly different contextual senses, denoting concepts such as the cosmos or the philosophical world.

8.8 References

[1] Wilson, Edward O. (1998). *Consilience: The Unity of Knowledge* (1st ed.). New York, NY: Vintage Books. pp. 49–71. ISBN 0-679-45077-7.

[2] "... modern science is a discovery as well as an invention. It was a discovery that nature generally acts regularly enough to be described by laws and even by mathematics; and required invention to devise the techniques, abstractions, apparatus, and organization for exhibiting the regularities and securing their law-like descriptions." —p.vii, J. L. Heilbron, (2003, editor-in-chief). *The Oxford Companion to the History of Modern Science*. New York: Oxford University Press. ISBN 0-19-511229-6.

[3] "science". *Merriam-Webster Online Dictionary*. Merriam-Webster, Inc. Retrieved 2011-10-16. **3 a:** knowledge or a system of knowledge covering general truths or the operation of general laws especially as obtained and tested through scientific method **b:** such knowledge or such a system of knowledge concerned with the physical world and its phenomena

[4] At the start of *The Feynman Lectures on Physics*, Richard Feynman offers the atomic hypothesis as the single most prolific scientific concept: "If, in some cataclysm, all [] scientific knowledge were to be destroyed [save] one sentence [...] what statement would contain the most information in the fewest words? I believe it is [...] that *all things are made up of atoms – little particles that move around in perpetual motion, attracting each other when they are a little distance apart, but repelling upon being squeezed into one another ...*" (Feynman, Leighton & Sands 1963, p. I-2)

[5] "Physical science is that department of knowledge which relates to the order of nature, or, in other words, to the regular succession of events." (Maxwell 1878, p. 9)

[6] Young & Freedman 2014, p. 9

[7] "Physics is the study of your world and the world and universe around you." (Holzner 2006, p. 7)

[8] "What is Chemistry?". Chemweb.ucc.ie. Retrieved 2011-06-12.

[9] Chemistry. (n.d.). Merriam-Webster's Medical Dictionary. Retrieved August 19, 2007.

[10] Scharringhausen, Britt. "What's the difference between astronomy and astrology? (Beginner) - Curious About Astronomy? Ask an Astronomer". *curious.astro.cornell.edu*. Retrieved 2017-05-27.

8.8.1 Works cited

- Feynman, R.P.; Leighton, R.B.; Sands, M. (1963). *The Feynman Lectures on Physics*. **1**. ISBN 0-201-02116-1.

- Holzner, S. (2006). *Physics for Dummies*. John Wiley & Sons. ISBN 0-470-61841-8. Physics is the study of your world and universe around you.

- Maxwell, J.C. (1878). *Matter and Motion*. D. Van Nostrand. ISBN 0-486-66895-9.

- Young, H.D.; Freedman, R.A. (2014). *Sears and Zemansky's University Physics with Modern Physics Technology Update* (13th ed.). Pearson Education. ISBN 978-1-292-02063-1.

8.9 External links

- Physical science topics and articles for school curricula (grades K-12)

Chapter 9

List of life sciences

"Life Sciences" redirects here. For the scientific journal, see Life Sciences (journal).

The **life sciences** comprise the branches of science that involve the scientific study of living organisms – such as microorganisms, plants, animals, and human beings – as well as related considerations like bioethics. While biology remains the centerpiece of the life sciences, technological advances in molecular biology and biotechnology have led to a burgeoning of specializations and interdisciplinary fields.[1][2]

Some life sciences focus on a specific type of life. For example, zoology is the study of animals, while botany is the study of plants. Other life sciences focus on aspects common to all or many life forms, such as anatomy and genetics. Yet other fields are interested in technological advances involving living things, such as bio-engineering. Another major, though more specific, branch of life sciences involves understanding the mind – neuroscience.

The life sciences are helpful in improving the quality and standard of life. They have applications in health, agriculture, medicine, and the pharmaceutical and food science industries.

There is considerable overlap between many of the topics of study in the life sciences.

9.1 Biology and its branches

Main article: Outline of biology § Branches of biology

Biology – burst and eclectic field, composed of many branches and subdisciplines. However, despite the broad scope of biology, there are certain general and unifying concepts within it that govern all study and research, consolidating it into a single, coherent field. In general, biology recognizes the cell as the basic unit of life, genes as the basic unit of heredity, and evolution as the engine that propels the synthesis and creation of new species. It is also understood today that all organisms survive by consuming and transforming energy and by regulating their internal environment to maintain a stable and vital condition. Here are some of biology's major branches:

- Anatomy – study of form and function, in plants, animals, and other organisms, or specifically in humans

- Biochemistry – study of the chemical reactions required for life to exist and function, usually a focus on the cellular level

- Bioengineering – study of biology through the means of engineering with an emphasis on applied knowledge and especially related to biotechnology

- Bioinformatics – interdisciplinary scientific field that develops methods for storing, retrieving, organizing and analyzing biological data. A major activity in bioinformatics is to develop software tools to generate useful biological knowledge.

- Biolinguistics – study of the biology and evolution of language

- Biological Anthropology - the study of the biological basis of humans, non-human primates, and hominins: a subfield of Anthropology. Also known as physical anthropology

- Biomechanics – the study of the mechanics of living beings

- Biomedical research – study of health and disease

- Biophysics – study of biological processes by applying the theories and methods traditionally used in the physical sciences

- Biotechnology – study of the manipulation of living matter, including genetic modification and synthetic biology

- Botany – study of plants
- Cell biology – study of the cell as a complete unit, and the molecular and chemical interactions that occur within a living cell
- Developmental biology – study of the processes through which an organism forms, from zygote to full structure
- Ecology – study of the interactions of living organisms with one another and with the non-living elements of their environment
- Entomology – study of insects
- Epidemiology – a major component of public health research, studying factors affecting the health of populations
- Ethology - the study of animal behaviour
- Evolutionary biology – study of the origin and descent of species over time
- Genetics – study of genes and heredity.
- Hematology (also known as Haematology) – study of blood and blood-forming organs.
- Marine biology – study of ocean ecosystems, plants, animals, and other living beings
- Microbiology – study of microscopic organisms (microorganisms) and their interactions with other living organisms
- Molecular biology – study of biology and biological functions at the molecular level, some cross over with biochemistry
- Neuroscience– study of the nervous system
- Physiology – study of the functioning of living organisms and the organs and parts of living organisms
- Population biology – study of groups of conspecific organisms
- Sociobiology – study of the biological bases of sociology
- Structural biology – a branch of molecular biology, biochemistry, and biophysics concerned with the molecular structure of biological macromolecules
- Synthetic biology - the design and construction of new biological entities such as enzymes, genetic circuits and cells, or the redesign of existing biological systems (LY)
- Systems biology - study of the integration and dependencies of various components within a biological system, with particular focus upon the role of metabolic pathways and cell-signaling strategies in physiology.
- Toxicology - study of the effects of chemicals on living organisms
- Zoology – study of animals, including classification, physiology, development, and behavior.

9.2 Medicine and its branches

See also: Outline of health sciences

Medicine – applied science or practice of the diagnosis, treatment, and prevention of disease. It encompasses a variety of health care practices evolved to maintain and restore health by the prevention and treatment of illness. Some of its branches are:

- Anesthesiology – branch of medicine that deals with life support and anesthesia during surgery.
- Cardiology – branch of medicine that deals with disorders of the heart and the blood vessels.
- Critical care medicine – focuses on life support and the intensive care of the seriously ill.
- Dermatology – branch of medicine that deals with the skin, its structure, functions and diseases.
- Emergency medicine – focuses on care provided in the emergency department.
- Endocrinology – branch of medicine that deals with disorders of the endocrine system.
- Gastroenterology – branch of medicine that deals with the study and care of the digestive system.
- General Practice (often called Family Medicine) is a branch of medicine that specializes in primary care.
- Geriatrics – branch of medicine that deals with the general health and well-being of the elderly.
- Gynecology – branch of medicine that deals with the health of the female reproductive systems and the breasts.
- Hematology – branch of medicine that deals with the blood and the circulatory system.
- Hepatology – branch of medicine that deals with the liver, gallbladder and the biliary system.

- Infectious disease – branch of medicine that deals with the diagnosis and management of infectious disease, especially for complex cases and immunocompromised patients.

- Neurology – branch of medicine that deals with the brain and the nervous system.

- Nephrology – branch of medicine which deals with the kidneys.

- Oncology – is the branch of medicine that studies of cancer.

- Ophthalmology – branch of medicine that deals with the eyes.

- Otolaryngology – branch of medicine that deals the ears, nose and throat.

- Pathology – study of diseases, and the causes, processes, nature, and development of disease

- Pediatrics – branch of medicine that deals with the general health and well-being of children.

- Pharmacology – study and practical application of preparation, use, and effects of drugs and synthetic medicines

- Pulmonology – branch of medicine that deals with the respiratory system.

- Psychiatry – branch of medicine that deals with the study, diagnosis, treatment, and prevention of mental disorders.

- Radiology – branch of medicine that employs medical imaging to diagnose and treat disease.

- Rheumatology – branch of medicine that deals with the diagnosis and treatment of rheumatic diseases.

- Splanchnology – branch of medicine that deals with visceral organs.

- Surgery – branch of medicine that uses operative techniques to investigate or treat both disease and injury, or to help improve bodily function or appearance.

- Urology – branch of medicine that deals with the urinary system and the male reproductive system.

- Veterinary medicine – branch of medicine that deals with the prevention, diagnosis and treatment of disease, disorder and injury in nonhuman animals.

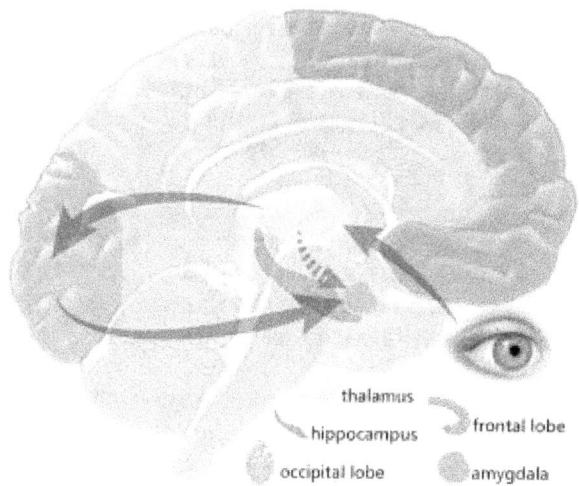

Brain parts involved with a fear amygdala hijack from optical stimulus

9.3 New and other life science types

- Affective neuroscience – study of the neural mechanisms of emotion. This interdisciplinary field combines neuroscience with the psychological study of personality, emotion, and mood.[3]

- Biocomputers – biocomputers use systems of biologically derived molecules, such as DNA and proteins, to perform computational calculations involving storing, retrieving, and processing data. The development of biocomputers has been made possible by the expanding new science of nanobiotechnology.

- Biocontrol – bioeffector-method of controlling pests

Encarsia formosa was one of the first biological control agents developed.

(including insects, mites, weeds and plant diseases) using other living organisms.[4]

- Biodynamics – method of organic farming originally developed by Rudolf Steiner that employs what proponents describe as "a holistic understanding of agricultural processes".[5]:145 One of the first sustainable agriculture movements.[6][7][8]

- Bioelectronics – the electrical state of biological matter significantly affects its structure and function, compare for instance the membrane potential, the signal transduction by neurons, the isoelectric point (IEP) and so on. Micro- and nano-electronic components and devices have increasingly been combined with biological systems[9] like medical implants, biosensors, lab-on-a-chip devices etc. causing the emergence of this new scientific field.

- Biomaterials – any matter, surface, or construct that interacts with biological systems. As a science, biomaterials is about fifty years old. The study of biomaterials is called biomaterials science. It has experienced steady and strong growth over its history, with many companies investing large amounts of money into the development of new products. Biomaterials science encompasses elements of medicine, biology, chemistry, tissue engineering and materials science.

- Biomedical science – healthcare science, also known as biomedical science, is a set of applied sciences applying portions of natural science or formal science, or both, to develop knowledge, interventions, or technology of use in healthcare or public health.[10] Such disciplines as medical microbiology, clinical virology, clinical epidemiology, genetic epidemiology, and biomedical engineering are medical sciences. Explaining physiological mechanisms operating in pathological processes, however, pathophysiology can be regarded as basic science.

- Biomedicine – branch of medical science that applies biological and other natural-science principles to clinical practice.[11] Biomedicine is related to the ability of humans to cope with environmental stress. The branch especially applies to biology and physiology.[12]

- Biomonitoring – measurement of the body burden[13] of toxic chemical compounds, elements, or their metabolites, in biological substances.[14][15] Often, these measurements are done in blood and urine.[16]

- Biopolymer – polymers produced by living organisms; in other words, they are polymeric biomolecules. Since they are polymers, biopolymers contain monomeric units that are covalently bonded to form larger structures. There are three main classes of biopolymers, classified according to the monomeric units used and the structure of the biopolymer formed: polynucleotides (RNA and DNA), which are long polymers composed of 13 or more nucleotide monomers; polypeptides, which are short polymers of amino acids; and polysaccharides, which are often linear bonded polymeric carbohydrate structures.[17][18][19][20]

- Cognitive neuroscience – academic field concerned with the scientific study of biological substrates underlying cognition,[21] with a specific focus on the neural substrates of mental processes. It addresses the questions of how psychological/cognitive functions are produced by the brain. Cognitive neuroscience is a branch of both psychology and neuroscience, overlapping with disciplines such as physiological psychology, cognitive psychology and neuropsychology.[22] Cognitive neuroscience relies upon theories in cognitive science coupled with evidence from neuropsychology, and computational modeling.[22]

- Computational neuroscience – study of brain function in terms of the information processing properties of the structures that make up the nervous system.[23] It is an interdisciplinary science that links the diverse fields of neuroscience, cognitive science, and psychology with electrical engineering, computer science, mathematics, and physics.

- Environmental health – multidisciplinary field concerned with environmental epidemiology, toxicology, and exposure science.

- Environmental science – multidisciplinary academic field that integrates physical and biological sciences, (including but not limited to ecology, physics, chemistry, zoology, mineralogy, oceanology, limnology, soil science, geology, atmospheric science, and geography) to the study of the environment, and the solution of environmental problems.

- Fermentation technology – study of use of microorganisms for industrial manufacturing of various products like vitamins, amino acids, antibiotics, beer, wine, etc.

- Food science – applied science devoted to the study of food. Activities of food scientists include the development of new food products, design of processes to produce and conserve these foods, choice of packaging materials, shelf-life studies, study of the effects of food on the human body, sensory evaluation of products using panels or potential consumers, as well as microbiological, physical (texture and rheology) and chemical testing.[24][25][26][27]

- Genomics – applies recombinant DNA, DNA sequencing methods, and bioinformatics to sequence, assemble, and analyze the function and structure of genomes (the *complete* set of DNA within a single cell of an organism).[28][29] The field includes efforts to determine the entire DNA sequence of organisms and fine-scale genetic mapping. The field also includes studies of intragenomic phenomena such as heterosis, epistasis, pleiotropy and other interactions between loci and alleles within the genome.[30] In contrast, the investigation of the roles and functions of single genes is a primary focus of molecular biology or genetics and is a common topic of modern medical and biological research. Research of single genes does not fall into the definition of genomics unless the aim of this genetic, pathway, and functional information analysis is to elucidate its effect on, place in, and response to the entire genome's networks.[31][32]

- Health sciences – The health sciences are a key branch of the life sciences, comprising all divisions of medicine and medical sciences.

- Immunogenetics – Immunogenetics or immungenetics is the branch of medical research that explores the relationship between the immune system and genetics. Autoimmune diseases, such as type 1 diabetes, are complex genetic traits which result from defects in the immune system. Identification of genes defining the immune defects may identify new target genes for therapeutic approaches. Alternatively, genetic variations can also help to define the immunological pathway leading to disease.

- Immunotherapy – is the "treatment of disease by inducing, enhancing, or suppressing an immune response".[33] Immunotherapies designed to elicit or amplify an immune response are classified as activation immunotherapies, while immunotherapies that reduce or suppress are classified as suppression immunotherapies.

- Kinesiology – Kinesiology, also known as human kinetics, is the scientific study of human movement. Kinesiology addresses physiological, mechanical, and psychological mechanisms. Applications of kinesiology to human health include: biomechanics and orthopedics; strength and conditioning; sport psychology; methods of rehabilitation, such as physical and occupational therapy; and sport and exercise.[34] Individuals who have earned degrees in kinesiology can work in research, the fitness industry, clinical settings, and in industrial environments.[35] Studies of human and animal motion include measures from motion tracking systems, electrophysiology of muscle and brain activity, various methods for monitoring physiological function, and other behavioral and cognitive research techniques.[36][37]

- Medical device – A medical device is an instrument, apparatus, implant, in vitro reagent, or similar or related article that is used to diagnose, prevent, or treat disease or other conditions, and does not achieve its purposes through chemical action within or on the body (which would make it a drug).[38] Whereas *medicinal products* (also called *pharmaceuticals*) achieve their principal action by pharmacological, metabolic or immunological means, *medical devices* act by other means like physical, mechanical, or thermal means.

- Medical imaging – Medical imaging is the technique and process used to create images of the human body (or parts and function thereof) for clinical purposes (medical procedures seeking to reveal, diagnose, or examine disease) or medical science (including the study of normal anatomy and physiology). Although imaging of removed organs and tissues can be performed for medical reasons, such procedures are not usually referred to as medical imaging, but rather are a part of pathology.[39] Examples of medical imaging include:[40]

  - X-rays
  - CT scans
  - Ultrasound
  - MRI scan

- Medical social work – Medical social work is a sub-discipline of social work, also known as hospital social work. Medical social workers typically work in a hospital, skilled nursing facility or hospice, have a graduate degree in the field, and work with patients and their families in need of psychosocial help. Medical social workers assess the psychosocial functioning of patients and families and intervene as necessary. Interventions may include connecting patients and families to necessary resources and supports in the community; providing psychotherapy, supportive counselling, or grief counselling; or helping a patient to expand and strengthen their network of social supports.

- Neuroethology – Neuroethology is the evolutionary and comparative approach study of animal behavior and the understanding of an animal's nervous system.[41]

- Optogenetics – Optogenetics is a neuromodulation technique employed in neuroscience that uses a combination of techniques from optics and genetics to control and monitor the activities of individual neurons in living tissue—even within freely-moving animals—and to precisely measure the effects of those manipulations in real-time.[42] The key reagents used in

- optogenetics are light-sensitive proteins. Spatially-precise neuronal control is achieved using optogenetic actuators like channelrhodopsin, halorhodopsin, and archaerhodopsin, while temporally-precise recordings can be made with the help of optogenetic sensors like Clomeleon, Mermaid, and SuperClomeleon.[43]

- Optometry – Optometry is a health care profession concerned with the health of the eyes and related structures, as well as vision, visual systems, and vision information processing in humans.

- Pharmacogenomics – Pharmacogenomics (a portmanteau of pharmacology and genomics) is the technology that analyses how genetic makeup affects an individual's response to drugs.[44] It deals with the influence of genetic variation on drug response in patients by correlating gene expression or single-nucleotide polymorphisms with a drug's efficacy or toxicity.[45]

- Pharmaceutical sciences – The pharmaceutical sciences are a group of interdisciplinary areas of study concerned with the design, action, delivery, disposition, inorganic, physical, biochemical and analytical biology (anatomy and physiology, biochemistry, cell biology, and molecular biology), epidemiology, statistics, chemometrics, mathematics, physics, and chemical engineering, and applies their principles to the study of drugs.

- Pharmacology – Pharmacology is the branch of

A variety of topics involved with pharmacology, including neuropharmacology, renal pharmacology, human metabolism, intracellular metabolism, and intracellular regulation

medicine and biology concerned with the study of drug action,[46] where a drug can be broadly defined as any man-made, natural, or endogenous (within the body) molecule which exerts a biochemical and/or physiological effect on the cell, tissue, organ, or organism. More specifically, it is the study of the interactions that occur between a living organism and chemicals that affect normal or abnormal biochemical function. If substances have medicinal properties, they are considered pharmaceuticals.

- Population dynamics – Population dynamics is the study of short-term and long-term changes in the size and age composition of populations, and the biological and environmental processes influencing those changes. Population dynamics deals with the way populations are affected by birth and death rates, and by immigration and emigration, and studies topics such as ageing populations or population decline.

- Proteomics – Proteomics is the large-scale study of proteins, particularly their structures and functions.[47][48] Proteins are vital parts of living organisms, as they are the main components of the physiological metabolic pathways of cells. The proteome is the entire set of proteins,[49] produced or modified by an organism or system. This varies with time and distinct requirements, or stresses, that a cell or organism undergoes.

- Psychiatric social work – Psychiatric social work is one of the oldest mental health professions.[50] Workers provide mental health services to the community, including psychotherapy and diagnosing mental illness.[51]

- Psychology – academic and applied discipline that involves the scientific study of mental functions and behaviors.[52][53] Psychology has the immediate goal of understanding individuals and groups by both establishing general principles and researching specific cases,[54][55] and by many accounts it ultimately aims to benefit society.[56][57] In this field, a professional practitioner or researcher is called a psychologist and can be classified as a social, behavioral, or cognitive scientist.

- Sports science – studies the application of treatment and prevention of injuries related to sports medicine. The study of sport science traditionally incorporates areas of physiology, psychology, and biomechanics but also includes other topics such as nutrition and diet.

9.4 References

[1] "Biocom Life Sciences Association California". *Biocom*.

[2] "Life Sciences". *Empire State Development Corporation*. Government of New York. Retrieved 3 February 2014.

[3] Panksepp J (1992). "A role for "affective neuroscience" in understanding stress: the case of separation distress circuitry". In Puglisi-Allegra S. Oliverio A. *Psychobiology of Stress*. Dordrecht, Netherlands: Kluwer Academic. pp. 41–58. ISBN 0-7923-0682-1.

9.4. REFERENCES

[4] Flint, Maria Louise & Dreistadt, Steve H. (1998). Clark, Jack K., ed. *Natural Enemies Handbook: The Illustrated Guide to Biological Pest Control*. University of California Press. ISBN 9780520218017.

[5] Florian Leiber, Nikolai Fuchs and Hartmut Spieß, "Biodynamic agriculture today", in Paul Kristiansen, Acram Taji, and John Reganold (2006), *Organic Agriculture: A global perspective*, Collingwood, AU: CSIRO Publishing

[6] Paull, John (2011). "Attending the First Organic Agriculture Course: Rudolf Steiner's Agriculture Course at Koberwitz, 1924" (PDF). *European Journal of Social Sciences*. **21** (1): 64–70.

[7] Lotter, D.W. 2003."Organic agriculture" J. Sustainable Agriculture 21(4)

[8] Richard Harwood, former C.S. Mott Chair for Sustainable Agriculture at Michigan State University, calls the biodynamic movement the "first organized and well-defined movement of growers and philosophies [in sustainable agriculture] (Harwood 1990; p.6).

[9] M. Birkholz; A. Mai; C. Wenger; C. Meliani; R. Scholz (2016). "Technology modules from micro- and nano-electronics for the life sciences". *WIREs Nanomed. Nanobiotech*. **8**: 355–377. doi:10.1002/wnan.1367.

[10] "The Future of the Healthcare Science Workforce. Modernising Scientific Careers: The Next Steps.". 26 Nov 2008. p. 2. Retrieved 1 June 2011.

[11] "biomedicine (applies biological, physiological) - Memidex dictionary/thesaurus". *memidex.com*. 2012-10-08. Retrieved 2012-10-20.

[12] "biomedicine - The Free Dictionary". 2013. Retrieved 2013-05-15.

[13] "What is body burden?". Chemicalbodyburden.org. Retrieved 9 February 2014.

[14] "Third National Report on Human Exposure to Environmental Chemicals" (PDF). Centers for Disease Control and Prevention – National Center for Environmental Health. Retrieved 9 August 2009.

[15] "What is Biomonitoring?" (PDF). American Chemistry Council. Retrieved 11 January 2009.

[16] Angerer, Jürgen; Ewers, Ulrich; Wilhelm, Michael (2007). "Human biomonitoring: State of the art". *International Journal of Hygiene and Environmental Health*. **210** (3–4): 201–28. PMID 17376741. doi:10.1016/j.ijheh.2007.01.024.

[17] Mohanty, A.K., et al., **Natural Fibers, Biopolymers, and Biocomposites** (CRC Press, 2005)

[18] Chandra, R., and Rustgi, R., "Biodegradable Polymers", Progress in Polymer Science, Vol. 23, p. 1273 (1998)

[19] Meyers, M.A., et al., "Biological Materials: Structure & Mechanical Properties", Progress in Materials Science, Vol. 53, p. 1 (2008)

[20] Kumar, A., et al., "Smart Polymers: Physical Forms & Bioengineering Applications", Progress in Polymer Science. Vol. 32, p.1205 (2007)

[21] Gazzaniga, Ivry and Mangun 2002, cf. title

[22] Gazzaniga 2002, p. xv

[23] What is computational neuroscience? Patricia S. Churchland, Christof Koch, Terrence J. Sejnowski. in Computational Neuroscience pp.46-55. Edited by Eric L. Schwartz. 1993. MIT Press "Archived copy". Archived from the original on 2011-06-04. Retrieved 2009-06-11.

[24] Geller, Martinne (22 January 2014). "Nestle teams up with Singapore for food science research". *Reuters*. Retrieved 9 February 2014.

[25] "Food science to fight obesity". *Euronews*. 9 December 2013. Retrieved 9 February 2014.

[26] Wood, David (31 August 2007). "Nothing Simple about Food Dating, Expiration Dates or 'Use-By' Dates". *ConsumerAffairs*. Retrieved 9 February 2014.

[27] Bhatia, Atish (16 November 2013). "A New Kind of Food Science: How IBM Is Using Big Data to Invent Creative Recipes". *Wired*. Retrieved 9 February 2014.

[28] National Human Genome Research Institute (2010-11-08). "A Brief Guide to Genomics". *Genome.gov*. Retrieved 2011-12-03.

[29] *Concepts of genetics* (10th ed.). San Francisco: Pearson Education. 2012. ISBN 9780321724120.

[30] Pevsner, Jonathan (2009). *Bioinformatics and functional genomics* (2nd ed.). Hoboken, NJ: Wiley-Blackwell. ISBN 9780470085851.

[31] National Human Genome Research Institute (2010-11-08). "FAQ About Genetic and Genomic Science". *Genome.gov*. Retrieved 2011-12-03.

[32] Culver, Kenneth W.; Mark A. Labow (2002-11-08). "Genomics". In Richard Robinson (ed.). *Genetics*. Macmillan Science Library. Macmillan Reference USA. ISBN 0028656067.

[33] "immunotherapies definition". *Dictionary.com*. Retrieved 2009-06-02.

[34] "Welcome to the Ontario Kinesiology Association". Oka.on.ca. Retrieved 2009-07-25.

[35] "CKA - Canadian Kinesiology Alliance - Alliance Canadienne de Kinésiologie". Cka.ca. Archived from the original on 2009-03-18. Retrieved 2009-07-25.

[36] Bodo Rosenhahn, Reinhard Klette and Dimitris Metaxas (eds.). Human Motion - Understanding, Modelling, Capture and Animation. Volume 36 in 'Computational Imaging and Vision', Springer, Dordrecht, 2007

[37] Ahmed Elgammal, Bodo Rosenhahn, and Reinhard Klette (eds.) Human Motion - Understanding, Modelling, Capture and Animation. 2nd Workshop, in conjunction with ICCV 2007, Rio de Janeiro, Lecture Notes in Computer Science, LNCS 4814, Springer, Berlin, 2007

[38] Summarised from the FDA's definition as per http://www.fda.gov/medicaldevices/deviceregulationandguidance/overview/classifyyourdevice/ucm051512.htm

[39] "Imaging". *Merriam-Webster*. Merriam-Webster, Incorporated. Retrieved 8 February 2014.

[40] "Diagnostic Imaging". *National Institutes of Health*. U.S. Department of Health and Human Services. Retrieved 8 February 2014.

[41] Hoyle, G. (1984) The scope of Neuroethology. The Behavioral and Brain Sciences. 7:367-412.

[42] Deisseroth, K.; Feng, G.; Majewska, A. K.; Miesenbock, G.; Ting, A.; Schnitzer, M. J. (2006). "Next-Generation Optical Technologies for Illuminating Genetically Targeted Brain Circuits". *Journal of Neuroscience*. **26** (41): 10380–6. PMC 2820367. PMID 17035522. doi:10.1523/JNEUROSCI.3863-06.2006.

[43] Mancuso, J. J.; Kim, J.; Lee, S.; Tsuda, S.; Chow, N. B. H.; Augustine, G. J. (2010). "Optogenetic probing of functional brain circuitry". *Experimental Physiology*. **96** (1): 26–33. PMID 21056968. doi:10.1113/expphysiol.2010.055731.

[44] Ermak G., Modern Science & Future Medicine (second edition). 164 p., 2013

[45] Wang L (2010). "Pharmacogenomics: a systems approach". *Wiley Interdiscip Rev Syst Biol Med*. **2** (1): 3–22. PMC 3894835. PMID 20836007. doi:10.1002/wsbm.42.

[46] Vallance P, Smart TG (January 2006). "The future of pharmacology". *British Journal of Pharmacology*. 147 Suppl 1 (S1): S304–7. PMC 1760753. PMID 16402118. doi:10.1038/sj.bjp.0706454.

[47] Anderson NL, Anderson NG (1998). "Proteome and proteomics: new technologies, new concepts, and new words". *Electrophoresis*. **19** (11): 1853–61. PMID 9740045. doi:10.1002/elps.1150191103.

[48] Blackstock WP, Weir MP (1999). "Proteomics: quantitative and physical mapping of cellular proteins". *Trends Biotechnol*. **17** (3): 121–7. PMID 10189717. doi:10.1016/S0167-7799(98)01245-1.

[49] Marc R. Wilkins; Christian Pasquali; Ron D. Appel; Keli Ou; Olivier Golaz; Jean-Charles Sanchez; Jun X. Yan; Andrew. A. Gooley; Graham Hughes; Ian Humphery-Smith; Keith L. Williams; Denis F. Hochstrasser (1996). "From Proteins to Proteomes: Large Scale Protein Identification by Two-Dimensional Electrophoresis and Amino Acid Analysis". *Nature Biotechnology*. **14** (1): 61–65. PMID 9636313. doi:10.1038/nbt0196-61.

[50] Silverman, Wade H. (1 September 1985). "The evolving mental health professions: Psychiatric social work, clinical psychology, psychiatry, and psychiatric nursing". *The journal of mental health administration*. **12** (2): 28–31. ISSN 0092-8623.

[51] "Psychiatric Social Workers and How to Become One". SocialWorkLicensure.org. Retrieved 9 February 2014.

[52] "How does the APA define "psychology"?". Retrieved 15 November 2011.

[53] "Definition of "Psychology (APA's Index Page)"". Retrieved 20 December 2011.

[54] Fernald LD (2008). *Psychology: Six perspectives* (pp. 12–15). Thousand Oaks, CA: Sage Publications.

[55] Hockenbury & Hockenbury. Psychology. Worth Publishers, 2010.

[56] O'Neil, H.F.; cited in Coon, D.; Mitterer, J.O. (2008). *Introduction to psychology: Gateways to mind and behavior* (12th ed., pp. 15–16). Stamford, CT: Cengage Learning.

[57] "The mission of the APA [American Psychological Association] is to advance the creation, communication and application of psychological knowledge to benefit society and improve people's lives": APA (2010). *About APA*. Retrieved 20 October 2010.

9.5 Further reading

- Magner, Lois N. (2002). *A history of the life sciences* (Rev. and expanded 3rd ed.). New York: M. Dekker. ISBN 0824708245.

Chapter 10

Social science

This article is about the science of studying social groups. For the integrated field of study intended to promote civic competence, see Social studies. For the social-political-economic theory first pioneered by Karl Marx, see Scientific socialism.

Social science is a major category of academic disciplines, concerned with society and the relationships among individuals within a society. It in turn has many branches, each of which is considered a "social science". The social sciences include, but is not limited to economics, political science, human geography, demography, management, psychology, sociology, anthropology, archaeology, jurisprudence, history, and linguistics. The term is also sometimes used to refer specifically to the field of sociology, the original 'science of society', established in the 19th century. A more detailed list of sub-disciplines within the social sciences can be found at Outline of social science.

Positivist social scientists use methods resembling those of the natural sciences as tools for understanding society, and so define science in its stricter modern sense. Interpretivist social scientists, by contrast, may use social critique or symbolic interpretation rather than constructing empirically falsifiable theories, and thus treat science in its broader sense. In modern academic practice, researchers are often eclectic, using multiple methodologies (for instance, by combining the quantitative and qualitative researchs). The term social research has also acquired a degree of autonomy as practitioners from various disciplines share in its aims and methods.

10.1 History

Main article: History of the social sciences

The history of the social sciences begins in the Age of Enlightenment after 1650, which saw a revolution within natural philosophy, changing the basic framework by which individuals understood what was "scientific". Social sciences came forth from the moral philosophy of the time and were influenced by the Age of Revolutions, such as the Industrial Revolution and the French Revolution.[1] The *social sciences* developed from the sciences (experimental and applied), or the systematic knowledge-bases or prescriptive practices, relating to the social improvement of a group of interacting entities.[2][3]

The beginnings of the social sciences in the 18th century are reflected in the grand encyclopedia of Diderot, with articles from Jean-Jacques Rousseau and other pioneers. The growth of the social sciences is also reflected in other specialized encyclopedias. The modern period saw "*social science*" first used as a distinct conceptual field.[4] Social science was influenced by positivism,[1] focusing on knowledge based on actual positive sense experience and avoiding the negative; metaphysical speculation was avoided. Auguste Comte used the term "*science sociale*" to describe the field, taken from the ideas of Charles Fourier; Comte also referred to the field as *social physics*.[1][5]

Following this period, there were five paths of development that sprang forth in the social sciences, influenced by Comte on other fields.[1] One route that was taken was the rise of social research. Large statistical surveys were undertaken in various parts of the United States and Europe. Another route undertaken was initiated by Émile Durkheim, studying "social facts", and Vilfredo Pareto, opening metatheoretical ideas and individual theories. A third means developed, arising from the methodological dichotomy present, in which social phenomena were identified with and understood; this was championed by figures such as Max Weber. The fourth route taken, based in economics, was developed and furthered economic knowledge as a hard science. The last path was the correlation of knowledge and social values; the antipositivism and verstehen sociology of Max Weber firmly demanded this distinction. In this route, theory (description) and prescription were non-overlapping formal discussions of a subject.

Around the start of the 20th century, Enlightenment

philosophy was challenged in various quarters. After the use of classical theories since the end of the scientific revolution, various fields substituted mathematics studies for experimental studies and examining equations to build a theoretical structure. The development of social science subfields became very quantitative in methodology. The interdisciplinary and cross-disciplinary nature of scientific inquiry into human behaviour, social and environmental factors affecting it, made many of the natural sciences interested in some aspects of social science methodology.[6] Examples of boundary blurring include emerging disciplines like social research of medicine, sociobiology, neuropsychology, bioeconomics and the history and sociology of science. Increasingly, quantitative research and qualitative methods are being integrated in the study of human action and its implications and consequences. In the first half of the 20th century, statistics became a free-standing discipline of applied mathematics. Statistical methods were used confidently.

In the contemporary period, Karl Popper and Talcott Parsons influenced the furtherance of the social sciences.[1] Researchers continue to search for a unified consensus on what methodology might have the power and refinement to connect a proposed "grand theory" with the various midrange theories that, with considerable success, continue to provide usable frameworks for massive, growing data banks; for more, see consilience. The social sciences will for the foreseeable future be composed of different zones in the research of, and sometime distinct in approach toward, the field.[1]

The term "social science" may refer either to the specific *sciences of society* established by thinkers such as Comte, Durkheim, Marx, and Weber, or more generally to all disciplines outside of "noble science" and arts. By the late 19th century, the academic social sciences were constituted of five fields: jurisprudence and amendment of the law, education, health, economy and trade, and art.[2]

Around the start of the 21st century, the expanding domain of economics in the social sciences has been described as economic imperialism.[7]

10.2 Branches

For a topical guide to this subject, see Outline of social science § Branches of social science.

The social science disciplines are branches of knowledge taught and researched at the college or university level. Social science disciplines are defined and recognized by the academic journals in which research is published, and the learned social science societies and academic departments or faculties to which their practitioners belong. Social science fields of study usually have several sub-disciplines or branches, and the distinguishing lines between these are often both arbitrary and ambiguous.

10.2.1 Anthropology

Main articles: Anthropology and Outline of anthropology

Anthropology is the holistic "science of man", a science of the totality of human existence. The discipline deals with the integration of different aspects of the social sciences, humanities, and human biology. In the twentieth century, academic disciplines have often been institutionally divided into three broad domains. The *natural sciences* seek to derive general laws through reproducible and verifiable experiments. The *humanities* generally study local traditions, through their history, literature, music, and arts, with an emphasis on understanding particular individuals, events, or eras. The *social sciences* have generally attempted to develop scientific methods to understand social phenomena in a generalizable way, though usually with methods distinct from those of the natural sciences.

The anthropological social sciences often develop nuanced descriptions rather than the general laws derived in physics or chemistry, or they may explain individual cases through more general principles, as in many fields of psychology. Anthropology (like some fields of history) does not easily fit into one of these categories, and different branches of anthropology draw on one or more of these domains.[8] Within the United States, anthropology is divided into four sub-fields: archaeology, physical or biological anthropology, anthropological linguistics, and cultural anthropology. It is an area that is offered at most undergraduate institutions. The word *anthropos* (ἄνθρωπος) is from the Greek for "human being" or "person". Eric Wolf described sociocultural anthropology as "the most scientific of the humanities, and the most humanistic of the sciences."

The goal of anthropology is to provide a holistic account of humans and human nature. This means that, though anthropologists generally specialize in only one sub-field, they always keep in mind the biological, linguistic, historic and cultural aspects of any problem. Since anthropology arose as a science in Western societies that were complex and industrial, a major trend within anthropology has been a methodological drive to study peoples in societies with more simple social organization, sometimes called "primitive" in anthropological literature, but without any connotation of "inferior".[9] Today, anthropologists use terms such as "less complex" societies or refer to specific modes of subsistence or production, such as "pastoralist" or "forager" or "horticulturalist" to refer to humans living in non-industrial, non-Western cultures, such people or folk (*ethnos*) remaining of great interest within anthropology.

The quest for holism leads most anthropologists to study a people in detail, using biogenetic, archaeological, and linguistic data alongside direct observation of contemporary customs.[10] In the 1990s and 2000s, calls for clarification of what constitutes a culture, of how an observer knows where his or her own culture ends and another begins, and other crucial topics in writing anthropology were heard. It is possible to view all human cultures as part of one large, evolving global culture. These dynamic relationships, between what can be observed on the ground, as opposed to what can be observed by compiling many local observations remain fundamental in any kind of anthropology, whether cultural, biological, linguistic or archaeological.[11]

10.2.2 Communication studies

Main articles: Communication studies and History of communication studies

Communication studies deals with processes of human communication, commonly defined as the sharing of symbols to create meaning. The discipline encompasses a range of topics, from face-to-face conversation to mass media outlets such as television broadcasting. Communication studies also examines how messages are interpreted through the political, cultural, economic, and social dimensions of their contexts. Communication is institutionalized under many different names at different universities, including "communication", "communication studies", "speech communication", "rhetorical studies", "communication science", "media studies", "communication arts", "mass communication", "media ecology", and "communication and media science".

Communication studies integrates aspects of both social sciences and the humanities. As a social science, the discipline often overlaps with sociology, psychology, anthropology, biology, political science, economics, and public policy, among others. From a humanities perspective, communication is concerned with rhetoric and persuasion (traditional graduate programs in communication studies trace their history to the rhetoricians of Ancient Greece). The field applies to outside disciplines as well, including engineering, architecture, mathematics, and information science.

10.2.3 Economics

Main articles: Economics and Outline of economics

Economics is a social science that seeks to analyze and describe the production, distribution, and consumption of wealth.[12] The word "economics" is from the Greek οἶκος [*oikos*], "family, household, estate", and νόμος [*nomos*], "custom, law", and hence means "household management" or "management of the state". An economist is a person using economic concepts and data in the course of employment, or someone who has earned a degree in the subject. The classic brief definition of economics, set out by Lionel Robbins in 1932, is "the science which studies human behavior as a relation between scarce means having alternative uses". Without scarcity and alternative uses, there is no economic problem. Briefer yet is "the study of how people seek to satisfy needs and wants" and "the study of the financial aspects of human behavior".

Economics has two broad branches: microeconomics, where the unit of analysis is the individual agent, such as a household or firm, and macroeconomics, where the unit of analysis is an economy as a whole. Another division of the subject distinguishes positive economics, which seeks to predict and explain economic phenomena, from normative economics, which orders choices and actions by some criterion; such orderings necessarily involve subjective value judgments. Since the early part of the 20th century, economics has focused largely on measurable quantities, employing both theoretical models and empirical analysis. Quantitative models, however, can be traced as far back as the physiocratic school. Economic reasoning has been increasingly applied in recent decades to other social situations such as politics, law, psychology, history, religion, marriage and family life, and other social interactions. This paradigm crucially assumes (1) that resources are scarce because they are not sufficient to satisfy all wants, and (2) that "economic value" is willingness to pay as revealed for instance by market (arms' length) transactions. Rival heterodox schools of thought, such as institutional economics, green economics, Marxist economics, and economic sociology, make other grounding assumptions. For example, Marxist economics assumes that economics primarily deals with the investigation of exchange value, of which human labour is the source.

The expanding domain of economics in the social sciences has been described as economic imperialism.[7][13]

10.2.4 Education

Main articles: Education and Outline of education

Education encompasses teaching and learning specific skills, and also something less tangible but more profound: the imparting of knowledge, positive judgement and well-developed wisdom. Education has as one of its fundamental aspects the imparting of culture from generation to gener-

A depiction of world's oldest university, the University of Bologna, in Italy

ation (see socialization). To educate means 'to draw out', from the Latin *educare*, or to facilitate the realization of an individual's potential and talents. It is an application of pedagogy, a body of theoretical and applied research relating to teaching and learning and draws on many disciplines such as psychology, philosophy, computer science, linguistics, neuroscience, sociology and anthropology.[14]

The education of an individual human begins at birth and continues throughout life. (Some believe that education begins even before birth, as evidenced by some parents' playing music or reading to the baby in the womb in the hope it will influence the child's development.) For some, the struggles and triumphs of daily life provide far more instruction than does formal schooling (thus Mark Twain's admonition to "never let school interfere with your education"). Family members may have a profound educational effect — often more profound than they realize — though family teaching may function very informally.

10.2.5 Geography

Main articles: Geography and Outline of geography

Geography as a discipline can be split broadly into two main sub fields: human geography and physical geography. The former focuses largely on the built environment and how space is created, viewed and managed by humans as well as the influence humans have on the space they occupy. This may involve cultural geography, transportation, health, military operations, and cities. The latter examines the natural environment and how the climate, vegetation and life, soil, oceans, water and landforms are produced and interact.[15] Physical geography examines phenomena related to the measurement of earth. As a result of the two subfields using different approaches a third field

Map of the Earth

has emerged, which is environmental geography. Environmental geography combines physical and human geography and looks at the interactions between the environment and humans.[16] Other branches of geography include social geography, regional geography, and geomatics.

Geographers attempt to understand the Earth in terms of physical and spatial relationships. The first geographers focused on the science of mapmaking and finding ways to precisely project the surface of the earth. In this sense, geography bridges some gaps between the natural sciences and social sciences. Historical geography is often taught in a college in a unified Department of Geography.

Modern geography is an all-encompassing discipline, closely related to GISc, that seeks to understand humanity and its natural environment. The fields of urban planning, regional science, and planetology are closely related to geography. Practitioners of geography use many technologies and methods to collect data such as GIS, remote sensing, aerial photography, statistics, and global positioning systems (GPS).

10.2.6 History

Main articles: History and Outline of history

History is the continuous, systematic narrative and research into past human events as interpreted through historiographical paradigms or theories.

History has a base in both the social sciences and the humanities. In the United States the National Endowment for the Humanities includes history in its definition of humanities (as it does for applied linguistics).[17] However, the National Research Council classifies history as a social science.[18] The historical method comprises the techniques and guidelines by which historians use primary sources and other evidence to research and then to write history. The Social Science History Association, formed in 1976, brings together scholars from numerous disciplines interested in social history.[19]

10.2.7 Law

Main articles: Law and Outline of law
The social science of law, jurisprudence, in common par-

A trial at a criminal court, the Old Bailey in London

lance, means a rule that (unlike a rule of ethics) is capable of enforcement through institutions.[20] However, many laws are based on norms accepted by a community and thus have an ethical foundation. The study of law crosses the boundaries between the social sciences and humanities, depending on one's view of research into its objectives and effects. Law is not always enforceable, especially in the international relations context. It has been defined as a "system of rules",[21] as an "interpretive concept"[22] to achieve justice, as an "authority"[23] to mediate people's interests, and even as "the command of a sovereign, backed by the threat of a sanction".[24] However one likes to think of law, it is a completely central social institution. Legal policy incorporates the practical manifestation of thinking from almost every social science and the humanities. Laws are politics, because politicians create them. Law is philosophy, because moral and ethical persuasions shape their ideas. Law tells many of history's stories, because statutes, case law and codifications build up over time. And law is economics, because any rule about contract, tort, property law, labour law, company law and many more can have long-lasting effects on the distribution of wealth. The noun *law* derives from the late Old English *lagu*, meaning something laid down or fixed[25] and the adjective *legal* comes from the Latin word *lex*.[26]

10.2.8 Linguistics

Main articles: Linguistics and Outline of linguistics
Linguistics investigates the cognitive and social aspects of

Ferdinand de Saussure, recognized as the father of modern linguistics

human language. The field is divided into areas that focus on aspects of the linguistic signal, such as syntax (the study of the rules that govern the structure of sentences), semantics (the study of meaning), morphology (the study of the structure of words), phonetics (the study of speech sounds) and phonology (the study of the abstract sound system of a particular language); however, work in areas like evolutionary linguistics (the study of the origins and evolution of language) and psycholinguistics (the study of psychological factors in human language) cut across these divisions.

The overwhelming majority of modern research in linguistics takes a predominantly synchronic perspective (focusing on language at a particular point in time), and a great deal of it—partly owing to the influence of Noam Chomsky—aims at formulating theories of the cognitive processing of language. However, language does not exist in a vacuum, or only in the brain, and approaches like contact linguistics, creole studies, discourse analysis, social interactional linguistics, and sociolinguistics explore language in its social context. Sociolinguistics often makes use of traditional quantitative analysis and statistics in investigating the frequency of features, while some disciplines, like contact linguistics, focus on qualitative analysis. While certain areas of linguistics can thus be understood as clearly falling within the social sciences, other areas, like acoustic phonetics and neurolinguistics, draw on the natural sciences. Linguistics draws only secondarily on the humanities, which played a rather greater role in linguistic inquiry in the 19th and early 20th centuries. Ferdinand Saussure is considered the father of modern linguistics.

10.2.9 Political science

Main articles: Political science, Outline of political science, and Politics

Political science is an academic and research discipline

Aristotle asserted that man is a political animal in his Politics.[27]

that deals with the theory and practice of politics and the description and analysis of political systems and political behaviour. Fields and subfields of political science include political economy, political theory and philosophy, civics and comparative politics, theory of direct democracy, apolitical governance, participatory direct democracy, national systems, cross-national political analysis, political development, international relations, foreign policy, international law, politics, public administration, administrative behaviour, public law, judicial behaviour, and public policy. Political science also studies power in international relations and the theory of great powers and superpowers.

Political science is methodologically diverse, although recent years have witnessed an upsurge in the use of the scientific method,[28] that is, the proliferation of formal-deductive model building and quantitative hypothesis testing. Approaches to the discipline include rational choice, classical political philosophy, interpretivism, structuralism, and behaviouralism, realism, pluralism, and institutionalism. Political science, as one of the social sciences, uses methods and techniques that relate to the kinds of inquiries sought: primary sources such as historical documents, interviews, and official records, as well as secondary sources such as scholarly articles are used in building and testing theories. Empirical methods include survey research, statistical analysis or econometrics, case studies, experiments, and model building. Herbert Baxter Adams is credited with coining the phrase "political science" while teaching history at Johns Hopkins University.

10.2.10 Psychology

Main articles: Psychology and Outline of psychology

Psychology is an academic and applied field involving

Wilhelm Maximilian Wundt was the founder of experimental psychology.

the study of behaviour and mental processes. Psychology also refers to the application of such knowledge to various spheres of human activity, including problems of individuals' daily lives and the treatment of mental illness. The word *psychology* comes from the ancient Greek ψυχή, *psyche* ("soul", "mind") and *logy* ("study").

Psychology differs from anthropology, economics, political science, and sociology in seeking to capture explanatory generalizations about the mental function and overt behaviour of individuals, while the other disciplines focus on creating descriptive generalizations about the functioning of social groups or situation-specific human behaviour. In practice, however, there is quite a lot of cross-fertilization that takes place among the various fields. Psychology differs from biology and neuroscience in that it is primarily concerned with the interaction of mental processes and behaviour, and of the overall processes of a system, and not simply the biological or neural processes themselves, though the subfield of neuropsychology combines the study of the actual neural processes with the study of the mental effects they have subjectively produced. Many people associate psychology with clinical psychology, which focuses on assessment and treatment of problems in living and psychopathology. In reality, psychology has myriad specialties including social psychology, developmental psychology, cognitive psychology, educational psychology, industrial-organizational psychology, mathematical psychology, neuropsychology, and quantitative analysis of behaviour.

Psychology is a very broad science that is rarely tackled as a whole, major block. Although some subfields encompass a natural science base and a social science application, others can be clearly distinguished as having little to do with the social sciences or having a lot to do with the social sciences. For example, biological psychology is considered a natural science with a social scientific application (as is clinical medicine), social and occupational psychology are, generally speaking, purely social sciences, whereas neuropsychology is a natural science that lacks application out of the scientific tradition entirely. In British universities, emphasis on what tenet of psychology a student has studied and/or concentrated is communicated through the degree conferred: B.Psy. indicates a balance between natural and social sciences, B.Sc. indicates a strong (or entire) scientific concentration, whereas a B.A. underlines a majority of social science credits. This is not always necessarily the case however, and in many UK institutions students studying the B.Psy, B.Sc. and B.A. follow the same curriculum as outlined by The British Psychological Society and have the same options of specialism open to them regardless of whether they choose a balance, a heavy science basis, or heavy social science basis to their degree. If they applied to read the B.A. for example, but specialized in heavily science-based modules, then they will still generally be awarded the B.A.

10.2.11 Sociology

Main articles: Sociology and Outline of sociology

Sociology is the systematic study of society and human social action. The meaning of the word comes from the suffix "-ology", which means "study of", derived from Greek, and the stem "soci-", which is from the Latin word socius, meaning "companion", or society in general.

Émile Durkheim is considered one of the founding fathers of sociology.

Sociology was originally established by Auguste Comte (1798–1857) in 1838.[29] Comte endeavoured to unify history, psychology and economics through the descriptive understanding of the social realm. He proposed that social ills could be remedied through sociological positivism, an epistemological approach outlined in *The Course in Positive Philosophy* [1830–1842] and *A General View of Positivism* (1844). Though Comte is generally regarded as the "Father of Sociology", the discipline was formally established by another French thinker, Émile Durkheim (1858–1917), who developed positivism as a foundation to practical social research. Durkheim set up the first European department of sociology at the University of Bordeaux in 1895, publishing his *Rules of the Sociological Method*. In 1896, he established the journal *L'Année Sociologique*. Durkheim's seminal monograph, *Suicide* (1897), a case study of suicide rates among Catholic and Protestant populations, distinguished sociological analysis from psychology or philosophy.[30]

Karl Marx rejected Comte's positivism but nevertheless aimed to establish a *science of society* based on historical materialism, becoming recognized as a founding figure of sociology posthumously as the term gained broader meaning. Around the start of the 20th century, the first wave of German sociologists, including Max Weber and Georg Simmel, developed sociological antipositivism. The field may be broadly recognized as an amalgam of three modes of social thought in particular: Durkheimian positivism and structural functionalism; Marxist historical materialism and conflict theory; and Weberian antipositivism and verstehen analysis. American sociology broadly arose on a separate trajectory, with little Marxist influence, an emphasis on rigorous experimental methodology, and a closer association with pragmatism and social psychology. In the 1920s, the Chicago school developed symbolic interactionism. Meanwhile, in the 1930s, the Frankfurt School pioneered the idea of critical theory, an interdisciplinary form of Marxist sociology drawing upon thinkers as diverse as Sigmund Freud and Friedrich Nietzsche. Critical theory would take on something of a life of its own after World War II, influencing literary criticism and the Birmingham School establishment of cultural studies.

Sociology evolved as an academic response to the challenges of modernity, such as industrialization, urbanization, secularization, and a perceived process of enveloping rationalization.[31] Because sociology is such a broad discipline, it can be difficult to define, even for professional sociologists. The field generally concerns the social rules and processes that bind and separate people not only as individuals, but as members of associations, groups, communities and institutions, and includes the examination of the organization and development of human social life. The sociological field of interest ranges from the analysis of short contacts between anonymous individuals on the street to the study of global social processes. In the terms of sociologists Peter L. Berger and Thomas Luckmann, social scientists seek an understanding of the *Social Construction of Reality*. Most sociologists work in one or more subfields. One useful way to describe the discipline is as a cluster of sub-fields that examine different dimensions of society. For example, social stratification studies inequality and class structure; demography studies changes in a population size or type; criminology examines criminal behaviour and deviance; and political sociology studies the interaction between society and state.

Since its inception, sociological epistemologies, methods, and frames of enquiry, have significantly expanded and diverged.[32] Sociologists use a diversity of research methods, drawing upon either empirical techniques or critical theory. Common modern methods include case studies, historical research, interviewing, participant observation, social network analysis, survey research, statistical analysis, and model building, among other approaches. Since the late 1970s, many sociologists have tried to make the discipline useful for non-academic purposes. The results of sociological research aid educators, lawmakers, administrators, developers, and others interested in resolving social problems and formulating public policy, through subdisciplinary areas such as evaluation research, methodological assessment, and public sociology.

New sociological sub-fields continue to appear — such as community studies, computational sociology, environmental sociology, network analysis, actor-network theory and a growing list, many of which are cross-disciplinary in nature.

10.3 Additional fields of study

Additional applied or interdisciplinary fields related to the social sciences include:

- Archaeology is the science that studies human cultures through the recovery, documentation, analysis, and interpretation of material remains and environmental data, including architecture, artifacts, features, biofacts, and landscapes.

- Area studies are interdisciplinary fields of research and scholarship pertaining to particular geographical, national/federal, or cultural regions.

- Behavioural science is a term that encompasses all the disciplines that explore the activities of and interactions among organisms in the natural world.

- Computational social science is an umbrella field encompassing computational approaches within the social sciences.

- Demography is the statistical study of all human populations.

- Development studies a multidisciplinary branch of social science that addresses issues of concern to developing countries.

- Environmental social science is the broad, transdisciplinary study of interrelations between humans and the natural environment.

- Environmental studies integrate social, humanistic, and natural science perspectives on the relation between humans and the natural environment.

- Information science is an interdisciplinary science primarily concerned with the collection, classification, manipulation, storage, retrieval and dissemination of information.

- International studies covers both International relations (the study of foreign affairs and global issues among states within the international system) and

International education (the comprehensive approach that intentionally prepares people to be active and engaged participants in an interconnected world).

- Legal management is a social sciences discipline that is designed for students interested in the study of state and legal elements.

- Library science is an interdisciplinary field that applies the practices, perspectives, and tools of management, information technology, education, and other areas to libraries; the collection, organization, preservation and dissemination of information resources; and the political economy of information.

- Management consists of various levels of leadership and administration of an organization in all business and human organizations. It is the effective execution of getting people together to accomplish desired goals and objectives through adequate planning, executing and controlling activities.

- Marketing the identification of human needs and wants, defines and measures their magnitude for demand and understanding the process of consumer buying behaviour to formulate products and services, pricing, promotion and distribution to satisfy these needs and wants through exchange processes and building long term relationships.

- Political economy is the study of production, buying and selling, and their relations with law, custom, and government.

- Public administration is one of the main branches of political science, and can be broadly described as the development, implementation and study of branches of government policy. The pursuit of the public good by enhancing civil society and social justice is the ultimate goal of the field. Though public administration has been historically referred to as government management, it increasingly encompasses non-governmental organizations (NGOs) that also operate with a similar, primary dedication to the betterment of humanity.

10.4 Methodology

10.4.1 Social research

Main article: Social research

The origin of the survey can be traced back at least early as the Domesday Book in 1086,[33][34] while some scholars pinpoint the origin of demography to 1663 with the publication of John Graunt's *Natural and Political Observations upon the Bills of Mortality*.[35] Social research began most intentionally, however, with the positivist philosophy of science in the 19th century.

In contemporary usage, "social research" is a relatively autonomous term, encompassing the work of practitioners from various disciplines that share in its aims and methods. Social scientists employ a range of methods in order to analyse a vast breadth of social phenomena; from census survey data derived from millions of individuals, to the in-depth analysis of a single agent's social experiences; from monitoring what is happening on contemporary streets, to the investigation of ancient historical documents. The methods originally rooted in classical sociology and statistical mathematics have formed the basis for research in other disciplines, such as political science, media studies, and marketing and market research.

Social research methods may be divided into two broad schools:

- Quantitative designs approach social phenomena through quantifiable evidence, and often rely on statistical analysis of many cases (or across intentionally designed treatments in an experiment) to create valid and reliable general claims.

- Qualitative designs emphasize understanding of social phenomena through direct observation, communication with participants, or analysis of texts, and may stress contextual and subjective accuracy over generality.

Social scientists will commonly combine quantitative and qualitative approaches as part of a multi-strategy design. Questionnaires, field-based data collection, archival database information and laboratory-based data collections are some of the measurement techniques used. It is noted the importance of measurement and analysis, focusing on the (difficult to achieve) goal of objective research or statistical hypothesis testing. A mathematical model uses mathematical language to describe a system. The process of developing a mathematical model is termed 'mathematical modelling' (also modeling). Eykhoff (1974) defined a *mathematical model* as 'a representation of the essential aspects of an existing system (or a system to be constructed) that presents knowledge of that system in usable form'.[36] Mathematical models can take many forms, including but not limited to dynamical systems, statistical models, differential equations, or game theoretic models.

These and other types of models can overlap, with a given model involving a variety of abstract structures. The system is a set of interacting or interdependent entities, real or abstract, forming an integrated whole. The concept of an *integrated whole* can also be stated in terms of a system embodying a set of relationships that are differentiated from relationships of the set to other elements, and from relationships between an element of the set and elements not a part of the relational regime. A dynamical system modeled as a mathematical formalization has a fixed "rule" that describes the time dependence of a point's position in its ambient space. Small changes in the state of the system correspond to small changes in the numbers. The *evolution rule* of the dynamical system is a fixed rule that describes what future states follow from the current state. The rule is deterministic: for a given time interval only one future state follows from the current state.

See also: Scholarly method, Teleology, Philosophy of science, and Philosophy of social science

10.4.2 Theory

Main article: Social theory

Other social scientists emphasize the subjective nature of research. These writers share social theory perspectives that include various types of the following:

- Critical theory is the examination and critique of society and culture, drawing from knowledge across social sciences and humanities disciplines.

- Dialectical materialism is the philosophy of Karl Marx, which he formulated by taking the dialectic of Hegel and joining it to the materialism of Feuerbach.

- Feminist theory is the extension of feminism into theoretical, or philosophical discourse; it aims to understand the nature of gender inequality.

- Marxist theories, such as revolutionary theory and class theory, cover work in philosophy that is strongly influenced by Karl Marx's materialist approach to theory or is written by Marxists.

- Phronetic social science is a theory and methodology for doing social science focusing on ethics and political power, based on a contemporary interpretation of Aristotelian phronesis.

- Post-colonial theory is a reaction to the cultural legacy of colonialism.

- Postmodernism refers to a point of departure for works of literature, drama, architecture, cinema, and design,

as well as in marketing and business and in the interpretation of history, law, culture and religion in the late 20th century.

- Rational choice theory is a framework for understanding and often formally modeling social and economic behaviour.

- Social constructionism considers how social phenomena develop in social contexts.

- Structuralism is an approach to the human sciences that attempts to analyze a specific field (for instance, mythology) as a complex system of interrelated parts.

- Structural functionalism is a sociological paradigm that addresses what social functions various elements of the social system perform in regard to the entire system.

Other fringe social scientists delve in alternative nature of research. These writers share social theory perspectives that include various types of the following:

- Intellectual critical-ism describes a sentiment of critique towards, or evaluation of, intellectuals and intellectual pursuits.

- Scientific criticalism is a position critical of science and the scientific method.

10.5 Education and degrees

Most universities offer degrees in social science fields.[37] The Bachelor of Social Science is a degree targeted at the social sciences in particular. It is often more flexible and in-depth than other degrees that include social science subjects.[38]

In the United States, a university may offer a student who studies a social sciences field a Bachelor of Arts degree, particularly if the field is within one of the traditional liberal arts such as history, or a BSc: Bachelor of Science degree such as those given by the London School of Economics, as the social sciences constitute one of the two main branches of science (the other being the natural sciences). In addition, some institutions have degrees for a particular social science, such as the Bachelor of Economics degree, though such specialized degrees are relatively rare in the United States.

10.6 See also

10.6.1 General

Outline of social science · Society · Culture · Structure and agency · Humanities (human science)

10.6.2 Methods

Historical method · Empiricism · Representation theory · Scientific method · Statistical hypothesis testing · Regression · Correlation · Terminology · Participatory Action Research

10.6.3 Areas

Political sciences · Natural sciences · Behavioural sciences · Geographic information science

10.6.4 History

History of science · History of technology

10.6.5 Lists

Fields of science · Outline of academic disciplines

10.6.6 People

Aristotle · Plato · Confucius · Augustine · Niccolò Machiavelli · Émile Durkheim · Max Weber · Karl Marx · Friedrich Engels · Herbert Spencer · Sir John Lubbock · Alfred Schutz · Adam Smith · David Ricardo · Jean-Baptiste Say · John Maynard Keynes · Robert Lucas · Milton Friedman · Sigmund Freud · Jean Piaget · Noam Chomsky · B.F. Skinner · John Stuart Mill · Thomas Hobbes · Jean-Jacques Rousseau · Montesquieu · John Locke · David Hume · Auguste Comte · Steven Pinker · John Rawls

10.6.7 Other

Behaviour · Ethology and Ethnology · Game theory · Gulbenkian commission · Labelling · "Periodic table of human sciences" (Tinbergen's four questions) · Social action · Philosophy of social sciences

10.7 Notes and references

[1] Kuper, A., and Kuper, J. (1985). *The Social Science Encyclopaedia*.

[2] Social sciences, *Columbian Cyclopedia*. (1897). Buffalo: Garretson, Cox & Company. Page 227.

[3] Peck, H. T., Peabody, S. H., and Richardson, C. F. (1897). *The International Cyclopedia, A Compendium of Human Knowledge*. Rev. with large additions. New York: Dodd, Mead and Company.

[4] William Thompson (1775–1833) (1824). *An Inquiry into the Principles of the Distribution of Wealth Most Conducive to Human Happiness; applied to the Newly Proposed System of Voluntary Equality of Wealth*.

[5] According to Comte, the *social physics* field was similar to that of natural sciences.

[6] Vessuri, H. (2002). "Ethical Challenges for the Social Sciences on the Threshold of the 21st Century". *Current Sociology*. **50**: 135–150. doi:10.1177/0011392102050001010.

[7] Lazear, E. P. (2000). "Economic Imperialism". *The Quarterly Journal of Economics*. **115**: 99–146. doi:10.1162/003355300554683.

[8] Wallerstein, I. (2003). "Anthropology, Sociology, and Other Dubious Disciplines". *Current Anthropology*. **44** (4): 453–465. doi:10.1086/375868.

[9] Lowie, Robert (1924). *Primitive Religion*. Routledge and Sons.; Tylor, Edward (1920). *Primitive Culture*. New York:: J. P. Putnam's Sons. Originally published 1871.

[10] Nanda, Serena and Richard Warms. *Culture Counts*. Wadsworth. 2008. Chapter One

[11] Rosaldo, Renato. *Culture and Truth: The remaking of social analysis*. Beacon Press. 1993; Inda, John Xavier and Renato Rosaldo. *The Anthropology of Globalization*. Wiley-Blackwell. 2007

[12] economics - Britannica Online Encyclopedia

[13] Becker, Gary S. (1976). *The Economic Approach to Human Behavior*. Links to arrow-page viewable chapter. University of Chicago Press.

[14] An overview of education

[15] "What is geography?". *AAG Career Guide: Jobs in Geography and Related Geographical Sciences*. Association of American Geographers. Archived from the original on October 6, 2006. Retrieved October 9, 2006.

[16] Hayes-Bohanan, James. "What is Environmental Geography, Anyway?". Retrieved October 9, 2006.

[17] "About NEH". National Endowment for the Humanities.

- [18] Research-Doctorate Programs in the United States: Continuity and Change
- [19] See the SSHA website
- [20] Robertson, Geoffrey (2006). *Crimes Against Humanity*. Penguin. p. 90. ISBN 978-0-14-102463-9.
- [21] Hart, H. L. A. (1961). *The Concept of Law*. Oxford University Press. ISBN 0-19-876122-8.
- [22] Dworkin, Ronald (1986). *Law's Empire*. Harvard University Press. ISBN 0-674-51836-5.
- [23] Raz, Joseph (1979). *The Authority of Law*. Oxford University Press. ISBN 0-19-956268-7.
- [24] Austin, John (1831). *The Providence of Jurisprudence Determined*.
- [25] see Etymonline Dictionary
- [26] see *Merriam-Webster's Dictionary*
- [27] Ebenstein, Alan (2002). Introduction to Political Thinkers. Boston, Massachusetts: Wadsworth.
- [28] https://www.amazon.com/dp/1403934223
- [29] *A Dictionary of Sociology*, Article: Comte, Auguste
- [30] Gianfranco Poggi (2000). *Durkheim*. Oxford: Oxford University Press. Chapter 1.
- [31] Habermas, Jürgen, *The Philosophical Discourse of Modernity: Modernity's Consciousness of Time*, Polity Press (1990), paperback, ISBN 0-7456-0830-2, p. 2.
- [32] Giddens, Anthony, Duneier, Mitchell, Applebaum, Richard. 2007. *Introduction to Sociology. Sixth Edition*. New York: W. W. Norton and Company. Chapter 1.
- [33] A. H. Halsey (2004), *A history of sociology in Britain: science, literature, and society*, p. 34
- [34] Geoffrey Duncan Mitchell (1970), *A new dictionary of sociology*, p. 201
- [35] Willcox, Walter (1938) *The Founder of Statistics*.
- [36] Eykhoff, Pieter *System Identification: Parameter and State Estimation*, Wiley & Sons, (1974). ISBN 0-471-24980-7
- [37] Peterson's (Firm : 2006-). (2007). Peterson's graduate programs in the humanities, arts, & social sciences, 2007. Lawrenceville, New Jersey: Peterson's.
- [38] A Bachelor of Social Science degree can be earned at the University of Adelaide, University of Waikato (Hamilton, New Zealand), University of Sydney, University of New South Wales, University of Hong Kong, University of Manchester, Lincoln University, New Zealand, National University of Malaysia and University of Queensland.

10.8 Bibliography

10.8.1 20th and 21st centuries sources

- Neil J. Smelser and Paul B. Baltes (2001). *International Encyclopedia of the Social & Behavioral Sciences*, Amsterdam: Elsevier.
- Byrne, D. S. (1998). *Complexity theory and the social sciences: an introduction*. Routledge. ISBN 0-415-16296-3
- Kuper, A., and Kuper, J. (1985). *The Social Science Encyclopedia*. London: Routledge & Kegan Paul. (ed., a limited preview of the 1996 version is available)
- Lave, C. A., and March, J. G. (1993). *An introduction to models in the social sciences*. Lanham, Md: University Press of America.
- Perry, John and Erna Perry. *Contemporary Society: An Introduction to Social Science* (12th Edition, 2008), college textbook
- Potter, D. (1988). *Society and the social sciences: An introduction*. London: Routledge [u.a.].
- David L. Sills and Robert K. Merton (1968). *International Encyclopedia of the Social Sciences*.
- Seligman, Edwin R. A. and Alvin Johnson (1934). *Encyclopedia of the Social Sciences*. (13 vol.)
- Ward, L. F. (1924). *Dynamic sociology, or applied social science: As based upon statical sociology and the less complex sciences*. New York: D. Appleton.
- Leavitt, F. M., and Brown, E. (1920). *Elementary social science*. New York: Macmillan.
- Bogardus, E. S. (1913). *Introduction to the social sciences: A textbook outline*. Los Angeles: Ralston Press.
- Small, A. W. (1910). *The meaning of social science*. Chicago, Ill: The University of Chicago Press.

10.8.2 19th century sources

- Andrews, S. P. (1888). *The science of society*. Boston, Mass: Sarah E. Holmes.
- Denslow, V. B. (1882). *Modern thinkers principally upon social science: What they think, and why*. Chicago: Belford, Clarke & Co.
- Harris, William Torrey (1879). *Method of Study in Social Science: A Lecture Delivered Before the St. Louis Social Science Association, March 4, 1879*. St. Louis: G.I. Jones and Co, 1879.

- Hamilton, R. S. (1873). *Present status of social science. A review, historical and critical, of the progress of thought in social philosophy*. New York: H. L. Hinton.

- Carey, H. C. (1867). *Principles of social science*. Philadelphia: J. B. Lippincott & Co. [etc.]. Volume I, Volume II, Volume III.

- Calvert, G. H. (1856). *Introduction to social science: A discourse in three parts*. New York: Redfield.

10.8.3 General sources

- Backhouse, Roger E., and Philippe Fontaine, eds. *A historiography of the modern social sciences* (Cambridge University Press, 2014).

- Backhouse, Roger E.; Fontaine, eds., Philippe, eds. (2010). *The History of the Social Sciences Since 1945*. Cambridge University Press.; covers the conceptual, institutional, and wider histories of economics, political science, sociology, social anthropology, psychology, and human geography.

- Delanty, G. (1997). *Social science: Beyond constructivism and realism*. Minneapolis: Univ. of Minnesota Press.

- Hargittai, E. (2009). *Research Confidential: Solutions to Problems Most Social Scientists Pretend They Never Have*. Ann Arbor: University of Michigan Press.

- Hunt, E. F.; Colander, D. C. (2008). *Social science: An introduction to the study of society*. Boston: Pearson/Allyn and Bacon.

- Carey, H. C.; McKean, K. (1883). *Manual of social science; Being a condensation of the Principles of social science*. Philadelphia: Baird.

- Galavotti, M. C. (2003). *Observation and experiment in the natural and social sciences. Boston studies in the philosophy of science*. **232**. Dordrecht: Kluwer Academic.

- Gorton, W. A. (2006). *Karl Popper and the social sciences*. SUNY series in the philosophy of the social sciences. Albany: State University of New York Press.

- Harris, F. R. (1973). *Social science and national policy*. New Brunswick, N.J.: Transaction Books. distributed by Dutton

- Krimerman, L. I. (1969). *The nature and scope of social science: A critical anthology*. New York: Appleton-Century-Crofts.

- Rule, J. B. (1997). *Theory and progress in social science*. Cambridge: Cambridge University Press.

- Shionoya, Y. (1997). *Schumpeter and the idea of social science: A metatheoretical study. Historical perspectives on modern economics*. Cambridge: Cambridge University Press.

- Singleton, Royce, A.; Straits, Bruce C. (1988). "Approaches to Social Research". Oxford University Press. ISBN 0-19-514794-4. Archived from the original on March 3, 2007.

- Thomas, D. (1979). *Naturalism and social science: a post-empiricist philosophy of social science*. CUP Archive. ISBN 978-0-521-29660-1.

- Trigg, R. (2001). *Understanding social science: A philosophical introduction to the social sciences*. Malden, Mass: Blackwell Publishers.

- Weber, M (1906) [1904]. *The Relations of the Rural Community to Other Branches of Social Science. Congress of Arts and Science: Universal Exposition*. St. Louis: Houghton, Mifflin and Company.

10.8.4 Academic resources

- *The ANNALS of the American Academy of Political and Social Science*, ISSN: 1552-3349 (electronic) ISSN 0002-7162 (paper), SAGE Publications

- Efferson, C. and Richerson, P. J. (In press). A prolegomenon to nonlinear empiricism in the human behavioral sciences. *Philosophy and Biology*. Full text

10.8.5 Opponents and critics

- George H. Smith (2014). *Intellectuals and Libertarianism: Thomas Sowell and Robert Nisbet*

- Phil Hutchinson, Rupert Read and Wes Sharrock (2008). *There's No Such Thing as a Social Science*. ISBN 978-0-7546-4776-8

- Sabia, D. R., and Wallulis, J. (1983). *Changing social science: Critical theory and other critical perspectives*. Albany: State University of New York Press.

10.9 External links

- Institute for Comparative Research in Human and Social Sciences (ICR) (JAPAN)

- Centre for Social Work Research

- Family Therapy and Systemic Research Centre
- International Conference on Social Sciences
- International Social Science Council
- Introduction to Hutchinson et al., *There's No Such Thing as a Social Science*
- Intute: Social Sciences (UK)
- Social Science Research Society
- Social Science Virtual Library
- Social Science Virtual Library: Canaktanweb (Turkish)
- Social Sciences And Humanities
- UC Berkeley Experimental Social Science Laboratory
- The Dialectic of Social Science by Paul A. Baran
- American Academy *Commission on the Humanities and Social Sciences*
- Social Phenomena by Teng Wang

Chapter 11

Applied science

Applied science is a discipline of science that applies existing scientific knowledge to develop more practical applications, like technology or inventions.

Within natural science, disciplines that are basic science, also called pure science, develop *information* to predict and perhaps explain—thus somehow understand—phenomena in the natural world. Applied science applies science to real world practice. This includes a broad range of applied science related fields from Engineering, Business, Medicine to Early Childhood Education.

Applied science can also apply formal science, such as statistics and probability theory, as in epidemiology. Genetic epidemiology is an applied science applying both biological and statistical methods.

11.1 Branches of applied science

For a topical guide to this subject, see Outline of applied science § Branches of applied science.

Engineering sciences include thermodynamics, heat transfer, fluid mechanics, statics, dynamics, mechanics of materials, kinematics, electromagnetism, materials science, earth sciences, engineering physics.

Medical sciences, for instance medical microbiology and clinical virology, are applied sciences that apply biology toward medical knowledge and inventions, but not necessarily medical technology, whose development is more specifically biomedicine or biomedical engineering.

11.2 In education

In Canada, the Netherlands and other places the Bachelor of Applied Science (BASc) is equivalent to the Bachelor of Engineering, and is classified as a professional degree. The BASc tends to focus more on the application of the engineering sciences. In Australia and New Zealand this degree is awarded in various fields of study and is considered a highly specialized professional degree.

In the United Kingdom's educational system, Applied Science refers to a suite of "vocational" science qualifications that run alongside "traditional" General Certificate of Secondary Education or A-Level Sciences.[1] Applied Science courses generally contain more coursework (also known as portfolio or internally assessed work) compared to their traditional counterparts. These are an evolution of the GNVQ qualifications that were offered up to 2005. These courses regularly come under scrutiny and are due for review following the Wolf Report 2011;[2] however, their merits are argued elsewhere.[3]

In the United States, The College of William & Mary offers an undergraduate minor as well as Master of Science and Doctor of Philosophy degrees in "applied science." Courses and research cover varied fields including neuroscience, optics, materials science and engineering, nondestructive testing, and nuclear magnetic resonance.[4] In New York City, the Bloomberg administration awarded the consortium of Cornell-Technion $100 million in City capital to construct the universities' proposed Applied Sciences campus on Roosevelt Island.[5]

11.3 See also

- Exact science
- Fundamental research
- Hard science vs. Soft science
- Invention

11.4 References

[1] Donnelly, Jim. "Applied Science - an invisible revolution?" (pdf). Nuffield Foundation. Retrieved 16 October 2015.

[2] Wolf, Alison (March 2011). Review of Vocational Education - The Wolf Report (Report). Department for Education and Department for Business, Innovation & Skills. DFE-00031-2011. Retrieved 16 October 2015.

[3] Bell, Jacqueline; Donnelly, Jim (2007). Positioning Applied Science In Schools: Uncertainty, Opportunity and Risk in Curriculum Reform (PDF) (Report). University of Leeds. Centre for Studies in Science & Mathematics Education. Archived from the original (pdf) on 3 October 2011. Retrieved 16 October 2015.

[4] "Applied Science". William & Mary. Retrieved 16 October 2015.

[5] "Mayor Bloomberg, Cornell President Skorton and Technion President Lavie announce historic partnership to build a new applied sciences campus on Roosevelt Island" (Press release). The City of New York. Office of the Mayor. 19 December 2011. Retrieved 16 October 2015.

Chapter 12

Interdisciplinarity

Interdisciplinarity involves the combining of two or more academic disciplines into one activity (e.g., a research project). It is about creating something new by thinking across boundaries. It is related to an **interdiscipline** or an **interdisciplinary field,** which is an organizational unit that crosses traditional boundaries between academic disciplines or schools of thought, as new needs and professions emerge. Large engineering teams are usually interdisciplinary, as a power station or mobile phone or other project requires the melding of several specialties. However, the term "interdisciplinary" is sometimes confined to academic settings.

The term *interdisciplinary* is applied within education and training pedagogies to describe studies that use methods and insights of several established disciplines or traditional fields of study. Interdisciplinarity involves researchers, students, and teachers in the goals of connecting and integrating several academic schools of thought, professions, or technologies—along with their specific perspectives—in the pursuit of a common task. The epidemiology of AIDS or global warming requires understanding of diverse disciplines to solve complex problems. *Interdisciplinary* may be applied where the subject is felt to have been neglected or even misrepresented in the traditional disciplinary structure of research institutions, for example, women's studies or ethnic area studies. Interdisciplinarity can likewise be applied to complex subjects that can only be understood by combining the perspectives of two or more fields.

The adjective *interdisciplinary* is most often used in educational circles when researchers from two or more disciplines pool their approaches and modify them so that they are better suited to the problem at hand, including the case of the team-taught course where students are required to understand a given subject in terms of multiple traditional disciplines. For example, the subject of land use may appear differently when examined by different disciplines, for instance, biology, chemistry, economics, geography, and politics.

12.1 Development

Although interdisciplinary and interdisciplinarity are frequently viewed as twentieth century terms, the concept has historical antecedents, most notably Greek philosophy.[1] Julie Thompson Klein attests that "the roots of the concepts lie in a number of ideas that resonate through modern discourse—the ideas of a unified science, general knowledge, synthesis and the integration of knowledge",[2] while Giles Gunn says that Greek historians and dramatists took elements from other realms of knowledge (such as medicine or philosophy) to further understand their own material.[3] Any broadminded humanist project involves interdisciplinarity, and history shows a crowd of cases, as seventeenth-century Leibniz's task to create a system of universal justice, which required linguistics, economics, management, ethics, law philosophy, politics, and even sinology.[4]

Interdisciplinary programs sometimes arise from a shared conviction that the traditional disciplines are unable or unwilling to address an important problem. For example, social science disciplines such as anthropology and sociology paid little attention to the social analysis of technology throughout most of the twentieth century. As a result, many social scientists with interests in technology have joined science, technology and society programs, which are typically staffed by scholars drawn from numerous disciplines. They may also arise from new research developments, such as nanotechnology, which cannot be addressed without combining the approaches of two or more disciplines. Examples include quantum information processing, an amalgamation of quantum physics and computer science, and bioinformatics, combining molecular biology with computer science. Sustainable development as a research area deals with problems requiring analysis and synthesis across economic, social and environmental spheres; often an integration of multiple social and natural science disciplines. Interdisciplinary research is also key to the study of health sciences, for example in studying optimal solutions to diseases.[5] Some institutions of higher education offer accredited degree programs in Interdisciplinary Studies.

At another level, interdisciplinarity is seen as a remedy to the harmful effects of excessive specialization. On some views, however, interdisciplinarity is entirely indebted to those who specialize in one field of study—that is, without specialists, interdisciplinarians would have no information and no leading experts to consult. Others place the focus of interdisciplinarity on the need to transcend disciplines, viewing excessive specialization as problematic both epistemologically and politically. When interdisciplinary collaboration or research results in new solutions to problems, much information is given back to the various disciplines involved. Therefore, both disciplinarians and interdisciplinarians may be seen in complementary relation to one another.

12.2 Barriers

Because most participants in interdisciplinary ventures were trained in traditional disciplines, they must learn to appreciate differing of perspectives and methods. For example, a discipline that places more emphasis on quantitative "rigor" may produce practitioners who think of themselves (and their discipline) as "more scientific" than others; in turn, colleagues in "softer" disciplines may associate quantitative approaches with an inability to grasp the broader dimensions of a problem. An interdisciplinary program may not succeed if its members remain stuck in their disciplines (and in disciplinary attitudes). On the other hand, and from the disciplinary perspective, much interdisciplinary work may be seen as "soft", lacking in rigor, or ideologically motivated; these beliefs place barriers in the career paths of those who choose interdisciplinary work. For example, interdisciplinary grant applications are often refereed by peer reviewers drawn from established disciplines; not surprisingly, interdisciplinary researchers may experience difficulty getting funding for their research. In addition, untenured researchers know that, when they seek promotion and tenure, it is likely that some of the evaluators will lack commitment to interdisciplinarity. They may fear that making a commitment to interdisciplinary research will increase the risk of being denied tenure.

Interdisciplinary programs may fail if they are not given sufficient autonomy. For example, interdisciplinary faculty are usually recruited to a joint appointment, with responsibilities in both an interdisciplinary program (such as women's studies) and a traditional discipline (such as history). If the traditional discipline makes the tenure decisions, new interdisciplinary faculty will be hesitant to commit themselves fully to interdisciplinary work. Other barriers include the generally disciplinary orientation of most scholarly journals, leading to the perception, if not the fact, that interdisciplinary research is hard to publish. In addition, since traditional budgetary practices at most universities channel resources through the disciplines, it becomes difficult to account for a given scholar or teacher's salary and time. During periods of budgetary contraction, the natural tendency to serve the primary constituency (i.e., students majoring in the traditional discipline) makes resources scarce for teaching and research comparatively far from the center of the discipline as traditionally understood. For these same reasons, the introduction of new interdisciplinary programs is often resisted because it is perceived as a competition for diminishing funds.

Due to these and other barriers, interdisciplinary research areas are strongly motivated to become disciplines themselves. If they succeed, they can establish their own research funding programs and make their own tenure and promotion decisions. In so doing, they lower the risk of entry. Examples of former interdisciplinary research areas that have become disciplines include neuroscience, cybernetics, biochemistry and biomedical engineering. These new fields are occasionally referred to as "interdisciplines". On the other hand, even though interdisciplinary activities are now a focus of attention for institutions promoting learning and teaching, as well as organizational and social entities concerned with education, they are practically facing complex barriers, serious challenges and criticism. The most important obstacles and challenges faced by interdisciplinary activities in the past two decades can be divided into "professional", "organizational", and "cultural" obstacles.[6]

12.3 Interdisciplinary studies and studies of interdisciplinarity

An initial distinction should be made between interdisciplinary studies, which can be found spread across the academy today, and the study of interdisciplinarity, which involves a much smaller group of researchers. The former is instantiated in thousands of research centers across the US and the world. The latter has one US organization, the Association for Interdisciplinary Studies[7] (founded in 1979), two international organizations, the International Network of Inter- and Transdisciplinarity[8] (founded in 2010) and the Philosophy of/as Interdisciplinarity Network[9] (founded in 2009), and one research institute devoted to the theory and practice of interdisciplinarity, the Center for the Study of Interdisciplinarity at the University of North Texas (founded in 2008).

An **interdisciplinary study** is an academic program or process seeking to synthesize broad perspectives, knowledge, skills, interconnections, and epistemology in an educational setting. Interdisciplinary programs may be founded in order to facilitate the study of subjects which have some coherence, but which cannot be adequately understood from a single disciplinary perspective (for example, women's studies or medieval studies). More rarely, and at a more advanced level, interdisciplinarity may itself become the focus of study, in a critique of institutionalized disciplines' ways of segmenting knowledge.

In contrast, **studies of interdisciplinarity** raise to self-consciousness questions about how interdisciplinarity works, the nature and history of disciplinarity, and the future of knowledge in post-industrial society. Researchers at the Center for the Study of Interdisciplinarity have made the distinction between philosophy 'of' and 'as' interdisciplinarity, the former identifying a new, discrete area within philosophy that raises epistemological and metaphysical questions about the status of interdisciplinary thinking, with the latter pointing toward a philosophical practice that is sometimes called 'field philosophy'.[10][11]

Perhaps the most common complaint regarding interdisciplinary programs, by supporters and detractors alike, is the lack of synthesis—that is, students are provided with multiple disciplinary perspectives, but are not given effective guidance in resolving the conflicts and achieving a coherent view of the subject. Others have argued that the very idea of synthesis or integration of disciplines presupposes questionable politico-epistemic commitments.[12] Critics of interdisciplinary programs feel that the ambition is simply unrealistic, given the knowledge and intellectual maturity of all but the exceptional undergraduate; some defenders concede the difficulty, but insist that cultivating interdisciplinarity as a habit of mind, even at that level, is both possible and essential to the education of informed and engaged citizens and leaders capable of analyzing, evaluating, and synthesizing information from multiple sources in order to render reasoned decisions.

While much has been written on the philosophy and promise of interdisciplinarity in academic programs and professional practice, social scientists are increasingly interrogating academic discourses on interdisciplinarity, as well as how interdisciplinarity actually works—and does not—in practice.[13][14][15] Some have shown, for example, that some interdisciplinary enterprises that aim to serve society can produce deleterious outcomes for which no one can be held to account.[16]

12.3.1 Politics of interdisciplinary studies

Since 1998, there has been an ascendancy in the value of interdisciplinary research and teaching and a growth in the number of bachelor's degrees awarded at U.S. universities classified as multi- or interdisciplinary studies. The number of interdisciplinary bachelor's degrees awarded annually rose from 7,000 in 1973 to 30,000 a year by 2005 according to data from the National Center of Educational Statistics (NECS). In addition, educational leaders from the Boyer Commission to Carnegie's President Vartan Gregorian to Alan I. Leshner, CEO of the American Association for the Advancement of Science have advocated for interdisciplinary rather than disciplinary approaches to problem solving in the 21st century. This has been echoed by federal funding agencies, particularly the National Institutes of Health under the Direction of Elias Zerhouni, who have advocated that grant proposals be framed more as interdisciplinary collaborative projects than single researcher, single discipline ones.

At the same time, many thriving longstanding bachelor's in interdisciplinary studies programs in existence for 30 or more years, have been closed down, in spite of healthy enrollment. Examples include Arizona International (formerly part of the University of Arizona), the School of Interdisciplinary Studies at Miami University, and the Department of Interdisciplinary Studies at Wayne State University; others such as the Department of Interdisciplinary Studies at Appalachian State University, and George Mason University's New Century College, have been cut back. Stuart Henry has seen this trend as part of the hegemony of the disciplines in their attempt to recolonize the experimental knowledge production of otherwise marginalized fields of inquiry. This is due to threat perceptions seemingly based on the ascendancy of interdisciplinary studies against traditional academia.

12.4 Historical examples

There are many examples of when a particular idea, almost on the same period, arises in different disciplines. One case is the shift from the approach of focusing on "specialized segments of attention" (adopting one particular perspective), to the idea of "instant sensory awareness of the whole", an attention to the "total field", a "sense of the whole pattern, of form and function as a unity", an "integral idea of structure and configuration". This has happened in painting (with cubism), physics, poetry, communication and educational theory. According to Marshall McLuhan, this paradigm shift was due to the passage from an era shaped by mechanization, which brought sequentiality, to the era shaped by the instant speed of electricity, which brought simultaneity.[17]

12.5 Efforts to simplify and defend the concept

An article in the *Social Science Journal*[18] attempts to provide a simple, common-sense, definition of interdisciplinarity, bypassing the difficulties of defining that concept and obviating the need for such related concepts as transdisciplinarity, pluridisciplinarity, and multidisciplinarity:

> "To begin with, a discipline can be conveniently defined as any comparatively self-contained and isolated domain of human experience which possesses its own community of experts. Interdisciplinarity is best seen as bringing together distinctive components of two or more disciplines. In academic discourse, interdisciplinarity typically applies to four realms: knowledge, research, education, and theory. Interdisciplinary knowledge involves familiarity with components of two or more disciplines. Interdisciplinary research combines components of two or more disciplines in the search or creation of new knowledge, operations, or artistic expressions. Interdisciplinary education merges components of two or more disciplines in a single program of instruction. Interdisciplinary theory takes interdisciplinary knowledge, research, or education as its main objects of study."

In turn, interdisciplinary *richness* of any two instances of knowledge, research, or education can be ranked by weighing four variables: number of disciplines involved, the "distance" between them, the novelty of any particular combination, and their extent of integration.[19]

Interdisciplinary knowledge and research are important because:

1. "Creativity often requires interdisciplinary knowledge.

2. Immigrants often make important contributions to their new field.

3. Disciplinarians often commit errors which can be best detected by people familiar with two or more disciplines.

4. Some worthwhile topics of research fall in the interstices among the traditional disciplines.

5. Many intellectual, social, and practical problems require interdisciplinary approaches.

6. Interdisciplinary knowledge and research serve to remind us of the unity-of-knowledge ideal.

7. Interdisciplinarians enjoy greater flexibility in their research.

8. More so than narrow disciplinarians, interdisciplinarians often treat themselves to the intellectual equivalent of traveling in new lands.

9. Interdisciplinarians may help breach communication gaps in the modern academy, thereby helping to mobilize its enormous intellectual resources in the cause of greater social rationality and justice.

10. By bridging fragmented disciplines, interdisciplinarians might play a role in the defense of academic freedom."[18]

12.6 Quotations

> "The modern mind divides, specializes, thinks in categories: the Greek instinct was the opposite, to take the widest view, to see things as an organic whole .... It was arete that the Olympic games were designed to test the arete of the whole man, not a merely specialized skill .... The great event was the pentathlon, if you won this, you were a man. Needless to say, the Marathon race was never heard of until modern times: the Greeks would have regarded it as a monstrosity."[20]

> "Previously, men could be divided simply into the learned and the ignorant, those more or less the one, and those more or less the other. But your specialist cannot be brought in under either of these two categories. He is not learned, for he is formally ignorant of all that does not enter into his specialty; but neither is he ignorant, because he is 'a scientist,' and 'knows' very well his own tiny portion of the universe. We shall have to say that he is a learned ignoramus, which is a very serious matter, as it implies that he is a person who is ignorant, not in the fashion of the ignorant man, but with all the petulance of one who is learned in his own special line."[21]

> "It is the custom among those who are called "practical" men to condemn any man capable of a wide survey as a visionary: no man is thought worthy of a voice in politics unless he ignores or does not know nine tenths of the most important relevant facts."[22]

12.7 See also

- Commensurability (philosophy of science)
- Crossdisciplinarity
- Encyclopedism
- Holism
- Holism in science
- Intellectual synthesis
- Integrative learning
- Interdiscipline
- Interprofessional education
- Multidisciplinarity
- Science of team science
- Social ecological model
- Synoptic philosophy
- Systems theory
- Systems thinking
- Periodic table of human sciences in Tinbergen's four questions
- Transdisciplinarity

12.8 References

[1] Ausburg, Tanya (2006). *Becoming Interdisciplinary: An Introduction to Interdisciplinary Studies* (2nd ed.). New York: Kendall/Hunt Publishing.

[2] Klein, Julie Thompson (1990). *Interdisciplinarity: History, Theory, and Practice*. Detroit: Wayne State University.

[3] Gunn, Giles (1992). "Interdisciplinary Studies". In Gibaldi, J. *Introduction to Scholarship in Modern Languages and Literatures*. New York: Modern Language Association. pp. 239–240. ISBN 978-0873523851.

[4] José Andrés-Gallego (2015). "Are Humanism and Mixed Methods Related? Leibniz's Universal (Chinese) Dream". *Journal of Mixed Methods Research*. **29** (2): 118–132. doi:10.1177/1558689813515332.

[5] J.S. Edge; S.J. Hoffman; C.L. Ramirez; S.J. Goldie (2013). "Research and Development Priorities to Achieve the "Grand Convergence": An Initial Scan of Priority Research Areas for Public Health, Implementation Science and Innovative Financing for Neglected Diseases: Working Paper for the Lancet Commission on Investing in Health" (PDF). London, UK: The Lancet.

[6] Khorsandi Taskoh, Ali (18 July 2011). *Interdisciplinary Higher Education: Criticism, Challenges and Obstacles*.

[7] Association for Interdisciplinary Studies

[8] International Network of Inter- and Transdisciplinarity

[9] Philosophy of/as Interdisciplinarity Network

[10] Frodeman, Robert (November 23, 2010). "Experiments of Field Psychology". *Opinionator*. Retrieved July 30, 2016.

[11] Frodeman, Robert; Briggle, Adam; Holbrook, J. Britt (2012). "Philosophy in the Age of Neoliberalism". *Social Epistemology*. **26** (3–4): 311–330. doi:10.1080/02691728.2012.722701.

[12] Holbrook, J. Britt. "What is interdisciplinary communication? Reflections on the very idea of disciplinary integration". *Synthese*. **190**: 1865–1879. doi:10.1007/s11229-012-0179-7.

[13] Barry, A.; G. Born & G. Weszkalnys (2008). "Logics of interdisciplinarity" (PDF). *Economy and Society*. **37** (1): 20–49. doi:10.1080/03085140701760841.

[14] Jacobs, J.A. & S. Frickel (2009). "Interdisciplinarity: a critical assessment" (PDF). *Annual Review of Sociology*. **35**: 43–65. doi:10.1146/annurev-soc-070308-115954.

[15] Strathern, M. (2004). *Commons and borderlands: working papers on interdisciplinarity, accountability and the flow of knowledge*. Wantage: Sean Kingston Publishing.

[16] Hall, E.F. & T. Sanders (2015). "Accountability and the academy: producing knowledge about the human dimensions of climate change". *Journal of the Royal Anthropological Institute*. **21** (2): 438–61. doi:10.1111/1467-9655.12162.

[17] Marshall McLuhan (1964) *Understanding Media*, p.13

[18] Nissani, M. (1997). "Ten cheers for interdisciplinarity: The Case for Interdisciplinary Knowledge and Research". *Social Science Journal*. **34** (2): 201–216. doi:10.1016/S0362-3319(97)90051-3.

[19] Nissani, M. (1995). "Fruits, Salads, and Smoothies: A Working Definition of Interdisciplinarity". *Journal of Educational Thought*. **29** (2): 119–126.

[20] Kitto, H.D.F. (1957). *The Greeks*. Middlesex: Penguin. pp. 173–4. ISBN 0140135219.

[21] Ortega y Gasset, José (1932). *The Revolt of the Masses*. New York: New American Library.

[22] Bertrand Russell, cited in: Nissani, M. (1992). *Lives in the Balance: the Cold War and American Politics, 1945-1991*. Hollowbrook. ISBN 978-0893416591.

12.9 Further reading

- Alderman, Harold; Chiappori, Pierre Andre; Haddad, Lawrence; Hoddinott, John. "Unitary Versus Collective Models of the Household: Time to Shift the Burden of Proof?". *World Bank Research Observer*. **10** (1): 1–19. doi:10.1093/wbro/10.1.1.

- Augsburg, Tanya (2005). *Becoming Interdisciplinary: An Introduction to Interdisciplinary Studies*. Kendall/Hunt.

- Association for Integrative Studies

- Bagchi, Amiya Kumar (1982). *The Political Economy of Underdevelopment*. New York: Cambridge University Press.

- Bernstein, Henry (1973). "Introduction: Development and The Social Sciences". In Henry Bernstein. *Underdevelopment and Development: The Third World Today*. Harmondsworth: Penguin. pp. 13–30.

- Center for the Study of Interdisciplinarity

- Centre for Interdisciplinary Research in the Arts (University of Manchester)

- Chambers, Robert (2001). "Qualitative approaches: self-criticism and what can be gained from quantitative approaches", in Kanbur, Ravi, *Qual-quant: qualitative and quantitative poverty appraisal - complementaries, tensions, and the way forward* (pdf), Ithaca, New York: Cornell University, pp. 22–25.

- Chubin, D. E. (1976). "The conceptualization of scientific specialties". *The Sociological Quarterly*. **17**: 448–476. doi:10.1111/j.1533-8525.1976.tb01715.x.

- College for Interdisciplinary Studies, University of British Columbia, Vancouver, British Columbia, Canada

- Callard, Felicity; Fitzgerald, Des (2015). *Rethinking Interdisciplinarity across the Social Sciences and Neurosciences*. Basingstoke: Palgrave Macmillan.

- Davies, M.; Devlin, M. (2007). "Interdisciplinary Higher Education: Implications for Teaching and Learning" (PDF). Centre for the Study of Higher Education, The University of Melbourne.

- Frodeman, R.; Mitcham, C. (Fall 2007). "New Directions in Interdisciplinarity: Broad, Deep, and Critical". *Bulletin of Science, Technology, and Society*. **27** (6): 506–514. doi:10.1177/0270467607308284.

- Franks, D.; Dale, P.; Hindmarsh, R.; Fellows, C.; Buckridge, M.; Cybinski, P. (2007). "Interdisciplinary foundations: reflecting on interdisciplinarity and three decades of teaching and research at Griffith University, Australia". *Studies in Higher Education*. **32** (2): 167–185. doi:10.1080/03075070701267228.

- Frodeman, R., Klein, J.T., and Mitcham, C. *Oxford Handbook of Interdisciplinarity*. Oxford University Press, 2010.

- The Evergreen State College, Olympia, Washington

- Gram Vikas (2007) Annual Report, p. 19.

- Granovetter, Mark (1985). "Economic Action and Social Structure: The Problem of Embeddedness". *The American Journal of Sociology*. **91** (3): 481–510. doi:10.1086/228311.

- Hang Seng Centre for Cognitive Studies

- Harriss, John (2002). "The Case for Cross-Disciplinary Approaches in International Development". *World Development*. **30** (3): 487–496. doi:10.1016/s0305-750x(01)00115-2.

- Henry, Stuart (2005). "Disciplinary hegemony meets interdisciplinary ascendancy: Can interdisciplinary/integrative studies survive, and if so how?". *Issues in Integrative Studies*. **23**: 1–37.

- Indiresan, P.V. (1990) *Managing Development: Decentralisation, Geographical Socialism And Urban Replication*. India: Sage

- Interdisciplinary Arts Department, Columbia College Chicago

- Interdisciplinarity and tenure

- Interdisciplinary Studies Project, Harvard University School of Education, Project Zero

- Jackson, Cecile (2002). "Disciplining Gender?". *World Development*. **30** (3): 497–509. doi:10.1016/s0305-750x(01)00113-9.

- Jacobs, J.A.; Frickel, S. (2009). "Interdisciplinarity: A Critical Assessment" (PDF). *Annual Review of Sociology*. **35**: 43–65. doi:10.1146/annurev-soc-070308-115954.

- Johnston, R (2003). "Integrating methodologists into teams of substantive experts" (PDF). *Studies in Intelligence*. **47** (1).

- Kanbur, Ravi (March 2002). "Economics, social science and development". *World Development*. Elsevier. **30** (3): 477–486. doi:10.1016/S0305-750X(01)00117-6.

- Kanbur, Ravi (2003), "Q-squared?: a commentry on qualitative and quantitative poverty appraisal", in Kanbur, Ravi, *Q-squared, combining qualitative and quantitative methods in poverty appraisal*, Delhi Bangalore: Permanent Black Distributed by Orient Longman, pp. 2–27. ISBN 9788178240534.

- Klein, Julie Thompson (1996) *Crossing Boundaries: Knowledge, Disciplinarities, and Interdisciplinarities* (University Press of Virginia)

- Klein, Julie Thompson (2006) "Resources for interdisciplinary studies." *Change*, (Mark/April). 52–58

- Kleinberg, Ethan (2008). "Interdisciplinary studies at the crossroads". *Liberal Education*. 94 (1): 6–11.

- Kockelmans, Joseph J. editor (1979) *Interdisciplinarity and Higher Education*. The Pennsylvania State University Press ISBN 9780271038261.

- Lipton, Michael (1970). "Interdisciplinary Studies in Less Developed Countries". *Journal of Development Studies*. 7 (1): 5–18. doi:10.1080/00220387008421343.

- Gerhard Medicus Interdisciplinarity in Human Sciences (Documents No. 6, 7 and 8 in English)

- Moran, Joe. (2002). Interdisciplinarity.

- NYU Gallatin School of Individualized Study, New York, NY

- Poverty Action Lab (accessed on 4 November 2008)

- Ravallion, Martin (2003), "Can qualitative methods help quantitative poverty", in Kanbur, Ravi, *Q-squared, combining qualitative and quantitative methods in poverty appraisal*, Delhi Bangalore: Permanent Black Distributed by Orient Longman, pp. 58–67. ISBN 9788178240534

- Rhoten, D. (2003). A multi-method analysis of the social and technical conditions for interdisciplinary collaboration.

- School of Social Ecology at the University of California, Irvine

- Schuurman, F.J. (2000). "Paradigms Lost, paradigms regained? Development studies in the twenty-first century". *Third World Quarterly*. 21 (1): 7–20. doi:10.1080/01436590013198.

- Sen, Amartya (1999). *Development as freedom*. New York: Oxford University Press. ISBN 9780198297581.

- Siskin, L.S. & Little, J.W. (1995). The Subjects in Question. Teachers College Press. about the departmental organization of high schools and efforts to change that.

- Stiglitz, Joseph (2002) Globalisation and its Discontents, United States of America, W.W. Norton and Company

- Sumner, A and M. Tribe (2008) International Development Studies: Theories and Methods in Research and Practice, London: Sage

- Thorbecke, Eric. (2006) "The Evolution of the Development Doctrine, 1950–2005". UNU-WIDER Research Paper No. 2006/155. United Nations University, World Institute for Development Economics Research

- Trans- & inter-disciplinary science approaches- A guide to on-line resources on integration and trans- and inter-disciplinary approaches.

- Truman State University's Interdisciplinary Studies Program

- Waldman, Amy (2003). "Distrust Opens the Door for Polio in India". Retrieved 4 November 2008.

- Peter Weingart and Nico Stehr, eds. 2000. *Practicing Interdisciplinarity* (University of Toronto Press)

- Peter Weingart and Britta Padberg, eds. 2014. "University Experiments in Interdisciplinarity - Obstacles and Opportunities", Bielefeld: transcript Verlag

- White, Howard (2002). "Combining Quantitative and Qualitative Approaches in Poverty Analysis". *World Development*. 30 (3): 511–522. doi:10.1016/s0305-750x(01)00114-0.

12.10 External links

- *National Science Foundation Workshop Report: Interdisciplinary Collaboration in Innovative Science and Engineering Fields*

- *Rethinking Interdisciplnarity* online conference, organized by the Institut Nicod, CNRS, Paris [broken]

- Center for the Study of Interdisciplinarity at the University of North Texas

- *Labyrinthe. Atelier interdisciplinaire*, a journal (in French), with a special issue on *La Fin des Disciplines?*

- *Rupkatha Journal on Interdisciplinary Studies in Humanities: An Online Open Access E-Journal*, publishing articles on a number of areas

- Article about interdisciplinary modeling (in French with an English abstract)

- Wolf, Dieter. Unity of Knowledge, an interdisciplinary project

- Soka University of America has no disciplinary departments and emphasizes interdisciplinary concentrations in the Humanities, Social and Behavioral Sciences, International Studies, and Environmental Studies.

- SystemsX.ch - The Swiss Initiative in Systems Biology

- Interdisciplinarity at Wikispaces - creative explorations of the term interdisciplinarity and its interactions with gender studies

Chapter 13

Philosophy of science

This article is about the concept. For the journal, see Philosophy of Science (journal).

Philosophy of science is a sub-field of philosophy concerned with the foundations, methods, and implications of science. The central questions of this study concern what qualifies as science, the reliability of scientific theories, and the ultimate purpose of science. This discipline overlaps with metaphysics, ontology, and epistemology, for example, when it explores the relationship between science and truth.

There is no consensus among philosophers about many of the central problems concerned with the philosophy of science, including whether science can reveal the truth about unobservable things and whether scientific reasoning can be justified at all. In addition to these general questions about science as a whole, philosophers of science consider problems that apply to particular sciences (such as biology or physics). Some philosophers of science also use contemporary results in science to reach conclusions about philosophy itself.

While philosophical thought pertaining to science dates back at least to the time of Aristotle, philosophy of science emerged as a distinct discipline only in the middle of the 20th century in the wake of the logical positivism movement, which aimed to formulate criteria for ensuring all philosophical statements' meaningfulness and objectively assessing them. Thomas Kuhn's landmark 1962 book *The Structure of Scientific Revolutions* was also formative, challenging the view of scientific progress as steady, cumulative acquisition of knowledge based on a fixed method of systematic experimentation and instead arguing that any progress is relative to a "paradigm," the set of questions, concepts, and practices that define a scientific discipline in a particular historical period.[1] Karl Popper and Charles Sanders Peirce moved on from positivism to establish a modern set of standards for scientific methodology.

Subsequently, the coherentist approach to science, in which a theory is validated if it makes sense of observations as part of a coherent whole, became prominent due to W. V. Quine and others. Some thinkers such as Stephen Jay Gould seek to ground science in axiomatic assumptions, such as the uniformity of nature. A vocal minority of philosophers, and Paul Feyerabend (1924–1994) in particular, argue that there is no such thing as the "scientific method", so all approaches to science should be allowed, including explicitly supernatural ones. Another approach to thinking about science involves studying how knowledge is created from a sociological perspective, an approach represented by scholars like David Bloor and Barry Barnes. Finally, a tradition in continental philosophy approaches science from the perspective of a rigorous analysis of human experience.

Philosophies of the particular sciences range from questions about the nature of time raised by Einstein's general relativity, to the implications of economics for public policy. A central theme is whether one scientific discipline can be reduced to the terms of another. That is, can chemistry be reduced to physics, or can sociology be reduced to individual psychology? The general questions of philosophy of science also arise with greater specificity in some particular sciences. For instance, the question of the validity of scientific reasoning is seen in a different guise in the foundations of statistics. The question of what counts as science and what should be excluded arises as a life-or-death matter in the philosophy of medicine. Additionally, the philosophies of biology, of psychology, and of the social sciences explore whether the scientific studies of human nature can achieve objectivity or are inevitably shaped by values and by social relations.

13.1 Introduction

13.1.1 Defining science

Main article: Demarcation problem

Distinguishing between science and non-science is referred to as the demarcation problem. For example, should psychoanalysis be considered science? How about so-

Karl Popper in the 1980s

13.1.2 Scientific explanation

Main article: Scientific explanation

A closely related question is what counts as a good scientific explanation. In addition to providing predictions about future events, society often takes scientific theories to provide explanations for events that occur regularly or have already occurred. Philosophers have investigated the criteria by which a scientific theory can be said to have successfully explained a phenomenon, as well as what it means to say a scientific theory has explanatory power.

One early and influential theory of scientific explanation is the deductive-nomological model. It says that a successful scientific explanation must deduce the occurrence of the phenomena in question from a scientific law.[10] This view has been subjected to substantial criticism, resulting in several widely acknowledged counterexamples to the theory.[11] It is especially challenging to characterize what is meant by an explanation when the thing to be explained cannot be deduced from any law because it is a matter of chance, or otherwise cannot be perfectly predicted from what is known. Wesley Salmon developed a model in which a good scientific explanation must be statistically relevant to the outcome to be explained.[12][13] Others have argued that the key to a good explanation is unifying disparate phenomena or providing a causal mechanism.[13]

13.1.3 Justifying science

Main article: Problem of induction
Although it is often taken for granted, it is not at all clear

called creation science, the inflationary multiverse hypothesis, or macroeconomics? Karl Popper called this the central question in the philosophy of science.[2] However, no unified account of the problem has won acceptance among philosophers, and some regard the problem as unsolvable or uninteresting.[3][4] Martin Gardner has argued for the use of a Potter Stewart standard ("I know it when I see it") for recognizing pseudoscience.[5]

Early attempts by the logical positivists grounded science in observation while non-science was non-observational and hence meaningless.[6] Popper argued that the central property of science is falsifiability. That is, every genuinely scientific claim is capable of being proven false, at least in principle.[7]

An area of study or speculation that masquerades as science in an attempt to claim a legitimacy that it would not otherwise be able to achieve is referred to as pseudoscience, fringe science, or junk science.[8] Physicist Richard Feynman coined the term "cargo cult science" for cases in which researchers believe they are doing science because their activities have the outward appearance of it but actually lack the "kind of utter honesty" that allows their results to be rigorously evaluated.[9]

The expectations chickens might form about farmer behavior illustrate the "problem of induction."

how one can infer the validity of a general statement from a number of specific instances or infer the truth of a theory from a series of successful tests.[14] For example, a chicken observes that each morning the farmer comes and gives it food, for hundreds of days in a row. The chicken may therefore use inductive reasoning to infer that the farmer will bring food *every* morning. However, one morning, the farmer comes and kills the chicken. How is scientific reasoning more trustworthy than the chicken's reasoning?

One approach is to acknowledge that induction cannot achieve certainty, but observing more instances of a general statement can at least make the general statement more probable. So the chicken would be right to conclude from all those mornings that it is likely the farmer will come with food again the next morning, even if it cannot be certain. However, there remain difficult questions about what precise probability any given evidence justifies putting on the general statement. One way out of these particular difficulties is to declare that all beliefs about scientific theories are subjective, or personal, and correct reasoning is merely about how evidence should change one's subjective beliefs over time.[14]

Some argue that what scientists do is not inductive reasoning at all but rather abductive reasoning, or inference to the best explanation. In this account, science is not about generalizing specific instances but rather about hypothesizing explanations for what is observed. As discussed in the previous section, it is not always clear what is meant by the "best explanation." Ockham's razor, which counsels choosing the simplest available explanation, thus plays an important role in some versions of this approach. To return to the example of the chicken, would it be simpler to suppose that the farmer cares about it and will continue taking care of it indefinitely or that the farmer is fattening it up for slaughter? Philosophers have tried to make this heuristic principle more precise in terms of theoretical parsimony or other measures. Yet, although various measures of simplicity have been brought forward as potential candidates, it is generally accepted that there is no such thing as a theory-independent measure of simplicity. In other words, there appear to be as many different measures of simplicity as there are theories themselves, and the task of choosing between measures of simplicity appears to be every bit as problematic as the job of choosing between theories.[15]

13.1.4 Observation inseparable from theory

When making observations, scientists look through telescopes, study images on electronic screens, record meter readings, and so on. Generally, on a basic level, they can agree on what they see, e.g., the thermometer shows 37.9 degrees C. But, if these scientists have different ideas about the theories that have been developed to explain these basic observations, they may disagree about what they are observ-

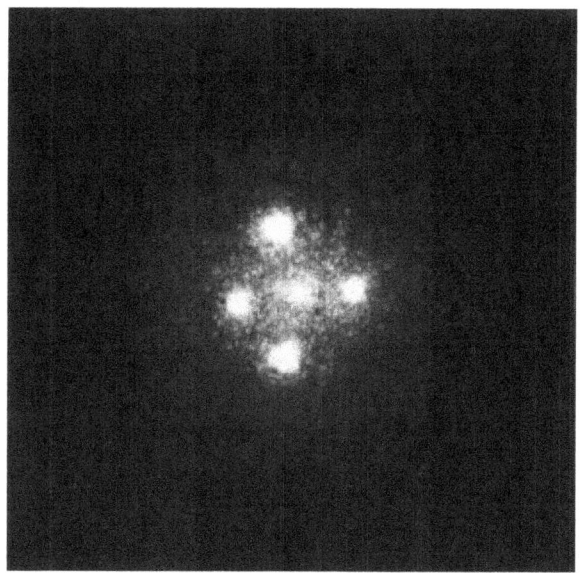

A celestial object known as the Einstein Cross.

ing. For example, before Albert Einstein's general theory of relativity, observers would have likely interpreted the image at left as five different objects in space. In light of that theory, however, astronomers will tell you that there are actually only two objects, one in the center and four different images of a second object around the sides. Alternatively, if other scientists suspect that something is wrong with the telescope and only one object is actually being observed, they are operating under yet another theory. Observations that cannot be separated from theoretical interpretation are said to be theory-laden.[16]

All observation involves both perception and cognition. That is, one does not make an observation passively, but rather is actively engaged in distinguishing the phenomenon being observed from surrounding sensory data. Therefore, observations are affected by one's underlying understanding of the way in which the world functions, and that understanding may influence what is perceived, noticed, or deemed worthy of consideration. In this sense, it can be argued that all observation is theory-laden.[16]

13.1.5 The purpose of science

See also: Scientific realism and Instrumentalism

Should science aim to determine ultimate truth, or are there questions that science cannot answer? *Scientific realists* claim that science aims at truth and that one ought to regard scientific theories as true, approximately true, or likely true. Conversely, *scientific anti-realists* argue that science does not aim (or at least does not succeed) at truth, especially truth about unobservables like electrons or other universes.[17] Instrumentalists argue that scientific theories should only be evaluated on whether they are useful. In their view, whether theories are true or not is beside the point, because the purpose of science is to make predictions and enable effective technology.

Realists often point to the success of recent scientific theories as evidence for the truth (or near truth) of current theories.[18][19] Antirealists point to either the many false theories in the history of science,[20][21] epistemic morals,[22] the success of false modeling assumptions,[23] or widely termed postmodern criticisms of objectivity as evidence against scientific realism.[18] Antirealists attempt to explain the success of scientific theories without reference to truth.[24] Some antirealists claim that scientific theories aim at being accurate only about observable objects and argue that their success is primarily judged by that criterion.[22]

13.1.6 Values and science

Values intersect with science in different ways. There are epistemic values that mainly guide the scientific research. The scientific enterprise is embedded in particular culture and values through individual practitioners. Values emerge from science, both as product and process and can be distributed among several cultures in the society.

If it is unclear what counts as science, how the process of confirming theories works, and what the purpose of science is, there is considerable scope for values and other social influences to shape science. Indeed, values can play a role ranging from determining which research gets funded to influencing which theories achieve scientific consensus.[25] For example, in the 19th century, cultural values held by scientists about race shaped research on evolution, and values concerning social class influenced debates on phrenology (considered scientific at the time).[26] Feminist philosophers of science, sociologists of science, and others explore how social values affect science.

13.2 History

See also: History of scientific method, History of science, and History of philosophy

13.2.1 Pre-modern

The origins of philosophy of science trace back to Plato and Aristotle[27] who distinguished the forms of approximate and exact reasoning, set out the threefold scheme of abductive, deductive, and inductive inference, and also analyzed reasoning by analogy. The eleventh century Arab polymath Ibn al-Haytham (known in Latin as Alhazen) conducted his research in optics by way of controlled experimental testing and applied geometry, especially in his investigations into the images resulting from the reflection and refraction of light. Roger Bacon (1214–1294), an English thinker and experimenter heavily influenced by al-Haytham, is recognized by many to be the father of modern scientific method.[28] His view that mathematics was essential to a correct understanding of natural philosophy was considered to be 400 years ahead of its time.[29]

13.2.2 Modern

Francis Bacon's statue at Gray's Inn, South Square, London

Francis Bacon (no direct relation to Roger, who lived 300 years earlier) was a seminal figure in philosophy of science at the time of the Scientific Revolution. In his work *Novum Organum* (1620) – a reference to Aristotle's *Organon* – Bacon outlined a new system of logic to improve upon the old philosophical process of syllogism. Bacon's method relied on experimental *histories* to eliminate alternative theories.[30] In 1637, René Descartes established a new framework for grounding scientific knowledge in his treatise, *Discourse on Method*, advocating the central role of reason as opposed to sensory experience. By contrast, in 1713, the 2nd edition of Isaac Newton's *Philosophiae Naturalis Principia Mathematica* argued that "... hypotheses ... have no place in experimental philosophy. In this philosophy[,] propositions are deduced from the phenomena and rendered general by induction."[31] This passage influenced a "later generation of philosophically-inclined readers to pronounce a ban on causal hypotheses in natural philosophy."[31] In particular, later in the 18th century, David Hume would famously articulate skepticism about the ability of science to determine causality and gave a definitive formulation of the problem of induction. The 19th century writings of John Stuart Mill are also considered important in the formation of current conceptions of the scientific method, as well as anticipating later accounts of scientific explanation.[32]

13.2.3 Logical positivism

Main article: Logical positivism

Instrumentalism became popular among physicists around the turn of the 20th century, after which logical positivism defined the field for several decades. Logical positivism accepts only testable statements as meaningful, rejects metaphysical interpretations, and embraces verificationism (a set of theories of knowledge that combines logicism, empiricism, and linguistics to ground philosophy on a basis consistent with examples from the empirical sciences). Seeking to overhaul all of philosophy and convert it to a new *scientific philosophy*,[33] the Berlin Circle and the Vienna Circle propounded logical positivism in the late 1920s.

Interpreting Ludwig Wittgenstein's early philosophy of language, logical positivists identified a verifiability principle or criterion of cognitive meaningfulness. From Bertrand Russell's logicism they sought reduction of mathematics to logic. They also embraced Russell's logical atomism, Ernst Mach's phenomenalism—whereby the mind knows only actual or potential sensory experience, which is the content of all sciences, whether physics or psychology—and Percy Bridgman's operationalism. Thereby, only the *verifiable* was scientific and *cognitively meaningful*, whereas the unverifiable was unscientific, cognitively meaningless "pseudostatements"—metaphysical, emotive, or such—not worthy of further review by philosophers, who were newly tasked to organize knowledge rather than develop new knowledge.

Logical positivism is commonly portrayed as taking the extreme position that scientific language should never refer to anything unobservable—even the seemingly core notions of causality, mechanism, and principles—but that is an exaggeration. Talk of such unobservables could be allowed as metaphorical—direct observations viewed in the abstract—or at worst metaphysical or emotional. *Theoretical laws* would be reduced to *empirical laws*, while *theoretical terms* would garner meaning from *observational terms* via *correspondence rules*. Mathematics in physics would reduce to symbolic logic via logicism, while rational reconstruction would convert ordinary language into standardized equivalents, all networked and united by a logical syntax. A scientific theory would be stated with its method of verification, whereby a logical calculus or empirical operation could verify its falsity or truth.

In the late 1930s, logical positivists fled Germany and Austria for Britain and America. By then, many had replaced Mach's phenomenalism with Otto Neurath's physicalism, and Rudolf Carnap had sought to replace *verification* with simply *confirmation*. With World War II's close in 1945, logical positivism became milder, *logical empiricism*, led largely by Carl Hempel, in America, who expounded the covering law model of scientific explanation as a way of identifying the logical form of explanations without any reference to the suspect notion of "causation". The logical positivist movement became a major underpinning of analytic philosophy,[34] and dominated Anglosphere philosophy, including philosophy of science, while influencing sciences, into the 1960s. Yet the movement failed to resolve its central problems,[35][36][37] and its doctrines were increasingly assaulted. Nevertheless, it brought about the establishment of philosophy of science as a distinct subdiscipline of philosophy, with Carl Hempel playing a key role.[38]

13.2.4 Thomas Kuhn

Main article: The Structure of Scientific Revolutions

In the 1962 book *The Structure of Scientific Revolutions*, Thomas Kuhn argued that the process of observation and evaluation takes place within a paradigm, a logically consistent "portrait" of the world that is consistent with observations made from its framing. A paradigm also encompasses the set of questions and practices that define a scientific discipline. He characterized *normal science* as the process of observation and "puzzle solving" which takes place within a paradigm, whereas *revolutionary science* occurs when one paradigm overtakes another in a paradigm shift.[39]

Kuhn denied that it is ever possible to isolate the hypothesis being tested from the influence of the theory in which the observations are grounded, and he argued that it is not possible to evaluate competing paradigms independently. More than one logically consistent construct can paint a usable likeness of the world, but there is no common ground from which to pit two against each other, theory against theory. Each paradigm has its own distinct questions, aims, and interpretations. Neither provides a standard by which

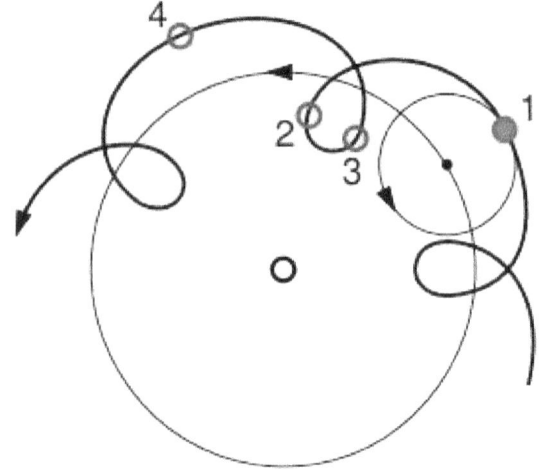

For Kuhn, the addition of epicycles in Ptolemaic astronomy was "normal science" within a paradigm, whereas the Copernican revolution was a paradigm shift.

the other can be judged, so there is no clear way to measure scientific progress across paradigms.

For Kuhn, the choice of paradigm was sustained by rational processes, but not ultimately determined by them. The choice between paradigms involves setting two or more "portraits" against the world and deciding which likeness is most promising. For Kuhn, acceptance or rejection of a paradigm is a social process as much as a logical process. Kuhn's position, however, is not one of relativism.[40] According to Kuhn, a paradigm shift occurs when a significant number of observational anomalies arise in the old paradigm and a new paradigm makes sense of them. That is, the choice of a new paradigm is based on observations, even though those observations are made against the background of the old paradigm.

13.3 Current approaches

13.3.1 Axiomatic assumptions

Some thinkers seek to articulate axiomatic assumptions on which science may be based, a form of foundationalism. This is typically the implicit philosophy of working scientists, that the following basic assumptions that are needed to justify the scientific method:

(1) that there is an objective reality shared by all rational observers;

(2) that this objective reality is governed by natural laws;

(3) that these laws can be discovered by means of systematic observation and experimentation.[41]

Proponents argue that these assumptions are reasonable and necessary for practicing science. For instance, Hugh Gauch argues that science presupposes that "the physical world is orderly and comprehensible."[42] Likewise, biologist Stephen Jay Gould cites the constancy of nature's laws as an assumption which a scientist should assume before proceeding to do geology.[43] In this view, the uniformity of scientific laws is an unprovable postulate which enables scientists to extrapolate into the unobservable past. In other words, the constancy of natural laws must be assumed in order to meaningfully study the past.[44]

13.3.2 Coherentism

Main article: Coherentism

In contrast to the view that science rests on foundational as-

Jeremiah Horrocks makes the first observation of the transit of Venus in 1639, as imagined by the artist W. R. Lavender in 1903

sumptions, coherentism asserts that statements are justified by being a part of a coherent system. Or, rather, individual statements cannot be validated on their own: only coherent systems can be justified.[45] A prediction of a transit of Venus is justified by its being coherent with broader beliefs about celestial mechanics and earlier observations. As explained above, observation is a cognitive act. That is, it relies on a pre-existing understanding, a systematic set of beliefs. An observation of a transit of Venus requires a huge range of auxiliary beliefs, such as those that describe the optics of telescopes, the mechanics of the telescope mount, and an understanding of celestial mechanics. If the prediction fails and a transit is not observed, that is likely to occasion an adjustment in the system, a change in some auxiliary assumption, rather than a rejection of the theoretical system.

In fact, according to the Duhem–Quine thesis, after Pierre Duhem and W. V. Quine, it is impossible to test a theory

13.3. CURRENT APPROACHES

in isolation.[46] One must always add auxiliary hypotheses in order to make testable predictions. For example, to test Newton's Law of Gravitation in the solar system, one needs information about the masses and positions of the Sun and all the planets. Famously, the failure to predict the orbit of Uranus in the 19th century led not to the rejection of Newton's Law but rather to the rejection of the hypothesis that the solar system comprises only seven planets. The investigations that followed led to the discovery of an eighth planet, Neptune. If a test fails, something is wrong. But there is a problem in figuring out what that something is: a missing planet, badly calibrated test equipment, an unsuspected curvature of space, or something else.

One consequence of the Duhem–Quine thesis is that one can make any theory compatible with any empirical observation by the addition of a sufficient number of suitable *ad hoc* hypotheses. Karl Popper accepted this thesis, leading him to reject naïve falsification. Instead, he favored a "survival of the fittest" view in which the most falsifiable scientific theories are to be preferred.[47]

13.3.3 Anything goes

Main article: Epistemological anarchism

Paul Feyerabend (1924–1994) argued that no description

Paul Karl Feyerabend

of scientific method could possibly be broad enough to include all the approaches and methods used by scientists, and that there are no useful and exception-free methodological rules governing the progress of science. He argued that "the only principle that does not inhibit progress is: *anything goes*".[48]

Feyerabend said that science started as a liberating movement, but that over time it had become increasingly dogmatic and rigid and had some oppressive features, and thus had become increasingly an ideology. Because of this, he said it was impossible to come up with an unambiguous way to distinguish science from religion, magic, or mythology. He saw the exclusive dominance of science as a means of directing society as authoritarian and ungrounded.[48] Promulgation of this epistemological anarchism earned Feyerabend the title of "the worst enemy of science" from his detractors.[49]

13.3.4 Sociology of scientific knowledge

Main article: Sociology of scientific knowledge

According to Kuhn, science is an inherently communal activity which can only be done as part of a community.[50] For him, the fundamental difference between science and other disciplines is the way in which the communities function. Others, especially Feyerabend and some postmodernist thinkers, have argued that there is insufficient difference between social practices in science and other disciplines to maintain this distinction. For them, social factors play an important and direct role in scientific method, but they do not serve to differentiate science from other disciplines. On this account, science is socially constructed, though this does not necessarily imply the more radical notion that reality itself is a social construct.

However, some (such as Quine) do maintain that scientific reality is a social construct:

> Physical objects are conceptually imported into the situation as convenient intermediaries not by definition in terms of experience, but simply as irreducible posits comparable, epistemologically, to the gods of Homer ... For my part I do, qua lay physicist, believe in physical objects and not in Homer's gods; and I consider it a scientific error to believe otherwise. But in point of epistemological footing, the physical objects and the gods differ only in degree and not in kind. Both sorts of entities enter our conceptions only as *cultural posits*.[51]

The public backlash of scientists against such views, particularly in the 1990s, became known as the science wars.[52]

A major development in recent decades has been the study of the formation, structure, and evolution of scientific communities by sociologists and anthropologists - including David Bloor, Harry Collins, Bruno Latour, and Anselm

Strauss. Concepts and methods (such as rational choice, social choice or game theory) from economics have also been applied for understanding the efficiency of scientific communities in the production of knowledge. This interdisciplinary field has come to be known as science and technology studies.[53] Here the approach to the philosophy of science is to study how scientific communities actually operate.

13.3.5 Continental philosophy

Philosophers in the continental philosophical tradition are not traditionally categorized as philosophers of science. However, they have much to say about science, some of which has anticipated themes in the analytical tradition. For example, Friedrich Nietzsche advanced the thesis in his "The Genealogy of Morals" that the motive for search of truth in sciences is a kind of ascetic ideal.[54]

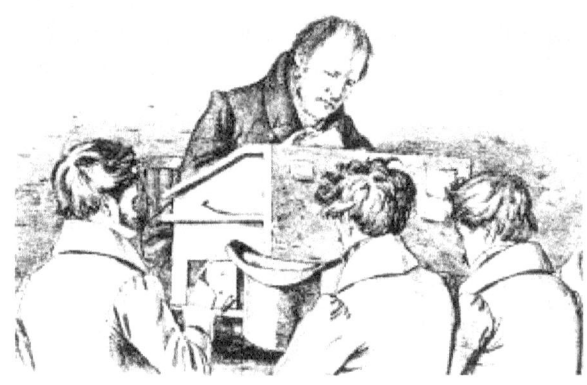

Hegel with his Berlin students
Sketch by Franz Kugler

In general, science in continental philosophy is viewed from a world-historical perspective. One of the first philosophers who supported this view was Georg Wilhelm Friedrich Hegel. Philosophers such as Pierre Duhem and Gaston Bachelard also wrote their works with this world-historical approach to science, predating Kuhn by a generation or more. All of these approaches involve a historical and sociological turn to science, with a priority on lived experience (a kind of Husserlian "life-world"), rather than a progress-based or anti-historical approach as done in the analytic tradition. This emphasis can be traced through Edmund Husserl's phenomenology, the late works of Merleau-Ponty (*Nature: Course Notes from the Collège de France*, 1956–1960), and Martin Heidegger's hermeneutics.[55]

The largest effect on the continental tradition with respect to science was Martin Heidegger's critique of the theoretical attitude in general which of course includes the scientific attitude.[56] For this reason the continental tradition has remained much more skeptical of the importance of science in human life and philosophical inquiry. Nonetheless, there have been a number of important works: especially a Kuhnian precursor, Alexandre Koyré. Another important development was that of Michel Foucault's analysis of the historical and scientific thought in *The Order of Things* and his study of power and corruption within the "science" of madness. Post-Heideggerian authors contributing to the continental philosophy of science in the second half of the 20th century include Jürgen Habermas (e.g., "Truth and Justification", 1998), Carl Friedrich von Weizsäcker ("The Unity of Nature", 1980), and Wolfgang Stegmüller ("Probleme und Resultate der Wissenschaftstheorie und Analytischen Philosophie", 1973–1986).

13.4 Other topics

13.4.1 Reductionism

Analysis is the activity of breaking an observation or theory down into simpler concepts in order to understand it. Reductionism can refer to one of several philosophical positions related to this approach. One type of reductionism is the belief that all fields of study are ultimately amenable to scientific explanation. Perhaps a historical event might be explained in sociological and psychological terms, which in turn might be described in terms of human physiology, which in turn might be described in terms of chemistry and physics.[57] Daniel Dennett distinguishes legitimate reductionism from what he calls *greedy reductionism*, which denies real complexities and leaps too quickly to sweeping generalizations.[58]

13.4.2 Social accountability

See also: The Mismeasure of Man

A broad issue affecting the neutrality of science concerns the areas which science chooses to explore, that is, what part of the world and man is studied by science. Philip Kitcher in his "Science, Truth, and Democracy"[59] argues that scientific studies that attempt to show one segment of the population as being less intelligent, successful or emotionally backward compared to others have a political feedback effect which further excludes such groups from access to science. Thus such studies undermine the broad consensus required for good science by excluding certain people, and so proving themselves in the end to be unscientific.

13.5 Philosophy of particular sciences

> *There is no such thing as philosophy-free science; there is only science whose philosophical baggage is taken on board without examination.*[60]
> — Daniel Dennett, *Darwin's Dangerous Idea*, 1995

In addition to addressing the general questions regarding science and induction, many philosophers of science are occupied by investigating foundational problems in particular sciences. They also examine the implications of particular sciences for broader philosophical questions. The late 20th and early 21st century has seen a rise in the number of practitioners of philosophy of a particular science.[61]

13.5.1 Philosophy of statistics

Main article: Philosophy of statistics

The problem of induction discussed above is seen in another form in debates over the foundations of statistics.[62] The standard approach to statistical hypothesis testing avoids claims about whether evidence supports a hypothesis or makes it more probable. Instead, the typical test yields a p-value, which is the probability of the *evidence* being such as it is, under the assumption that the hypothesis being tested is true. If the *p*-value is too low, the hypothesis is rejected, in a way analogous to falsification. In contrast, Bayesian inference seeks to assign probabilities to hypotheses. Related topics in philosophy of statistics include probability interpretations, overfitting, and the difference between correlation and causation.

13.5.2 Philosophy of mathematics

Main article: Philosophy of mathematics

Philosophy of mathematics is concerned with the philosophical foundations and implications of mathematics.[63] The central questions are whether numbers, triangles, and other mathematical entities exist independently of the human mind and what is the nature of mathematical propositions. Is asking whether "1+1=2" is true fundamentally different from asking whether a ball is red? Was calculus invented or discovered? A related question is whether learning mathematics requires experience or reason alone. What does it mean to prove a mathematical

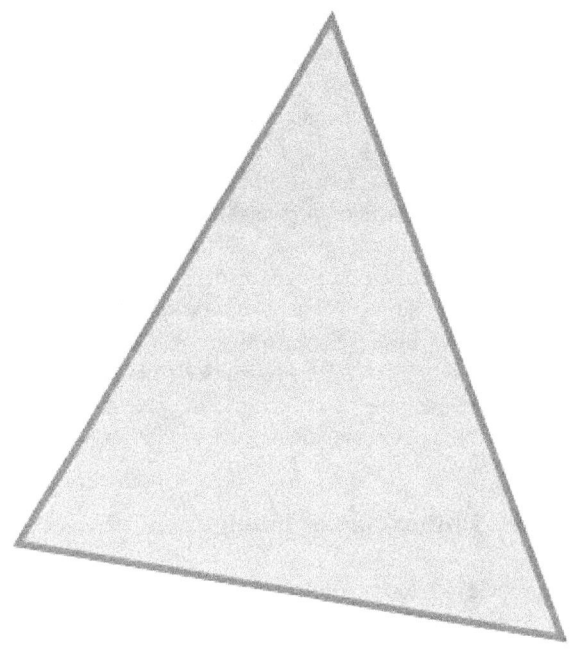

A triangle.

theorem and how does one know whether a mathematical proof is correct? Philosophers of mathematics also aim to clarify the relationships between mathematics and logic, human capabilities such as intuition, and the material universe.

13.5.3 Philosophy of physics

Main article: Philosophy of physics

Philosophy of physics is the study of the fundamental, philosophical questions underlying modern physics, the study of matter and energy and how they interact. The main questions concern the nature of space and time, atoms and atomism. Also included are the predictions of cosmology, the interpretation of quantum mechanics, the foundations of statistical mechanics, causality, determinism, and the nature of physical laws.[64] Classically, several of these questions were studied as part of metaphysics (for example, those about causality, determinism, and space and time).

13.5.4 Philosophy of chemistry

Main article: Philosophy of chemistry

Philosophy of chemistry is the philosophical study of the methodology and content of the science of chemistry. It is explored by philosophers, chemists, and philosopher-chemist teams. It includes research on general philosophy of science issues as applied to chemistry. For example, can all chemical phenomena be explained by quantum mechanics or is it not possible to reduce chemistry to physics? For another example, chemists have discussed the philosophy of how theories are confirmed in the context of confirming reaction mechanisms. Determining reaction mechanisms is difficult because they cannot be observed directly. Chemists can use a number of indirect measures as evidence to rule out certain mechanisms, but they are often unsure if the remaining mechanism is correct because there are many other possible mechanisms that they have not tested or even thought of.[65] Philosophers have also sought to clarify the meaning of chemical concepts which do not refer to specific physical entities, such as chemical bonds.

13.5.5 Philosophy of biology

Main article: Philosophy of biology

Philosophy of biology deals with epistemological, metaphysical, and ethical issues in the biological and biomedical sciences. Although philosophers of science and philosophers generally have long been interested in biology (e.g., Aristotle, Descartes, Leibniz and even Kant), philosophy of biology only emerged as an independent field of philosophy in the 1960s and 1970s.[66] Philosophers of science began to pay increasing attention to developments in biology, from the rise of the modern synthesis in the 1930s and 1940s to the discovery of the structure of deoxyribonucleic acid (DNA) in 1953 to more recent advances in genetic engineering. Other key ideas such as the reduction of all life processes to biochemical reactions as well as the incorporation of psychology into a broader neuroscience are also addressed. Research in current philosophy of biology includes investigation of the foundations of evolutionary theory,[67] and the role of viruses as persistent symbionts in host genomes. As a consequence the evolution of genetic content order is seen as the result of competent genome editors in contrast to former narratives in which error replication events (mutations) dominated.[68]

13.5.6 Philosophy of medicine

Main article: Philosophy of medicine

Beyond medical ethics and bioethics, the philosophy of medicine is a branch of philosophy that includes the epistemology and ontology/metaphysics of medicine. Within the epistemology of medicine, evidence-based medicine (EBM) (or evidence-based practice (EBP)) has attracted attention, most notably the roles of

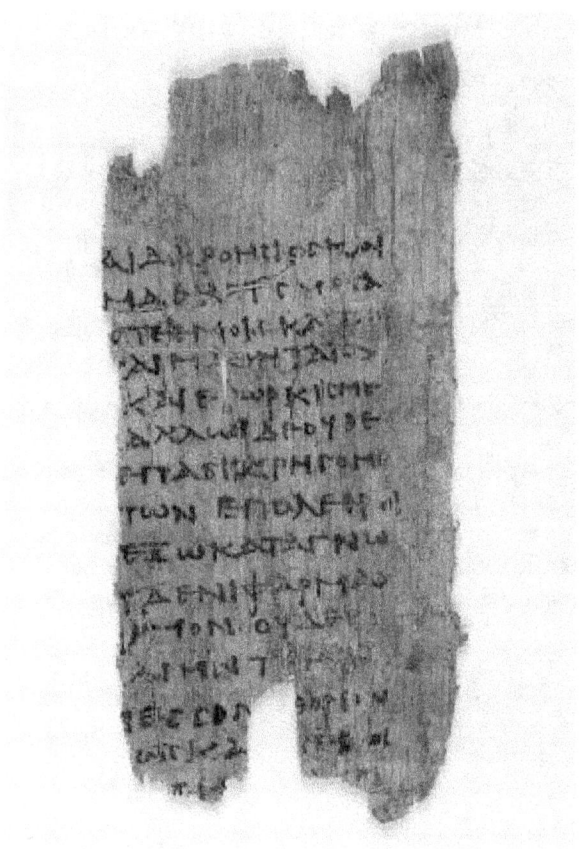

A fragment of the Hippocratic Oath from the third century.

randomisation,[69][70][71] blinding and placebo controls. Related to these areas of investigation, ontologies of specific interest to the philosophy of medicine include Cartesian dualism, the monogenetic conception of disease[72] and the conceptualization of 'placebos' and 'placebo effects'.[73][74][75][76] There is also a growing interest in the metaphysics of medicine,[77] particularly the idea of causation. Philosophers of medicine might not only be interested in how medical knowledge is generated, but also in the nature of such phenomena. Causation is of interest because the purpose of much medical research is to establish causal relationships, e.g. what causes disease, or what causes people to get better.[78]

13.5.7 Philosophy of psychology

Main article: Philosophy of psychology

Philosophy of psychology refers to issues at the theoretical foundations of modern psychology. Some of these issues are epistemological concerns about the methodology of psychological investigation. For example, is the best method for studying psychology to focus only on the response of behavior to external stimuli or should psycholo-

Wilhelm Wundt (seated) with colleagues in his psychological laboratory, the first of its kind.

gists focus on mental perception and thought processes?[79] If the latter, an important question is how the internal experiences of others can be measured. Self-reports of feelings and beliefs may not be reliable because, even in cases in which there is no apparent incentive for subjects to intentionally deceive in their answers, self-deception or selective memory may affect their responses. Then even in the case of accurate self-reports, how can responses be compared across individuals? Even if two individuals respond with the same answer on a Likert scale, they may be experiencing very different things.

Other issues in philosophy of psychology are philosophical questions about the nature of mind, brain, and cognition, and are perhaps more commonly thought of as part of cognitive science, or philosophy of mind. For example, are humans rational creatures?[79] Is there any sense in which they have free will, and how does that relate to the experience of making choices? Philosophy of psychology also closely monitors contemporary work conducted in cognitive neuroscience, evolutionary psychology, and artificial intelligence, questioning what they can and cannot explain in psychology.

Philosophy of psychology is a relatively young field, because psychology only became a discipline of its own in the late 1800s. In particular, neurophilosophy has just recently become its own field with the works of Paul Churchland and Patricia Churchland.[61] Philosophy of mind, by contrast, has been a well-established discipline since before psychology was a field of study at all. It is concerned with questions about the very nature of mind, the qualities of experience, and particular issues like the debate between dualism and monism. Another related field is philosophy of language.

A notable recent development in Philosophy of Psychology is Functional Contextualism or Contextual Behavioural Science (CBS). Functional Contextualism is a modern philosophy of science rooted in philosophical pragmatism and contextualism. It is most actively developed in behavioral science in general, the field of behavior analysis, and contextual behavioral science in particular (see the entry for the Association for Contextual Behavioral Science). Functional contextualism serves as the basis of a theory of language known as relational frame theory[1] and its most prominent application, acceptance and commitment therapy (ACT).[2] It is an extension and contextualistic interpretation of B.F. Skinner's radical behaviorism first delineated by Steven C. Hayes which emphasizes the importance of predicting and influencing psychological events (including thoughts, feelings, and behaviors) with precision, scope, and depth, by focusing on manipulable variables in their context.

13.5.8 Philosophy of psychiatry

Main article: Philosophy of psychiatry

Philosophy of psychiatry explores philosophical questions relating to psychiatry and mental illness. The philosopher of science and medicine Dominic Murphy identifies three areas of exploration in the philosophy of psychiatry. The first concerns the examination of psychiatry as a science, using the tools of the philosophy of science more broadly. The second entails the examination of the concepts employed in discussion of mental illness, including the experience of mental illness, and the normative questions it raises. The third area concerns the links and discontinuities between the philosophy of mind and psychopathology.[80]

13.5.9 Philosophy of economics

Main article: Philosophy and economics

Philosophy of economics is the branch of philosophy which studies philosophical issues relating to economics. It can also be defined as the branch of economics which studies its own foundations and morality. It can be categorized into three central topics.[82] The first concerns the definition and scope of economics and by what methods it should be studied and whether these methods rise to the level of epistemic reliability associated with the other special sciences. For example, is it possible to research economics in such a way that it is value-free, establishing facts that are independent of the normative views of the researcher? The second topic is the meaning and implications of rationality. For example, can buying lottery tickets (increasing the riskiness of your income) at the same time as buying insurance (decreasing the riskiness of your income) be rational? The third topic is the normative evaluation of economic policies and outcomes. What criteria should be used to determine whether

Amartya Sen was awarded the Nobel Prize in Economics for "combining tools from economics and philosophy."[81]

a given public policy is beneficial for society?

13.5.10 Philosophy of social science

Main article: Philosophy of social science

The philosophy of social science is the study of the logic and method of the social sciences, such as sociology, anthropology, and political science.[83] Philosophers of social science are concerned with the differences and similarities between the social and the natural sciences, causal relationships between social phenomena, the possible existence of social laws, and the ontological significance of structure and agency.

The French philosopher, Auguste Comte (1798–1857), established the epistemological perspective of positivism in *The Course in Positivist Philosophy*, a series of texts published between 1830 and 1842. The first three volumes of the *Course* dealt chiefly with the physical sciences already in existence (mathematics, astronomy, physics, chemistry, biology), whereas the latter two emphasised the inevitable coming of social science: *"sociologie"*.[84] For Comte, the physical sciences had necessarily to arrive first, before humanity could adequately channel its efforts into the most challenging and complex "Queen science" of human society itself. Comte offers an evolutionary system proposing that society undergoes three phases in its quest for the truth according to a general 'law of three stages'. These are (1) the *theological*, (2) the *metaphysical*, and (3) the *positive*.[85]

Comte's positivism established the initial philosophical foundations for formal sociology and social research. Durkheim, Marx, and Weber are more typically cited as the fathers of contemporary social science. In psychology, a positivistic approach has historically been favoured in behaviourism. Positivism has also been espoused by 'technocrats' who believe in the inevitability of social progress through science and technology.[86]

The positivist perspective has been associated with 'scientism'; the view that the methods of the natural sciences may be applied to all areas of investigation, be it philosophical, social scientific, or otherwise. Among most social scientists and historians, orthodox positivism has long since lost popular support. Today, practitioners of both social and physical sciences instead take into account the distorting effect of observer bias and structural limitations. This scepticism has been facilitated by a general weakening of deductivist accounts of science by philosophers such as Thomas Kuhn, and new philosophical movements such as critical realism and neopragmatism. The philosopher-sociologist Jürgen Habermas has critiqued pure instrumental rationality as meaning that scientific-thinking becomes something akin to ideology itself.[87]

13.6 See also

13.7 References

[1] Encyclopædia Britannica: Thomas S. Kuhn. "Instead, he argued that the paradigm determines the kinds of experiments scientists perform, the types of questions they ask, and the problems they consider important."

[2] Thornton, Stephen (2006). "Karl Popper". *Stanford Encyclopedia of Philosophy*. Retrieved 2007-12-01.

[3] "Science and Pseudo-science" (2008) in Stanford Encyclopedia of Philosophy

[4] Laudan, Larry (1983). "The Demise of the Demarcation Problem". In Adolf Grünbaum; Robert Sonné Cohen; Larry Laudan. *Physics, Philosophy, and Psychoanalysis: Essays in Honor of Adolf Grünbaum*. Springer. ISBN 90-277-1533-5.

13.7. REFERENCES

[5] Gordin, Michael D. (2012). *The Pseudoscience Wars: Immanuel Velikovsky and the Birth of the Modern Fringe*. University of Chicago Press. pp. 12–13. ISBN 9780226304427.

[6] Uebel, Thomas (2006). "Vienna Circle". *Stanford Encyclopedia of Philosophy*. Retrieved 2007-12-01.

[7] Popper, Karl (2004). *The logic of scientific discovery* (reprint ed.). London & New York: Routledge Classics. ISBN 0-415-27844-9 First published 1959 by Hutchinson & Co.

[8] "*Pseudoscientific – pretending to be scientific, falsely represented as being scientific*", from the *Oxford American Dictionary*, published by the Oxford English Dictionary; Hansson, Sven Ove (1996)."Defining Pseudoscience". Philosophia Naturalis. 33: 169–176, as cited in "Science and Pseudoscience" (2008) in Stanford Encyclopedia of Philosophy. The Stanford article states: "Many writers on pseudoscience have emphasized that pseudoscience is non-science posing as science. The foremost modern classic on the subject (Gardner 1957) bears the title Fads and Fallacies in the Name of Science. According to Brian Baigrie (1988, 438), "[w]hat is objectionable about these beliefs is that they masquerade as genuinely scientific ones." These and many other authors assume that to be pseudoscientific, an activity or a teaching has to satisfy the following two criteria (Hansson 1996): (1) it is not scientific, and (2) its major proponents try to create the impression that it is scientific".

- For example, Hewitt et al. *Conceptual Physical Science* Addison Wesley; 3 edition (July 18, 2003) ISBN 0-321-05173-4, Bennett et al. *The Cosmic Perspective* 3e Addison Wesley; 3 edition (July 25, 2003) ISBN 0-8053-8738-2; *See also*, e.g., Gauch HG Jr. *Scientific Method in Practice* (2003).

- A 2006 National Science Foundation report on Science and engineering indicators quoted Michael Shermer's (1997) definition of pseudoscience: "'claims presented so that they appear [to be] scientific even though they lack supporting evidence and plausibility"(p. 33). In contrast, science is "a set of methods designed to describe and interpret observed and inferred phenomena, past or present, and aimed at building a testable body of knowledge open to rejection or confirmation"(p. 17)'.Shermer M. (1997). *Why People Believe Weird Things: Pseudoscience, Superstition, and Other Confusions of Our Time*. New York: W. H. Freeman and Company. ISBN 0-7167-3090-1. as cited by National Science Foundation; Division of Science Resources Statistics (2006). "Science and Technology: Public Attitudes and Understanding". *Science and engineering indicators 2006*.

- "A pretended or spurious science; a collection of related beliefs about the world mistakenly regarded as being based on scientific method or as having the status that scientific truths now have," from the *Oxford English Dictionary*, second edition 1989.

[9] Cargo Cult Science by Feynman, Richard. Retrieved 2015-10-25.

[10] Hempel, Carl G.; Paul Oppenheim (1948). "Studies in the Logic of Explanation". *Philosophy of Science*. **15** (2): 135–175. doi:10.1086/286983.

[11] Salmon, Merrilee; John Earman, Clark Glymour, James G. Lenno, Peter Machamer, J.E. McGuire, John D. Norton, Wesley C. Salmon, Kenneth F. Schaffner (1992). *Introduction to the Philosophy of Science*. Prentice-Hall. ISBN 0-13-663345-5.

[12] Salmon, Wesley (1971). *Statistical Explanation and Statistical Relevance*. Pittsburgh: University of Pittsburgh Press.

[13] Woodward, James (2003). "Scientific Explanation". *Stanford Encyclopedia of Philosophy*. Retrieved 2007-12-07.

[14] Vickers, John (2013). "The Problem of Induction". *Stanford Encyclopedia of Philosophy*. Retrieved 2014-02-25.

[15] Baker, Alan (2013). "Simplicity". *Stanford Encyclopedia of Philosophy*. Retrieved 2014-02-25.

[16] Bogen, Jim (2013). "Theory and Observation in Science". *Stanford Encyclopedia of Philosophy*. Retrieved 2014-02-25.

[17] Levin, Michael (1984). "What Kind of Explanation is Truth?". In Jarrett Leplin. *Scientific Realism*. Berkeley: University of California Press. pp. 124–1139. ISBN 0-520-05155-6.

[18] Boyd, Richard (2002). "Scientific Realism". *Stanford Encyclopedia of Philosophy*. Retrieved 2007-12-01.

[19] Specific examples include:

- Popper, Karl (2002). *Conjectures and Refutations*. London & New York: Routledge Classics. ISBN 0-415-28594-1 First published 1963 by Routledge and Kegan Paul
- Smart, J. J. C. (1968). *Between Science and Philosophy*. New York: Random House.
- Putnam, Hilary (1975). *Mathematics, Matter and Method (Philosophical Papers, Vol. 1)*. London: Cambridge University Press.
- Putnam, Hilary (1978). *Meaning and the Moral Sciences*. London: Routledge and Kegan Paul.
- Boyd, Richard (1984). "The Current Status of Scientific Realism". In Jarrett Leplin. *Scientific Realism*. Berkeley: University of California Press. pp. 41–82. ISBN 0-520-05155-6.

[20] Stanford, P. Kyle (2006). *Exceeding Our Grasp: Science, History, and the Problem of Unconceived Alternatives*. Oxford University Press. ISBN 978-0-19-517408-3.

[21] Laudan, Larry (1981). "A Confutation of Convergent Realism". *Philosophy of Science*. **48**: 218–249. doi:10.1086/288975.

[22] van Fraassen, Bas (1980). *The Scientific Image*. Oxford: The Clarendon Press. ISBN 0-19-824424-X.

[23] Winsberg, Eric (September 2006). "Models of Success Versus the Success of Models: Reliability without Truth". *Synthese*. **152**: 1–19. doi:10.1007/s11229-004-5404-6.

[24] Stanford, P. Kyle (June 2000). "An Antirealist Explanation of the Success of Science". *Philosophy of Science*. **67** (2): 266–284. doi:10.1086/392775.

[25] Longino, Helen (2013). "The Social Dimensions of Scientific Knowledge". *Stanford Encyclopedia of Philosophy*. Retrieved 2014-03-06.

[26] Douglas Allchin, "Values in Science and in Science Education," in International Handbook of Science Education, B.J. Fraser and K.G. Tobin (eds.), 2:1083–1092, Kluwer Academic Publishers (1988).

[27] Aristotle, "Prior Analytics", Hugh Tredennick (trans.), pp. 181–531 in *Aristotle, Volume 1*, Loeb Classical Library, William Heinemann, London, UK, 1938.

[28] Lindberg, David C. (1980). *Science in the Middle Ages*. University of Chicago Press. pp. 350–351. ISBN 978-0-226-48233-0.

[29] Clegg, Brian. "The First Scientist: A Life of Roger Bacon". Carroll and Graf Publishers, NY, 2003, p. 2.

[30] Bacon, Francis *Novum Organum (The New Organon)*, 1620. Bacon's work described many of the accepted principles, underscoring the importance of empirical results, data gathering and experiment. *Encyclopædia Britannica* (1911), "Bacon, Francis" states: [In Novum Organum, we] "proceed to apply what is perhaps the most valuable part of the Baconian method, the process of exclusion or rejection. This elimination of the non-essential, ... is the most important of Bacon's contributions to the logic of induction, and that in which, as he repeatedly says, his method differs from all previous philosophies."

[31] McMullin, Ernan. "The Impact of Newton's Principia on the Philosophy of Science". www.paricenter.com. Pari Center for New Learning. Retrieved 29 October 2015.

[32] "John Stuart Mill (Stanford Encyclopedia of Philosophy)". plato.stanford.edu. Retrieved 2009-07-31.

[33] Michael Friedman, *Reconsidering Logical Positivism* (New York: Cambridge University Press, 1999), p xiv.

[34] See "Vienna Circle" in *Stanford Encyclopedia of Philosophy*.

[35] Smith, L.D. (1986). *Behaviorism and Logical Positivism: A Reassessment of the Alliance*. Stanford University Press. p. 314. ISBN 978-0-8047-1301-6. LCCN 85030366. The secondary and historical literature on logical positivism affords substantial grounds for concluding that logical positivism failed to solve many of the central problems it generated for itself. Prominent among the unsolved problems was the failure to find an acceptable statement of the verifiability (later confirmability) criterion of meaningfulness. Until a competing tradition emerged (about the late 1950's), the problems of logical positivism continued to be attacked from within that tradition. But as the new tradition in the philosophy of science began to demonstrate its effectiveness—by dissolving and rephrasing old problems as well as by generating new ones—philosophers began to shift allegiances to the new tradition, even though that tradition has yet to receive a canonical formulation.

[36] Bunge, M.A. (1996). *Finding Philosophy in Social Science*. Yale University Press. p. 317. ISBN 978-0-300-06606-7. LCCN lc96004399. To conclude, logical positivism was progressive compared with the classical positivism of Ptolemy, Hume, d'Alembert, Compte, John Stuart Mill, and Ernst Mach. It was even more so by comparison with its contemporary rivals—neo-Thomisism, neo-Kantianism, intuitionism, dialectical materialism, phenomenology, and existentialism. However, neo-positivism failed dismally to give a faithful account of science, whether natural or social. It failed because it remained anchored to sense-data and to a phenomenalist metaphysics, overrated the power of induction and underrated that of hypothesis, and denounced realism and materialism as metaphysical nonsense. Although it has never been practiced consistently in the advanced natural sciences and has been criticized by many philosophers, notably Popper (1959 [1935], 1963), logical positivism remains the tacit philosophy of many scientists. Regrettably, the anti-positivism fashionable in the metatheory of social science is often nothing but an excuse for sloppiness and wild speculation.

[37] "Popper, Falsifiability, and the Failure of Positivism", 7 August 2000. Archived from the original on January 7, 2014. Retrieved 7 January 2014. The upshot is that the positivists seem caught between insisting on the V.C. [Verifiability Criterion]—but for no defensible reason—or admitting that the V.C. requires a background language, etc., which opens the door to relativism, etc. In light of this dilemma, many folk—especially following Popper's "last-ditch" effort to "save" empiricism/positivism/realism with the falsifiability criterion—have agreed that positivism is a dead-end.

[38] Friedman, *Reconsidering Logical Positivism* (Cambridge U P, 1999), p xii.

[39] Bird, Alexander (2013). Zalta, Edward N., ed. "Thomas Kuhn". *Stanford Encyclopedia of Philosophy*. Retrieved 2015-10-26.

[40] T. S. Kuhn, *The Structure of Scientific Revolutions*, 2nd. ed., Chicago: Univ. of Chicago Pr., 1970, p. 206. ISBN 0-226-45804-0

[41] Heilbron 2003, p. vii

[42] Gauch 2002, p. 154. "Expressed as a single grand statement, science presupposes that the physical world is orderly and comprehensible. The most obvious components of this comprehensive presupposition are that the physical world exists and that our sense perceptions are generally reliable."

[43] Gould 1987, p. 120. "You cannot go to a rocky outcrop and observe either the constancy of nature's laws or the working of known processes. It works the other way around." You first assume these propositions and "then you go to the outcrop of rock."

[44] Simpson 1963, pp. 24–48. "Uniformity is an unprovable postulate justified, or indeed required, on two grounds. First, nothing in our incomplete but extensive knowledge of history disagrees with it. Second, only with this postulate is a rational interpretation of history possible and we are justified in seeking—as scientists we must seek—such a rational interpretation."

[45] Olsson, Erik (2014). Zalta, Edward N., ed. "Coherentist Theories of Epistemic Justification". *Stanford Encyclopedia of Philosophy*. Retrieved 2015-10-26.

[46] Sandra Harding (1976). *Can theories be refuted?: essays on the Dunhem–Quine thesis*. Springer Science & Business Media. pp. 9–. ISBN 978-90-277-0630-0.

[47] Popper, Karl (2005). *The Logic of Scientific Discovery* (Taylor & Francis e-Library ed.). London and New York: Routledge / Taylor & Francis e-Library. chapters 3–4. ISBN 0-203-99462-0.

[48] Paul Feyerabend. *Against Method: Outline of an Anarchistic Theory of Knowledge* (1975), ISBN 0-391-00381-X, ISBN 0-86091-222-1, ISBN 0-86091-481-X, ISBN 0-86091-646-4, ISBN 0-86091-934-X, ISBN 0-902308-91-2

[49] Preston, John (2007-02-15). "Paul Feyerabend". *Stanford Encyclopedia of Philosophy*.

[50] Kuhn, T. S. (1996). "[Postscript]". *The Structure of Scientific Revolutions*, 3rd. ed. [Univ. of Chicago Pr]. p. 176. ISBN 0-226-45808-3. A paradigm is what the members of a community of scientists share, *and*, conversely, a scientific community consists of men who share a paradigm.

[51] Quine, Willard Van Orman (1980). "Two Dogmas of Empiricism". *From a Logical Point of View*. Harvard University Press. ISBN 0-674-32351-3.

[52] Ashman, Keith M.; Barringer, Philip S., eds. (2001). *After the Science Wars*. London, UK: Routledge. ISBN 0-415-21209-X. Retrieved 29 October 2015. The 'war' is between scientists who believe that science and its methods are objective, and an increasing number of social scientists, historians, philosophers, and others gathered under the umbrella of Science Studies.

[53] Woodhouse, Edward. Science Technology and Society. Spring 2015 ed. N.p.: U Readers, 2014. Print.

[54] Hatab, Lawrence J. (2008). "How Does the Ascetic Ideal Function in Nietzsche's *Genealogy*?". *The Journal of Nietzsche Studies* (35/36): 107.

[55] Gutting, Gary (2004). *Continental Philosophy of Science*. Blackwell Publishers, Cambridge, MA.

[56] Wheeler, Michael (2015). "Martin Heidegger". *Stanford Encyclopedia of Philosophy*. Retrieved 2015-10-29.

[57] Cat, Jordi (2013). "The Unity of Science". *Stanford Encyclopedia of Philosophy*. Retrieved 2014-03-01.

[58] Levine, George (2008). *Darwin Loves You: Natural Selection and the Re-enchantment of the World*. Princeton University Press. p. 104. ISBN 978-0-691-13639-4. Retrieved 28 October 2015.

[59] Kitcher, P. Science, Truth, and Democracy. Oxford: Oxford University Press, 2001

[60] Dennett, Daniel (1995). *Darwin's Dangerous Idea: Evolution and the Meanings of Life*. Simon and Schuster. p. 21. ISBN 978-1-4391-2629-5.

[61] Bickle, John, Mandik, Peter and Landreth, Anthony, "The Philosophy of Neuroscience", The Stanford Encyclopedia of Philosophy (Summer 2010 Edition), Edward N. Zalta (ed.), URL = <http://plato.stanford.edu/archives/sum2010/entries/neuroscience/

[62] Romeijn, Jan-Willem (2014). Zalta, Edward N., ed. "Philosophy of Statistics". *Stanford Encyclopedia of Philosophy*. Retrieved 2015-10-29.

[63] Horsten, Leon (2015). Zalta, Edward N., ed. "Philosophy of Mathematics". *Stanford Encyclopedia of Philosophy*. Retrieved 2015-10-29.

[64] Ismael, Jenann (2015). Zalta, Edward N., ed. "Quantum Mechanics". *Stanford Encyclopedia of Philosophy*. Retrieved 2015-10-29.

[65] Weisberg, Michael; Needham, Paul; Hendry, Robin (2011). "Philosophy of Chemistry". *Stanford Encyclopedia of Philosophy*. Retrieved 2014-02-14.

[66] Hull D. (1969). What philosophy of biology is not. Journal of the History of Biology, 2, p. 241–268.

[67] Recent examples include Okasha S. (2006), *Evolution and the Levels of Selection*. Oxford: Oxford University Press. and Godfrey-Smith P. (2009), *Darwinian Populations and Natural Selection*. Oxford: Oxford University Press.

[68] Witzany G. (2010). "Biocommunication and Natural Genome Editing". Dortrecht: Springer Sciences and Business Media.

[69] Papineau, D (1994). "The Virtues of Randomization". *British Journal for the Philosophy of Science*. **45** (2): 437–450. doi:10.1093/bjps/45.2.437.

[70] Jstor, Worrall, J., 2002. What Evidence in Evidence-Based Medicine?" *Philosophy of Science* 69(3), p.S316-S330.

[71] Worrall, J (2007). "Why there's no cause to randomize". *British Journal for the Philosophy of Science*. **58**: 451–488. doi:10.1093/bjps/axm024.

[72] Lee, K., 2012. *The Philosophical Foundations of Modern Medicine*, London/New York, Palgrave/Macmillan.

[73] Grünbaum, A (1981). "The Placebo Concept". *Behavioural Research & Therapy*. **19** (2): 157–167. doi:10.1016/0005-7967(81)90040-1.

[74] Gøtzsche, P.C. (1994). "Is there logic in the placebo?". *Lancet*. **344** (8927): 925–926. doi:10.1016/s0140-6736(94)92273-x.

[75] Nunn, R., 2009. It's time to put the placebo out of our misery" *British Medical Journal* 338, b1568.

[76] Turner, A (2012). "Placebos" and the logic of placebo comparison". *Biology & Philosophy*. **27** (3): 419–432. doi:10.1007/s10539-011-9289-8.

[77] Worrall, J (2011). "Causality in medicine: getting back to the Hill top". *Preventive Medicine*. **53** (4–5): 235–238. PMID 21888926. doi:10.1016/j.ypmed.2011.08.009.

[78] Cartwright, N (2009). "What are randomised controlled trials good for?". *Philosophical Studies*. **147** (1): 59–70. doi:10.1007/s11098-009-9450-2.

[79] Mason, Kelby; Sripada, Chandra Sekhar; Stich, Stephen (2010). "Philosophy of Psychology" (PDF). In Moral, Dermot. *Routledge Companion to Twentieth-Century Philosophy*. London: Routledge.

[80] Murphy, Dominic (Spring 2015). "Philosophy of Psychiatry". *The Stanford Encyclopedia of Philosophy*. edited by Edward N. Zalta. Accessed 18 August 2016.

[81] https://www.nobelprize.org/nobel_prizes/economic-sciences/laureates/1998/press.html

[82] Hausman, Daniel (December 18, 2012). "Philosophy of Economics". *Stanford Encyclopedia of Philosophy*. Stanford University. Retrieved 20 February 2014.

[83] • Hollis, Martin (1994). *The Philosophy of Social Science: An Introduction*. Cambridge. ISBN 0-521-44780-1.

[84] Stanford Encyclopaedia: Auguste Comte

[85] Giddens, *Positivism and Sociology*, 1

[86] Schunk, *Learning Theories: An Educational Perspective*, 5th, 315

[87] Outhwaite, William, 1988 *Habermas: Key Contemporary Thinkers*, Polity Press (Second Edition 2009), ISBN 978-0-7456-4328-1 p.68

13.8 Cited texts

- Gauch, Hugh G. (2002). *Scientific Method in Practice*. Cambridge University Press.

- Heilbron, J. L. (editor-in-chief) (2003). *The Oxford Companion to the History of Modern Science*. New York: Oxford University Press. ISBN 0-19-511229-6.

- Kneale, William; Martha Kneale (1962). *The Development of Logic*. London: Oxford University Press. p. 243. ISBN 0-19-824183-6.

- Simpson, G. G. (1963). "Historical science". In Albritton, Jr., C. C. *Fabric of geology*. Stanford, California: Freeman, Cooper, and Company. pp. 24–48.

- Gould, Stephen J (1987). *Time's Arrow, Time's Cycle: Myth and Metaphor in the Discovery of Geological Time*. Cambridge, MA: Harvard University Press. p. 120. ISBN 0-674-89199-6.

- Whitehead, A.N. (1997) [1920]. *Science and the Modern World*. Lowell Lectures. Free Press. p. 135. ISBN 978-0-684-83639-3. LCCN 67002244.

13.9 Further reading

- Bovens, L. and Hartmann, S. (2003), *Bayesian Epistemology*, Oxford University Press, Oxford.

- Gutting, Gary (2004), *Continental Philosophy of Science*, Blackwell Publishers, Cambridge, MA.

- Kuhn, T. S. (1970). *The Structure of Scientific Revolutions*, 2nd. ed. [Univ. of Chicago Pr]. ISBN 0-226-45804-0.

- Losee, J. (1998), *A Historical Introduction to the Philosophy of Science*, Oxford University Press, Oxford, UK.

- Papineau, David (2005) *Science, problems of the philosophy of*. Oxford Companion to Philosophy. Oxford.

- Salmon, Merrilee; John Earman, Clark Glymour, James G. Lenno, Peter Machamer, J.E. McGuire, John D. Norton, Wesley C. Salmon, Kenneth F. Schaffner (1992). *Introduction to the Philosophy of Science*. Prentice-Hall. ISBN 0-13-663345-5.

- Popper, Karl, (1963) *Conjectures and Refutations: The Growth of Scientific Knowledge*, ISBN 0-415-04318-2

- van Fraassen, Bas (1980). *The Scientific Image*. Oxford: The Clarendon Press. ISBN 0-19-824424-X.
- Ziman, John (2000). *Real Science: what it is, and what it means*. Cambridge, Uk: Cambridge University Press.

13.10 External links

- Philosophy of science at PhilPapers
- Philosophy of science at the Indiana Philosophy Ontology Project
- "Philosophy of science". *Internet Encyclopedia of Philosophy*.

Chapter 14

History of science

For the academic journal, see History of Science (journal). "New science" redirects here. For the treatise about history, see The New Science.

The **history of science** is the study of the development of science and scientific knowledge, including both the natural sciences and social sciences. (The history of the arts and humanities is termed the history of scholarship.) Science is a body of empirical, theoretical, and practical knowledge about the natural world, produced by scientists who emphasize the observation, explanation, and prediction of real world phenomena. Historiography of science, in contrast, studies the methods by which historians study the history of science.

The English word *scientist* is relatively recent—first coined by William Whewell in the 19th century.[1] Previously, people investigating nature called themselves "natural philosophers". While empirical investigations of the natural world have been described since classical antiquity (for example, by Thales and Aristotle), and scientific method has been employed since the Middle Ages (for example, by Ibn al-Haytham and Roger Bacon), modern science began to develop in the early modern period, and in particular in the scientific revolution of 16th- and 17th-century Europe.[2] Traditionally, historians of science have defined science sufficiently broadly to include those earlier inquiries.[3]

From the 18th century through late 20th century, the history of science, especially of the physical and biological sciences, was often presented in a progressive narrative in which true theories replaced false beliefs.[4] Some more recent historical interpretations, such as those of Thomas Kuhn, tend to portray the history of science in different terms, such as that of competing paradigms or conceptual systems in a wider matrix that includes intellectual, cultural, economic and political themes outside of science.[5]

14.1 Early cultures

Main article: History of science in early cultures
See also: Protoscience and Alchemy

In prehistoric times, advice and knowledge was passed from generation to generation in an oral tradition. For example, the domestication of maize for agriculture has been dated to about 9,000 years ago in southern Mexico, before the development of writing systems.[6][7][8] Similarly, archaeological evidence indicates the development of astronomical knowledge in preliterate societies.[9][10] The development of writing enabled knowledge to be stored and communicated across generations with much greater fidelity.

Many ancient civilizations collected astronomical information in a systematic manner through simple observation. Though they had no knowledge of the real physical structure of the planets and stars, many theoretical explanations were proposed. Basic facts about human physiology were known in some places, and alchemy was practiced in several civilizations.[11][12] Considerable observation of macroscopic flora and fauna was also performed.

14.1.1 Africa

Ancient Egypt made significant advances in astronomy, mathematics and medicine.[13] Their development of geometry was a necessary outgrowth of surveying to preserve the layout and ownership of farmland, which was flooded annually by the Nile river. The 3-4-5 right triangle and other rules of geometry were used to build rectilinear structures, and the post and lintel architecture of Egypt. Egypt was also a center of alchemy research for much of the Mediterranean. The Edwin Smith papyrus is one of the first medical documents still extant, and perhaps the earliest document that attempts to describe and analyse the brain: it might be seen as the very beginnings of modern neuroscience. However, while Egyptian medicine had some effective practices, it was not without its ineffective and sometimes harmful practices. Medical historians believe that ancient Egyptian pharmacology, for example, was largely ineffective.[14] Nevertheless, it applies the following components to the treatment of disease: examination, diagnosis, treatment, and prognosis,[15] which display strong parallels to the basic empirical method of science and according to G. E. R. Lloyd[16] played a significant role in the development of this methodology. The Ebers papyrus (c. 1550 BC) also contains evidence of traditional empiricism.

14.1.2 Ancient Near East

Further information: Babylonian astronomy, Babylonian mathematics, Babylonian medicine, Egyptian astronomy, Egyptian mathematics, and Egyptian medicine
From their beginnings in Sumer (now Iraq) around 3500

Mesopotamian clay tablet, 492 BC. Writing allowed the recording of astronomical information.

BC, the Mesopotamian people began to attempt to record some observations of the world with numerical data. But their observations and measurements were seemingly taken for purposes other than for elucidating scientific laws. A concrete instance of Pythagoras' law was recorded, as early as the 18th century BC: the Mesopotamian cuneiform tablet Plimpton 322 records a number of Pythagorean triplets (3,4,5) (5,12,13), ... dated 1900 BC, possibly millennia before Pythagoras, but an abstract formulation of the Pythagorean theorem was not.[17]

In Babylonian astronomy, records of the motions of the stars, planets, and the moon are left on thousands of clay tablets created by scribes. Even today, astronomical periods identified by Mesopotamian proto-scientists are still widely used in Western calendars such as the solar year and the lunar month. Using these data they developed arithmetical methods to compute the changing length of daylight in the course of the year and to predict the appearances and disappearances of the Moon and planets and eclipses of the Sun and Moon. Only a few astronomers' names are known, such as that of Kidinnu, a Chaldean astronomer and mathematician. Kiddinu's value for the solar year is in use for today's calendars. Babylonian astronomy was "the first and highly successful attempt at giving a refined mathematical description of astronomical phenomena." According to the historian A. Aaboe, "all subsequent varieties of scientific astronomy, in the Hellenistic world, in India, in Islam, and in the West—if not indeed all subsequent endeavour in the exact sciences—depend upon Babylonian astronomy in decisive and fundamental ways."[18]

14.1.3 Greco-Roman world

Main article: History of science in classical antiquity
In Classical Antiquity, the inquiry into the workings of

Plato's Academy. 1st century mosaic from Pompeii

the universe took place both in investigations aimed at such practical goals as establishing a reliable calendar or determining how to cure a variety of illnesses and in those abstract investigations known as natural philosophy. The ancient people who are considered the first *scientists* may have thought of themselves as *natural philosophers*, as practitioners of a skilled profession (for example, physicians), or as followers of a religious tradition (for example, temple healers).

The earliest Greek philosophers, known as the pre-Socratics,[19] provided competing answers to the question found in the myths of their neighbors: "How did the ordered cosmos in which we live come to be?"[20] The pre-Socratic philosopher Thales (640-546 BC), dubbed the "father of science", was the first to postulate non-supernatural explanations for natural phenomena. For example, that land floats on water and that earthquakes are caused by the agitation of the water upon which the land floats, rather than the god Poseidon.[21] Thales' student Pythagoras of Samos founded the Pythagorean school, which investigated mathematics for its own sake, and was the first to postulate that the Earth is spherical in shape.[22] Leucippus (5th century BC) introduced atomism, the theory that all matter is made of indivisible, imperishable units called atoms. This was greatly expanded on by his pupil Democritus and later Epicurus.

Subsequently, Plato and Aristotle produced the first systematic discussions of natural philosophy, which did much to shape later investigations of nature. Their development of deductive reasoning was of particular importance and usefulness to later scientific inquiry. Plato founded the Platonic Academy in 387 BC, whose motto was "Let none unversed in geometry enter here", and turned out many notable philosophers. Plato's student Aristotle introduced empiricism and the notion that universal truths can be arrived at via observation and induction, thereby laying the foundations of the scientific method.[23] Aristotle also produced many biological writings that were empirical in nature, focusing on biological causation and the diversity of life. He made countless observations of nature, especially the habits and attributes of plants and animals in the world around him, classified more than 540 animal species, and dissected at least 50. Aristotle's writings profoundly influenced subsequent Islamic and European scholarship, though they were eventually superseded in the Scientific Revolution.

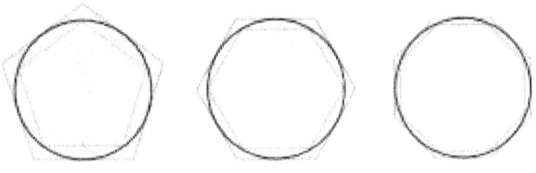

Archimedes used the method of exhaustion to approximate the value of π.

The important legacy of this period included substantial advances in factual knowledge, especially in anatomy, zoology, botany, mineralogy, geography, mathematics and astronomy; an awareness of the importance of certain scientific problems, especially those related to the problem of change and its causes; and a recognition of the methodological importance of applying mathematics to natural phenomena and of undertaking empirical research.[24] In the Hellenistic age scholars frequently employed the principles developed in earlier Greek thought: the application of mathematics and deliberate empirical research, in their scientific investigations.[25] Thus, clear unbroken lines of influence lead from ancient Greek and Hellenistic philosophers, to medieval Muslim philosophers and scientists, to the European Renaissance and Enlightenment, to the secular sciences of the modern day. Neither reason nor inquiry began with the Ancient Greeks, but the Socratic method did, along with the idea of Forms, great advances in geometry, logic, and the natural sciences. According to Benjamin Farrington, former Professor of Classics at Swansea University:

> "Men were weighing for thousands of years before Archimedes worked out the laws of equilibrium; they must have had practical and intuitional knowledge of the principles involved. What Archimedes did was to sort out the theoretical implications of this practical knowledge and present the resulting body of knowledge as a logically coherent system."

and again:

> "With astonishment we find ourselves on the threshold of modern science. Nor should it be supposed that by some trick of translation the extracts have been given an air of modernity. Far from it. The vocabulary of these writings and their style are the source from which our own vocabulary and style have been derived."[26]

The astronomer Aristarchus of Samos was the first known person to propose a heliocentric model of the solar system, while the geographer Eratosthenes accurately calculated the circumference of the Earth. Hipparchus (c. 190 – c. 120 BC) produced the first systematic star catalog. The level of achievement in Hellenistic astronomy and engineering is impressively shown by the Antikythera mechanism (150-100 BC), an analog computer for calculating the position of planets. Technological artifacts of similar complexity did not reappear until the 14th century, when mechanical astronomical clocks appeared in Europe.[27]

In medicine, Hippocrates (c. 460 BC – c. 370 BC) and his followers were the first to describe many diseases and medical conditions and developed the Hippocratic Oath for physicians, still relevant and in use today. Herophilos (335–280 BC) was the first to base his conclusions on dissection of the human body and to describe the nervous system. Galen (129 – c. 200 AD) performed many audacious operations—including brain and eye surgeries— that were not tried again for almost two millennia.

In Hellenistic Egypt, the mathematician Euclid laid down the foundations of mathematical rigor and introduced the concepts of definition, axiom, theorem and proof still in use today in his *Elements*, considered the most influential

14.1. EARLY CULTURES

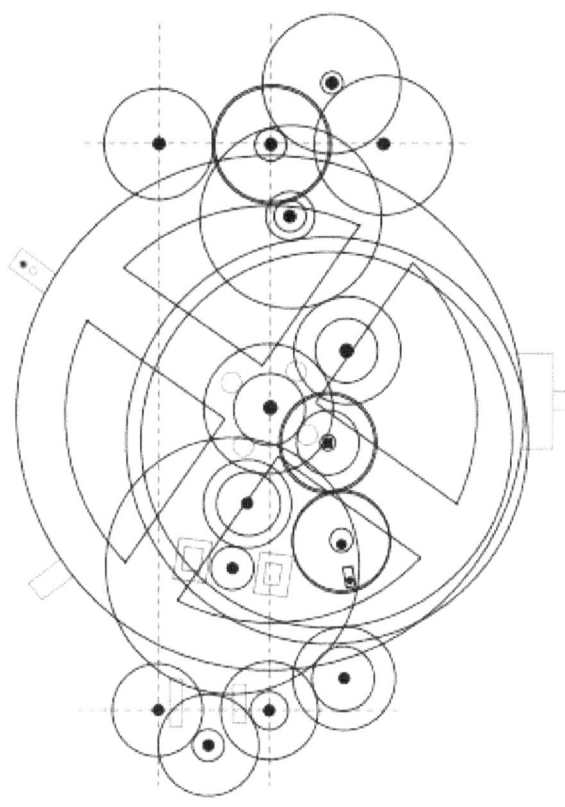

Schematic of the Antikythera mechanism (150-100 BC).

Octahedral shape of a diamond.

textbook ever written.[29] Archimedes, considered one of the greatest mathematicians of all time,[30] is credited with using the method of exhaustion to calculate the area under the arc of a parabola with the summation of an infinite series, and gave a remarkably accurate approximation of Pi.[31] He is also known in physics for laying the foundations of hydrostatics, statics, and the explanation of the principle of the lever.

One of the oldest surviving fragments of Euclid's Elements, *found at Oxyrhynchus and dated to c. 100 AD.*[28]

Theophrastus wrote some of the earliest descriptions of plants and animals, establishing the first taxonomy and looking at minerals in terms of their properties such as hardness. Pliny the Elder produced what is one of the largest encyclopedias of the natural world in 77 AD, and must be regarded as the rightful successor to Theophrastus. For example, he accurately describes the octahedral shape of the diamond, and proceeds to mention that diamond dust is used by engravers to cut and polish other gems owing to its great hardness. His recognition of the importance of crystal shape is a precursor to modern crystallography, while mention of numerous other minerals presages mineralogy. He also recognises that other minerals have characteristic crystal shapes, but in one example, confuses the crystal habit with the work of lapidaries. He was also the first to recognise that amber was a fossilized resin from pine trees because he had seen samples with trapped insects within them.

14.1.4 India

Main article: Science and technology in ancient India

Mathematics: The earliest traces of mathematical knowledge in the Indian subcontinent appear with the Indus Valley Civilization (c. 4th millennium BC – c. 3rd millennium BC). The people of this civilization made bricks whose dimensions were in the proportion 4:2:1, considered favorable for the stability of a brick structure.[32] They also tried to standardize measurement of length to a high degree of accuracy. They designed a ruler—the *Mohenjo-daro ruler*—whose unit of length (approximately 1.32 inches or 3.4 centimetres) was divided into ten equal parts. Bricks manufactured in ancient Mohenjo-daro often had dimensions that were integral multiples of this unit of length.[33]

Indian astronomer and mathematician Aryabhata (476-550), in his *Aryabhatiya* (499) introduced a number of trigonometric functions (including sine, versine, cosine and inverse sine), trigonometric tables, and techniques and

Ancient India was an early leader in metallurgy, as evidenced by the wrought-iron Pillar of Delhi.

algorithms of algebra. In 628 AD, Brahmagupta suggested that gravity was a force of attraction.[34][35] He also lucidly explained the use of zero as both a placeholder and a decimal digit, along with the Hindu-Arabic numeral system now used universally throughout the world. Arabic translations of the two astronomers' texts were soon available in the Islamic world, introducing what would become Arabic numerals to the Islamic World by the 9th century.[36][37] During the 14th–16th centuries, the Kerala school of astronomy and mathematics made significant advances in astronomy and especially mathematics, including fields such as trigonometry and analysis. In particular, Madhava of Sangamagrama is considered the "founder of mathematical analysis".[38]

Astronomy: The first textual mention of astronomical concepts comes from the Vedas, religious literature of India.[39] According to Sarma (2008): "One finds in the Rigveda intelligent speculations about the genesis of the universe from nonexistence, the configuration of the universe, the spherical self-supporting earth, and the year of 360 days divided into 12 equal parts of 30 days each with a periodical intercalary month.".[39] The first 12 chapters of the *Siddhanta Shiromani*, written by Bhāskara in the 12th century, cover topics such as: mean longitudes of the planets; true longitudes of the planets; the three problems of diurnal rotation; syzygies; lunar eclipses; solar eclipses; latitudes of the planets; risings and settings; the moon's crescent; conjunctions of the planets with each other; conjunctions of the planets with the fixed stars; and the patas of the sun and moon. The 13 chapters of the second part cover the nature of the sphere, as well as significant astronomical and trigonometric calculations based on it.

Nilakantha Somayaji's astronomical treatise the Tantrasangraha similar in nature to the Tychonic system proposed by Tycho Brahe had been the most accurate astronomical model until the time of Johannes Kepler in the 17th century.[40]

Linguistics: Some of the earliest linguistic activities can be found in Iron Age India (1st millennium BC) with the analysis of Sanskrit for the purpose of the correct recitation and interpretation of Vedic texts. The most notable grammarian of Sanskrit was Pāṇini (c. 520–460 BC), whose grammar formulates close to 4,000 rules which together form a compact generative grammar of Sanskrit. Inherent in his analytic approach are the concepts of the phoneme, the morpheme and the root.

Medicine: Findings from Neolithic graveyards in what is now Pakistan show evidence of proto-dentistry among an early farming culture.[41] Ayurveda is a system of traditional medicine that originated in ancient India before 2500 BC,[42] and is now practiced as a form of alternative medicine in other parts of the world. Its most famous text is the Suśrutasamhitā of Suśruta, which is notable for describing procedures on various forms of surgery, including rhinoplasty, the repair of torn ear lobes, perineal lithotomy, cataract surgery, and several other excisions and other surgical procedures.

Metallurgy: The wootz, crucible and stainless steels were invented in India, and were widely exported in Classic Mediterranean world. It was known from Pliny the Elder as *ferrum indicum*. Indian Wootz steel was held in high regard in Roman Empire, was often considered to be the best. After in Middle Age it was imported in Syria to produce with special techniques the "Damascus steel" by the year 1000.[43]

> The Hindus excel in the manufacture of iron, and in the preparations of those ingredients along with which it is fused to obtain that kind of soft iron which is usually styled Indian steel (Hindiah). They also have workshops wherein are forged the most famous sabres in the world.
>
> —Henry Yule quoted the 12th-century Arab Edrizi.[44]

14.1.5 China

Main articles: History of science and technology in China and List of Chinese discoveries

Further information: Chinese mathematics and List of Chinese inventions

Mathematics: From the earliest the Chinese used a po-

Lui Hui's Survey of sea island

sitional decimal system on counting boards in order to calculate. To express 10, a single rod is placed in the second box from the right. The spoken language uses a similar system to English: e.g. four thousand two hundred seven. No symbol was used for zero. By the 1st century BC, negative numbers and decimal fractions were in use and *The Nine Chapters on the Mathematical Art* included methods for extracting higher order roots by Horner's method and solving linear equations and by Pythagoras' theorem. Cubic equations were solved in the Tang dynasty and solutions of equations of order higher than 3 appeared in print in 1245 AD by Ch'in Chiu-shao. Pascal's triangle for binomial coefficients was described around 1100 by Jia Xian.

Although the first attempts at an axiomatisation of geometry appear in the Mohist canon in 330 BC, Liu Hui developed algebraic methods in geometry in the 3rd century AD and also calculated pi to 5 significant figures. In 480, Zu Chongzhi improved this by discovering the ratio $\frac{355}{113}$ which remained the most accurate value for 1200 years.

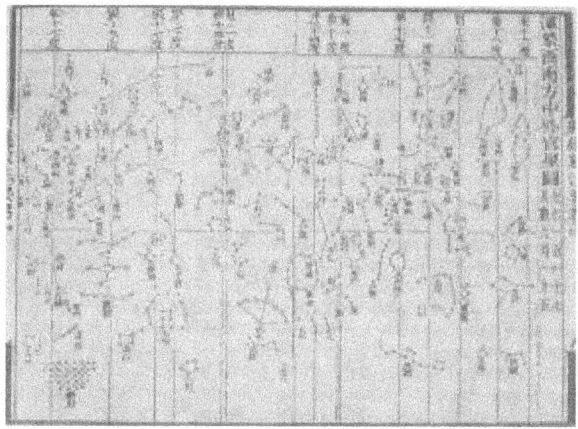

One of the star maps from Su Song's Xin Yi Xiang Fa Yao *published in 1092, featuring a cylindrical projection similar to Mercator projection and the corrected position of the pole star thanks to Shen Kuo's astronomical observations.*[45]

Astronomy: Astronomical observations from China constitute the longest continuous sequence from any civilisation and include records of sunspots (112 records from 364 BC), supernovas (1054), lunar and solar eclipses. By the 12th century, they could reasonably accurately make predictions of eclipses, but the knowledge of this was lost during the Ming dynasty, so that the Jesuit Matteo Ricci gained much favour in 1601 by his predictions.[46] By 635 Chinese astronomers had observed that the tails of comets always point away from the sun.

From antiquity, the Chinese used an equatorial system for describing the skies and a star map from 940 was drawn using a cylindrical (Mercator) projection. The use of an armillary sphere is recorded from the 4th century BC and a sphere permanently mounted in equatorial axis from 52 BC. In 125 AD Zhang Heng used water power to rotate the sphere in real time. This included rings for the meridian and ecliptic. By 1270 they had incorporated the principles of the Arab torquetum.

Seismology: To better prepare for calamities, Zhang Heng invented a seismometer in 132 CE which provided instant alert to authorities in the capital Luoyang that an earthquake had occurred in a location indicated by a specific cardinal or ordinal direction.[47] Although no tremors could be felt in the capital when Zhang told the court that an earthquake had just occurred in the northwest, a message came soon afterwards that an earthquake had indeed struck 400 km (248 mi) to 500 km (310 mi) northwest of Luoyang (in what is now modern Gansu).[48] Zhang called his device the 'instrument for measuring the seasonal winds and the movements of the Earth' (Houfeng didong yi 候風地動儀), so-

A modern replica of Zhang Heng's seismometer of 132 CE

named because he and others thought that earthquakes were most likely caused by the enormous compression of trapped air.[49] See Zhang's seismometer for further details.

There are many notable contributors to the field of Chinese science throughout the ages. One of the best examples would be Shen Kuo (1031–1095), a polymath scientist and statesman who was the first to describe the magnetic-needle compass used for navigation, discovered the concept of true north, improved the design of the astronomical gnomon, armillary sphere, sight tube, and clepsydra, and described the use of drydocks to repair boats. After observing the natural process of the inundation of silt and the find of marine fossils in the Taihang Mountains (hundreds of miles from the Pacific Ocean), Shen Kuo devised a theory of land formation, or geomorphology. He also adopted a theory of gradual climate change in regions over time, after observing petrified bamboo found underground at Yan'an, Shaanxi province. If not for Shen Kuo's writing,[50] the architectural works of Yu Hao would be little known, along with the inventor of movable type printing, Bi Sheng (990-1051). Shen's contemporary Su Song (1020–1101) was also a brilliant polymath, an astronomer who created a celestial atlas of star maps, wrote a pharmaceutical treatise with related subjects of botany, zoology, mineralogy, and metallurgy, and had erected a large astronomical clocktower in Kaifeng city in 1088. To operate the crowning armillary sphere, his clocktower featured an escapement mechanism and the world's oldest known use of an endless power-transmitting chain drive.

The Jesuit China missions of the 16th and 17th centuries "learned to appreciate the scientific achievements of this ancient culture and made them known in Europe. Through their correspondence European scientists first learned about the Chinese science and culture."[51] Western academic thought on the history of Chinese technology and science was galvanized by the work of Joseph Needham and the Needham Research Institute. Among the technological accomplishments of China were, according to the British scholar Needham, early seismological detectors (Zhang Heng in the 2nd century), the water-powered celestial globe (Zhang Heng), matches, the independent invention of the decimal system, dry docks, sliding calipers, the double-action piston pump, cast iron, the blast furnace, the iron plough, the multi-tube seed drill, the wheelbarrow, the suspension bridge, the winnowing machine, the rotary fan, the parachute, natural gas as fuel, the raised-relief map, the propeller, the crossbow, and a solid fuel rocket, the multistage rocket, the horse collar, along with contributions in logic, astronomy, medicine, and other fields.

However, cultural factors prevented these Chinese achievements from developing into what we might call "modern science". According to Needham, it may have been the religious and philosophical framework of Chinese intellectuals which made them unable to accept the ideas of laws of nature:

> It was not that there was no order in nature for the Chinese, but rather that it was not an order ordained by a rational personal being, and hence there was no conviction that rational personal beings would be able to spell out in their lesser earthly languages the divine code of laws which he had decreed aforetime. The Taoists, indeed, would have scorned such an idea as being too naïve for the subtlety and complexity of the universe as they intuited it.[52]

14.2 Science in the Middle Ages

With the division of the Roman Empire, the Western Roman Empire lost contact with much of its past. In the Middle East, Greek philosophy was able to find some support under the newly created Arab Empire. With the spread of Islam in the 7th and 8th centuries, a period of Muslim scholarship, known as the Islamic Golden Age, lasted until the 13th century. This scholarship was aided by several factors. The use of a single language, Arabic, allowed communication without need of a translator. Access to Greek texts from the Byzantine Empire, along with Indian sources of learning, provided Muslim scholars a knowledge base to build upon.

While the Byzantine Empire still held learning centers such as Constantinople, Western Europe's knowledge was concentrated in monasteries until the development of medieval universities in the 12th and 13th centuries. The curriculum of monastic schools included the study of the few available ancient texts and of new works on practical subjects like medicine[53] and timekeeping.[54]

14.2.1 Islamic world

Main articles: Science in the medieval Islamic world and Timeline of science and engineering in the Islamic world
See also: Alchemy and chemistry in medieval Islam, Islamic astronomy, Islamic mathematics, Islamic medicine, Islamic physics, Islamic psychological thought, and Early Muslim sociology

Scientific method began developing in the Muslim world,

15th-century manuscript of Avicenna's The Canon of Medicine.

where significant progress in methodology was made, beginning with the experiments of Ibn al-Haytham (Alhazen) on optics from c. 1000, in his *Book of Optics*.[55] The most important development of the scientific method was the use of experiments to distinguish between competing scientific theories set within a generally empirical orientation, which began among Muslim scientists. Ibn al-Haytham is also regarded as the father of optics, especially for his empirical proof of the intromission theory of light. Some have also described Ibn al-Haytham as the "first scientist" for his development of the modern scientific method.[56]

In mathematics, the mathematician Muhammad ibn Musa al-Khwarizmi gave his name to the concept of the algorithm, while the term algebra is derived from *al-jabr*, the beginning of the title of one of his publications. What is now known as Arabic numerals originally came from India, but Muslim mathematicians did make several refinements to the number system, such as the introduction of decimal point notation. Mathematician Al-Battani (850-929) contributed to astronomy and mathematics, while scholar Al-Razi contributed to chemistry and medicine.

In astronomy, Al-Battani improved the measurements of Hipparchus, preserved in the translation of Ptolemy's *Hè Megalè Syntaxis* (*The great treatise*) translated as *Almagest*. Al-Battani also improved the precision of the measurement of the precession of the Earth's axis. The corrections made to the geocentric model by al-Battani, Ibn al-Haytham,[57] Averroes and the Maragha astronomers such as Nasir al-Din al-Tusi, Mo'ayyeduddin Urdi and Ibn al-Shatir are similar to Copernican heliocentric model.[58][59] Heliocentric theories may have also been discussed by several other Muslim astronomers such as Ja'far ibn Muhammad Abu Ma'shar al-Balkhi,[60] Abu-Rayhan Biruni, Abu Said al-Sijzi,[61] Qutb al-Din al-Shirazi, and Najm al-Dīn al-Qazwīnī al-Kātibī.[62]

Muslim chemists and alchemists played an important role in the foundation of modern chemistry. Scholars such as Will Durant[63] and Fielding H. Garrison[64] considered Muslim chemists to be the founders of chemistry. In particular, Jābir ibn Hayyān is "considered by many to be the father of chemistry".[65][66] The works of Arabic scientists influenced Roger Bacon (who introduced the empirical method to Europe, strongly influenced by his reading of Persian writers),[67] and later Isaac Newton.[68]

Ibn Sina (Avicenna) is regarded as the most influential philosopher of Islam.[69] He pioneered the science of experimental medicine[70] and was the first physician to conduct clinical trials.[71] His two most notable works in medicine are the *Kitāb al-shifā'* ("Book of Healing") and The Canon of Medicine, both of which were used as standard medicinal texts in both the Muslim world and in Europe well into the 17th century. Amongst his many contributions are the discovery of the contagious nature of infectious diseases,[70] and the introduction of clinical pharmacology.[72]

Some of the other famous scientists from the Islamic world include al-Farabi (polymath), Abu al-Qasim al-Zahrawi (pioneer of surgery),[73] Abū Rayhān al-Bīrūnī (pioneer of Indology,[74] geodesy and anthropology),[75] Nasīr al-Dīn al-Tūsī (polymath), and Ibn Khaldun (forerunner of social sciences[76] such as demography,[77] cultural history,[78] historiography,[79] philosophy of history and sociology),[80] among many others.

Islamic science began its decline in the 12th or 13th century, before the Renaissance in Europe, and due in part to the 11th–13th century Mongol conquests, during which libraries, observatories, hospitals and universities were destroyed.[81] The end of the Islamic Golden Age is marked by the destruction of the intellectual center of Baghdad, the capital of the Abbasid caliphate in 1258.[81]

14.2.2 Europe

Main articles: European science in the Middle Ages and Byzantine science
Further information: Renaissance of the 12th century, Scholasticism, Medieval technology, List of medieval European scientists, and Islamic contributions to Medieval Europe

An intellectual revitalization of Europe started with the

Map of medieval universities.

birth of medieval universities in the 12th century. The contact with the Islamic world in Spain and Sicily, and during the Reconquista and the Crusades, allowed Europeans access to scientific Greek and Arabic texts, including the works of Aristotle, Ptolemy, Jābir ibn Hayyān, al-Khwarizmi, Alhazen, Avicenna, and Averroes. European scholars had access to the translation programs of Raymond of Toledo, who sponsored the 12th century Toledo School of Translators from Arabic to Latin. Later translators like Michael Scotus would learn Arabic in order to study these texts directly. The European universities aided materially in the translation and propagation of these texts and started a new infrastructure which was needed for scientific communities. In fact, European university put many works about the natural world and the study of nature at the center of its curriculum,[82] with the result that the "medieval university laid far greater emphasis on science than does its modern counterpart and descendent."[83]

As well as this, Europeans began to venture further and further east (most notably, perhaps, Marco Polo) as a result of the Pax Mongolica. This led to the increased awareness of Indian and even Chinese culture and civilization within the European tradition. Technological advances were also made, such as the early flight of Eilmer of Malmesbury (who had studied Mathematics in 11th century England),[84] and the metallurgical achievements of the Cistercian blast furnace at Laskill.[85][86]

Statue of Roger Bacon at the Oxford University Museum.

At the beginning of the 13th century, there were reasonably accurate Latin translations of the main works of almost all the intellectually crucial ancient authors, allowing a sound transfer of scientific ideas via both the universities and the monasteries. By then, the natural philosophy contained in these texts began to be extended by notable scholastics such as Robert Grosseteste, Roger Bacon, Albertus Magnus and Duns Scotus. Precursors of the modern scientific method, influenced by earlier contributions of the Islamic world, can be seen already in Grosseteste's emphasis on mathematics as a way to understand nature, and in the empirical approach admired by Bacon, particularly in his *Opus Majus*. Pierre Duhem's provocative thesis of the Catholic Church's Condemnation of 1277 led to the study of medieval science as a serious discipline, "but no one in the field any longer endorses his view that modern science started in 1277".[87] However, many scholars agree with Duhem's view that the Middle Ages were a period of important scientific developments.[88][89][90][91]

The first half of the 14th century saw much important scientific work being done, largely within the framework of scholastic commentaries on Aristotle's scientific writings.[92] William of Ockham introduced the principle of parsimony: natural philosophers should not postulate unnecessary entities, so that motion is not a distinct thing but is only the moving object[93] and an intermediary "sensible species" is not needed to transmit an image of an object to the eye.[94] Scholars such as Jean Buridan and Nicole Oresme started to reinterpret elements of Aristotle's mechanics. In particular, Buridan developed the theory that impetus was the cause of the motion of projectiles, which was a first step towards the modern concept of inertia.[95] The Oxford Calculators began to mathematically analyze the kinematics of motion, making this analysis without considering the causes of motion.[96]

In 1348, the Black Death and other disasters sealed a sudden end to the previous period of massive philosophic and scientific development. Yet, the rediscovery of ancient texts was improved after the Fall of Constantinople in 1453, when many Byzantine scholars had to seek refuge in the West. Meanwhile, the introduction of printing was to have great effect on European society. The facilitated dissemination of the printed word democratized learning and allowed a faster propagation of new ideas. New ideas also helped to influence the development of European science at this point: not least the introduction of Algebra. These developments paved the way for the Scientific Revolution, which may also be understood as a resumption of the process of scientific inquiry, halted at the start of the Black Death.

14.3 Impact of science in Europe

Main articles: Scientific revolution and Age of Enlightenment
See also: Continuity thesis, Decline of Western alchemy, and Natural magic

The renewal of learning in Europe, that began with 12th century Scholasticism, came to an end about the time of the Black Death, and the initial period of the subsequent Italian Renaissance is sometimes seen as a lull in scientific activity. The Northern Renaissance, on the other hand, showed a decisive shift in focus from Aristoteleian natural philosophy to chemistry and the biological sciences (botany, anatomy, and medicine).[98] Thus modern science in Europe was resumed in a period of great upheaval: the Protestant Reformation and Catholic Counter-Reformation; the discovery of the Americas by Christopher Columbus; the Fall of Constantinople; but also the re-discovery of Aristotle during the Scholastic period presaged large social and political changes. Thus, a suitable environment was created in which

Isaac Newton initiated classical mechanics in physics.

it became possible to question scientific doctrine, in much the same way that Martin Luther and John Calvin questioned religious doctrine. The works of Ptolemy (astronomy) and Galen (medicine) were found not always to match everyday observations. Work by Vesalius on human cadavers found problems with the Galenic view of anatomy.[99]

The willingness to question previously held truths and search for new answers resulted in a period of major scientific advancements, now known as the Scientific Revolution. The Scientific Revolution is traditionally held by most historians to have begun in 1543, when the books *De humani corporis fabrica* (*On the Workings of the Human Body*) by Andreas Vesalius, and also *De Revolutionibus*, by the astronomer Nicolaus Copernicus, were first printed. The thesis of Copernicus' book was that the Earth moved around the Sun. The period culminated with the publication of the *Philosophiæ Naturalis Principia Mathematica* in 1687 by Isaac Newton, representative of the unprecedented growth of scientific publications throughout Europe.

Other significant scientific advances were made during this time by Galileo Galilei, Edmond Halley, Robert Hooke, Christiaan Huygens, Tycho Brahe, Johannes Kepler, Gottfried Leibniz, and Blaise Pascal. In philosophy, major contributions were made by Francis Bacon, Sir

Galileo Galilei, father of modern science.[97]

Thomas Browne, René Descartes, and Thomas Hobbes. The scientific method was also better developed as the modern way of thinking emphasized experimentation and reason over traditional considerations.

14.3.1 Age of Enlightenment

Main article: Science in the Age of Enlightenment
Further information: Age of Enlightenment
Distinguished Men of Science.[1] Use your cursor to see who is who.[2]

1. ^ Engraving after 'Men of Science Living in 1807-8', John Gilbert engraved by George Zobel and William Walker, ref. NPG 1075a, National Portrait Gallery, London, accessed February 2010
2. ^ Smith, HM (May 1941). "Eminent men of science living in 1807-8". *J. Chem. Educ.* **18** (5): 203. doi:10.1021/ed018p203.

The Age of Enlightenment was a European affair. The 17th century brought decisive steps towards modern science, which accelerated during the 18th century. Directly based on the works[100] of Newton, Descartes, Pascal and Leibniz, the way was now clear to the development of

Alessandro Volta demonstrates the first electrical cell to Napoleon in 1801.

modern mathematics, physics and technology by the generation of Benjamin Franklin (1706–1790), Leonhard Euler (1707–1783), Mikhail Lomonosov (1711–1765) and Jean le Rond d'Alembert (1717–1783). Denis Diderot's *Encyclopédie*, published between 1751 and 1772 brought this new understanding to a wider audience. The impact of this process was not limited to science and technology, but affected philosophy (Immanuel Kant, David Hume), religion (the increasingly significant impact of science upon religion), and society and politics in general (Adam Smith, Voltaire). The early modern period is seen as a flowering of the European Renaissance, in what is often known as the Scientific Revolution, viewed as a foundation of modern science.[101]

14.3.2 Romanticism in science

Main article: Romanticism in science

The Romantic Movement of the early 19th century reshaped science by opening up new pursuits unexpected in the classical approaches of the Enlightenment. Major breakthroughs came in biology, especially in Darwin's theory of evolution, as well as physics (electromagnetism), mathematics (non-Euclidean geometry, group theory) and chemistry (organic chemistry). The decline of Romanticism occurred because a new movement, Positivism, began to take hold of the ideals of the intellectuals after 1840 and lasted until about 1880.

14.4 Modern science

The scientific revolution established science as a source for the growth of knowledge.[102] During the 19th century, the practice of science became professionalized and institutionalized in ways that continued through the 20th century. As the role of scientific knowledge grew in society, it became incorporated with many aspects of the functioning of nation-states.

14.4.1 Natural sciences

Physics

Main article: History of physics

The scientific revolution is a convenient boundary between

James Clerk Maxwell

Albert Einstein

ancient thought and classical physics. Nicolaus Copernicus revived the heliocentric model of the solar system described by Aristarchus of Samos. This was followed by the first known model of planetary motion given by Johannes Kepler in the early 17th century, which proposed that the planets follow elliptical orbits, with the Sun at one focus of the ellipse. Galileo ("*Father of Modern Physics*") also made use of experiments to validate physical theories, a key element of the scientific method. William Gilbert did some of the earliest experiments with electricity and magnetism, establishing that the Earth itself is magnetic.

In 1687, Isaac Newton published the *Principia Mathematica*, detailing two comprehensive and successful physical theories: Newton's laws of motion, which led to classical mechanics; and Newton's Law of Gravitation, which describes the fundamental force of gravity.

During the early 19th century, the behavior of electricity and magnetism was studied by Faraday, Ohm, and others. These studies led to the unification of the two phenomena into a single theory of electromagnetism, by James Clerk Maxwell (known as Maxwell's equations).

The beginning of the 20th century brought the start of a revolution in physics. The long-held theories of Newton were shown not to be correct in all circumstances. Beginning in 1900, Max Planck, Albert Einstein, Niels Bohr and others developed quantum theories to explain various anomalous experimental results, by introducing discrete energy levels. Not only did quantum mechanics show that the laws of motion did not hold on small scales, but even more disturbingly, the theory of general relativity, proposed by Einstein in 1915, showed that the fixed background of spacetime, on which both Newtonian mechanics and special relativity depended, could not exist. In 1925, Werner Heisenberg and Erwin Schrödinger formulated quantum mechanics, which explained the preceding quantum theories. The observation by Edwin Hubble in 1929 that the speed at which galaxies recede positively correlates with their distance, led to the understanding that the universe is expanding, and the formulation of the Big Bang theory by Georges Lemaître.

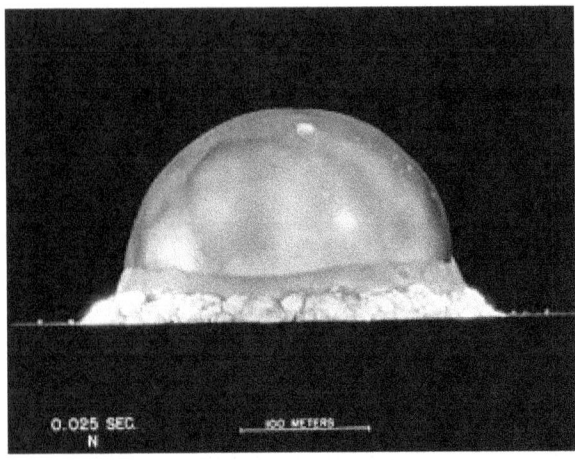

The atomic bomb ushered in "Big Science" in physics.

Dmitri Mendeleev

In 1938 Otto Hahn and Fritz Strassmann discovered nuclear fission with radiochemical methods, and in 1939 Lise Meitner and Otto Robert Frisch wrote the first theoretical interpretation of the fission process, which was later improved by Niels Bohr and John A. Wheeler. Further developments took place during World War II, which led to the practical application of radar and the development and use of the atomic bomb. Though the process had begun with the invention of the cyclotron by Ernest O. Lawrence in the 1930s, physics in the postwar period entered into a phase of what historians have called "Big Science", requiring massive machines, budgets, and laboratories in order to test their theories and move into new frontiers. The primary patron of physics became state governments, who recognized that the support of "basic" research could often lead to technologies useful to both military and industrial applications. Currently, general relativity and quantum mechanics are inconsistent with each other, and efforts are underway to unify the two.

Chemistry

Main article: History of chemistry

Modern chemistry emerged from the sixteenth through the eighteenth centuries through the material practices and theories promoted by alchemy, medicine, manufacturing and mining.[103] A decisive moment came when 'chemistry' was distinguished from alchemy by Robert Boyle in his work *The Sceptical Chymist*, in 1661; although the alchemical tradition continued for some time after his work. Other important steps included the gravimetric experimental practices of medical chemists like William Cullen, Joseph Black, Torbern Bergman and Pierre Macquer and through the work of Antoine Lavoisier (*Father of Modern Chemistry*) on oxygen and the law of conservation of mass, which refuted phlogiston theory. The theory that all matter is made of atoms, which are the smallest constituents of matter that cannot be broken down without losing the basic chemical and physical properties of that matter, was provided by John Dalton in 1803, although the question took a hundred years to settle as proven. Dalton also formulated the law of mass relationships. In 1869, Dmitri Mendeleev composed his periodic table of elements on the basis of Dalton's discoveries.

The synthesis of urea by Friedrich Wöhler opened a new research field, organic chemistry, and by the end of the 19th century, scientists were able to synthesize hundreds of organic compounds. The later part of the 19th century saw the exploitation of the Earth's petrochemicals, after the exhaustion of the oil supply from whaling. By the 20th century, systematic production of refined materials provided a ready supply of products which provided not only energy, but also synthetic materials for clothing, medicine, and everyday disposable resources. Application of the techniques of organic chemistry to living organisms resulted in physiological chemistry, the precursor to biochemistry. The 20th century also saw the integration of physics and chemistry, with chemical properties explained as the result of the electronic structure of the atom. Linus Pauling's book on *The Nature of the Chemical Bond* used the principles of quantum mechanics to deduce bond angles in ever-more complicated molecules. Pauling's work culminated in the physical modelling of DNA, *the secret of life* (in the words of Francis Crick, 1953). In the same year, the Miller–Urey experiment demonstrated in a simulation of primordial processes, that basic constituents of proteins, simple amino acids, could themselves be built up from simpler molecules.

Geology

Main article: History of geology

Geology existed as a cloud of isolated, disconnected ideas about rocks, minerals, and landforms long before it became a coherent science. Theophrastus' work on rocks, *Peri lithōn*, remained authoritative for millennia: its interpretation of fossils was not overturned until after the Scientific Revolution. Chinese polymath Shen Kua (1031–1095) first formulated hypotheses for the process of land formation. Based on his observation of fossils in a geological stratum in a mountain hundreds of miles from the ocean, he deduced that the land was formed by erosion of the mountains and by deposition of silt.

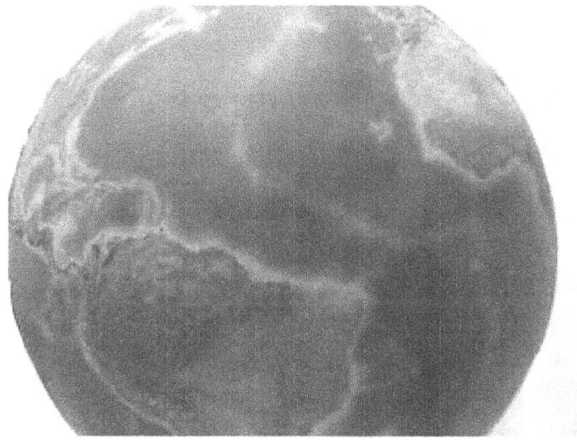

Plate tectonics—seafloor spreading and continental drift illustrated on a relief globe

Geology did not undergo systematic restructuring during the Scientific Revolution, but individual theorists made important contributions. Robert Hooke, for example, formulated a theory of earthquakes, and Nicholas Steno developed the theory of superposition and argued that fossils were the remains of once-living creatures. Beginning with Thomas Burnet's *Sacred Theory of the Earth* in 1681, natural philosophers began to explore the idea that the Earth had changed over time. Burnet and his contemporaries interpreted Earth's past in terms of events described in the Bible, but their work laid the intellectual foundations for secular interpretations of Earth history.

James Hutton, the father of modern geology

Modern geology, like modern chemistry, gradually evolved during the 18th and early 19th centuries. Benoît de Maillet and the Comte de Buffon saw the Earth as much older than the 6,000 years envisioned by biblical scholars. Jean-Étienne Guettard and Nicolas Desmarest hiked central France and recorded their observations on some of the first geological maps. Aided by chemical experimentation, naturalists such as Scotland's John Walker,[104] Sweden's Torbern Bergman, and Germany's Abraham Werner created comprehensive classification systems for rocks and minerals—a collective achievement that transformed geology into a cutting edge field by the end of the eighteenth century. These early geologists also proposed a generalized interpretations of Earth history that led James Hutton, Georges Cuvier and Alexandre Brongniart, following in the steps of Steno, to argue that layers of rock could be dated by the fossils they contained: a principle first applied to the geology of the Paris Basin. The use of index fossils became a powerful tool for making geological maps, because it allowed geologists to correlate the rocks in one locality with those of similar age in other, distant localities. Over the first half of the 19th century, geologists such as Charles Lyell, Adam Sedgwick, and Roderick Murchison applied the new technique to rocks throughout Europe and eastern North America, setting the stage for more detailed, government-funded mapping projects in later decades.

Midway through the 19th century, the focus of geology shifted from description and classification to attempts to understand *how* the surface of the Earth had changed. The first comprehensive theories of mountain building were proposed during this period, as were the first modern theories of earthquakes and volcanoes. Louis Agassiz and others established the reality of continent-covering ice ages, and "fluvialists" like Andrew Crombie Ramsay argued that river valleys were formed, over millions of years by the rivers that flow through them. After the discovery of radioactivity, radiometric dating methods were developed, starting in the 20th century. Alfred Wegener's theory of "continental drift" was widely dismissed when he proposed it in the 1910s, but new data gathered in the 1950s and 1960s led to the theory of plate tectonics, which provided a plausible mechanism for it. Plate tectonics also provided a unified explanation for a wide range of seemingly unrelated geological phenomena. Since 1970 it has served as the unifying principle in geology.

Geologists' embrace of plate tectonics became part of a broadening of the field from a study of rocks into a study of the Earth as a planet. Other elements of this transformation include: geophysical studies of the interior of the Earth, the grouping of geology with meteorology and oceanography as one of the "earth sciences", and comparisons of Earth and the solar system's other rocky planets.

Astronomy

Main article: History of astronomy

Aristarchus of Samos published work on how to determine the sizes and distances of the Sun and the Moon, and Eratosthenes used this work to figure the size of the Earth. Hipparchus later discovered the precession of the Earth.

Advances in astronomy and in optical systems in the 19th century resulted in the first observation of an asteroid (1 Ceres) in 1801, and the discovery of Neptune in 1846.

George Gamow, Ralph Alpher, and Robert Herman had calculated that there should be evidence for a Big Bang in the background temperature of the universe.[105] In 1964, Arno Penzias and Robert Wilson[106] discovered a 3 Kelvin background hiss in their Bell Labs radiotelescope (the Holmdel Horn Antenna), which was evidence for this hypothesis, and formed the basis for a number of results that helped determine the age of the universe.

Supernova SN1987A was observed by astronomers on Earth both visually, and in a triumph for neutrino astronomy, by the solar neutrino detectors at Kamiokande. But the solar neutrino flux was a fraction of its theoretically expected value. This discrepancy forced a change in some values in the standard model for particle physics.

William Harvey published *De Motu Cordis* in 1628, which revealed his conclusions based on his extensive studies of vertebrate circulatory systems. He identified the central role of the heart, arteries, and veins in producing blood movement in a circuit, and failed to find any confirmation of Galen's pre-existing notions of heating and cooling functions.[107]

The British Royal Society had received a letter from Antonie van Leeuwenhoek and published it in 1673, bringing to light the scientist's observations of microscopic organisms with his custom crafted lenses.[108]

In 1847, Hungarian physician Ignác Fülöp Semmelweis dramatically reduced the occurrency of puerperal fever by simply requiring physicians to wash their hands before attending to women in childbirth. This discovery predated the germ theory of disease. However, Semmelweis' findings were not appreciated by his contemporaries and came into use only with discoveries by British surgeon Joseph Lister, who in 1865 proved the principles of antisepsis. Lister's work was based on the important findings by French biologist Louis Pasteur. Pasteur was able to link microorganisms with disease, revolutionizing medicine. He also devised one of the most important methods in preventive medicine, when in 1880 he produced a vaccine against rabies. Pasteur invented the process of pasteurization, to help prevent the spread of disease through milk and other foods.[109]

Perhaps the most prominent, controversial and far-reaching theory in all of science has been the theory of evolution by natural selection put forward by the British naturalist Charles Darwin in his book On the Origin of Species in 1859. Darwin proposed that the features of all living things, including humans, were shaped by natural processes over long periods of time. The theory of evolution in its current form affects almost all areas of biology.[110] Implications of evolution on fields outside of pure science have led to both opposition and support from different parts of society, and profoundly influenced the popular understanding

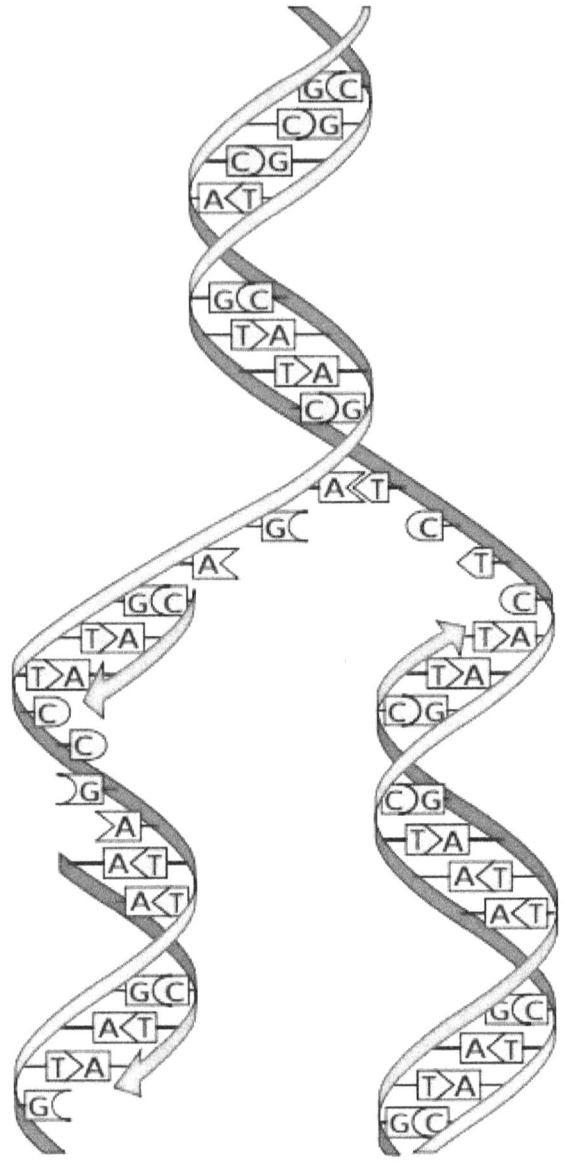

Semi-conservative DNA replication

Ecology

Main article: History of ecology

The discipline of ecology typically traces its origin to

Earthrise over the Moon, Apollo 8, NASA. This image helped create awareness of the finiteness of Earth, and the limits of its natural resources.

the synthesis of Darwinian evolution and Humboldtian biogeography, in the late 19th and early 20th centuries. Equally important in the rise of ecology, however, were microbiology and soil science—particularly the cycle of life concept, prominent in the work Louis Pasteur and Ferdinand Cohn. The word *ecology* was coined by Ernst Haeckel, whose particularly holistic view of nature in general (and Darwin's theory in particular) was important in the spread of ecological thinking. In the 1930s, Arthur Tansley and others began developing the field of ecosystem ecology, which combined experimental soil science with physiological concepts of energy and the techniques of field biology. The history of ecology in the 20th century is closely tied to that of environmentalism; the Gaia hypothesis, first formulated in the 1960s, and spreading in the 1970s, and more recently the scientific-religious movement of Deep Ecology have brought the two closer together.

14.4.2 Social sciences

Main article: History of the social sciences

Successful use of the scientific method in the physical sciences led to the same methodology being adapted to better understand the many fields of human endeavor. From this

of "man's place in the universe". In the early 20th century, the study of heredity became a major investigation after the rediscovery in 1900 of the laws of inheritance developed by the Moravian[111] monk Gregor Mendel in 1866. Mendel's laws provided the beginnings of the study of genetics, which became a major field of research for both scientific and industrial research. By 1953, James D. Watson, Francis Crick and Maurice Wilkins clarified the basic structure of DNA, the genetic material for expressing life in all its forms.[112] In the late 20th century, the possibilities of genetic engineering became practical for the first time, and a massive international effort began in 1990 to map out an entire human genome (the Human Genome Project).

effort the social sciences have been developed.

Political science

Main article: History of political science

Political science is a late arrival in terms of social sciences. However, the discipline has a clear set of antecedents such as moral philosophy, political philosophy, political economy, history, and other fields concerned with normative determinations of what ought to be and with deducing the characteristics and functions of the ideal form of government. The roots of politics are in prehistory. In each historic period and in almost every geographic area, we can find someone studying politics and increasing political understanding.

In Western culture, the study of politics is first found in Ancient Greece. The antecedents of European politics trace their roots back even earlier than Plato and Aristotle, particularly in the works of Homer, Hesiod, Thucydides, Xenophon, and Euripides. Later, Plato analyzed political systems, abstracted their analysis from more literary- and history- oriented studies and applied an approach we would understand as closer to philosophy. Similarly, Aristotle built upon Plato's analysis to include historical empirical evidence in his analysis.

An ancient Indian treatise on statecraft, economic policy and military strategy by Kautilya[113] and Vishnugupta,[114] who are traditionally identified with Chāṇakya (c. 350– –283 BCE). In this treatise, the behaviors and relationships of the people, the King, the State, the Government Superintendents, Courtiers, Enemies, Invaders, and Corporations are analysed and documented. Roger Boesche describes the *Arthaśāstra* as "a book of political realism, a book analysing how the political world does work and not very often stating how it ought to work, a book that frequently discloses to a king what calculating and sometimes brutal measures he must carry out to preserve the state and the common good."[115]

During the rule of Rome, famous historians such as Polybius, Livy and Plutarch documented the rise of the Roman Republic, and the organization and histories of other nations, while statesmen like Julius Caesar, Cicero and others provided us with examples of the politics of the republic and Rome's empire and wars. The study of politics during this age was oriented toward understanding history, understanding methods of governing, and describing the operation of governments.

With the fall of the Western Roman Empire, there arose a more diffuse arena for political studies. The rise of monotheism and, particularly for the Western tradition, Christianity, brought to light a new space for politics and political action. During the Middle Ages, the study of politics was widespread in the churches and courts. Works such as Augustine of Hippo's *The City of God* synthesized current philosophies and political traditions with those of Christianity, redefining the borders between what was religious and what was political. Most of the political questions surrounding the relationship between Church and State were clarified and contested in this period.

In the Middle East and later other Islamic areas, works such as the Rubaiyat of Omar Khayyam and Epic of Kings by Ferdowsi provided evidence of political analysis, while the Islamic Aristotelians such as Avicenna and later Maimonides and Averroes, continued Aristotle's tradition of analysis and empiricism, writing commentaries on Aristotle's works.

During the Italian Renaissance, Niccolò Machiavelli established the emphasis of modern political science on direct empirical observation of political institutions and actors. Later, the expansion of the scientific paradigm during the Enlightenment further pushed the study of politics beyond normative determinations. In particular, the study of statistics, to study the subjects of the state, has been applied to polling and voting.

In the 20th century, the study of ideology, behaviouralism and international relations led to a multitude of 'pol-sci' subdisciplines including rational choice theory, voting theory, game theory (also used in economics), psephology, political geography/geopolitics, political psychology/political sociology, political economy, policy analysis, public administration, comparative political analysis and peace studies/conflict analysis.

Linguistics

Main article: History of linguistics

Historical linguistics emerged as an independent field of study at the end of the 18th century. Sir William Jones proposed that Sanskrit, Persian, Greek, Latin, Gothic, and Celtic languages all shared a common base. After Jones, an effort to catalog all languages of the world was made throughout the 19th century and into the 20th century. Publication of Ferdinand de Saussure's *Cours de linguistique générale* created the development of descriptive linguistics. Descriptive linguistics, and the related structuralism movement caused linguistics to focus on how language changes over time, instead of just describing the differences between languages. Noam Chomsky further diversified linguistics with the development of generative linguistics in the 1950s. His effort is based upon a mathematical model of language that allows for the description and prediction of valid syntax. Additional specialties such as sociolinguistics, cognitive linguistics, and computational linguistics have emerged from collaboration between linguistics and other disciplines.

Economics

Main article: History of economics

The basis for classical economics forms Adam Smith's *An*

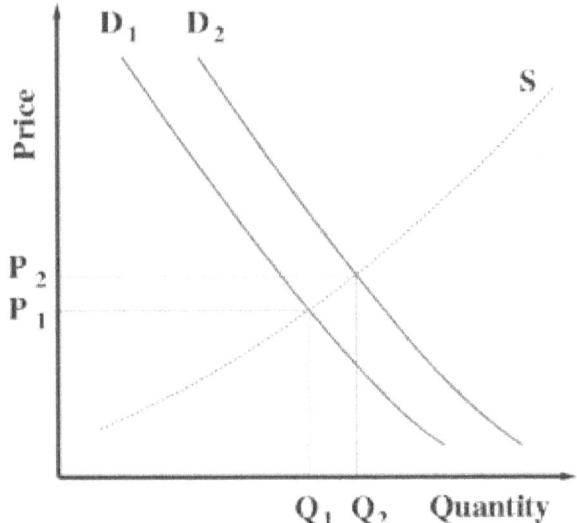

The supply and demand model

Adam Smith wrote The Wealth of Nations, *the first modern work of economics*

Inquiry into the Nature and Causes of the Wealth of Nations, published in 1776. Smith criticized mercantilism, advocating a system of free trade with division of labour. He postulated an "invisible hand" that regulated economic systems made up of actors guided only by self-interest. Karl Marx developed an alternative economic theory, called Marxian economics. Marxian economics is based on the labor theory of value and assumes the value of good to be based on the amount of labor required to produce it. Under this assumption, capitalism was based on employers not paying the full value of workers labor to create profit. The Austrian school responded to Marxian economics by viewing entrepreneurship as driving force of economic development. This replaced the labor theory of value by a system of supply and demand.

In the 1920s, John Maynard Keynes prompted a division between microeconomics and macroeconomics. Under Keynesian economics macroeconomic trends can overwhelm economic choices made by individuals. Governments should promote aggregate demand for goods as a means to encourage economic expansion. Following World War II, Milton Friedman created the concept of monetarism. Monetarism focuses on using the supply and demand of money as a method for controlling economic ac-

tivity. In the 1970s, monetarism has adapted into supply-side economics which advocates reducing taxes as a means to increase the amount of money available for economic expansion.

Other modern schools of economic thought are New Classical economics and New Keynesian economics. New Classical economics was developed in the 1970s, emphasizing solid microeconomics as the basis for macroeconomic growth. New Keynesian economics was created partially in response to New Classical economics, and deals with how inefficiencies in the market create a need for control by a central bank or government.

The above "history of economics" reflects modern economic textbooks and this means that the last stage of a science is represented as the culmination of its history (Kuhn, 1962). The "invisible hand" mentioned in a lost page in the middle of a chapter in the middle of the "Wealth of Nations", 1776, advances as Smith's central message. It is played down that this "invisible hand" acts only "frequently" and that it is "no part of his [the individual's] intentions" because competition leads to lower prices by imitating "his" invention. That this "invisible hand" prefers "the support of domestic to foreign industry" is cleansed—often without indication that part of the citation is truncated.[116] The opening passage of the "Wealth" containing Smith's message is never mentioned as it cannot be integrated into modern theory: "Wealth" depends on the division of labour which changes with market volume and on the proportion of productive to Unproductive labor.

Psychology

Main article: History of psychology

The end of the 19th century marks the start of psychology as a scientific enterprise. The year 1879 is commonly seen as the start of psychology as an independent field of study. In that year Wilhelm Wundt founded the first laboratory dedicated exclusively to psychological research (in Leipzig). Other important early contributors to the field include Hermann Ebbinghaus (a pioneer in memory studies), Ivan Pavlov (who discovered classical conditioning), William James, and Sigmund Freud. Freud's influence has been enormous, though more as cultural icon than a force in scientific psychology.

The 20th century saw a rejection of Freud's theories as being too unscientific, and a reaction against Edward Titchener's atomistic approach of the mind. This led to the formulation of behaviorism by John B. Watson, which was popularized by B.F. Skinner. Behaviorism proposed epistemologically limiting psychological study to overt behavior, since that could be reliably measured. Scientific knowledge of the "mind" was considered too metaphysical, hence impossible to achieve.

The final decades of the 20th century have seen the rise of a new interdisciplinary approach to studying human psychology, known collectively as cognitive science. Cognitive science again considers the mind as a subject for investigation, using the tools of psychology, linguistics, computer science, philosophy, and neurobiology. New methods of visualizing the activity of the brain, such as PET scans and CAT scans, began to exert their influence as well, leading some researchers to investigate the mind by investigating the brain, rather than cognition. These new forms of investigation assume that a wide understanding of the human mind is possible, and that such an understanding may be applied to other research domains, such as artificial intelligence.

Sociology

Main article: History of sociology

Ibn Khaldun can be regarded as the earliest scientific systematic sociologist.[117] The modern sociology, emerged in the early 19th century as the academic response to the modernization of the world. Among many early sociologists (e.g., Émile Durkheim), the aim of sociology was in structuralism, understanding the cohesion of social groups, and developing an "antidote" to social disintegration. Max Weber was concerned with the modernization of society through the concept of rationalization, which he believed would trap individuals in an "iron cage" of rational thought. Some sociologists, including Georg Simmel and W. E. B. Du Bois, utilized more microsociological, qualitative analyses. This microlevel approach played an important role in American sociology, with the theories of George Herbert Mead and his student Herbert Blumer resulting in the creation of the symbolic interactionism approach to sociology.

American sociology in the 1940s and 1950s was dominated largely by Talcott Parsons, who argued that aspects of society that promoted structural integration were therefore "functional". This structural functionalism approach was questioned in the 1960s, when sociologists came to see this approach as merely a justification for inequalities present in the status quo. In reaction, conflict theory was developed, which was based in part on the philosophies of Karl Marx. Conflict theorists saw society as an arena in which different groups compete for control over resources. Symbolic interactionism also came to be regarded as central to sociological thinking. Erving Goffman saw social interactions as a stage performance, with individuals preparing "backstage" and attempting to control their audience through impression management. While these theories are currently prominent in sociological thought, other approaches exist, including feminist theory, post-structuralism, rational choice theory, and postmodernism.

Anthropology

Main article: History of anthropology

Anthropology can best be understood as an outgrowth of the Age of Enlightenment. It was during this period that Europeans attempted systematically to study human behaviour. Traditions of jurisprudence, history, philology and sociology developed during this time and informed the development of the social sciences of which anthropology was a part.

At the same time, the romantic reaction to the Enlightenment produced thinkers such as Johann Gottfried Herder and later Wilhelm Dilthey whose work formed the basis for the culture concept which is central to the discipline. Traditionally, much of the history of the subject was based on colonial encounters between Western Europe and the rest of the world, and much of 18th- and 19th-century anthropology is now classed as forms of scientific racism.

During the late 19th-century, battles over the "study of man" took place between those of an "anthropological" persuasion (relying on anthropometrical techniques) and those of an "ethnological" persuasion (looking at cultures and traditions), and these distinctions became part of the later divide between physical anthropology and cultural anthropology, the latter ushered in by the students of Franz Boas.

In the mid-20th century, much of the methodologies of earlier anthropological and ethnographical study were reevaluated with an eye towards research ethics, while at the same time the scope of investigation has broadened far beyond the traditional study of "primitive cultures" (scientific practice itself is often an arena of anthropological study).

The emergence of paleoanthropology, a scientific discipline which draws on the methodologies of paleontology, physical anthropology and ethology, among other disciplines, and increasing in scope and momentum from the mid-20th century, continues to yield further insights into human origins, evolution, genetic and cultural heritage, and perspectives on the contemporary human predicament as well.

14.4.3 Emerging disciplines

During the 20th century, a number of interdisciplinary scientific fields have emerged. Examples include:

Communication studies combines animal communication, information theory, marketing, public relations, telecommunications and other forms of communication.

Computer science, built upon a foundation of theoretical linguistics, discrete mathematics, and electrical engineering, studies the nature and limits of computation. Subfields include computability, computational complexity, database design, computer networking, artificial intelligence, and the design of computer hardware. One area in which advances in computing have contributed to more general scientific development is by facilitating large-scale archiving of scientific data. Contemporary computer science typically distinguishes itself by emphasising mathematical 'theory' in contrast to the practical emphasis of software engineering.

Environmental science is an interdisciplinary field. It draws upon the disciplines of biology, chemistry, earth sciences, ecology, geography, mathematics, and physics.

Materials science has its roots in metallurgy, mineralogy, and crystallography. It combines chemistry, physics, and several engineering disciplines. The field studies metals, ceramics, glass, plastics, semiconductors, and composite materials.

14.5 Academic study

Main article: History of science and technology

As an academic field, history of science and technology began with the publication of William Whewell's *History of the Inductive Sciences* (first published in 1837). A more formal study of the history of science as an independent discipline was launched by George Sarton's publications, *Introduction to the History of Science* (1927) and the *Isis* journal (founded in 1912). Sarton exemplified the early 20th-century view of the history of science as the history of great men and great ideas. He shared with many of his contemporaries a Whiggish belief in history as a record of the advances and delays in the march of progress. The history of science was not a recognized subfield of American history in this period, and most of the work was carried out by interested scientists and physicians rather than professional historians.[118] With the work of I. Bernard Cohen at Harvard, the history of science became an established sub-discipline of history after 1945.[119]

The history of mathematics, history of technology, and history of philosophy are distinct areas of research and are covered in other articles. Mathematics is closely related to but distinct from natural science (at least in the modern conception). Technology is likewise closely related to but clearly differs from the search for empirical truth.

History of science is an academic discipline, with an international community of specialists. Main professional organizations for this field include the History of Science Society, the British Society for the History of Science, and the European Society for the History of Science.

14.5.1 Theories and sociology of the history of science

Main article: Theories and sociology of the history of science

Much of the study of the history of science has been devoted to answering questions about what science *is*, how it *functions*, and whether it exhibits large-scale patterns and trends.[120] The sociology of science in particular has focused on the ways in which scientists work, looking closely at the ways in which they "produce" and "construct" scientific knowledge. Since the 1960s, a common trend in science studies (the study of the sociology and history of science) has been to emphasize the "human component" of scientific knowledge, and to de-emphasize the view that scientific data are self-evident, value-free, and context-free.[121] The field of Science and Technology Studies, an area that overlaps and often informs historical studies of science, focuses on the social context of science in both contemporary and historical periods.

Humboldtian science refers to the early 19th century approach of combining scientific field work with the age of Romanticism sensitivity, ethics and aesthetic ideals.[122] It helped to install natural history as a separate field, gave base for ecology and was based on the role model of scientist, naturalist and explorer Alexander von Humboldt.[123] The later 19th century positivism asserted that all authentic knowledge allows verification and that all authentic knowledge assumes that the only valid knowledge is scientific.[124]

A major subject of concern and controversy in the philosophy of science has been the nature of *theory change* in science. Karl Popper argued that scientific knowledge is progressive and cumulative; Thomas Kuhn, that scientific knowledge moves through "paradigm shifts" and is not necessarily progressive; and Paul Feyerabend, that scientific knowledge is not cumulative or progressive and that there can be no demarcation in terms of method between science and any other form of investigation.[125]

The mid 20th century saw a series of studies relying to the role of science in a social context, starting from Thomas Kuhn's *The Structure of Scientific Revolutions* in 1962. It opened the study of science to new disciplines by suggesting that the evolution of science was in part sociologically determined and that positivism did not explain the actual interactions and strategies of the human participants in science. As Thomas Kuhn put it, the history of science may be seen in more nuanced terms, such as that of competing paradigms or conceptual systems in a wider matrix that includes intellectual, cultural, economic and political themes outside of science. "Partly by selection and partly by distortion, the scientists of earlier ages are implicitly presented as having worked upon the same set of fixed problems and in accordance with the same set of fixed canons that the most recent revolution in scientific theory and method made seem scientific."[126]

Further studies, e.g. Jerome Ravetz 1971 *Scientific Knowledge and its Social Problems* referred to the role of the scientific community, as a social construct, in accepting or rejecting (objective) scientific knowledge.[127] The Science wars of the 1990 were about the influence of especially French philosophers, which denied the objectivity of science in general or seemed to do so. They described as well differences between the idealized model of a pure science and the actual scientific practice; while scientism, a revival of the positivism approach, saw in precise measurement and rigorous calculation the basis for finally settling enduring metaphysical and moral controversies.[128][129] However, more recently some of the leading critical theorists have recognized that their postmodern deconstructions have at times been counter-productive, and are providing intellectual ammunition for reactionary interests. Bruno Latour noted that "dangerous extremists are using the very same argument of social construction to destroy hard-won evidence that could save our lives. Was I wrong to participate in the invention of this field known as science studies? Is it enough to say that we did not really mean what we meant?"[130]

14.5.2 The Plight of Many Scientific Innovators

See also: List of overlooked scientific innovators

One recurring observation in the history of science involves the struggle for recognition of first-rate scientists working on the periphery of the scientific establishment. For instance, the great physicist Lord Rayleigh looked back (cited here) on John James Waterston's seminal paper on the kinetic theory of gases. The history of the neglect of Waterston's path-breaking article, Rayleigh felt, suggests that "a young author who believes himself capable of great things would usually do well to secure favourable recognition of the scientific world . . . before embarking upon higher flights."

William Harvey's experiences led him to an even more pessimistic view:[131]

> "But what remains to be said about the quantity and source of the blood which thus passes, is of so novel and unheard-of character that I not only fear injury to myself from the envy of a few, but I tremble lest I have mankind at large for my enemies, so much doth wont and custom, that become as another nature, and doctrine once sown and that hath struck deep root, and respect for antiquity, influence all men."

In more general terms, Robert K. Merton[132] remarks that "the history of science abounds in instances[133] of basic papers having been written by comparatively unknown scientists, only to be rejected or neglected for years."

14.6 See also

- History
  - 2000s in science and technology
  - History of mathematics
  - History of physics
  - History of philosophy

- History of science and technology
- History of science and technology in China
- History of technology
- Science and technology in Canada
- Science and technology in India
- Women in science
- Timeline of science and technology in the Islamic world
- History of science policy
- History and Philosophy of Science
- List of discoveries
- List of famous experiments
- List of multiple discoveries
- List of Nobel laureates
- List of scientists
- List of years in science
- Multiple discovery
- Philosophy of history
- Science
  - Fields of science
    - Behavioural sciences
    - Natural sciences
      - Natural Sciences Tripos University of Cambridge, UK
    - Social sciences
  - History of technology
- History of scholarship
  - Philosophy of science
    - Imre Lakatos
    - Naïve empiricism
  - Science studies
- Theories and sociology of the history of science
- Timelines of science
  - Timeline of scientific discoveries
  - Timeline of scientific experiments
  - Timeline of scientific thought
  - Timeline of the history of scientific method

14.7 Notes and references

[1] "Whewell and the coining of 'scientist' in the Quarterly Review » Science Comma". *blogs.kent.ac.uk*. Retrieved 2016-10-19.

[2] Hendrix, Scott E. (2011). "Natural Philosophy or Science in Premodern Epistemic Regimes? The Case of the Astrology of Albert the Great and Galileo Galilei". *Teorie vědy / Theory of Science*. 33 (1): 111–132. Retrieved 20 February 2012.

[3] "For our purpose, science may be defined as ordered knowledge of natural phenomena and of the relations between them." William C. Dampier-Whetham, "Science", in *Encyclopædia Britannica*, 11th ed. (New York: 1911); "Science comprises, first, the orderly and systematic comprehension, description and/or explanation of natural phenomena and, secondly, the [mathematical and logical] tools necessary for the undertaking." Marshall Clagett, *Greek Science in Antiquity* (New York: Collier Books, 1955); "Science is a systematic explanation of perceived or imaginary phenomena, or else is based on such an explanation. Mathematics finds a place in science only as one of the symbolical languages in which scientific explanations may be expressed." David Pingree, "Hellenophilia versus the History of Science", *Isis* 83, 559 (1982); Pat Munday, entry "History of Science", *New Dictionary of the History of Ideas* (Charles Scribner's Sons, 2005).

[4] Golinski, Jan (2001). *Making Natural Knowledge: Constructivism and the History of Science* (reprint ed.). University of Chicago Press. p. 2. ISBN 9780226302324. When [history of science] began, during the eighteenth century, it was practiced by scientists (or "natural philosophers") with an interest in validating and defending their enterprise. They wrote histories in which ... the science of the day was exhibited as the outcome of the progressive accumulation of human knowledge, which was an integral part of moral and cultural development.

[5] Kuhn, T., 1962, "The Structure of Scientific Revolutions", University of Chicago Press, p. 137: "Partly by selection and partly by distortion, the scientists of earlier ages are implicitly presented as having worked upon the same set of fixed problems and in accordance with the same set of fixed canons that the most recent revolution in scientific theory and method made seem scientific."

[6] Matsuoka, Yoshihiro; Vigouroux, Yves; Goodman, Major M.; Sanchez G., Jesus; Buckler, Edward; Doebley, John (30 April 2002). "A single domestication for maize shown by multilocus microsatellite genotyping". *Proceedings of the National Academy of Sciences*. 99 (9): 6080–6084. Bibcode:2002PNAS...99.6080M. PMC 122905. PMID 11983901. doi:10.1073/pnas.052125199.

[7] Sean B. Carroll (24 May 2010)."Tracking the Ancestry of Corn Back 9,000 Years" *New York Times*.

[8] Francesca Bray (1984), *Science and Civilisation in China* **VI.2 Agriculture** pp 299, 453 writes that teosinte, 'the father of corn' helps the success and vitality of corn when planted between the rows of its 'children', maize.

[9] Hoskin, Michael (2001). *Tombs, Temples and their Orientations: a New Perspective on Mediterranean Prehistory*. Bognor Regis, UK: Ocarina Books. ISBN 0-9540867-1-6.

[10] Ruggles, Clive (1999). *Astronomy in Prehistoric Britain and Ireland*. New Haven: Yale University Press. ISBN 0-300-07814-5.

[11] See Homer's *Odyssey* 4.227–232 '[The Egyptians] are of the race of Paeeon [(physician to the gods)]'

[12] See, for example Joseph Needham (1974, 1976, 1980, 1983) and his co-authors, *Science and Civilisation in China*, **V**. Cambridge University Press, specifically:

- Joseph Needham and Lu Gwei-djen (1974), **V.2 Spagyrical Discovery and Invention: Magisteries of Gold and Immortality**
- Joseph Needham, Ho Ping-Yu [Ho Peng-Yoke], and Lu Gwei-djen (1976), **V.3 Spagyrical Discovery and Invention: Historical Survey, from Cinnabar Elixirs to Synthetic Insulin**
- Joseph Needham, Lu Gwei-djen, and Nathan Sivin (1980), **V.4 Spagyrical Discovery and Invention: Apparatus and Theory**
- Joseph Needham and Lu Gwei-djen (1983), **V.5 Spagyrical Discovery and Invention: Physiological Alchemy**

[13] Homer (May 1998). *The Odyssey*. Translated by Walter Shewring. Oxford University Press. p. 40. ISBN 0-19-283375-8. In Egypt, more than in other lands, the bounteous earth yields a wealth of drugs, healthful and baneful side by side; and every man there is a physician; the rest of the world has no such skill, for these are all of the family of Paeon.

[14] Microsoft Word - Proceedings-2001.doc Archived 7 April 2008 at the Wayback Machine.

[15] *Edwin Smith papyrus: Egyptian medical book*, Encyclopædia Britannica, retrieved 21 December 2016

[16] Lloyd, G. E. R. "The development of empirical research", in his *Magic, Reason and Experience: Studies in the Origin and Development of Greek Science*.

[17] Paul Hoffman, *The man who loved only numbers: the story of Paul Erdös and the search for mathematical truth*, (New York: Hyperion), 1998, p.187. ISBN 0-7868-6362-5

[18] A. Aaboe (2 May 1974). "Scientific Astronomy in Antiquity". *Philosophical Transactions of the Royal Society*. **276** (1257): 21–42. Bibcode:1974RSPTA.276...21A. JSTOR 74272. doi:10.1098/rsta.1974.0007.

[19] Sambursky 1974, pp. 3,37 called the pre-Socratics the transition from *mythos* to *logos*

[20] F. M. Cornford, *Principium Sapientiae: The Origins of Greek Philosophical Thought*, (Gloucester, Massachusetts, Peter Smith, 1971), p. 159.

[21] Arieti, James A. *Philosophy in the ancient world: an introduction*, p. 45 . Rowman & Littlefield, 2005. 386 pages. ISBN 978-0-7425-3329-5.

[22] Dicks, D.R. (1970). *Early Greek Astronomy to Aristotle*. Ithaca, N.Y.: Cornell University Press. pp. 72–198. ISBN 978-0-8014-0561-7.

[23] O'Leary, De Lacy (1949). *How Greek Science Passed to the Arabs*. London: Routledge & Kegan Paul Ltd. ISBN 0-7100-1903-3.

[24] G. E. R. Lloyd, *Early Greek Science: Thales to Aristotle*, (New York: W. W. Norton, 1970), pp. 144-6.

[25] Lloyd (1973), p. 177.

[26] *Greek Science*, many editions, such as the paperback by Penguin Books. Copyrights in 1944, 1949, 1953, 1961, 1963. The first quote above comes from Part 1, Chapter 1; the second, from Part 2, Chapter 4.

[27] Marchant, Jo (2006). "In search of lost time". *Nature*. **444**: 534–538. Bibcode:2006Natur.444..534M. PMID 17136067. doi:10.1038/444534a.

[28] Bill Casselman. "One of the Oldest Extant Diagrams from Euclid". University of British Columbia. Retrieved 26 September 2008.

[29] Boyer (1991). "Euclid of Alexandria". *A History of Mathematics*. p. 119. The *Elements* of Euclid not only was the earliest major Greek mathematical work to come down to us, but also the most influential textbook of all times. [...]The first printed versions of the *Elements* appeared at Venice in 1482, one of the very earliest of mathematical books to be set in type; it has been estimated that since then at least a thousand editions have been published. Perhaps no book other than the Bible can boast so many editions, and certainly no mathematical work has had an influence comparable with that of Euclid's *Elements*.

[30] Calinger, Ronald (1999). *A Contextual History of Mathematics*. Prentice-Hall. p. 150. ISBN 0-02-318285-7. Shortly after Euclid, compiler of the definitive textbook, came Archimedes of Syracuse (c. 287–212 BC.), the most original and profound mathematician of antiquity.

[31] O'Connor, J.J.; Robertson, E.F. (February 1996). "A history of calculus". University of St Andrews. Retrieved 7 August 2007.

[32] "3: Early Indian culture - Indus civilisation". *st-and.ac.uk*.

[33] Bisht, R. S. (1982). "Excavations at Banawali: 1974-77". In Possehl, Gregory L. *Harappan Civilization: A Contemporary Perspective*. New Delhi: Oxford and IBH Publishing Co. pp. 113–124.

[34] Pickover, Clifford (2008). *Archimedes to Hawking: laws of science and the great minds behind them*. Oxford University Press US. p. 105. ISBN 978-0-19-533611-5.

[35] Mainak Kumar Bose, *Late Classical India*. A. Mukherjee & Co., 1988, p. 277.

[36] Ifrah, Georges. 1999. *The Universal History of Numbers : From Prehistory to the Invention of the Computer*. Wiley. ISBN 0-471-37568-3.

[37] O'Connor, J.J. and E.F. Robertson. 2000. 'Indian Numerals', *MacTutor History of Mathematics Archive*, School of Mathematics and Statistics, University of St. Andrews, Scotland.

[38] George G. Joseph (1991). *The crest of the peacock*. London.

[39] Sarma (2008), *Astronomy in India*

[40] George G. Joseph (2000). *The Crest of the Peacock: Non-European Roots of Mathematics*, p. 408. Princeton University Press.

[41] Coppa, A.; et al. (6 April 2006). "Early Neolithic tradition of dentistry: Flint tips were surprisingly effective for drilling tooth enamel in a prehistoric population" (PDF). *Nature*. **440** (7085): 755–6. Bibcode:2006Natur.440..755C. PMID 16598247. doi:10.1038/440755a.

[42] Pullaiah (2006). *Biodiversity in India, Volume 4*. Daya Books. p. 83. ISBN 978-81-89233-20-4.

[43] C. S. Smith, A History of Metallography, University Press, Chicago (1960); Juleff 1996; Srinivasan, Sharda and Srinivasa Rangnathan 2004

[44] Srinivasan, Sharda and Srinivasa Rangnathan. 2004. *India's Legendary Wootz Steel*. Bangalore: Tata Steel.

[45] Needham, Joseph (1986). *Science and Civilization in China: Volume 3, Mathematics and the Sciences of the Heavens and the Earth*. Taipei: Caves Books Ltd. Page 208.

[46] Needham p422

[47] de Crespigny (2007), 1050; Morton & Lewis (2005), 70.

[48] Minford & Lau (2002), 307; Balchin (2003), 26–27; Needham (1986a), 627; Needham (1986c), 484; Krebs (2003), 31.

[49] Needham (1986a), 626.

[50] Shen Kuo 沈括 (1086, last supplement dated 1091), *Meng Ch'i Pi T'han* (夢溪筆談, *Dream Pool Essays*) as cited in Needham, Robinson & Huang 2004 p.244

[51] Agustín Udías, *Searching the Heavens and the Earth: The History of Jesuit Observatories*. (Dordrecht, The Netherlands: Kluwer Academic Publishers, 2003). p.53

[52] Needham & Wang 1954 581.

[53] Linda E. Voigts, "Anglo-Saxon Plant Remedies and the Anglo-Saxons", *Isis*, 70 (1979): 250-268; reprinted in Michael H. Shank, *The Scientific Enterprise in Antiquity and the Middle Ages*, Chicago: Univ. of Chicago Pr., 2000, pp. 163-181. ISBN 0-226-74951-7.

[54] Faith Wallis, *Bede: The Reckoning of Time*, Liverpool: Liverpool Univ. Pr., 2004, pp. xviii-xxxiv. ISBN 0-85323-693-3.

[55] Sameen Ahmed Khan, Arab Origins of the Discovery of the Refraction of Light; Roshdi Hifni Rashed (Picture) Awarded the 2007 King Faisal International Prize, Optics & Photonics News (OPN, Logo), Vol. 18, No. 10, pp. 22-23 (October 2007).

[56] Al-Khalili, Jim (4 January 2009). "BBC News". BBC News. Retrieved 11 April 2014.

[57] Rosen, Edward (1985). "The Dissolution of the Solid Celestial Spheres". *Journal of the History of Ideas*. **46** (1): 19–20 & 21. doi:10.2307/2709773.

[58] Rabin, Sheila (2004). "Nicolaus Copernicus". *Stanford Encyclopedia of Philosophy*. Metaphysics Research Lab, CSLI, Stanford University. Retrieved 24 June 2012.

[59] Saliba, George (1994). *A History of Arabic Astronomy: Planetary Theories During the Golden Age of Islam*. New York University Press. pp. 254 & 256–257. ISBN 0-8147-8023-7.

[60] Bartel, B. L. (1987). "The Heliocentric System in Greek, Persian and Hindu Astronomy". *Annals of the New York Academy of Sciences*. **500** (1): 525–545 [534–537]. Bibcode:1987NYASA.500..525V. doi:10.1111/j.1749-6632.1987.tb37224.x.

[61] Nasr, Seyyed H. (1993). "An Introduction to Islamic Cosmological Doctrines" (2nd ed.). State University of New York Press: 135–136. ISBN 0-7914-1516-3.

[62] Baker, A.; Chapter, L. (2002). "Part 4: The Sciences". Missing or empty |title= (help), in Sharif, M. M. "A History of Muslim Philosophy". *Philosophia Islamica*.

[63] Will Durant (1980). *The Age of Faith (The Story of Civilization, Volume 4)*, pp. 162-186. Simon & Schuster. ISBN 0-671-01200-2.

[64] Fielding H. Garrison, *An Introduction to the History of Medicine with Medical Chronology, Suggestions for Study and Bibliographic Data*, p. 86

[65] Derewenda, Zygmunt S.; Derewenda, ZS (2007). "On wine, chirality and crystallography". *Acta Crystallographica Section A*. **64** (Pt 1): 246–258 [247]. Bibcode:2008AcCrA..64..246D. PMID 18156689. doi:10.1107/S0108767307054293.

[66] Warren, John (2005). "War and the Cultural Heritage of Iraq: a sadly mismanaged affair". *Third World Quarterly*. **26** (4–5): 815–830. doi:10.1080/01436590500128048.

14.7. NOTES AND REFERENCES

[67] Lindberg, David C. (1967). "Alhazen's Theory of Vision and Its Reception in the West". *Isis.* **58** (3): 321–341. PMID 4867472. doi:10.1086/350266.

[68] Faruqi, Yasmeen M. (2006). "Contributions of Islamic scholars to the scientific enterprise". *International Education Journal.* **7** (4): 391–396.

[69] Nasr, Seyyed Hossein (2007). "Avicenna". *Encyclopædia Britannica Online.* Retrieved 3 June 2010.

[70] Jacquart, Danielle (2008). "Islamic Pharmacology in the Middle Ages: Theories and Substances". European Review (Cambridge University Press) 16: 219–27.

[71] David W. Tschanz, MSPH, PhD (August 2003). "Arab Roots of European Medicine", Heart Views 4 (2).

[72] Brater, D. Craig; Daly, Walter J. (2000). "Clinical pharmacology in the Middle Ages: Principles that presage the 21st century". *Clinical Pharmacology & Therapeutics.* **67** (5): 447–450 [448]. PMID 10824622. doi:10.1067/mcp.2000.106465.

[73] Martin-Araguz, A.; Bustamante-Martinez, C.; Fernández-Armayor Ajo, V.; Moreno-Martínez, J. M. (2002). "Neuroscience in al-Andalus and its influence on medieval scholastic medicine". *Revista de neurología.* **34** (9): 877–892. PMID 12134355.

[74] Zafarul-Islam Khan, At The Threshold Of A New Millennium – II, *The Milli Gazette.*

[75] Ahmed, Akbar S. (1984). "Al-Beruni: The First Anthropologist". *RAIN.* **60** (60): 9–10. doi:10.2307/3033407.

[76] Ahmed, Akbar (2002). "Ibn Khaldun's Understanding of Civilizations and the Dilemmas of Islam and the West Today". *Middle East Journal.* **56** (1): 25.

[77] H. Mowlana (2001). "Information in the Arab World", *Co-operation South Journal* **1**.

[78] Abdalla, Mohamad (2007). "Ibn Khaldun on the Fate of Islamic Science after the 11th Century". *Islam & Science.* **5** (1): 61–70.

[79] Salahuddin Ahmed (1999). *A Dictionary of Muslim Names.* C. Hurst & Co. Publishers. ISBN 1-85065-356-9.

[80] Dr. Akhtar, S. W. (1997). "The Islamic Concept of Knowledge". *Al-Tawhid: A Quarterly Journal of Islamic Thought & Culture.* **12**: 3.

[81] Erica Fraser. The Islamic World to 1600. University of Calgary.

[82] Toby Huff, *Rise of early modern science* 2nd ed. p. 180-181

[83] Edward Grant, "Science in the Medieval University", in James M. Kittleson and Pamela J. Transue, ed., *Rebirth, Reform and Resilience: Universities in Transition, 1300-1700,* Columbus: Ohio State University Press, 1984, p. 68

[84] William of Malmesbury, *Gesta Regum Anglorum / The history of the English kings,* ed. and trans. R. A. B. Mynors, R. M. Thomson, and M. Winterbottom, 2 vols., Oxford Medieval Texts (1998–99)

[85] R. W. Vernon, G. McDonnell and A. Schmidt, 'An integrated geophysical and analytical appraisal of early iron-working: three case studies' Historical Metallurgy 31(2) (1998), 72-5 79.

[86] David Derbyshire, Henry "Stamped Out Industrial Revolution", The Daily Telegraph (21 June 2002)

[87] Hans Thijssen (30 January 2003). "Condemnation of 1277". *Stanford Encyclopedia of Philosophy.* University of Stanford. Retrieved 14 September 2009.

[88] "Rediscovering the Science of the Middle Ages". *BioLogos.org.*

[89] "023-A03: The Middle Ages and the Birth of Science – International Catholic University". *International Catholic University.*

[90] "History: A medieval multiverse". *Nature News & Comment.*

[91] http://www.rae.org/pdf/jaki.pdf

[92] Edward Grant, *The Foundations of Modern Science in the Middle Ages: Their Religious, Institutional, and Intellectual Contexts,* (Cambridge: Cambridge Univ. Pr., 1996), pp. 127-31.

[93] Edward Grant, *A Source Book in Medieval Science,* (Cambridge: Harvard Univ. Pr., 1974), p. 232

[94] David C. Lindberg, *Theories of Vision from al-Kindi to Kepler,* (Chicago: Univ. of Chicago Pr., 1976), pp. 140-2.

[95] Edward Grant, *The Foundations of Modern Science in the Middle Ages: Their Religious, Institutional, and Intellectual Contexts,* (Cambridge: Cambridge Univ. Pr., 1996), pp. 95-7.

[96] Edward Grant, *The Foundations of Modern Science in the Middle Ages: Their Religious, Institutional, and Intellectual Contexts,* (Cambridge: Cambridge Univ. Pr., 1996), pp. 100-3.

[97] Weidhorn, Manfred (2005). *The Person of the Millennium: The Unique Impact of Galileo on World History.* iUniverse. p. 155. ISBN 0-595-36877-8.

[98] Allen Debus, *Man and Nature in the Renaissance,* (Cambridge: Cambridge Univ. Pr., 1978).

[99] Precise titles of these landmark books can be found in the collections of the Library of Congress. A list of these titles can be found in Bruno 1989

[100] Heilbron 2003, 741

[101] See, for example, pp 741-744 of Heilbron 2003

[102] Heilbron 2003, 741-743

[103] Matthew Daniel Eddy; Seymour Mauskopf; William R. Newman, eds. (2014). *Chemical Knowledge in the Early Modern World*. Chicago: University of Chicago Press. pp. 1–15.

[104] Matthew Daniel Eddy (2008). *The Language of Mineralogy: John Walker, Chemistry and the Edinburgh Medical School 1750-1800*. Ashgate.

[105] Alpher, Ralph A.; Herman, Robert (1948). "Evolution of the Universe". *Nature*. **162** (4124): 774–775. Bibcode:1948Natur.162..774A. doi:10.1038/162774b0. Gamow, G. (1948). "The Evolution of the Universe". *Nature*. **162** (4122): 680–682. Bibcode:1948Natur.162..680G. PMID 18893719. doi:10.1038/162680a0.

[106] "Wilson's 1978 Nobel lecture" (PDF). *nobelprize.org*.

[107] Power, d'Arcey. Life of Harvey. Longmans, Green, & co.

[108] Dobell, Clifford (1923). "A Protozoological Bicentenary: Antony van Leeuwenhoek (1632–1723) and Louis Joblot (1645–1723)". Parasitology. 15: 308–19.

[109] Campbell, Neil A.; Brad Williamson; Robin J. Heyden (2006). *Biology: Exploring Life*. Boston, Massachusetts: Pearson Prentice Hall. ISBN 0-13-250882-6. OCLC 75299209.

[110] Dobzhansky, Theodosius (1964). "Biology, Molecular and Organismic" (PDF). *American Zoologist*. **4**: 443–452. doi:10.1093/icb/4.4.443.

[111] Henig, Robin Marantz (2000). *The Monk in the Garden : The Lost and Found Genius of Gregor Mendel, the Father of Genetics*. Houghton Mifflin. ISBN 0-395-97765-7. OCLC 43648512. The article, written by an obscure Moravian monk named Gregor Mendel...

[112] James D. Watson and Francis H. Crick. "Letters to *Nature*: Molecular structure of Nucleic Acid." *Nature* **171**, 737–738 (1953).

[113] Mabbett, I. W. (1 April 1964). "The Date of the Arthaśāstra". *Journal of the American Oriental Society*. **84** (2): 162–169. JSTOR 597102. doi:10.2307/597102. Trautmann, Thomas R. (1971). *Kauṭilya and the Arthaśāstra: A Statistical Investigation of the Authorship and Evolution of the Text*. Leiden: E.J. Brill. p. 10. while in his character as author of an *arthaśāstra* he is generally referred to by his *gotra* name, Kauṭilya.

[114] Mabbett 1964
Trautmann 1971:5 "the very last verse of the work...is the unique instance of the personal name Viṣṇugupta rather than the *gotra* name Kauṭilya in the *Arthaśāstra*.

[115] Boesche, Roger (2002). *The First Great Political Realist: Kautilya and His Arthashastra*. Lanham: Lexington Books. p. 17. ISBN 0-7391-0401-2.

[116] Compare Smith's original phrase with Samuelson's quotation of it. In brackets what Samuelson curtailed without indication and without giving a reference: "[As] every individual ... [therefore, endeavours as much as he can, both to employ his capital in the support of domestic industry, and so to direct that industry that its produce maybe of the greatest value; every individual necessarily labours to render the annual revenue of the society as great as he can. He generally, indeed,] neither intends to promote the general [Smith said "public"] interest, nor knows how much he is promoting it. [By preferring the support of domestic to that of foreign industry,] he intends only his own security, [and by directing that industry in such a manner as its produce may be of the greatest value, he intends only] his own gain; and he is in this, [as in many other cases,] led by an invisible hand to promote an end which was no part of his intention. [Nor is it always the worse for the society that it was no part of it.] By pursuing his own interest, he frequently promotes that of the society more effectually than when he really intends to promote it" Samuelson, Paul A./Nordhaus, William D., 1989, Economics, 13th edition, N.Y. et al.: McGraw-Hill, page 825; Smith, Adam, 1937, The Wealth of Nations, N.Y.: Random House, page 423

[117] Muhammed Abdullah Enan, *Ibn Khaldun: His Life and Works*, The Other Press, 2007, pp. 104–105. ISBN 983-9541-53-6.

[118] Reingold, Nathan (1986). "History of Science Today, 1. Uniformity as Hidden Diversity: History of Science in the United States, 1920-1940". *British Journal for the History of Science*. **19** (3): 243–262. doi:10.1017/S0007087400023268.

[119] Dauben JW, Gleason ML, Smith GE (2009). "Seven Decades of History of Science". *ISIS: Journal of the History of Science in Society*. **100** (1): 4–35. PMID 19554868. doi:10.1086/597575.

[120] *What is this thing called science?*. Hackett Pub. 1999. ISBN 978-0-87220-452-2.

[121] King Merton, Robert (1979). *The Sociology of Science: Theoretical and Empirical Investigations*. University of Chicago Press. ISBN 978-0-226-52092-6.

[122] Böhme, Hartmut: Ästhetische Wissenschaft, in: Matices. Nr. 23, 1999, S. 37-41

[123] Jardine et al., *Cultures of Natural History*, p. 304

[124] Jorge Larrain (1979) *The Concept of Ideology* p.197, quotation:

> one of the features of positivism is precisely its postulate that scientific knowledge is the paradigm of valid knowledge, a postulate that indeed is never proved nor intended to be proved.

[125] Matthews, Michael Robert (1994). *Science Teaching: The Role of History and Philosophy of Science*. Routledge. ISBN 978-0-415-90899-3.

[126] Kuhn, T., 1962, "The Structure of Scientific Revolutions", University of Chicago Press, p. 137

[127] Ravetz, Jerome R. (1979). *Scientific knowledge and its social problems*. Oxford: Oxford Univ. Press. ISBN 0-19-519721-6.

[128] Lears, T.J. Jackson. "Get Happy!!". The Nation. Retrieved 21 December 2013. ...scientism is a revival of the nineteenth-century positivist faith that a reified "science" has discovered (or is about to discover) all the important truths about human life. Precise measurement and rigorous calculation, in this view, are the basis for finally settling enduring metaphysical and moral controversies—explaining consciousness and choice, replacing ambiguity with certainty.

[129] Sorell, Thomas 'Tom' (1994). *Scientism: Philosophy and the Infatuation with Science*, Routledge, pp. 1ff.

[130] Latour, B (2004). "Why Has Critique Run Out of Steam? From Matters of Fact to Matters of Concern" (PDF). *Critical Inquiry*. 30: 225–48. doi:10.1086/421123.

[131] Moran, Gordon (1998). *Silencing Scientists and Scholars in Other Fields: Power, Paradigm Controls, Peer Review, and Scholarly Communication*. Santa Barbara, California: Ablex. pp. (cited on page) 38. ISBN 978-1567503432.

[132] Merton, Robert K. (1973). *The Sociology of Science*. Chicago: University of Chicago Press. pp. 456–457.

[133] Nissani, Moti (1995). "The Plight of the Obscure Innovator in Science: A Few Reflections on Campanario's Note". *Social Studies of Science*. 25: 165–183. doi:10.1177/030631295025001008.

14.8 Further reading

- Agar, Jon (2012) *Science in the Twentieth Century and Beyond* (Polity Press, Cambridge, 2012. ISBN 978-0-7456-3469-2.)

- Agassi, Joseph (2007) *Science and Its History: A Reassessment of the Historiography of Science* (Boston Studies in the Philosophy of Science, 253) Springer. ISBN 1-4020-5631-1, 2008.

- Boorstin, Daniel (1983). *The Discoverers : A History of Man's Search to Know His World and Himself*. New York: Random House. ISBN 0-394-40229-4. OCLC 9645583.

- Bowler, Peter J. *The Norton History of the Environmental Sciences* (1993)

- Brock, W. H. *The Norton History of Chemistry* (1993)

- Bronowski, J. *The Common Sense of Science* (Heinemann Educational Books Ltd., London, 1951. ISBN 84-297-1380-8.) (Includes a description of the history of science in England.)

- Bruno, Leonard C. (1989). *The Landmarks of Science*. ISBN 0-8160-2137-6

- Byers, Nina and Gary Williams, ed. (2006) *Out of the Shadows: Contributions of Twentieth-Century Women to Physics*. Cambridge University Press ISBN 978-0-521-82197-1

- Heilbron, John L., ed. (2003). *The Oxford Companion to the History of Modern Science*. New York: Oxford University Press. ISBN 0-19-511229-6.

- Herzenberg, Caroline L. 1986. *Women Scientists from Antiquity to the Present* Locust Hill Press ISBN 0-933951-01-9

- Kuhn, Thomas S. (1996). *The Structure of Scientific Revolutions*. University of Chicago Press. ISBN 0-226-45807-5. (3rd ed.)

- Kumar, Deepak (2006). *Science and the Raj: A Study of British India*. 2nd edition. Oxford University Press. ISBN 0-19-568003-0

- Lakatos, Imre *History of Science and its Rational Reconstructions* published in *The Methodology of Scientific Research Programmes: Philosophical Papers Volume 1*. Cambridge: Cambridge University Press 1978

- Ilizarov Simon S., Sobisevich Alexey V. New feature: History of science now. The Russian Federation // CENTAURUS. — 2015. — Vol. 57, no. 4. — P. 301–306.

- Levere, Trevor Harvey. *Transforming Matter: A History of Chemistry from Alchemy to the Buckyball* (2001)

- Lindberg, David C.; Shank, Michael H., eds. (2013). *The Cambridge History of Science*. 2, Medieval Science. Cambridge University Press. ISBN 978-0-521-59448-6.

- Margolis, Howard (2002). *It Started with Copernicus*. New York: McGraw-Hill. ISBN 0-07-138507-X

- Mayr, Ernst. *The Growth of Biological Thought: Diversity, Evolution, and Inheritance* (1985)

- Needham, Joseph. *Science and Civilisation in China*. Multiple volumes (1954–2004).

  - Needham, Joseph; Wang, Ling (??) (1954). "Science and Civilisation in China". 1 *Introductory Orientations*. Cambridge University Press.

- Needham, Joseph; Robinson, Kenneth G.; Huang, Jen-Yü (2004). "Science and Civilisation in China". 7, part II *General Conclusions and Reflections*. Cambridge University Press.

- North, John. *The Norton History of Astronomy and Cosmology* (1995)

- Nye, Mary Jo, ed. *The Cambridge History of Science, Volume 5: The Modern Physical and Mathematical Sciences* (2002)

- Park, Katharine, and Lorraine Daston, eds. *The Cambridge History of Science, Volume 3: Early Modern Science* (2006)

- Porter, Roy, ed. *The Cambridge History of Science, Volume 4: The Eighteenth Century* (2003)

- Rousseau, George and Roy Porter, eds., *The Ferment of Knowledge: Studies in the Historiography of Science* (Cambridge: Cambridge University Press, 1980). ISBN 0-521-22599-X

- Sambursky, Shmuel (1974). *Physical Thought from the Presocratics to the Quantum Physicists: an anthology selected, introduced and edited by Shmuel Sambursky*. New York: Pica Press. p. 584. ISBN 0-87663-712-8.

- Slotten, Hugh Richard, ed., *The Oxford Encyclopedia of the History of American Science, Medicine, and Technology* (2014), 1456 pp

14.9 External links

- International Academy of the History of Science

- Division of History of Science and Technology of the International Union of History and Philosophy of Science

- A History of Science, Vols 1–4, online text

- History of Science Society ("HSS")

- IsisCB Explore: History of Science Index An open access discovery tool

- (in French) The CNRS History of Science and Technology Research Center in Paris (France)

- The official site of the Nobel Foundation. Features biographies and info on Nobel laureates

- Museo Galileo - Institute and Museum of the History of Science in Florence, Italy

- The Royal Society, trailblazing science from 1650 to date

- The Vega Science Trust Free to view videos of scientists including Feynman, Perutz, Rotblat, Born and many Nobel Laureates.

- National Center for Atmospheric Research (NCAR) Archives

- History of Science Digital Collection: Utah State University - Contains primary sources by such major figures in the history of scientific inquiry as Otto Brunfels, Charles Darwin, Erasmus Darwin, Carolus Linnaeus Antony van Leeuwenhoek, Jan Swammerdam, James Sowerby, Andreas Vesalius, and others.

- Inter-Divisional Teaching Commission (IDTC) of the International Union for the History and Philosophy of Science (IUHPS)

- International History, Philosophy and Science Teaching Group

- Digital facsimiles of books from the History of Science Collection, Linda Hall Library Digital Collections

- ""Scientific Change"". *Internet Encyclopedia of Philosophy*.

Chapter 15

Outline of science

The following outline is provided as a topical overview of science:

Science – systematic effort of acquiring knowledge—through observation and experimentation coupled with logic and reasoning to find out what can be proved or not proved—and the knowledge thus acquired. The word "science" comes from the Latin word "scientia" meaning knowledge. A practitioner of science is called a "scientist". Modern science respects objective logical reasoning, and follows a set of core procedures or rules in order to determine the nature and underlying natural laws of the universe and everything in it. Some scientists do not know of the rules themselves, but follow them through research policies. These procedures are known as the scientific method.

15.1 Essence of science

- Research – systematic investigation into existing or new knowledge.

- Scientific discovery – observation of new phenomena, new actions, or new events and providing new reasoning to explain the knowledge gathered through such observations with previously acquired knowledge from abstract thought and everyday experiences.

- Laboratory – facility that provides controlled conditions in which scientific research, experiments, and measurement may be performed.

- Objectivity – the idea that scientists, in attempting to uncover truths about the natural world, must aspire to eliminate personal or cognitive biases, a priori commitments, emotional involvement, etc.

- Inquiry – any process that has the aim of augmenting knowledge, resolving doubt, or solving a problem.

15.2 Scientific method

Outline of scientific method Scientific method – body of techniques for investigating phenomena and acquiring new knowledge, as well as for correcting and integrating previous knowledge. It is based on observable, empirical, measurable evidence, and subject to laws of reasoning, both deductive and inductive.

- Empirical method –

- Experimental method – The steps involved in order to produce a reliable and logical conclusion include:

    1. Asking a question about a natural phenomenon
    2. Making observations of the phenomenon
    3. Forming a hypothesis – proposed explanation for a phenomenon. For a hypothesis to be a scientific hypothesis, the scientific method requires that one can test it. Scientists generally base scientific hypotheses on previous observations that cannot satisfactorily be explained with the available scientific theories.
    4. Predicting a logical consequence of the hypothesis
    5. Testing the hypothesis through an experiment – methodical procedure carried out with the goal of verifying, falsifying, or establishing the validity of a hypothesis. The 3 types of scientific experiments are:
        - Controlled experiment – experiment that compares the results obtained from an experimental sample against a control sample, which is practically identical to the experimental sample except for the one aspect the effect of which is being tested (the independent variable).
        - Natural experiment – empirical study in which the experimental conditions (i.e., which units receive which treatment) are

determined by nature or by other factors out of the control of the experimenters and yet the treatment assignment process is arguably exogenous. Thus, natural experiments are observational studies and are not controlled in the traditional sense of a randomized experiment.

- Observational study – draws inferences about the possible effect of a treatment on subjects, where the assignment of subjects into a treated group versus a control group is outside the control of the investigator.
- Field experiment – applies the scientific method to experimentally examine an intervention in the real world (or as many experimentalists like to say, naturally occurring environments) rather than in the laboratory. See also field research.

6. Gather and analyze data from experiments or observations, including indicators of uncertainty.

7. Draw conclusions by comparing data with predictions. Possible outcomes:
    - Conclusive:
        - The hypothesis is falsified by the data.
        - Data are consistent with the hypothesis.
        - Data are consistent with alternative hypotheses.
    - Inconclusive:
        - Data are not relevant to the hypothesis, or data and predictions are incommensurate.
        - There is too much uncertainty in the data to draw any conclusion.

- Deductive-nomological model
- Scientific modelling –
- Models of scientific method
    - Hypothetico-deductive model – proposed description of scientific method. According to it, scientific inquiry proceeds by formulating a hypothesis in a form that could conceivably be falsified by a test on observable data. A test that could and does run contrary to predictions of the hypothesis is taken as a falsification of the hypothesis. A test that could but does not run contrary to the hypothesis corroborates the theory.

15.3 Branches of science

See also: Index of branches of science

Branches of science – divisions within science with respect to the entity or system concerned, which typically embodies its own terminology and nomenclature.

15.3.1 Natural science

Outline of natural science

See also: Outline of science § Social science

Natural science – major branch of science, that tries to explain and predict nature's phenomena, based on empirical evidence. In natural science, hypotheses must be verified scientifically to be regarded as scientific theory. Validity, accuracy, and social mechanisms ensuring quality control, such as peer review and repeatability of findings, are amongst the criteria and methods used for this purpose. Natural science can be broken into two main branches: biology, and physical science. Each of these branches, and all of their sub-branches, are referred to as natural sciences.

- Branches of natural science (also known as the natural sciences)

15.3.2 Formal science

Formal science – branches of knowledge that are concerned with formal systems, such as those under the branches of: logic, mathematics, computer science, statistics, and some aspects of linguistics. Unlike other sciences, the formal sciences are not concerned with the validity of theories based on observations in the real world, but instead with the properties of formal systems based on definitions and rules.

- Computer science – study of the theoretical foundations of information and computation and their implementation and application in computer systems. *(See also Branches of Computer Science and ACM Computing Classification System)*
    - Theory of computation – branch that deals with whether and how efficiently problems can be solved on a model of computation, using an algorithm
        - Automata theory – study of mathematical objects called abstract machines or automata and the computational problems that can be solved using them.

- Formal languages – set of strings of symbols.
- Computability theory – branch of mathematical logic and computer science that originated in the 1930s with the study of computable functions and Turing degrees.
- Computational complexity theory – branch of the theory of computation in theoretical computer science and mathematics that focuses on classifying computational problems according to their inherent difficulty, and relating those classes to each other
- Concurrency theory – In computer science, concurrency is a property of systems in which several computations are executing simultaneously, and potentially interacting with each other
- Algorithms – step-by-step procedure for calculations
  - Randomized algorithms – algorithm which employs a degree of randomness as part of its logic.
  - Distributed algorithms – algorithm designed to run on computer hardware constructed from interconnected processors
  - Parallel algorithms – algorithm which can be executed a piece at a time on many different processing devices, and then put back together again at the end to get the correct result.
- Data structures – particular way of storing and organizing data in a computer so that it can be used efficiently.
- Computer architecture – In computer science and engineering, computer architecture is the practical art of selecting and interconnecting hardware components to create computers that meet functional, performance and cost goals and the formal modeling of those systems.
  - VLSI design – process of creating integrated circuits by combining thousands of transistors into a single chip
- Operating systems – set of software that manages computer hardware resources and provides common services for computer programs
- Computer communications (networks) – collection of hardware components and computers interconnected by communication channels that allow sharing of resources and information
  - Information theory – branch of applied mathematics and electrical engineering involving the quantification of information
  - Internet – global system of interconnected computer networks that use the standard Internet protocol suite (often called TCP/IP, although not all applications use TCP) to serve billions of users worldwide.
    - World wide web – part of the Internet; system of interlinked hypertext documents accessed via the Internet.
  - Wireless computing – any type of computer network that is not connected by cables of any kind.
    - Mobile computing – form of human–computer interaction by which a computer is expected to be transported during normal usage.
- Computer security – branch of computer technology known as information security as applied to computers and networks.
  - reliability – system design approach and associated service implementation that ensures a prearranged level of operational performance will be met during a contractual measurement period.
  - Cryptography – practice and study of hiding information.
  - Fault-tolerant computing – property that enables a system (often computer-based) to continue operating properly in the event of the failure of (or one or more faults within) some of its components
- Distributed computing – field of computer science that studies distributed systems
  - Grid computing – federation of computer resources from multiple administrative domains to reach a common goal
- Parallel computing – form of computation in which many calculations are carried out simultaneously, operating on the principle that large problems can often be divided into smaller ones, which are then solved concurrently ("in parallel").
  - High-performance computing – computer at the frontline of current processing capacity, particularly speed of calculation
- Quantum computing – device for computation that makes direct use of quantum mechanical phenomena, such as superposition and entanglement, to perform operations on data
- Computer graphics – graphics created using computers and, more generally, the representation and manipulation of image data by a com-

puter with help from specialized software and hardware.

- Image processing – any form of signal processing for which the input is an image, such as a photograph or video frame; the output of image processing may be either an image or a set of characteristics or parameters related to the image
- Scientific visualization – interdisciplinary branch of science according to Friendly (2008) "primarily concerned with the visualization of three-dimensional phenomena (architectural, meteorological, medical, biological, etc.), where the emphasis is on realistic renderings of volumes, surfaces, illumination sources, and so forth, perhaps with a dynamic (time) component".
- Computational geometry – branch of computer science devoted to the study of algorithms which can be stated in terms of geometry

- Software engineering – application of a systematic, disciplined, quantifiable approach to the development, operation, and maintenance of software; that is the application of engineering to software
    - Formal methods – particular kind of mathematically based techniques for the specification, development and verification of software and hardware systems
        - Formal verification – act of proving or disproving the correctness of intended algorithms underlying a system with respect to a certain formal specification or property, using formal methods of mathematics
- Programming languages – artificial language designed to communicate instructions to a machine, particularly a computer
    - Programming paradigms – fundamental style of computer programming
        - Object-oriented programming – programming paradigm using "objects" – data structures consisting of data fields and methods together with their interactions – to design applications and computer programs
        - Functional programming – programming paradigm that treats computation as the evaluation of mathematical functions and avoids state and mutable data

- Program semantics – field concerned with the rigorous mathematical study of the meaning of programming languages
- Type theory – any of several formal systems that can serve as alternatives to naive set theory, or the study of such formalisms in general
- Compilers – computer program (or set of programs) that transforms source code written in a programming language (the source language) into another computer language (the target language, often having a binary form known as object code)
- Concurrent programming languages – form of computing in which programs are designed as collections of interacting computational processes that may be executed in parallel

- Information science – interdisciplinary field primarily concerned with the analysis, collection, classification, manipulation, storage, retrieval and dissemination of information
    - Database – organized collection of data, today typically in digital form
        - Relational database – collection of data items organized as a set of formally described tables from which data can be accessed easily
        - Distributed database – database in which storage devices are not all attached to a common CPU.
        - Object database – database management system in which information is represented in the form of objects as used in object-oriented programming
    - Multimedia – media and content that uses a combination of different content forms.
    - hypermedia – computer-based information retrieval system that enables a user to gain or provide access to texts, audio and video recordings, photographs and computer graphics related to a particular subject.
    - Data mining – process that results in the discovery of new patterns in large data sets
    - Information retrieval – area of study concerned with searching for documents, for information within documents, and for metadata about documents, as well as that of searching structured storage, relational databases, and the World Wide Web.

15.3. BRANCHES OF SCIENCE

- Artificial intelligence – branch of computer science that deals with intelligent behavior, learning, and adaptation in machines.
    - Automated reasoning – area of computer science and mathematical logic dedicated to understand different aspects of reasoning.
    - Computer vision – field that includes methods for acquiring, processing, analysing, and understanding images and, in general, high-dimensional data from the real world in order to produce numerical or symbolic information, e.g., in the forms of decisions.
    - Machine learning – scientific discipline concerned with the design and development of algorithms that allow computers to evolve behaviors based on empirical data, such as from sensor data or databases
        - Artificial neural network – mathematical model or computational model that is inspired by the structure and/or functional aspects of biological neural networks
    - Natural language processing – field of computer science, artificial intelligence (also called machine learning), and linguistics concerned with the interactions between computers and human (natural) languages.
        - Computational linguistics – interdisciplinary field dealing with the statistical or rule-based modeling of natural language from a computational perspective.
    - Expert systems – computer system that emulates the decision-making ability of a human expert
    - Robotics – branch of technology that deals with the design, construction, operation, structural disposition, manufacture and application of robots
- Human-computer interaction – study, planning, and design of the interaction between people (users) and computers.
    - Numerical analysis – study of algorithms that use numerical approximation (as opposed to general symbolic manipulations) for the problems of mathematical analysis (as distinguished from discrete mathematics).
    - Algebraic (symbolic) computation – relates to algorithms and software for manipulating mathematical expressions and equations in symbolic form, as opposed to manipulating the approximations of specific numerical quantities represented by those symbols. Software applications that perform symbolic calculations are called computer algebra systems.
- Computational number theory – study of algorithms for performing number theoretic computations
- Computational mathematics – involves mathematical research in areas of science where computing plays a central and essential role, emphasizing algorithms, numerical methods, and symbolic methods
- Scientific computing (Computational science) –
- Computational biology (bioinformatics) – involves the development and application of data-analytical and theoretical methods, mathematical modeling and computational simulation techniques to the study of biological, behavioral, and social systems.
- Computational science – subfield of computer science concerned with constructing mathematical models and quantitative analysis techniques and using computers to analyze and solve scientific problems
- Computational chemistry – branch of chemistry that uses principles of computer science to assist in solving chemical problems
- Computational neuroscience – study of brain function in terms of the information processing properties of the structures that make up the nervous system.
- Computer-aided engineering – broad usage of computer software to aid in engineering tasks.
    - Finite element analysis – numerical technique for finding approximate solutions of partial differential equations (PDE) as well as integral equations.
    - Computational fluid dynamics – branch of fluid mechanics that uses numerical methods and algorithms to solve and analyze problems that involve fluid flows.
- Computational economics – research discipline at the interface between computer science and economic and management science
- Computational sociology – branch of sociology that uses computationally intensive

methods to analyze and model social phenomena.
- Computational finance – cross-disciplinary field which relies on computational intelligence, mathematical finance, numerical methods and computer simulations to make trading, hedging and investment decisions, as well as facilitating the risk management of those decisions
- Humanities computing (Digital Humanities) – area of research, teaching, and creation concerned with the intersection of computing and the disciplines of the humanities
- Information systems – study of complementary networks of hardware and software that people and organizations use to collect, filter, process, create, and distribute data
    - Business informatics – discipline combining information technology (IT), informatics and management concepts.
    - Information technology –
    - Management information systems – provides information that is needed to manage organizations efficiently and effectively
    - Health informatics – discipline at the intersection of information science, computer science, and health care.
- Mathematics – search for fundamental truths in pattern, quantity, and change. *(See also Branches of Mathematics and AMS Mathematics Subject Classification)*
    - Algebra – one of the main branches of mathematics, it concerns the study of structure, relation and quantity.
        - Group theory – studies the algebraic structures known as groups.
            - Group representation – describe abstract groups in terms of linear transformations of vector spaces
        - Ring theory – study of ring–algebraic structures in which addition and multiplication are defined and have similar properties to those familiar from the integers
        - Field theory – branch of mathematics which studies the properties of fields
        - Linear algebra – branch of mathematics concerning finite or countably infinite dimensional vector spaces, as well as linear mappings between such spaces.
            - Vector space – mathematical structure formed by a collection of vectors: objects that may be added together and multiplied ("scaled") by numbers, called scalars in this context.
        - Multilinear algebra – extends the methods of linear algebra
        - Lie algebra – algebraic structure whose main use is in studying geometric objects such as Lie groups and differentiable manifolds
        - Associative algebra – associative ring that has a compatible structure of a vector space over a certain field K or, more generally, of a module over a commutative ring R.
        - Non-associative algebra – K-vector space (or more generally a module) A equipped with a K-bilinear map
        - Universal algebra – field of mathematics that studies algebraic structures themselves, not examples ("models") of algebraic structures
        - Homological algebra – branch of mathematics which studies homology in a general algebraic setting
        - Category theory – area of study in mathematics that examines in an abstract way the properties of particular mathematical concepts, by formalising them as collections of objects and arrows (also called morphisms, although this term also has a specific, non-category-theoretical sense), where these collections satisfy some basic conditions
        - Lattice theory – partially ordered set in which any two elements have a unique supremum (also called a least upper bound or join) and a unique infimum (also called a greatest lower bound or meet).
            - Order theory – branch of mathematics which investigates our intuitive notion of order using binary relations.
        - Differential algebra – algebras equipped with a derivation, which is a unary function that is linear and satisfies the Leibniz product rule.
    - Analysis – branch of pure mathematics that includes the theories of differentiation, integration and measure, limits, infinite series, and analytic functions
        - Real analysis – branch of mathematical analysis dealing with the set of real numbers and functions of a real variable.

- Calculus – branch of mathematics focused on limits, functions, derivatives, integrals, and infinite series.
- Complex analysis – branch of mathematical analysis that investigates functions of complex numbers
- Functional analysis – branch of mathematical analysis, the core of which is formed by the study of vector spaces endowed with some kind of limit-related structure (e.g. inner product, norm, topology, etc.) and the linear operators acting upon these spaces and respecting these structures in a suitable sense
    - Operator theory – branch of functional analysis that focuses on bounded linear operators, but which includes closed operators and nonlinear operators.
- Non-standard analysis – branch of classical mathematics that formulates analysis using a rigorous notion of an infinitesimal number.
- Harmonic analysis – branch of mathematics concerned with the representation of functions or signals as the superposition of basic waves, and the study of and generalization of the notions of Fourier series and Fourier transforms.
- p-adic analysis – branch of number theory that deals with the mathematical analysis of functions of p-adic numbers.
- Ordinary differential equations – ordinary differential equation (ODE) is an equation in which there is only one independent variable and one or more derivatives of a dependent variable with respect to the independent variable, so that all the derivatives occurring in the equation are ordinary derivatives.
- Partial differential equations – differential equation that contains unknown multivariable functions and their partial derivatives.
- Probability theory – branch of mathematics concerned with probability, the analysis of random phenomena.
    - Measure theory – systematic way to assign a number to each suitable subset of that set, intuitively interpreted as its size.
    - Ergodic theory – branch of mathematics that studies dynamical systems with an invariant measure and related problems.
    - Stochastic process – collection of random variables; this is often used to represent the evolution of some random value, or system, over time.
- Geometry – branch of mathematics concerned with questions of shape, size, relative position of figures, and the properties of space. Geometry is one of the oldest mathematical sciences.
    - Topology – major area of mathematics concerned with properties that are preserved under continuous deformations of objects, such as deformations that involve stretching, but no tearing or gluing.
    - General topology – branch of topology which studies properties of topological spaces and structures defined on them.
    - Algebraic topology – branch of mathematics which uses tools from abstract algebra to study topological spaces
    - Geometric topology – study of manifolds and maps between them, particularly embeddings of one manifold into another.
    - Differential topology – field dealing with differentiable functions on differentiable manifolds
    - Algebraic geometry – branch of mathematics which combines techniques of abstract algebra, especially commutative algebra, with the language and the problems of geometry
    - Differential geometry – mathematical discipline that uses the techniques of differential calculus and integral calculus, as well as linear algebra and multilinear algebra, to study problems in geometry
    - Projective geometry – study of geometric properties that are invariant under projective transformations
    - Affine geometry – study of geometric properties which remain unchanged by affine transformations
    - Non-Euclidean geometry – either of two specific geometries that are, loosely speaking, obtained by negating the Euclidean parallel postulate, namely hyperbolic and elliptic geometry.
    - Convex geometry – branch of geometry studying convex sets, mainly in Euclidean space.
    - Discrete geometry – branch of geometry that studies combinatorial properties and constructive methods of discrete geometric objects.

- Trigonometry – branch of mathematics that studies relationships involving lengths and angles of triangles
- Number theory – branch of pure mathematics devoted primarily to the study of the integers
  - Analytic number theory – branch of number theory that uses methods from mathematical analysis to solve problems about the integers
  - Algebraic number theory – major branch of number theory which studies algebraic structures related to algebraic integers
  - Geometric number theory – studies convex bodies and integer vectors in n-dimensional space
- Logic and Foundations of mathematics – subfield of mathematics with close connections to the foundations of mathematics, theoretical computer science and philosophical logic.
  - Set theory – branch of mathematics that studies sets, which are collections of objects
  - Proof theory – branch of mathematical logic that represents proofs as formal mathematical objects, facilitating their analysis by mathematical techniques
  - Model theory – study of (classes of) mathematical structures (e.g. groups, fields, graphs, universes of set theory) using tools from mathematical logic
  - Recursion theory – branch of mathematical logic and computer science that originated in the 1930s with the study of computable functions and Turing degrees
  - Modal logic – type of formal logic primarily developed in the 1960s that extends classical propositional and predicate logic to include operators expressing modality
  - Intuitionistic logic – symbolic logic system differing from classical logic in its definition of the meaning of a statement being true
- Applied mathematics – branch of mathematics that concerns itself with mathematical methods that are typically used in science, engineering, business, and industry.
  - Mathematical statistics – study of statistics from a mathematical standpoint, using probability theory as well as other branches of mathematics such as linear algebra and analysis
    - Probability – likelihood or chance that something is the case or will happen
- Econometrics – application of mathematics and statistical methods to economic data
- Actuarial science – discipline that applies mathematical and statistical methods to assess risk in the insurance and finance industries.
- Demography – statistical study of human populations and sub-populations.
- Approximation theory – study of how functions can best be approximated with simpler functions, and with quantitatively characterizing the errors introduced thereby.
- Numerical analysis – study of algorithms that use numerical approximation (as opposed to general symbolic manipulations) for the problems of mathematical analysis (as distinguished from discrete mathematics).
- Optimization (Mathematical programming) – selection of a best element from some set of available alternatives.
  - Operations research – study of the application of advanced analytical methods to help make better decisions
  - Linear programming – mathematical method for determining a way to achieve the best outcome (such as maximum profit or lowest cost) in a given mathematical model for some list of requirements represented as linear relationships
- Dynamical systems – concept in mathematics where a fixed rule describes the time dependence of a point in a geometrical space
  - Chaos theory – study of the behavior of dynamical systems that are highly sensitive to initial conditions, an effect which is popularly referred to as the butterfly effect.
  - Fractal geometry – mathematical set that has a fractal dimension that usually exceeds its topological dimension and may fall between the integers.
- Mathematical physics – development of mathematical methods for application to problems in physics
  - Quantum field theory – theoretical framework for constructing quantum mechanical models of systems classically parametrized (represented) by an infinite number of degrees of freedom,

that is. fields and (in a condensed matter context) many-body systems.
- Statistical mechanics – branch of physics that applies probability theory, which contains mathematical tools for dealing with large populations, to the study of the thermodynamic behavior of systems composed of a large number of particles.
- Information theory – branch of applied mathematics and electrical engineering involving the quantification of information.
- Cryptography – study of means of obscuring information, such as codes and ciphers
- Combinatorics – branch of mathematics concerning the study of finite or countable discrete structures
  - Coding theory – study of the properties of codes and their fitness for a specific application
- Graph theory – study of graphs, mathematical structures used to model pairwise relations between objects from a certain collection
- Game theory – study of strategic decision making. More formally, it is "the study of mathematical models of conflict and cooperation between intelligent rational decision-makers."

- Statistics – collection, analysis, interpretation, and presentation of data.
  - Computational statistics – interface between statistics and computer science.
    - Data mining – process that results in the discovery of new patterns in large data sets
    - Regression – estimates the conditional expectation of the dependent variable given the independent variables – that is, the average value of the dependent variable when the independent variables are held fixed.
    - Simulation – Simulation is the imitation of the operation of a real-world process or system over time. The act of simulating something first requires that a model be developed; this model represents the key characteristics or behaviors of the selected physical or abstract system or process. The model represents the system itself, whereas the simulation represents the operation of the system over time.
    - Bootstrap (statistics) – method for assigning measures of accuracy to sample estimates (Efron and Tibshirani 1993).
  - Design of experiments – design of any information-gathering exercises where variation is present, whether under the full control of the experimenter or not
    - Block design – set together with a family of subsets (repeated subsets are allowed at times) whose members are chosen to satisfy some set of properties that are deemed useful for a particular application.
    - Analysis of variance – collection of statistical models, and their associated procedures, in which the observed variance in a particular variable is partitioned into components attributable to different sources of variation.
    - Response surface methodology – explores the relationships between several explanatory variables and one or more response variables.
  - Engineering statistics – Engineering statistics combines engineering and statistics
  - Spatial statistics – any of the formal techniques which study entities using their topological, geometric, or geographic properties.
  - Social statistics – use of statistical measurement systems to study human behavior in a social environment
  - Statistical modelling – formalization of relationships between variables in the form of mathematical equations
    - Biostatistics – application of statistics to a wide range of topics in biology.
      - Epidemiology – study of the distribution and patterns of health-events, health-characteristics and their causes or influences in well-defined populations.
    - Multivariate analysis – observation and analysis of more than one statistical variable at a time.
      - Structural equation model – statistical technique for testing and estimating causal relations using a combination of statistical data and qualitative causal assumptions.
      - Time series – sequence of data points, measured typically at successive time instants spaced at uniform time intervals.

- Reliability theory – describes the probability of a system completing its expected function during an interval of time.
- Quality control – process by which entities review the quality of all factors involved in production.
- Statistical theory – provides a basis for the whole range of techniques, in both study design and data analysis, that are used within applications of statistics.
    - Decision theory – identifies the values, uncertainties and other issues relevant in a given decision, its rationality, and the resulting optimal decision.
    - Mathematical statistics – study of statistics from a mathematical standpoint, using probability theory as well as other branches of mathematics such as linear algebra and analysis.
        - Probability – likelihood or chance that something is the case or will happen.
- Sample Survey – process of selecting a sample of elements from a target population in order to conduct a survey.
    - Sampling theory – study of the collection, organization, analysis, and interpretation of data.
    - Survey methodology – field that studies the sampling of individuals from a population with a view towards making statistical inferences about the population using the sample.
- Systems science – interdisciplinary field of science that studies the nature of complex systems in nature, society, and science.
    - Chaos theory – field of study in mathematics, with applications in several disciplines including physics, engineering, economics, biology, and philosophy; studies the behavior of dynamical systems that are highly sensitive to initial conditions.
    - Complex systems and Complexity Theory – studies how relationships between parts give rise to the collective behaviors of a system and how the system interacts and forms relationships with its environment.
    - Cybernetics – interdisciplinary study of the structure of regulatory systems.
        - Biocybernetics – application of cybernetics to biological science, composed of biological disciplines that benefit from the application of cybernetics: neurology, multicellular systems and others.
        - Engineering cybernetics – field of cybernetics, which deals with the question of control engineering of mechatronic systems as well as chemical or biological systems.
        - Management cybernetics – field of cybernetics concerned with management and organizations.
        - Medical cybernetics – branch of cybernetics which has been heavily affected by the development of the computer, which applies the concepts of cybernetics to medical research and practice.
        - New Cybernetics – study of self-organizing systems according to Peter Harries-Jones (1988), "looking beyond the issues of the "first", "old" or "original" cybernetics and their politics and sciences of control, to the autonomy and self-organization capabilities of complex systems".
        - Second-order cybernetics – investigates the construction of models of cybernetic systems.
    - Control theory – Control theory is an interdisciplinary branch of engineering and mathematics that deals with the behavior of dynamical systems. The external input of a system is called the reference. When one or more output variables of a system need to follow a certain reference over time, a controller manipulates the inputs to a system to obtain the desired effect on the output of the system.
        - Control engineering – engineering discipline that applies control theory to design systems with desired behaviors.
        - Control systems – device, or set of devices to manage, command, direct or regulate the behavior of other devices or system.
        - Dynamical systems – concept in mathematics where a fixed rule describes the time dependence of a point in a geometrical space.
    - Operations research – study of the use of advanced analytical methods to help make better decisions.
    - Systems dynamics – approach to understanding the behaviour of complex systems over time.
        - Systems analysis – study of sets of interacting entities, including computer systems analysis.

- Systems theory – interdisciplinary study of systems in general, with the goal of elucidating principles that can be applied to all types of systems at all nesting levels in all fields of research.
    - Developmental systems theory – overarching theoretical perspective on biological development, heredity, and evolution
    - General systems theory – interdisciplinary study of systems in general, with the goal of elucidating principles that can be applied to all types of systems at all nesting levels in all fields of research.
    - Linear time-invariant systems – investigates the response of a linear and time-invariant system to an arbitrary input signal.
    - Mathematical system theory – area of mathematics used to describe the behavior of complex dynamical systems, usually by employing differential equations or difference equations.
    - Systems biology – several related trends in bioscience research, and a movement that draws on those trends.
    - Systems ecology – interdisciplinary field of ecology, taking a holistic approach to the study of ecological systems, especially ecosystems.
    - Systems engineering – interdisciplinary field of engineering focusing on how complex engineering projects should be designed and managed over their life cycles.
    - Systems neuroscience – subdiscipline of neuroscience and systems biology that studies the function of neural circuits and systems.
    - Systems psychology – branch of applied psychology that studies human behaviour and experience in complex systems.

15.3.3 Social science

Social science – study of the social world constructed between humans. The social sciences usually limit themselves to an anthropomorphically centric view of these interactions with minimal emphasis on the inadvertent impact of social human behavior on the external environment (physical, biological, ecological, etc.). 'Social' is the concept of exchange/influence of ideas, thoughts, and relationship interactions (resulting in harmony, peace, self enrichment, favoritism, maliciousness, justice seeking, etc.) between humans. The scientific method is utilized in many social sciences, albeit adapted to the needs of the social construct being studied.

- Branches of social science (also known as the social sciences)

15.3.4 Applied science

Applied science – branch of science that applies existing scientific knowledge to develop more practical applications, including inventions and other technological advancements.

- Branches of applied science (also known as the applied sciences)

15.4 How scientific fields differ

- Exact science – any field of science capable of accurate quantitative expression or precise predictions and rigorous methods of testing hypotheses, especially reproducible experiments involving quantifiable predictions and measurements.

- Fundamental science – science that describes the most basic objects, forces, relations between them and laws governing them, such that all other phenomena may be in principle derived from them following the logic of scientific reductionism.

- Hard and soft science – colloquial terms often used when comparing scientific fields of academic research or scholarship, with hard meaning perceived as being more scientific, rigorous, or accurate.

15.5 Politics of science

- Disruptive technology – innovation that helps create a new market and value network, and eventually goes on to disrupt an existing market and value network (over a few years or decades), displacing an earlier technology.

- Kansas evolution hearings – series of hearings held in Topeka, Kansas, United States 5 to 12 May 2005 by the Kansas State Board of Education and its State Board Science Hearing Committee to change how evolution and the origin of life would be taught in the state's public high school science classes.

- List of books about the politics of science – list of books about the politics of science.

- Politicization of science – politicization of science is the manipulation of science for political gain.
- Science by press release – refers to scientists who put an unusual focus on publicizing results of research in the media.

15.6 History of science

See also: Outline_of_history § History_of_Science

- History of science - History of science in general
  - History of scientific method – history of scientific method is a history of the methodology of scientific inquiry, as differentiated from a history of science in general.
  - Theories/sociology of science – sociology and philosophy of science, as well as the entire field of science studies, have in the 20th century been occupied with the question of large-scale patterns and trends in the development of science, and asking questions about how science "works" both in a philosophical and practical sense.
  - Historiography – study of the history and methodology of the sub-discipline of history, known as the history of science, including its disciplinary aspects and practices (methods, theories, schools) and to the study of its own historical development ("History of History of Science", i.e., the history of the discipline called History of Science).
  - History of pseudoscience – history of pseudoscience is the study of pseudoscientific theories over time. A pseudoscience is a set of ideas that presents itself as science, while it does not meet the criteria to properly be called such.
  - Timeline of scientific discoveries – shows the date of publication of major scientific theories and discoveries, along with the discoverer. In many cases, the discoveries spanned several years.
  - Timeline of scientific thought – lists the major landmarks across all scientific philosophy and methodological sciences.

15.6.1 By period

- History of science in early cultures – history of science in early cultures refers to the study of protoscience in ancient history, prior to the development of science in the Middle Ages.
- History of science in Classical Antiquity – history of science in classical antiquity encompasses both those inquiries into the workings of the universe aimed at such practical goals as establishing a reliable calendar or determining how to cure a variety of illnesses and those abstract investigations known as natural philosophy.
- History of science in the Middle Ages – Science in the Middle Ages comprised the study of nature, including practical disciplines, the mathematics and natural philosophy in medieval Europe.
- History of science in the Renaissance – During the Renaissance, great advances occurred in geography, astronomy, chemistry, physics, mathematics, manufacturing, and engineering.
  - Science and inventions of Leonardo da Vinci – Italian polymath, regarded as the epitome of the "Renaissance Man", displaying skills in numerous diverse areas of study.
- Scientific revolution – scientific revolution is an era associated primarily with the 16th and 17th centuries during which new ideas and knowledge in physics, astronomy, biology, medicine and chemistry transformed medieval and ancient views of nature and laid the foundations for modern science.
- Governmental impact on science during WWII – Governmental impact on science during World War II represents the effect of public administration on technological development that provided many advantages to the armed forces, economies and societies in their strategies during the war.

By date

- List of years in science – events related to science or technology which occurred in the listed year.
- Timeline of scientific discoveries – shows the date of publication of major scientific theories and discoveries, along with the discoverer. In many cases, the discoveries spanned several years.
- Timeline of scientific experiments – shows the date of publication of major scientific experiments.
- Timeline of the history of scientific method – shows an overview of the cultural inventions that have contributed to the development of the scientific method.

15.6.2 By field

- History of natural science – study of nature and the physical universe that was dominant before the development of modern science.

  - Natural philosophy – study of nature and the physical universe that was dominant before the development of modern science.
  - Natural history – scientific research of plants or animals, leaning more towards observational rather than experimental methods of study, and encompasses more research published in magazines than in academic journals.
  - History of biology – traces the study of the living world from ancient to modern times.
  - History of ecology – history of the science of ecology.
  - History of molecular biology – begins in the 1930s with the convergence of various, previously distinct biological disciplines: biochemistry, genetics, microbiology, and virology.
  - History of astronomy – Timeline
  - History of chemistry – By 1000 BC, ancient civilizations used technologies that would eventually form the basis of the various branches of chemistry.
  - History of geography
  - History of geology – Timeline
  - History of meteorology – Timeline
  - History of physics – As forms of science historically developed out of philosophy, physics was originally referred to as natural philosophy, a field of study concerned with "the workings of nature."
  - History of science and technology

- History of the social sciences – has origin in the common stock of Western philosophy and shares various precursors, but began most intentionally in the early 19th century with the positivist philosophy of science.

  - History of archaeology – Timeline
  - History of cognitive science
  - History of criminal justice – Throughout the history of criminal justice, evolving forms of punishment, added rights for offenders and victims, and policing reforms have reflected changing customs, political ideals, and economic conditions.
  - History of economics – study of different thinkers and theories in the subject that became political economy and economics from the ancient world to the present day.
  - History of education – development of systematic methods of teaching and learning.
  - History of law – study of how law has evolved and why it changed.
  - History of linguistics – endeavors to describe and explain the human faculty of language.
  - History of marketing – recognized discipline, along with concomitant changes in marketing theory and practice.
  - History of parapsychology
  - History of political science – social science discipline concerned with the study of the state, government, and politics.
  - History of psychology – Timeline
  - History of sociology – Timeline

See also: Outline of technology:: History of technology

15.6.3 By region

History of science in present states, by continent

See – Category:Science and technology by continent

History of science in historic states

- Science and technology of the Han Dynasty
- Science and technology in the Ottoman Empire
- Science and technology of the Song Dynasty
- Science and technology in the Soviet Union
- Science and technology of the Tang Dynasty

15.7 Philosophy of science

See also: Outline_of_philosophy § Philosophy_of_science

- Philosophy of science – questions the assumptions, foundations, methods and implications of science.

15.8 Scientific community

- Scientific community – group of all interacting scientists.
- Big Science – a series of changes in science that occurred in industrial nations during and after World War II.

15.8.1 Scientific organizations

- Academy of Sciences – national academy or another learned society dedicated to sciences.

15.8.2 Scientists

- Scientist – practitioner of science; an individual who uses scientific method to objectively inquire into the nature of reality—be it the fundamental laws of physics or how people behave. There are many names for scientists, often named in relation to the job that they do. One example of this is a biologist, a scientist who studies biology (the study of living organisms and their environments).

Types of scientist

By field The scientific fields mentioned below are generally described by the science they study.

- Agricultural scientist – broad multidisciplinary field that encompasses the parts of exact, natural, economic and social sciences that are used in the practice and understanding of agriculture.
- Archaeologist – study of human activity, primarily through the recovery and analysis of the material culture and environmental data that they have left behind, which includes artifacts, architecture, biofacts and cultural landscapes (the archaeological record).
- Astronomer – astronomer is a scientist who studies celestial bodies such as planets, stars and galaxies.
  - Astrophysicist – branch of astronomy that deals with the physics of the universe, including the physical properties of celestial objects, as well as their interactions and behavior.
- Biologist – scientist devoted to the study of living organisms and their relationship to their environment.
  - Astrobiologist – study of the origin, evolution, distribution, and future of extraterrestrial life.
- Biophysicist – interdisciplinary science that uses the methods of physical science to study biological systems.
- Biotechnologist – field of applied biology that involves the use of living organisms and bioprocesses in engineering, technology, medicine and other fields requiring bioproducts.
- Botanist – discipline of biology, is the science of plant life.
- Cognitive scientists – scientific study of the mind and its processes.
- Ecologist – scientific study of the relations that living organisms have with respect to each other and their natural environment.
- Entomologist – scientific study of insects, a branch of arthropodology.
- Evolutionary biologist – sub-field of biology concerned with the study of the evolutionary processes that have given rise to the diversity of life on Earth.
- Geneticist – biologist who studies genetics, the science of genes, heredity, and variation of organisms.
- Herpetologist – branch of zoology concerned with the study of amphibians (including frogs, toads, salamanders, newts, and gymnophiona) and reptiles (including snakes, lizards, amphisbaenids, turtles, terrapins, tortoises, crocodilians, and the tuataras).
- Immunologist – branch of biomedical science that covers the study of all aspects of the immune system in all organisms.
- Ichthyologist – study of fish.
- Lepidopterist – person who specialises in the study of Lepidoptera, members of an order encompassing moths and the three superfamilies of butterflies, skipper butterflies, and moth-butterflies.
- Marine biologist – scientific study of organisms in the ocean or other marine or brackish bodies of water.
- Medical scientist – basic research, applied research, or translational research conducted to aid and support the body of knowledge in the field of medicine.
- Microbiologist – study of microscopic organisms.
- Mycologist – branch of biology concerned with the study of fungi, including their genetic and

15.8. SCIENTIFIC COMMUNITY

biochemical properties, their taxonomy and their use to humans as a source for tinder, medicinals (e.g., penicillin), food (e.g., beer, wine, cheese, edible mushrooms) and entheogens, as well as their dangers, such as poisoning or infection.

- Neuroscientist – individual who studies the scientific field of neuroscience or any of its related sub-fields.
- Ornithologist – branch of zoology that concerns the study of birds.
- Paleontologist – study of prehistoric life.
- Pathologist – precise study and diagnosis of disease.
- Pharmacologist – branch of medicine and biology concerned with the study of drug action.
- Physiologist – science of the function of living systems.
- Zoologist – branch of biology that relates to the animal kingdom, including the structure, embryology, evolution, classification, habits, and distribution of all animals, both living and extinct.

- Chemist – scientist trained in the study of chemistry.

  - Analytical chemist – study of the separation, identification, and quantification of the chemical components of natural and artificial materials.
  - Biochemist – study of chemical processes in living organisms, including, but not limited to, living matter.
  - Inorganic chemist – branch of chemistry concerned with the properties and behavior of inorganic compounds.
  - Organic chemist – subdiscipline within chemistry involving the scientific study of the structure, properties, composition, reactions, and preparation (by synthesis or by other means) of carbon-based compounds, hydrocarbons, and their derivatives.
  - Physical chemist – study of macroscopic, atomic, subatomic, and particulate phenomena in chemical systems in terms of physical laws and concepts.

- Earth scientist – all-embracing term for the sciences related to the planet Earth.

  - Geologist – scientist who studies the solid and liquid matter that constitutes the Earth as well as the processes and history that has shaped it.
  - Glaciologist – study of glaciers, or more generally ice and natural phenomena that involve ice.
  - Hydrologist – study of the movement, distribution, and quality of water on Earth and other planets, including the hydrologic cycle, water resources and environmental watershed sustainability.
  - Limnologist – study of inland waters
  - Meteorologist – study of weather
  - Mineralogist – study of chemistry, crystal structure, and physical (including optical) properties of minerals.
  - Oceanographer – branch of Earth science that studies the ocean
  - Paleontologist – study of prehistoric life
  - Seismologist – scientific study of earthquakes and the propagation of elastic waves through the Earth or through other planet-like bodies.
  - Volcanologist – study of volcanoes, lava, magma, and related geological, geophysical and geochemical phenomena.

- Informatician – science of information, the practice of information processing, and the engineering of information systems.

  - Computer scientist – scientist who has acquired knowledge of computer science, the study of the theoretical foundations of information and computation

- Library scientist – interdisciplinary or multidisciplinary field that applies the practices, perspectives, and tools of management, information technology, education, and other areas to libraries; the collection, organization, preservation, and dissemination of information resources; and the political economy of information.

- Management scientist – study of advanced analytical methods to help make better decisions.

- Mathematician – person with an extensive knowledge of mathematics, a field that has been informally defined as being concerned with numbers, data, collection, quantity, structure, space, and change.

  - Statistician – someone who works with theoretical or applied statistics.

- Military scientist – process of translating national defence policy to produce military capability by employing military scientists, including theorists, researchers, experimental scientists, applied scientists, designers, engineers, test technicians, and military personnel responsible for prototyping.

- Physicist – scientist who does research in physics
- Psychologist – professional or academic title used by individuals who practice psychology
    - Abnormal psychologist – branch of psychology that studies unusual patterns of behavior, emotion and thought, which may or may not be understood as precipitating a mental disorder.
    - Educational psychologist – psychologist whose differentiating functions may include diagnostic and psycho-educational assessment, psychological counseling in educational communities (students, teachers, parents and academic authorities), community-type psycho-educational intervention, and mediation, coordination, and referral to other professionals, at all levels of the educational system.
    - Biopsychologist – application of the principles of biology (in particular neurobiology), to the study of physiological, genetic, and developmental mechanisms of behavior in human and non-human animals.
    - Clinical psychologist – integration of science, theory and clinical knowledge for the purpose of understanding, preventing, and relieving psychologically based distress or dysfunction and to promote subjective well-being and personal development.
    - Comparative psychologist – scientific study of the behavior and mental processes of non-human animals, especially as these relate to the phylogenetic history, adaptive significance, and development of behavior.
    - Cognitive psychologist – subdiscipline of psychology exploring internal mental processes. It is the study of how people perceive, remember, think, speak, and solve problems.
    - Developmental psychologist – scientific study of systematic psychological changes, emotional changes, and perception changes that occur in human beings over the course of their life span.
    - Evolutionary psychologist – approach in the social and natural sciences that examines psychological traits such as memory, perception, and language from a modern evolutionary perspective.
    - Experimental psychologist – study of behavior and the processes that underlie it, by means of experiment
    - Neuropsychologist – studies the structure and function of the brain as they relate to specific psychological processes and behaviors.
    - Social psychologist – scientific study of how people's thoughts, feelings, and behaviors are influenced by the actual, imagined, or implied presence of others.
- Social scientist – field of study concerned with society and human behaviours.
    - Anthropologist – study of humanity.
        - Ethnologist – branch of anthropology that compares and analyzes the origins, distribution, technology, religion, language, and social structure of the ethnic, racial, and/or national divisions of humanity.
    - Communication scientist – academic field that deals with processes of human communication, commonly defined as the sharing of symbols to create meaning.
    - Criminologist – study of criminal behavior
    - Demographer – statistical study of populations
    - Economist – professional in the social science discipline of economics.
    - Geographer – geographer is a scholar whose area of study is geography, the study of Earth's natural environment and human society.
    - Political economist – study of production, buying, and selling, and their relations with law, custom, and government, as well as with the distribution of national income and wealth, including through the budget process.
    - Political scientist – social science discipline concerned with the study of the state, government, and politics.
    - Sociologist –
- Technologist
    - Architectural technologist – specialist in the technology of building design and construction
    - Educational technologist – specialist in tools to enhance learning
    - Engineering technologist – specialist who implements technology within a field of engineering
    - Industrial technologist – specialist in the management, operation, and maintenance of complex operation systems
    - Medical Technologist – healthcare professional who performs diagnostic analysis on a variety of body fluids
    - Radiologic technologist – medical professional who applies doses of radiation for imaging and treatment

- Surgical technologist – health specialist who facilitates the conduct of invasive surgical procedures

By employment status

- Academic – community of students and scholars engaged in higher education and research.
- Corporate Scientist – someone who is employed by a business to do research and development for the benefit of that business
- Layperson – someone who is not an expert or someone who has not had professional training
- Gentleman scientist – financially independent scientist who pursues scientific study as a hobby.
- Government scientist – scientist employed by a country's government

Famous scientists

Main list: Lists of scientists

- Aristotle – Greek philosopher and polymath, a student of Plato and teacher of Alexander the Great.
- Archimedes – Greek mathematician, physicist, engineer, inventor, and astronomer.
- Andreas Vesalius – Flemish anatomist, physician, and author of one of the most influential books on human anatomy. De humani corporis fabrica (On the Structure of the Human Body).
- Nicolaus Copernicus – Renaissance astronomer and the first person to formulate a comprehensive heliocentric cosmology which displaced the Earth from the center of the universe.
- Galileo Galilei – Italian physicist, mathematician, astronomer, and philosopher who played a major role in the Scientific Revolution.
- Johannes Kepler – German mathematician, astronomer and astrologer. A key figure in the 17th century scientific revolution, he is best known for his eponymous laws of planetary motion, codified by later astronomers, based on his works Astronomia nova, Harmonices Mundi, and Epitome of Copernican Astronomy.
- René Descartes – French philosopher, mathematician, and writer who spent most of his adult life in the Dutch Republic.
- Isaac Newton – English physicist, mathematician, astronomer, natural philosopher, alchemist, and theologian, who has been "considered by many to be the greatest and most influential scientist who ever lived."
- Leonhard Euler – pioneering Swiss mathematician and physicist.
- Pierre-Simon Laplace – French mathematician and astronomer whose work was pivotal to the development of mathematical astronomy and statistics.
- Alexander von Humboldt – German geographer, naturalist and explorer, and the younger brother of the Prussian minister, philosopher and linguist Wilhelm von Humboldt.
- Charles Darwin – Charles Robert Darwin FRS (12 February 1809 -[?] 19 April 1882) was an English naturalist.[1] He established that all species of life have descended over time from common ancestors, and proposed the scientific theory that this branching pattern of evolution resulted from a process that he called natural selection.
- James Clerk Maxwell – Scottish physicist and mathematician.
- Marie Curie – Polish physicist and chemist famous for her pioneering research on radioactivity.
- Albert Einstein – German-born theoretical physicist who developed the theory of general relativity, effecting a revolution in physics
- Linus Pauling – American chemist, biochemist, peace activist, author, and educator. He was one of the most influential chemists in history and ranks among the most important scientists of the 20th century.
- John Bardeen – American physicist and electrical engineer, the only person to have won the Nobel Prize in Physics twice
- Frederick Sanger – English biochemist and a two-time Nobel laureate in chemistry, the only person to have been so.
- Stephen Hawking – British theoretical physicist, cosmologist, and author.

15.9 Science education

Science education

- Scientific literacy – encompasses written, numerical, and digital literacy as they pertain to understanding science, its methodology, observations, and theories.

15.10 See also

- Sci-Mate – open collaboration of scientists using Web 2.0 software to address well known challenges in academic publishing and technology transfer

- Science Daily – news website for topical science articles

- Science.tv – virtual community for people interested in science

15.11 References

Chapter 16

Objectivity (philosophy)

Not to be confused with Objectivism (Ayn Rand).

Objectivity is a central philosophical concept, related to reality and truth, which has been variously defined by sources. Generally, objectivity means the state or quality of being true even outside of a subject's individual biases, interpretations, feelings, and imaginings. A proposition is generally considered objectively true (to have **objective truth**) when its truth conditions are met without biases caused by feelings, ideas, opinions, etc., of a sentient subject. A second, broader meaning of the term refers to the ability in any context to judge fairly, without partiality or external influence. This second meaning of *objectivity* is sometimes used synonymously with *neutrality*.

16.1 Objectivism

"Objectivism" is a branch of philosophy that originated in the early nineteenth century. Gottlob Frege was the first to apply it, when he expounded an epistemological and metaphysical theory contrary to that of Immanuel Kant. Kant's rationalism attempted to reconcile the failures he perceived in philosophical realism.

Stronger versions of this claim hold that there is only one correct description of this reality. If it is true that reality is mind-independent, then reality might include objects that are unknown to consciousness and thus might include objects not the subject of intensionality. Objectivity in referring requires a definition of truth. According to metaphysical objectivists, an object may truthfully be said to have this or that attribute, as in the statement "This object exists," whereas the statement "This object is true" or "false" is meaningless. For them, only propositions have truth values. The terms "objectivity" and "objectivism" are not synonymous, with objectivism being an ontological theory that incorporates a commitment to the objectivity of objects.

Plato's idealism was a form of metaphysical objectivism, holding that the Ideas exist objectively and independently. Berkeley's empiricist idealism, on the other hand, could be called a subjectivism: he held that things only exist to the extent that they are perceived. Both theories claim methods of objectivity. Plato's definition of objectivity can be found in his epistemology, which takes as a model mathematics, and his metaphysics, where knowledge of the ontological status of objects and ideas is resistant to change.

Plato considered knowledge of geometry a condition of philosophical knowledge, both being concerned with universal truths. Plato's opposition between objective knowledge and *doxa* (opinions) became the basis for later philosophies intent on resolving the problem of reality, knowledge, and human existence. Personal opinions belong to the changing sphere of the sensible, opposed to a fixed and eternal incorporeal realm that is mutually intelligible.

Where Plato distinguishes between what and how we know things (epistemology), and their ontological status as things (metaphysics), subjectivism such as Berkeley's and a mind dependence of knowledge and reality fails to distinguish between what one knows and what is to be known, or at least explains the distinction superficially. In Platonic terms, a criticism of subjectivism is that it is difficult to distinguish between knowledge, *doxa*, and subjective knowledge (true belief), distinctions that Plato makes.

The importance of perception in evaluating and understanding objective reality is debated. Realists argue that perception is key in directly observing objective reality, while instrumentalists hold that perception is not necessarily useful in directly observing objective reality, but is useful in interpreting and predicting reality. The concepts that encompasses these ideas are important in the philosophy of science.

16.2 Objectivity in ethics

16.2.1 Ethical subjectivism

See also: David Hume, Non-cognitivism, and Subjectivism

The term, "ethical subjectivism", covers two distinct theories in ethics. According to cognitive versions of ethical subjectivism, the truth of moral statements depends upon people's values, attitudes, feelings, or beliefs. Some forms of cognitivist ethical subjectivism can be counted as forms of realism, others are forms of anti-realism. David Hume is a foundational figure for cognitive ethical subjectivism. On a standard interpretation of his theory, a trait of character counts as a moral virtue when it evokes a sentiment of approbation in a sympathetic, informed, and rational human observer. Similarly, Roderick Firth's ideal observer theory held that right acts are those that an impartial, rational observer would approve of. William James, another ethical subjectivist, held that an end is good (to or for a person) just in the case it is desired by that person (see also ethical egoism). According to non-cognitive versions of ethical subjectivism, such as emotivism, prescriptivism, and expressivism, ethical statements cannot be true or false, at all; rather, they are expressions of personal feelings or commands. For example, on A. J. Ayer's emotivism, the statement, "Murder is wrong" is equivalent in meaning to the emotive, "Murder, Boo!"

16.2.2 Ethical objectivism

Main article: Moral realism

According to the ethical objectivist, the truth or falsehood of typical moral judgments does not depend upon the beliefs or feelings of any person or group of persons. This view holds that moral propositions are analogous to propositions about chemistry, biology, or history, in so much as they are true despite what anyone believes, hopes, wishes, or feels. When they fail to describe this mind-independent moral reality, they are false—no matter what anyone believes, hopes, wishes, or feels.

There are many versions of ethical objectivism, including various religious views of morality, Platonistic intuitionism, Kantianism, utilitarianism, and certain forms of ethical egoism and contractualism. Note that Platonists define ethical objectivism in an even more narrow way, so that it requires the existence of intrinsic value. Consequently, they reject the idea that contractualists or egoists could be ethical objectivists. Objectivism, in turn, places primacy on the origin of the frame of reference—and, as such, considers any arbitrary frame of reference ultimately a form of ethical subjectivism by a transitive property, even when the frame incidentally coincides with reality and can be used for measurements.

16.3 See also

- Factual relativism
- Journalistic objectivity
- Naïve realism
- Objectivity (science)

16.4 Further reading

- Bachelard, Gaston. *La formation de l'esprit scientifique : contribution à une psychanalyse de la connaissance*. Paris: Vrin, 2004. ISBN 2-7116-1150-7.
- Castillejo, David. *The Formation of Modern Objectivity*. Madrid: Ediciones de Arte y Bibliofilia, 1982.
- Kuhn, Thomas S. *The Structure of Scientific Revolutions*. Chicago: University of Chicago Press, 1996. 3rd ed. ISBN 0-226-45808-3.
- Megill, Allan. *Rethinking Objectivity*. London: Duke UP, 1994.
- Nagel, Ernest. *The Structure of Science*. New York: Brace and World, 1961.
- Nagel, Thomas. *The View from Nowhere*. Oxford: Oxford UP, 1986
- Nozick, Robert. *Invariances: the structure of the objective world*. Cambridge: Harvard UP, 2001.
- Popper, Karl. R. *Objective Knowledge: An Evolutionary Approach*. Oxford University Press, 1972. ISBN 0-19-875024-2.
- Rescher, Nicholas. *Objectivity: the obligations of impersonal reason*. Notre Dame: Notre Dame Press, 1977.
- Rorty, Richard. *Objectivity, Relativism, and Truth*. Cambridge: Cambridge University Press, 1991
- Rousset, Bernard. *La théorie kantienne de l'objectivité*, Paris: Vrin, 1967.
- Schaeffler, Israel. *Science and Subjectivity*. Hackett, 1982. Voices of Wisdom: a multicultural philosophy reader. kessler

16.5 External links

- Mulder, Dwayne H. "Objectivity". *Internet Encyclopedia of Philosophy.*
- Subjectivity and Objectivity—by Pete Mandik

Chapter 17

Logical reasoning

Informally, two kinds of **logical reasoning** can be distinguished in addition to formal deduction: induction and abduction. Given a precondition or *premise*, a conclusion or *logical consequence* and a rule or *material conditional* that implies the *conclusion* given the *precondition*, one can explain that:

- **Deductive reasoning** determines whether the truth of a *conclusion* can be determined for that *rule*, based solely on the truth of the premises. Example: "When it rains, things outside get wet. The grass is outside, therefore: when it rains, the grass gets wet." Mathematical logic and philosophical logic are commonly associated with this type of reasoning.

- **Inductive reasoning** attempts to support a determination of the *rule*. It hypothesizes a *rule* after numerous examples are taken to be a *conclusion* that follows from a *precondition* in terms of such a *rule*. Example: "The grass got wet numerous times when it rained, therefore: the grass always gets wet when it rains." While they may be persuasive, these arguments are not deductively valid, see the problem of induction. Science is associated with this type of reasoning.

- **Abductive reasoning**, a.k.a. *inference to the best explanation*, selects a cogent set of *preconditions*. Given a true *conclusion* and a *rule*, it attempts to select some possible *premises* that, if true also, can support the *conclusion*, though not uniquely. Example: "When it rains, the grass gets wet. The grass is wet. Therefore, it might have rained." This kind of reasoning can be used to develop a hypothesis, which in turn can be tested by additional reasoning or data. Diagnosticians, detectives, and scientists often use this type of reasoning.

17.1 See also

- Analogy
- Argument
- Defeasible reasoning
- Fallacy
- Inference
- Reason

17.2 References

- Menzies, T. *Applications of Abduction: Knowledge-Level Modeling.* November 1996

Chapter 18

Scientific method

"Scientific research" redirects here. For the publisher, see Scientific Research Publishing.
Compare Observational study and Experiment
For a broader coverage related to this topic, see Research.
For other uses, see Scientific method (disambiguation).

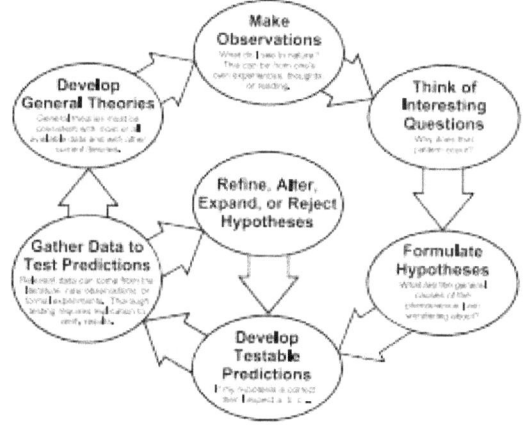

The scientific method as a cyclic or iterative process[1]

The **scientific method** is a body of techniques for investigating phenomena, acquiring new knowledge, or correcting and integrating previous knowledge.[2] To be termed scientific, a method of inquiry is commonly based on empirical or measurable evidence subject to specific principles of reasoning.[3] The Oxford Dictionaries Online defines the scientific method as "a method or procedure that has characterized natural science since the 17th century, consisting in systematic observation, measurement, and experiment, and the formulation, testing, and modification of hypotheses".[4] Experiments need to be designed to test hypotheses. Experiments are an important tool of the scientific method.[5]

The method is a continuous process that begins with observations about the natural world. People are naturally inquisitive, so they often come up with questions about things they see or hear, and they often develop ideas or hypotheses about why things are the way they are. The best hypotheses lead to predictions that can be tested in various ways. The strongest tests of hypotheses come from carefully controlled experiments that gather empirical data. Depending on how well additional tests match the predictions, the original hypothesis may require refinement, alteration, expansion or even rejection. If a particular hypothesis becomes very well supported, a general theory may be developed.[1]

Although procedures vary from one field of inquiry to another, they are frequently the same from one to another. The process of the scientific method involves making conjectures (hypotheses), deriving predictions from them as logical consequences, and then carrying out experiments based on those predictions.[6][7] A hypothesis is a conjecture, based on knowledge obtained while seeking answers to the question. The hypothesis might be very specific, or it might be broad. Scientists then test hypotheses by conducting experiments. A scientific hypothesis must be falsifiable, implying that it is possible to identify a possible outcome of an experiment that conflicts with predictions deduced from the hypothesis; otherwise, the hypothesis cannot be meaningfully tested.[8]

The purpose of an experiment is to determine whether observations agree with or conflict with the predictions derived from a hypothesis.[9] Experiments can take place anywhere from a college lab to CERN's Large Hadron Collider. There are difficulties in a formulaic statement of method, however. Though the scientific method is often presented as a fixed sequence of steps, it represents rather a set of general principles.[10] Not all steps take place in every scientific inquiry (nor to the same degree), and they are not always in the same order.[11] Some philosophers and scientists have argued that there is no scientific method; they include physicist Lee Smolin[12] and philosopher Paul Feyerabend (in his *Against Method*). Nola and Sankey remark that "For some, the whole idea of a theory of scientific method is yesteryear's debate".[13]

18.1 Overview

The DNA example below is a synopsis of this method

The scientific method is the process by which science is carried out.[18] As in other areas of inquiry, science (through the scientific method) can build on previous knowledge and develop a more sophisticated understanding of its topics of study over time.[19][20][21][22][23][24] This model can be seen to underlay the scientific revolution.[25] One thousand years ago, Alhazen argued the importance of forming questions and subsequently testing them,[26] an approach which was advocated by Galileo in 1638 with the publication of *Two New Sciences*.[27] The current method is based on a hypothetico-deductive model[28] formulated in the 20th century, although it has undergone significant revision since first proposed (for a more formal discussion, see below).

18.1.1 Process

The overall process involves making conjectures (hypotheses), deriving predictions from them as logical consequences, and then carrying out experiments based on those predictions to determine whether the original conjecture was correct.[6] There are difficulties in a formulaic statement of method, however. Though the scientific method is often presented as a fixed sequence of steps, these actions are better considered as general principles.[29] Not all steps take place in every scientific inquiry (nor to the same degree), and they are not always done in the same order. As noted by scientist and philosopher William Whewell (1794–1866), "invention, sagacity, [and] genius"[11] are required at every step.

Formulation of a question

The question can refer to the explanation of a specific observation, as in "Why is the sky blue?" but can also be open-ended, as in "How can I design a drug to cure this particular disease?" This stage frequently involves finding and evaluating evidence from previous experiments, personal scientific observations or assertions, as well as the work of other scientists. If the answer is already known, a different question that builds on the evidence can be posed. When applying the scientific method to research, determining a good question can be very difficult and it will affect the outcome of the investigation.[30]

Hypothesis

A hypothesis is a conjecture, based on knowledge obtained while formulating the question, that may explain any given behavior. The hypothesis might be very specific; for example, Einstein's equivalence principle or Francis Crick's "DNA makes RNA makes protein",[31] or it might be broad; for example, unknown species of life dwell in the unexplored depths of the oceans. A statistical hypothesis is a conjecture about a given statistical population. For example, the population might be *people with a particular disease*. The conjecture might be that a new drug will cure the disease in some of those people. Terms commonly associated with statistical hypotheses are null hypothesis and alternative hypothesis. A null hypothesis is the conjecture that the statistical hypothesis is false; for example, that the new drug does nothing and that any cure is caused by chance. Researchers normally want to show that the null hypothesis is false. The alternative hypothesis is the desired outcome, that the drug does better than chance. A final point: a scientific hypothesis must be falsifiable, meaning that one can identify a possible outcome of an experiment that conflicts with predictions deduced from the hypothesis; otherwise, it cannot be meaningfully tested.

Prediction

This step involves determining the logical consequences of the hypothesis. One or more predictions are then selected for further testing. The more unlikely that a prediction would be correct simply by coincidence, then the more convincing it would be if the prediction were fulfilled; evidence is also stronger if the answer to the prediction is not already known, due to the effects of hindsight bias (see also postdiction). Ideally, the prediction must also distinguish the hypothesis from likely alternatives; if two hypotheses make the same prediction, observing the prediction to be correct is not evidence for either one over the other. (These statements about the relative strength of evidence can be mathematically derived using Bayes' Theorem).[32]

Testing

This is an investigation of whether the real world behaves as predicted by the hypothesis. Scientists (and other people) test hypotheses by conducting experiments. The purpose of an experiment is to determine whether observations of the real world agree with or conflict with the predictions derived from a hypothesis. If they agree, confidence in the hypothesis increases; otherwise, it decreases. Agreement does not assure that the hypothesis is true; future experiments may reveal problems. Karl Popper advised scientists to try to falsify hypotheses, i.e., to search for and test those experiments that seem most doubtful. Large numbers of successful confirmations are not convincing if they arise from experiments that avoid risk.[9] Experiments should be designed to minimize possible errors, especially through the use of appropriate scientific controls. For example, tests of medical treatments are commonly run as double-blind tests. Test personnel, who might unwittingly reveal to test subjects which samples are the desired test drugs and which are placebos, are kept ignorant of which are which. Such hints can bias the responses of the test subjects. Furthermore, failure of an experiment does not necessarily mean the hypothesis is false. Experiments always depend on several hypotheses, e.g., that the test equipment is working properly, and a failure may be a failure of one of the auxiliary hypotheses. (See the Duhem–Quine thesis.) Experiments can be conducted in a college lab, on a kitchen table, at CERN's Large Hadron Collider, at the bottom of an ocean, on Mars (using one of the working rovers), and so on. Astronomers do experiments, searching for planets around distant stars. Finally, most individual experiments address highly specific topics for reasons of practicality. As a result, evidence about broader topics is usually accumulated gradually.

Analysis

This involves determining what the results of the experiment show and deciding on the next actions to take. The predictions of the hypothesis are compared to those of the null hypothesis, to determine which is better able to explain the data. In cases where an experiment is repeated many times, a statistical analysis such as a chi-squared test may be required. If the evidence has falsified the hypothesis, a new hypothesis is required; if the experiment supports the hypothesis but the evidence is not strong enough for high confidence, other predictions from the hypothesis must be tested. Once a hypothesis is strongly supported by evidence, a new question can be asked to provide further insight on the same topic. Evidence from other scientists and experience are frequently incorporated at any stage in the process. Depending on the complexity of the experiment, many iterations may be required to gather sufficient evidence to answer a question with confidence, or to build up many answers to highly specific questions in order to answer a single broader question.

18.1.2 DNA example

The discovery became the starting point for many further studies involving the genetic material, such as the field of molecular genetics, and it was awarded the Nobel Prize in 1962. Each step of the example is examined in more detail later in the article.

18.1.3 Other components

The scientific method also includes other components required even when all the iterations of the steps above have been completed:[41]

Replication

If an experiment cannot be repeated to produce the same results, this implies that the original results might have been in error. As a result, it is common for a single experiment to be performed multiple times, especially when there are uncontrolled variables or other indications of experimental error. For significant or surprising results, other scientists may also attempt to replicate the results for themselves, especially if those results would be important to their own work.[42]

External review

The process of peer review involves evaluation of the experiment by experts, who typically give their opinions anonymously. Some journals request that the experimenter provide lists of possible peer reviewers, especially if the field is highly specialized. Peer review does not certify correctness of the results, only that, in the opinion of the reviewer, the experiments themselves were sound (based on the description supplied by the experimenter). If the work passes peer review, which occasionally may require new experiments requested by the reviewers, it will be published in a peer-reviewed scientific journal. The specific journal that publishes the results indicates the perceived quality of the work.[43]

Data recording and sharing

Scientists typically are careful in recording their data, a requirement promoted by Ludwik Fleck (1896–1961) and others.[44] Though not typically required, they might be requested to supply this data to other scientists who wish to replicate their original results (or parts of their original results), extending to the sharing of any experimental samples that may be difficult to obtain.[45]

18.2 Scientific inquiry

Scientific inquiry generally aims to obtain knowledge in the form of testable explanations that scientists can use to predict the results of future experiments. This allows scientists to gain a better understanding of the topic under study, and later to use that understanding to intervene in its causal mechanisms (such as to cure disease). The better an explanation is at making predictions, the more useful it frequently can be, and the more likely it will continue to explain a body of evidence better than its alternatives. The most successful explanations – those which explain and make accurate predictions in a wide range of circumstances – are often called scientific theories.

Most experimental results do not produce large changes in human understanding; improvements in theoretical scientific understanding typically result from a gradual process of development over time, sometimes across different domains of science.[46] Scientific models vary in the extent to which they have been experimentally tested and for how long, and in their acceptance in the scientific community. In general, explanations become accepted over time as evidence accumulates on a given topic, and the explanation in question proves more powerful than its alternatives at explaining the evidence. Often subsequent researchers reformulate the explanations over time, or combined explanations to produce new explanations.

Tow sees the scientific method in terms of an evolutionary algorithm applied to science and technology.[47]

18.2.1 Properties of scientific inquiry

Scientific knowledge is closely tied to empirical findings, and can remain subject to falsification if new experimental observation incompatible with it is found. That is, no theory can ever be considered final, since new problematic evidence might be discovered. If such evidence is found, a new theory may be proposed, or (more commonly) it is found that modifications to the previous theory are sufficient to explain the new evidence. The strength of a theory can be argued to relate to how long it has persisted without major alteration to its core principles.

Theories can also become subsumed by other theories. For example, Newton's laws explained thousands of years of scientific observations of the planets almost perfectly. However, these laws were then determined to be special cases of a more general theory (relativity), which explained both the (previously unexplained) exceptions to Newton's laws and predicted and explained other observations such as the deflection of light by gravity. Thus, in certain cases independent, unconnected, scientific observations can be connected to each other, unified by principles of increasing explanatory power.[48]

Since new theories might be more comprehensive than what preceded them, and thus be able to explain more than previous ones, successor theories might be able to meet a higher standard by explaining a larger body of observations than their predecessors.[48] For example, the theory of evolution explains the diversity of life on Earth, how species adapt to their environments, and many other patterns observed in the natural world;[49][50] its most recent major modification was unification with genetics to form the modern evolutionary synthesis. In subsequent modifications, it has also subsumed aspects of many other fields such as biochemistry and molecular biology.

18.2.2 Beliefs and biases

Flying gallop falsified; see image below

Muybridge's photographs of The Horse in Motion, *1878, were used to answer the question whether all four feet of a galloping horse are ever off the ground at the same time. This demonstrates a use of photography as an experimental tool in science.*

Scientific methodology often directs that hypotheses be tested in controlled conditions wherever possible. This is frequently possible in certain areas, such as in the biological sciences, and more difficult in other areas, such as in astronomy. The practice of experimental control and reproducibility can have the effect of diminishing the potentially harmful effects of circumstance, and to a degree, personal bias. For example, pre-existing beliefs can alter the interpretation of results, as in confirmation bias; this is a heuristic that leads a person with a particular belief to see things as reinforcing their belief, even if another observer might disagree (in other words, people tend to observe what they expect to observe).

A historical example is the belief that the legs of a galloping horse are splayed at the point when none of the horse's legs touches the ground, to the point of this image being included in paintings by its supporters. However, the first stop-action pictures of a horse's gallop by Eadweard Muybridge showed this to be false, and that the legs are instead gathered together.[51] Another important human bias that plays a role is a preference for new, surprising statements (see appeal to novelty), which can result in a search for evidence that the new is true.[2] In contrast to this standard in the scientific method, poorly attested beliefs can be believed and acted upon via a less rigorous heuristic,[52] sometimes taking advantage of the narrative fallacy that when narrative is constructed its elements become easier to believe.[53][54] Sometimes, these have their elements assumed *a priori*, or contain some other logical or methodological flaw in the process that ultimately produced them.[55]

18.3 Elements of the scientific method

There are different ways of outlining the basic method used for scientific inquiry. The scientific community and philosophers of science generally agree on the following classification of method components. These methodological elements and organization of procedures tend to be more characteristic of natural sciences than social sciences. Nonetheless, the cycle of formulating hypotheses, testing and analyzing the results, and formulating new hypotheses, will resemble the cycle described below.

Four essential elements[56][57][58] of the scientific method[59] are iterations,[60][61] recursions,[62] interleavings, or orderings of the following:

- Characterizations (observations,[63] definitions, and measurements of the subject of inquiry)
- Hypotheses[64][65] (theoretical, hypothetical explanations of observations and measurements of the subject)[66]
- Predictions (reasoning including deductive reasoning[67] from the hypothesis or theory)
- Experiments[68] (tests of all of the above)

Each element of the scientific method is subject to peer review for possible mistakes. These activities do not describe all that scientists do (see below) but apply mostly to experimental sciences (e.g., physics, chemistry, and biology). The elements above are often taught in the educational system as "the scientific method".[69]

The scientific method is not a single recipe: it requires intelligence, imagination, and creativity.[70] In this sense, it is not a mindless set of standards and procedures to follow, but is rather an ongoing cycle, constantly developing more useful, accurate and comprehensive models and methods. For example, when Einstein developed the Special and General Theories of Relativity, he did not in any way refute or discount Newton's *Principia*. On the contrary, if the astronomically large, the vanishingly small, and the extremely fast are removed from Einstein's theories – all phenomena Newton could not have observed – Newton's equations are what remain. Einstein's theories are expansions and refinements of Newton's theories and, thus, increase confidence in Newton's work.

A linearized, pragmatic scheme of the four points above is sometimes offered as a guideline for proceeding:[71]

1. Define a question
2. Gather information and resources (observe)
3. Form an explanatory hypothesis
4. Test the hypothesis by performing an experiment and collecting data in a reproducible manner
5. Analyze the data
6. Interpret the data and draw conclusions that serve as a starting point for new hypothesis
7. Publish results
8. Retest (frequently done by other scientists)

The iterative cycle inherent in this step-by-step method goes from point 3 to 6 back to 3 again.

While this schema outlines a typical hypothesis/testing method,[72] it should also be noted that a number of philosophers, historians, and sociologists of science, including Paul Feyerabend, claim that such descriptions of scientific method have little relation to the ways that science is actually practiced.

18.3.1 Characterizations

The scientific method depends upon increasingly sophisticated characterizations of the subjects of investigation. (The *subjects* can also be called *unsolved problems* or the *unknowns*.) For example, Benjamin Franklin conjectured, correctly, that St. Elmo's fire was electrical in nature, but it has taken a long series of experiments and theoretical changes to establish this. While seeking the pertinent properties of the subjects, careful thought may also entail some definitions and observations; the observations often demand careful measurements and/or counting.

The systematic, careful collection of measurements or counts of relevant quantities is often the critical difference between pseudo-sciences, such as alchemy, and science, such as chemistry or biology. Scientific measurements are usually tabulated, graphed, or mapped, and statistical manipulations, such as correlation and regression, performed on them. The measurements might be made in a controlled setting, such as a laboratory, or made on more or less inaccessible or unmanipulatable objects such as stars or human populations. The measurements often require specialized scientific instruments such as thermometers, spectroscopes, particle accelerators, or voltmeters, and the progress of a scientific field is usually intimately tied to their invention and improvement.

> I am not accustomed to saying anything with certainty after only one or two observations.
> — Andreas Vesalius, (1546)[73]

Uncertainty

Measurements in scientific work are also usually accompanied by estimates of their uncertainty. The uncertainty is often estimated by making repeated measurements of the desired quantity. Uncertainties may also be calculated by consideration of the uncertainties of the individual underlying quantities used. Counts of things, such as the number of people in a nation at a particular time, may also have an uncertainty due to data collection limitations. Or counts may represent a sample of desired quantities, with an uncertainty that depends upon the sampling method used and the number of samples taken.

Definition

Measurements demand the use of *operational definitions* of relevant quantities. That is, a scientific quantity is described or defined by how it is measured, as opposed to some more vague, inexact or "idealized" definition. For example, electric current, measured in amperes, may be operationally defined in terms of the mass of silver deposited in a certain time on an electrode in an electrochemical device that is described in some detail. The operational definition of a thing often relies on comparisons with standards: the operational definition of "mass" ultimately relies on the use of an artifact, such as a particular kilogram of platinum-iridium kept in a laboratory in France.

The scientific definition of a term sometimes differs substantially from its natural language usage. For example, mass and weight overlap in meaning in common discourse, but have distinct meanings in mechanics. Scientific quantities are often characterized by their units of measure which can later be described in terms of conventional physical units when communicating the work.

New theories are sometimes developed after realizing certain terms have not previously been sufficiently clearly defined. For example, Albert Einstein's first paper on relativity begins by defining simultaneity and the means for determining length. These ideas were skipped over by Isaac Newton with, "I do not define time, space, place and motion, as being well known to all." Einstein's paper then demonstrates that they (viz., absolute time and length independent of motion) were approximations. Francis Crick cautions us that when characterizing a subject, however, it can be premature to define something when it remains ill-understood.[74] In Crick's study of consciousness, he actually found it easier to study awareness in the visual system, rather than to study free will, for example. His cautionary example was the gene; the gene was much more poorly understood before Watson and Crick's pioneering discovery of the structure of DNA; it would have been counterproductive to spend much time on the definition of the gene, before them.

DNA-characterizations

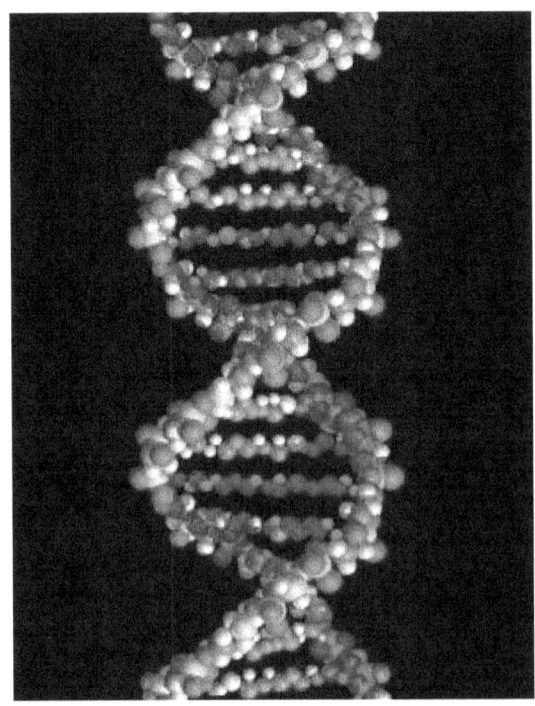

18.3. ELEMENTS OF THE SCIENTIFIC METHOD

The history of the discovery of the structure of DNA is a classic example of the elements of the scientific method: in 1950 it was known that genetic inheritance had a mathematical description, starting with the studies of Gregor Mendel, and that DNA contained genetic information (Oswald Avery's *transforming principle*).[33] But the mechanism of storing genetic information (i.e., genes) in DNA was unclear. Researchers in Bragg's laboratory at Cambridge University made X-ray diffraction pictures of various molecules, starting with crystals of salt, and proceeding to more complicated substances. Using clues painstakingly assembled over decades, beginning with its chemical composition, it was determined that it should be possible to characterize the physical structure of DNA, and the X-ray images would be the vehicle.[75] ..2. DNA-hypotheses

Another example: precession of Mercury

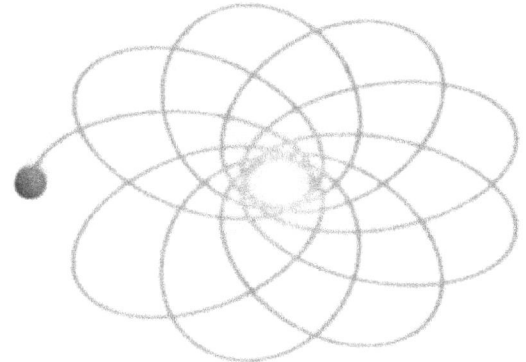

Precession of the perihelion (exaggerated)

The characterization element can require extended and extensive study, even centuries. It took thousands of years of measurements, from the Chaldean, Indian, Persian, Greek, Arabic and European astronomers, to fully record the motion of planet Earth. Newton was able to include those measurements into consequences of his laws of motion. But the perihelion of the planet Mercury's orbit exhibits a precession that cannot be fully explained by Newton's laws of motion (see diagram to the right), as Leverrier pointed out in 1859. The observed difference for Mercury's precession between Newtonian theory and observation was one of the things that occurred to Einstein as a possible early test of his theory of General Relativity. His relativistic calculations matched observation much more closely than did Newtonian theory. The difference is approximately 43 arc-seconds per century.

18.3.2 Hypothesis development

Main article: Hypothesis formation

A hypothesis is a suggested explanation of a phenomenon, or alternately a reasoned proposal suggesting a possible correlation between or among a set of phenomena.

Normally hypotheses have the form of a mathematical model. Sometimes, but not always, they can also be formulated as existential statements, stating that some particular instance of the phenomenon being studied has some characteristic and causal explanations, which have the general form of universal statements, stating that every instance of the phenomenon has a particular characteristic.

Scientists are free to use whatever resources they have – their own creativity, ideas from other fields, inductive reasoning, Bayesian inference, and so on – to imagine possible explanations for a phenomenon under study. Charles Sanders Peirce, borrowing a page from Aristotle (*Prior Analytics*, 2.25) described the incipient stages of inquiry, instigated by the "irritation of doubt" to venture a plausible guess, as *abductive reasoning*. The history of science is filled with stories of scientists claiming a "flash of inspiration", or a hunch, which then motivated them to look for evidence to support or refute their idea. Michael Polanyi made such creativity the centerpiece of his discussion of methodology.

William Glen observes that

> the success of a hypothesis, or its service to science, lies not simply in its perceived "truth", or power to displace, subsume or reduce a predecessor idea, but perhaps more in its ability to stimulate the research that will illuminate ... bald suppositions and areas of vagueness.[76]

In general scientists tend to look for theories that are "elegant" or "beautiful". In contrast to the usual English use of these terms, they here refer to a theory in accordance with the known facts, which is nevertheless relatively simple and easy to handle. Occam's Razor serves as a rule of thumb for choosing the most desirable amongst a group of equally explanatory hypotheses.

To minimize the confirmation bias which results from entertaining a single hypothesis, strong inference emphasizes the need for entertaining multiple alternative hypotheses.[77]

DNA-hypotheses

Linus Pauling proposed that DNA might be a triple helix.[78] This hypothesis was also considered by Francis Crick

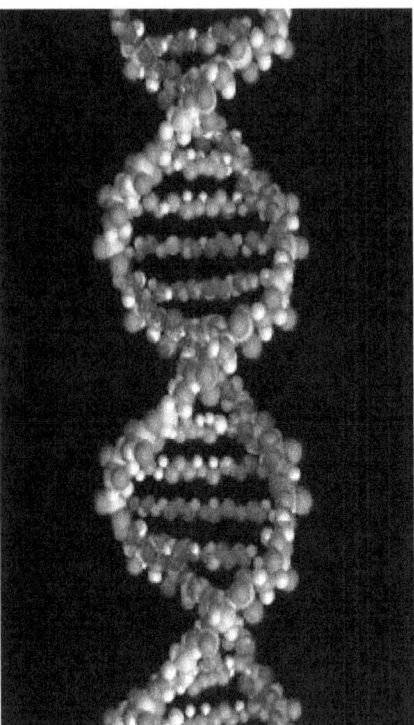

and James D. Watson but discarded. When Watson and Crick learned of Pauling's hypothesis, they understood from existing data that Pauling was wrong[79] and that Pauling would soon admit his difficulties with that structure. So, the race was on to figure out the correct structure (except that Pauling did not realize at the time that he was in a race) ...3. DNA-predictions

18.3.3 Predictions from the hypothesis

Main article: Prediction in science

Any useful hypothesis will enable predictions, by reasoning including deductive reasoning. It might predict the outcome of an experiment in a laboratory setting or the observation of a phenomenon in nature. The prediction can also be statistical and deal only with probabilities.

It is essential that the outcome of testing such a prediction be currently unknown. Only in this case does a successful outcome increase the probability that the hypothesis is true. If the outcome is already known, it is called a consequence and should have already been considered while formulating the hypothesis.

If the predictions are not accessible by observation or experience, the hypothesis is not yet testable and so will remain to that extent unscientific in a strict sense. A new technology or theory might make the necessary experiments feasible. Thus, much scientifically based speculation might convince one (or many) that the hypothesis that other intelligent species exist is true. But since there no experiment now known which can test this hypothesis, science itself can have little to say about the possibility. In future, some new technique might lead to an experimental test and the speculation would then become part of accepted science.

DNA-predictions

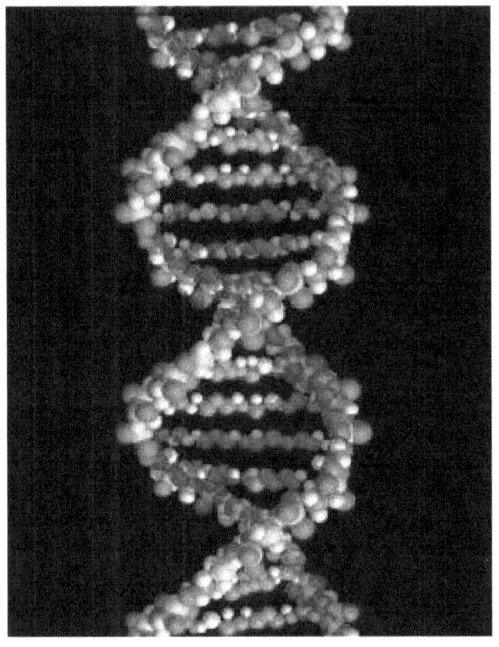

James D. Watson, Francis Crick, and others hypothesized that DNA had a helical structure. This implied that DNA's X-ray diffraction pattern would be 'x shaped'.[36][80] This prediction followed from the work of Cochran, Crick and Vand[37] (and independently by Stokes). The Cochran-Crick-Vand-Stokes theorem provided a mathematical explanation for the empirical observation that diffraction from helical structures produces x shaped patterns.

In their first paper, Watson and Crick also noted that the double helix structure they proposed provided a simple mechanism for DNA replication, writing, "It has not escaped our notice that the specific pairing we have postulated immediately suggests a possible copying mechanism for the genetic material".[81] ...4. DNA-experiments

Another example: general relativity

Einstein's theory of General Relativity makes several specific predictions about the observable structure of spacetime, such as that light bends in a gravitational field, and that the amount of bending depends in a precise way on the strength of that gravitational field. Arthur Eddington's observations made during a 1919 solar eclipse supported General Relativity rather than Newtonian gravitation.[82]

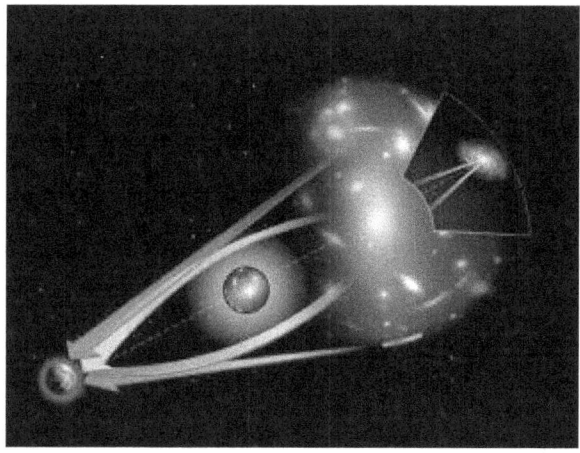

Einstein's prediction (1907): Light bends in a gravitational field

18.3.4 Experiments

Main article: Experiment

Once predictions are made, they can be sought by experiments. If the test results contradict the predictions, the hypotheses which entailed them are called into question and become less tenable. Sometimes the experiments are conducted incorrectly or are not very well designed, when compared to a crucial experiment. If the experimental results confirm the predictions, then the hypotheses are considered more likely to be correct, but might still be wrong and continue to be subject to further testing. The experimental control is a technique for dealing with observational error. This technique uses the contrast between multiple samples (or observations) under differing conditions to see what varies or what remains the same. We vary the conditions for each measurement, to help isolate what has changed. Mill's canons can then help us figure out what the important factor is.[83] Factor analysis is one technique for discovering the important factor in an effect.

Depending on the predictions, the experiments can have different shapes. It could be a classical experiment in a laboratory setting, a double-blind study or an archaeological excavation. Even taking a plane from New York to Paris is an experiment which tests the aerodynamical hypotheses used for constructing the plane.

Scientists assume an attitude of openness and accountability on the part of those conducting an experiment. Detailed record keeping is essential, to aid in recording and reporting on the experimental results, and supports the effectiveness and integrity of the procedure. They will also assist in reproducing the experimental results, likely by others. Traces of this approach can be seen in the work of Hipparchus (190–120 BCE), when determining a value for the precession of the Earth, while controlled experiments can be seen in the works of Jābir ibn Hayyān (721–815 CE), al-Battani (853–929) and Alhazen (965–1039).[84]

DNA-experiments

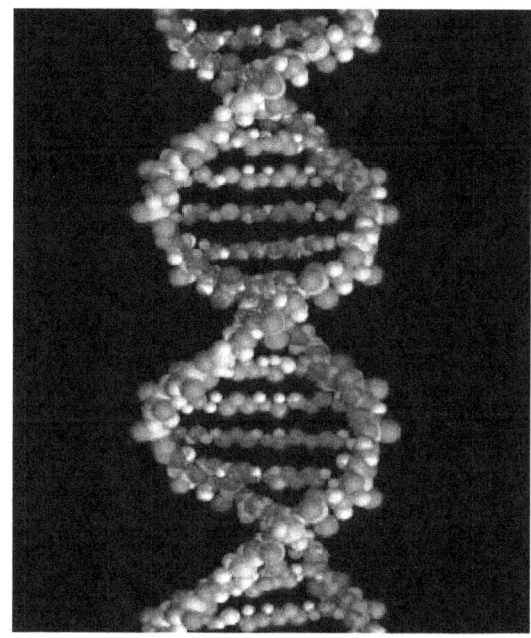

Watson and Crick showed an initial (and incorrect) proposal for the structure of DNA to a team from Kings College – Rosalind Franklin, Maurice Wilkins, and Raymond Gosling. Franklin immediately spotted the flaws which concerned the water content. Later Watson saw Franklin's detailed X-ray diffraction images which showed an X-shape and was able to confirm the structure was helical.[38][39] This rekindled Watson and Crick's model building and led to the correct structure. ..1. DNA-characterizations

18.3.5 Evaluation and improvement

The scientific method is iterative. At any stage it is possible to refine its accuracy and precision, so that some consideration will lead the scientist to repeat an earlier part of the process. Failure to develop an interesting hypothesis may lead a scientist to re-define the subject under consideration. Failure of a hypothesis to produce interesting and testable predictions may lead to reconsideration of the hypothesis or of the definition of the subject. Failure of an experiment to produce interesting results may lead a scientist to reconsider the experimental method, the hypothesis, or the definition of the subject.

Other scientists may start their own research and enter the process at any stage. They might adopt the characterization and formulate their own hypothesis, or they might adopt the hypothesis and deduce their own predictions. Often the experiment is not done by the person who made the prediction, and the characterization is based on experiments done by someone else. Published results of experiments can also serve as a hypothesis predicting their own reproducibility.

DNA-iterations

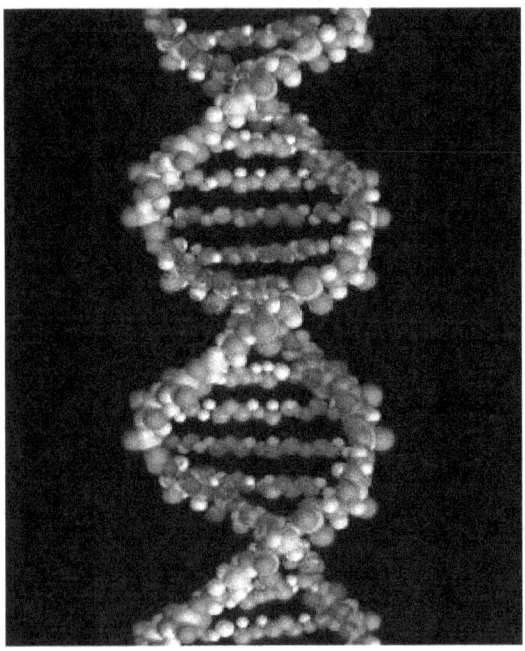

After considerable fruitless experimentation, being discouraged by their superior from continuing, and numerous false starts,[85][86][87] Watson and Crick were able to infer the essential structure of DNA by concrete modeling of the physical shapes of the nucleotides which comprise it.[40][88] They were guided by the bond lengths which had been deduced by Linus Pauling and by Rosalind Franklin's X-ray diffraction images. ..*DNA Example*

18.3.6 Confirmation

Science is a social enterprise, and scientific work tends to be accepted by the scientific community when it has been confirmed. Crucially, experimental and theoretical results must be reproduced by others within the scientific community. Researchers have given their lives for this vision; Georg Wilhelm Richmann was killed by ball lightning (1753) when attempting to replicate the 1752 kite-flying experiment of Benjamin Franklin.[89]

To protect against bad science and fraudulent data, government research-granting agencies such as the National Science Foundation, and science journals, including *Nature* and *Science*, have a policy that researchers must archive their data and methods so that other researchers can test the data and methods and build on the research that has gone before. Scientific data archiving can be done at a number of national archives in the U.S. or in the World Data Center.

18.4 Models of scientific inquiry

Main article: Models of scientific inquiry

18.4.1 Classical model

The classical model of scientific inquiry derives from Aristotle,[90] who distinguished the forms of approximate and exact reasoning, set out the threefold scheme of abductive, deductive, and inductive inference, and also treated the compound forms such as reasoning by analogy.

18.4.2 Pragmatic model

See also: Pragmatic theory of truth

In 1877,[19] Charles Sanders Peirce (/ˈpɜːrs/ like "purse"; 1839–1914) characterized inquiry in general not as the pursuit of truth *per se* but as the struggle to move from irritating, inhibitory doubts born of surprises, disagreements, and the like, and to reach a secure belief, belief being that on which one is prepared to act. He framed scientific inquiry as part of a broader spectrum and as spurred, like inquiry generally, by actual doubt, not mere verbal or hyperbolic doubt, which he held to be fruitless.[91] He outlined four methods of settling opinion, ordered from least to most successful:

1. The method of tenacity (policy of sticking to initial belief) – which brings comforts and decisiveness but leads to trying to ignore contrary information and others' views as if truth were intrinsically private, not public. It goes against the social impulse and easily falters since one may well notice when another's opinion is as good as one's own initial opinion. Its successes can shine but tend to be transitory.[92]

2. The method of authority – which overcomes disagreements but sometimes brutally. Its successes can be majestic and long-lived, but it cannot operate thoroughly enough to suppress doubts indefinitely, especially when people learn of other societies present and past.

3. The method of the *a priori* – which promotes conformity less brutally but fosters opinions as something

like tastes, arising in conversation and comparisons of perspectives in terms of "what is agreeable to reason." Thereby it depends on fashion in paradigms and goes in circles over time. It is more intellectual and respectable but, like the first two methods, sustains accidental and capricious beliefs, destining some minds to doubt it.

4. **The scientific method** – the method wherein inquiry regards itself as fallible and purposely tests itself and criticizes, corrects, and improves itself.

Peirce held that slow, stumbling ratiocination can be dangerously inferior to instinct and traditional sentiment in practical matters, and that the scientific method is best suited to theoretical research,[93] which in turn should not be trammeled by the other methods and practical ends; reason's "first rule" is that, in order to learn, one must desire to learn and, as a corollary, must not block the way of inquiry.[94] The scientific method excels the others by being deliberately designed to arrive – eventually – at the most secure beliefs, upon which the most successful practices can be based. Starting from the idea that people seek not truth *per se* but instead to subdue irritating, inhibitory doubt, Peirce showed how, through the struggle, some can come to submit to truth for the sake of belief's integrity, seek as truth the guidance of potential practice correctly to its given goal, and wed themselves to the scientific method.[19][22]

For Peirce, rational inquiry implies presuppositions about truth and the real; to reason is to presuppose (and at least to hope), as a principle of the reasoner's self-regulation, that the real is discoverable and independent of our vagaries of opinion. In that vein he defined truth as the correspondence of a sign (in particular, a proposition) to its object and, pragmatically, not as actual consensus of some definite, finite community (such that to inquire would be to poll the experts), but instead as that final opinion which all investigators *would* reach sooner or later but still inevitably, if they were to push investigation far enough, even when they start from different points.[95] In tandem he defined the real as a true sign's object (be that object a possibility or quality, or an actuality or brute fact, or a necessity or norm or law), which is what it is independently of any finite community's opinion and, pragmatically, depends only on the final opinion destined in a sufficient investigation. That is a destination as far, or near, as the truth itself to you or me or the given finite community. Thus, his theory of inquiry boils down to "Do the science." Those conceptions of truth and the real involve the idea of a community both without definite limits (and thus potentially self-correcting as far as needed) and capable of definite increase of knowledge.[96] As inference, "logic is rooted in the social principle" since it depends on a standpoint that is, in a sense, unlimited.[97]

Paying special attention to the generation of explanations,

Peirce outlined the scientific method as a coordination of three kinds of inference in a purposeful cycle aimed at settling doubts, as follows (in §III–IV in "A Neglected Argument"[6] except as otherwise noted):

1. **Abduction** (or retroduction). Guessing, inference to explanatory hypotheses for selection of those best worth trying. From abduction, Peirce distinguishes induction as inferring, on the basis of tests, the proportion of truth in the hypothesis. Every inquiry, whether into ideas, brute facts, or norms and laws, arises from surprising observations in one or more of those realms (and for example at any stage of an inquiry already underway). All explanatory content of theories comes from abduction, which guesses a new or outside idea so as to account in a simple, economical way for a surprising or complicative phenomenon. Oftenest, even a well-prepared mind guesses wrong. But the modicum of success of our guesses far exceeds that of sheer luck and seems born of attunement to nature by instincts developed or inherent, especially insofar as best guesses are optimally plausible and simple in the sense, said Peirce, of the "facile and natural", as by Galileo's natural light of reason and as distinct from "logical simplicity". Abduction is the most fertile but least secure mode of inference. Its general rationale is inductive: it succeeds often enough and, without it, there is no hope of sufficiently expediting inquiry (often multi-generational) toward new truths.[98] Coordinative method leads from abducing a plausible hypothesis to judging it for its testability[99] and for how its trial would economize inquiry itself.[100] Peirce calls his pragmatism "the logic of abduction".[101] His pragmatic maxim is: "Consider what effects that might conceivably have practical bearings you conceive the objects of your conception to have. Then, your conception of those effects is the whole of your conception of the object".[95] His pragmatism is a method of reducing conceptual confusions fruitfully by equating the meaning of any conception with the conceivable practical implications of its object's conceived effects – a method of experimentational mental reflection hospitable to forming hypotheses and conducive to testing them. It favors efficiency. The hypothesis, being insecure, needs to have practical implications leading at least to mental tests and, in science, lending themselves to scientific tests. A simple but unlikely guess, if uncostly to test for falsity, may belong first in line for testing. A guess is intrinsically worth testing if it has instinctive plausibility or reasoned objective probability, while subjective likelihood, though reasoned, can be misleadingly seductive. Guesses can be chosen for trial strategically, for their caution (for which Peirce gave as example the game

of Twenty Questions), breadth, and incomplexity.[102] One can hope to discover only that which time would reveal through a learner's sufficient experience anyway, so the point is to expedite it; the economy of research is what demands the leap, so to speak, of abduction and governs its art.[100]

2. **Deduction**. Two stages:

    (a) Explication. Unclearly premissed, but deductive, analysis of the hypothesis in order to render its parts as clear as possible.

    (b) Demonstration: Deductive Argumentation, Euclidean in procedure. Explicit deduction of hypothesis's consequences as predictions, for induction to test, about evidence to be found. Corollarial or, if needed, theorematic.

3. **Induction**. The long-run validity of the rule of induction is deducible from the principle (presuppositional to reasoning in general[95]) that the real is only the object of the final opinion to which adequate investigation would lead;[103] anything to which no such process would ever lead would not be real. Induction involving ongoing tests or observations follows a method which, sufficiently persisted in, will diminish its error below any predesignate degree. Three stages:

    (a) Classification. Unclearly premissed, but inductive, classing of objects of experience under general ideas.

    (b) Probation: direct inductive argumentation. Crude (the enumeration of instances) or gradual (new estimate of proportion of truth in the hypothesis after each test). Gradual induction is qualitative or quantitative; if qualitative, then dependent on weightings of qualities or characters;[104] if quantitative, then dependent on measurements, or on statistics, or on countings.

    (c) Sentential Induction. "...which, by inductive reasonings, appraises the different probations singly, then their combinations, then makes self-appraisal of these very appraisals themselves, and passes final judgment on the whole result".

18.5 Communication and community

See also: Scientific community and Scholarly communication

Frequently the scientific method is employed not only by a single person, but also by several people cooperating directly or indirectly. Such cooperation can be regarded as an important element of a scientific community. Various standards of scientific methodology are used within such an environment.

18.5.1 Peer review evaluation

Scientific journals use a process of *peer review*, in which scientists' manuscripts are submitted by editors of scientific journals to (usually one to three) fellow (usually anonymous) scientists familiar with the field for evaluation. In certain journals, the journal itself selects the referees; while in others (especially journals that are extremely specialized), the manuscript author might recommend referees. The referees may or may not recommend publication, or they might recommend publication with suggested modifications, or sometimes, publication in another journal. This standard is practiced to various degrees by different journals, and can have the effect of keeping the literature free of obvious errors and to generally improve the quality of the material, especially in the journals who use the standard most rigorously. The peer review process can have limitations when considering research outside the conventional scientific paradigm: problems of "groupthink" can interfere with open and fair deliberation of some new research.[105]

18.5.2 Documentation and replication

Main article: Reproducibility

Sometimes experimenters may make systematic errors during their experiments, veer from standard methods and practices (Pathological science) for various reasons, or, in rare cases, deliberately report false results. Occasionally because of this then, other scientists might attempt to repeat the experiments in order to duplicate the results.

Archiving

Researchers sometimes practice scientific data archiving, such as in compliance with the policies of government funding agencies and scientific journals. In these cases, detailed records of their experimental procedures, raw data, statistical analyses and source code can be preserved in order to provide evidence of the methodology and practice of the procedure and assist in any potential future attempts to reproduce the result. These procedural records may also assist in the conception of new experiments to test the hypothesis, and may prove useful to engineers who might examine the potential practical applications of a discovery.

Data sharing

When additional information is needed before a study can be reproduced, the author of the study might be asked to provide it. They might provide it, or if the author refuses to share data, appeals can be made to the journal editors who published the study or to the institution which funded the research.

Limitations

Since it is impossible for a scientist to record *everything* that took place in an experiment, facts selected for their apparent relevance are reported. This may lead, unavoidably, to problems later if some supposedly irrelevant feature is questioned. For example, Heinrich Hertz did not report the size of the room used to test Maxwell's equations, which later turned out to account for a small deviation in the results. The problem is that parts of the theory itself need to be assumed in order to select and report the experimental conditions. The observations are hence sometimes described as being 'theory-laden'.

18.5.3 Dimensions of practice

Further information: Rhetoric of science

The primary constraints on contemporary science are:

- Publication, i.e. Peer review

- Resources (mostly funding)

It has not always been like this: in the old days of the "gentleman scientist" funding (and to a lesser extent publication) were far weaker constraints.

Both of these constraints indirectly require scientific method – work that violates the constraints will be difficult to publish and difficult to get funded. Journals require submitted papers to conform to "good scientific practice" and to a degree this can be enforced by peer review. Originality, importance and interest are more important – see for example the author guidelines for *Nature*.

Smaldino and McElreath 2016 have noted that our need to reward scientific understanding is being nullified by poor research design and poor data analysis, which is leading to false-positive findings.[106]

18.6 Philosophy and sociology of science

See also: Philosophy of science and Sociology of science

Philosophy of science looks at the underpinning logic of the scientific method, at what separates science from non-science, and the ethic that is implicit in science. There are basic assumptions, derived from philosophy by at least one prominent scientist, that form the base of the scientific method – namely, that reality is objective and consistent, that humans have the capacity to perceive reality accurately, and that rational explanations exist for elements of the real world.[107] These assumptions from methodological naturalism form a basis on which science may be grounded. Logical Positivist, empiricist, falsificationist, and other theories have criticized these assumptions and given alternative accounts of the logic of science, but each has also itself been criticized. More generally, the scientific method can be recognized as an idealization.[108]

Thomas Kuhn examined the history of science in his *The Structure of Scientific Revolutions*, and found that the actual method used by scientists differed dramatically from the then-espoused method. His observations of science practice are essentially sociological and do not speak to how science is or can be practiced in other times and other cultures.

Norwood Russell Hanson, Imre Lakatos and Thomas Kuhn have done extensive work on the "theory laden" character of observation. Hanson (1958) first coined the term for the idea that all observation is dependent on the conceptual framework of the observer, using the concept of gestalt to show how preconceptions can affect both observation and description.[109] He opens Chapter 1 with a discussion of the Golgi bodies and their initial rejection as an artefact of staining technique, and a discussion of Brahe and Kepler observing the dawn and seeing a "different" sun rise despite the same physiological phenomenon. Kuhn[110] and Feyerabend[111] acknowledge the pioneering significance of his work.

Kuhn (1961) said the scientist generally has a theory in mind before designing and undertaking experiments so as to make empirical observations, and that the "route from theory to measurement can almost never be traveled backward". This implies that the way in which theory is tested is dictated by the nature of the theory itself, which led Kuhn (1961, p. 166) to argue that "once it has been adopted by a profession ... no theory is recognized to be testable by any quantitative tests that it has not already passed".[112]

Paul Feyerabend similarly examined the history of science, and was led to deny that science is genuinely a methodological process. In his book *Against Method* he argues that scientific progress is *not* the result of applying any particular method. In essence, he says that for any specific method or norm of science, one can find a historic episode where violating it has contributed to the progress of science. Thus, if believers in scientific method wish to express a single universally valid rule, Feyerabend jokingly suggests, it should be 'anything goes'.[113] Criticisms such as his led to the strong programme, a radical approach to the sociology of science.

The postmodernist critiques of science have themselves been the subject of intense controversy. This ongoing debate, known as the science wars, is the result of conflicting values and assumptions between the postmodernist and realist camps. Whereas postmodernists assert that scientific knowledge is simply another discourse (note that this term has special meaning in this context) and not representative of any form of fundamental truth, realists in the scientific community maintain that scientific knowledge does reveal real and fundamental truths about reality. Many books have been written by scientists which take on this problem and challenge the assertions of the postmodernists while defending science as a legitimate method of deriving truth.[114]

18.6.1 Role of chance in discovery

Main article: Role of chance in scientific discoveries

Somewhere between 33% and 50% of all scientific discoveries are estimated to have been *stumbled upon*, rather than sought out. This may explain why scientists so often express that they were lucky.[115] Louis Pasteur is credited with the famous saying that "Luck favours the prepared mind", but some psychologists have begun to study what it means to be 'prepared for luck' in the scientific context. Research is showing that scientists are taught various heuristics that tend to harness chance and the unexpected.[115][116] This is what Nassim Nicholas Taleb calls "Anti-fragility"; while some systems of investigation are fragile in the face of human error, human bias, and randomness, the scientific method is more than resistant or tough – it actually benefits from such randomness in many ways (it is anti-fragile). Taleb believes that the more anti-fragile the system, the more it will flourish in the real world.[23]

Psychologist Kevin Dunbar says the process of discovery often starts with researchers finding bugs in their experiments. These unexpected results lead researchers to try to fix what they *think* is an error in their method. Eventually, the researcher decides the error is too persistent and systematic to be a coincidence. The highly controlled, cautious and curious aspects of the scientific method are thus what make it well suited for identifying such persistent systematic errors. At this point, the researcher will begin to think of theoretical explanations for the error, often seeking the help of colleagues across different domains of expertise.[115][116]

18.7 History

Main article: History of scientific method
See also: Timeline of the history of scientific method
The history of scientific method considers changes in

Aristotle, 384–322 BCE. "As regards his method, Aristotle is recognized as the inventor of scientific method because of his refined analysis of logical implications contained in demonstrative discourse, which goes well beyond natural logic and does not owe anything to the ones who philosophized before him." – Riccardo Pozzo[117]

the methodology of scientific inquiry, as distinct from the history of science itself. The development of rules for scientific reasoning has not been straightforward; scientific method has been the subject of intense and recurring debate throughout the history of science, and eminent natural philosophers and scientists have argued for the primacy of one or another approach to establishing scientific knowledge. Despite the disagreements about approaches, scientific method has advanced in definite steps. Rationalist explanations of nature, including atomism, appeared both in ancient Greece in the thought of Leucippus and

Democritus, and in ancient India, in the Nyaya, Vaisesika and Buddhist schools, while Charvaka materialism rejected inference as a source of knowledge in favour of an empiricism that was always subject to doubt. Aristotle pioneered scientific method in ancient Greece alongside his empirical biology and his work on logic, rejecting a purely deductive framework in favour of generalisations made from observations of nature. Important debates in the history of scientific method center on rationalism, especially as advocated by René Descartes, inductivism, which rose to particular prominence with Isaac Newton and his followers, and hypothetico-deductivism, which came to the fore in the early 19th century. In the late 19th and early 20th centuries, a debate over realism vs. antirealism was conducted as powerful scientific theories extended beyond the realm of the observable, while in the mid-20th century, prominent philosophers such as Paul Feyerabend argued against any universal rules of science at all.[118]

18.8 Relationship with mathematics

Science is the process of gathering, comparing, and evaluating proposed models against observables. A model can be a simulation, mathematical or chemical formula, or set of proposed steps. Science is like mathematics in that researchers in both disciplines can clearly distinguish what is *known* from what is *unknown* at each stage of discovery. Models, in both science and mathematics, need to be internally consistent and also ought to be *falsifiable* (capable of disproof). In mathematics, a statement need not yet be proven; at such a stage, that statement would be called a conjecture. But when a statement has attained mathematical proof, that statement gains a kind of immortality which is highly prized by mathematicians, and for which some mathematicians devote their lives.[119]

Mathematical work and scientific work can inspire each other.[120] For example, the technical concept of time arose in science, and timelessness was a hallmark of a mathematical topic. But today, the Poincaré conjecture has been proven using time as a mathematical concept in which objects can flow (see Ricci flow).

Nevertheless, the connection between mathematics and reality (and so science to the extent it describes reality) remains obscure. Eugene Wigner's paper, *The Unreasonable Effectiveness of Mathematics in the Natural Sciences*, is a very well known account of the issue from a Nobel Prize-winning physicist. In fact, some observers (including some well known mathematicians such as Gregory Chaitin, and others such as Lakoff and Núñez) have suggested that mathematics is the result of practitioner bias and human limitation (including cultural ones), somewhat like the post-modernist view of science.

George Pólya's work on problem solving,[121] the construction of mathematical proofs, and heuristic[122][123] show that the mathematical method and the scientific method differ in detail, while nevertheless resembling each other in using iterative or recursive steps.

In Pólya's view, *understanding* involves restating unfamiliar definitions in your own words, resorting to geometrical figures, and questioning what we know and do not know already; *analysis*, which Pólya takes from Pappus,[124] involves free and heuristic construction of plausible arguments, working backward from the goal, and devising a plan for constructing the proof; *synthesis* is the strict Euclidean exposition of step-by-step details[125] of the proof; *review* involves reconsidering and re-examining the result and the path taken to it.

Gauss, when asked how he came about his theorems, once replied "durch planmässiges Tattonieren" (through systematic palpable experimentation).[126]

Imre Lakatos argued that mathematicians actually use contradiction, criticism and revision as principles for improving their work.[127] In like manner to science, where truth is sought, but certainty is not found, in *Proofs and refutations* (1976), what Lakatos tried to establish was that no theorem of informal mathematics is final or perfect. This means that we should not think that a theorem is ultimately true, only that no counterexample has yet been found. Once a counterexample, i.e. an entity contradicting/not explained by the theorem is found, we adjust the theorem, possibly extending the domain of its validity. This is a continuous way our knowledge accumulates, through the logic and process of proofs and refutations. (If axioms are given for a branch of mathematics, however, Lakatos claimed that proofs from those axioms were tautological, i.e. logically true, by rewriting them, as did Poincaré (*Proofs and Refutations*, 1976).)

Lakatos proposed an account of mathematical knowledge based on Polya's idea of heuristics. In *Proofs and Refutations*, Lakatos gave several basic rules for finding proofs and counterexamples to conjectures. He thought that mathematical 'thought experiments' are a valid way to discover mathematical conjectures and proofs.[128]

18.9 Relationship with statistics

The scientific method has been extremely successful in bringing the world out of medieval thinking, especially once it was combined with industrial processes.[129] However, when the scientific method employs statistics as part of its arsenal, there are mathematical and practical issues that can have a deleterious effect on the reliability of the output of scientific methods. This is described in a popular 2005 scientific paper "Why Most Published Research Findings Are False" by John Ioannidis.[130]

The particular points raised are statistical ("The smaller the studies conducted in a scientific field, the less likely the research findings are to be true" and "The greater the flexibility in designs, definitions, outcomes, and analytical modes in a scientific field, the less likely the research findings are to be true.") and economical ("The greater the financial and other interests and prejudices in a scientific field, the less likely the research findings are to be true" and "The hotter a scientific field (with more scientific teams involved), the less likely the research findings are to be true.") Hence: "Most research findings are false for most research designs and for most fields" and "As shown, the majority of modern biomedical research is operating in areas with very low pre- and poststudy probability for true findings." However: "Nevertheless, most new discoveries will continue to stem from hypothesis-generating research with low or very low pre-study odds," which means that *new* discoveries will come from research that, when that research started, had low or very low odds (a low or very low chance) of succeeding. Hence, if the scientific method is used to expand the frontiers of knowledge, research into areas that are outside the mainstream will yield most new discoveries.

18.10 See also

- Armchair theorizing
- Contingency
- Empirical limits in science
- Evidence-based medicine
- Fuzzy logic
- Information theory
- Logic
  - Historical method
  - Philosophical methodology
  - Scholarly method
- Operationalization
- Quantitative research
- Replication crisis
- Social research
- Strong inference
- Testability
- Verificationism

18.10.1 Problems and issues

- Holism in science
- Junk science
- List of cognitive biases
- Normative science
- Philosophical skepticism
- Poverty of the stimulus
- Problem of induction
- Reference class problem
- Skeptical hypotheses
- Underdetermination

18.10.2 History, philosophy, sociology

- Epistemology
- Epistemic truth
- Mertonian norms
- Normal science
- Post-normal science
- Science studies
- Sociology of scientific knowledge

18.11 Notes

[1] Garland, Jr., Theodore (20 March 2015). "The Scientific Method as an Ongoing Process". U C Riverside. Archived from the original on 19 August 2016.

[2] Goldhaber & Nieto 2010, p. 940

[3] "[4] Rules for the study of natural philosophy", Newton transl 1999, pp. 794–96, after Book 3, *The System of the World*.

[4] "scientific method", *Oxford Dictionaries: British and World English*, 2016, retrieved 28 May 2016

[5] "The Scientific Method is a process used to design and perform experiments. It helps to minimize experimental errors and bias, and increase confidence in the accuracy of your results." —The Scientific Method by Science Made Simple. (n.d.). Retrieved November 08, 2016, from http://www.sciencemadesimple.com/scientific_method.html

[6] Peirce, Charles Sanders (1908). *A Neglected Argument for the Reality of God*. **7**. Wikisource. pp. 90–112. with added notes. Reprinted with previously unpublished part, *Collected Papers* v. 6, paragraphs 452–85, *The Essential Peirce* v. 2, pp. 434–50, and elsewhere.

[7] See, for example, Galileo 1638. His thought experiments disprove Aristotle's physics of falling bodies, in *Two New Sciences*.

[8] Popper 1959[273]

[9] Karl R. Popper, *Conjectures and Refutations: The Growth of Scientific Knowledge*, Routledge, 2003 ISBN 0-415-28594-1

[10] Gauch, Hugh G. (2003). *Scientific Method in Practice* (Reprint ed.). Cambridge University Press. p. 3. ISBN 9780521017084. The scientific method 'is often misrepresented as a fixed sequence of steps,' rather than being seen for what it truly is, 'a highly variable and creative process' (AAAS 2000:18). The claim here is that science has general principles that must be mastered to increase productivity and enhance perspective, not that these principles provide a simple and automated sequence of steps to follow.

[11] *History of Inductive Science* (1837), and in *Philosophy of Inductive Science* (1840)

[12] Smolin, Lee. "There is No Scientific Method". Retrieved 2016-06-07.

[13] Nola, Robert (2001). *After Popper, Kuhn and Feyerabend. Recent Issues in Theories of Scientific Method*. Springer Science & Business Media. p. "Introduction". ISBN 1402002467.

[14] Jim Al-Khalili (4 January 2009). "The 'first true scientist'". BBC News.

[15] Tracey Tokuhama-Espinosa (2010). *Mind, Brain, and Education Science: A Comprehensive Guide to the New Brain-Based Teaching*. W. W. Norton & Company. p. 39. ISBN 9780393706079. Alhazen (or Al-Haytham; 965–1039 CE) was perhaps one of the greatest physicists of all times and a product of the Islamic Golden Age or Islamic Renaissance (7th–13th centuries). He made significant contributions to anatomy, astronomy, engineering, mathematics, medicine, ophthalmology, philosophy, physics, psychology, and visual perception and is primarily attributed as the inventor of the scientific method, for which author Bradley Steffens (2006) describes him as the "first scientist".

[16] Peirce, C. S., *Collected Papers* v. 1, paragraph 74.

[17] Albert Einstein, "On the Method of Theoretical Physics", in Essays in Science (Dover, 2009 [1934]), pp. 12-21.

[18] " The thesis of this book, as set forth in Chapter One, is that there are general principles applicable to all the sciences." __ Gauch 2003, p. xv

[19] Peirce, Charles Sanders (1877). *The Fixation of Belief*. **12**. Wikisource. pp. 1–15..

[20] Gauch 2003, p. 1 The scientific method can function in the same way; This is the principle of noncontradiction.

[21] Francis Bacon(1629) *New Organon*, lists 4 types of error: Idols of the tribe (error due to the entire human race), the cave (errors due to an individual's own intellect), the marketplace (errors due to false words), and the theater (errors due to incredulous acceptance).

[22] Peirce, C. S., *Collected Papers* v. 5, in paragraph 582, from 1898:

> ... [rational] inquiry of every type, fully carried out, has the vital power of self-correction and of growth. This is a property so deeply saturating its inmost nature that it may truly be said that there is but one thing needful for learning the truth, and that is a hearty and active desire to learn what is true.

[23] Taleb contributes a brief description of anti-fragility. http://www.edge.org/q2011/q11_3.html

[24] For example, the concept of falsification (first proposed in 1934) formalizes the attempt to *disprove* hypotheses rather than prove them. *Karl R. Popper (1963), 'The Logic of Scientific Discovery'. The Logic of Scientific Discovery pp. 17–20, 249–52, 437–38, and elsewhere.*

- Leon Lederman, for teaching physics first, illustrates how to avoid confirmation bias: Ian Shelton, in Chile, was initially skeptical that supernova 1987a was real, but possibly an artifact of instrumentation (null hypothesis), so he went outside and disproved his null hypothesis by observing SN 1987a with the naked eye. The Kamiokande experiment, in Japan, independently observed neutrinos from SN 1987a at the same time.

[25] Lindberg 2007, pp. 2–3: "There is a danger that must be avoided. ... If we wish to do justice to the historical enterprise, we must take the past for what it was. And that means we must resist the temptation to scour the past for examples or precursors of modern science. ...My concern will be with the beginnings of scientific *theories*, the methods by which they were formulated, and the uses to which they were put: ... "

[26] "How does light travel through transparent bodies? Light travels through transparent bodies in straight lines only.... We have explained this exhaustively in our *Book of Optics*. But let us now mention something to prove this convincingly: the fact that light travels in straight lines is clearly observed in the lights which enter into dark rooms through holes.... [T]he entering light will be clearly observable in the dust which fills the air. – Alhazen, *Treatise on Light* (رسالة في الضوء), translated into English from German by M. Schwarz, from "Abhandlung über das Licht", J. Baarmann (editor and translator from Arabic to German, 1882) *Zeitschrift der Deutschen Morgenländischen Gesellschaft* Vol **36** as quoted in Sambursky 1974, p. 136.

- He demonstrated his conjecture that "light travels through transparent bodies in straight lines only" by placing a straight stick or a taut thread next to the light beam, as quoted in Sambursky 1974, p. 136 to prove that light travels in a straight line.

- David Hockney, (2001, 2006) in *Secret Knowledge: rediscovering the lost techniques of the old masters* ISBN 0-14-200512-6 (expanded edition) cites Alhazen several times as the likely source for the portraiture technique using the camera obscura, which Hockney rediscovered with the aid of an optical suggestion from Charles M. Falco. *Kitab al-Manazir*, which is Alhazen's *Book of Optics*, at that time denoted *Opticae Thesaurus, Alhazen Arabis*, was translated from Arabic into Latin for European use as early as 1270. Hockney cites Friedrich Risner's 1572 Basle edition of *Opticae Thesaurus*. Hockney quotes Alhazen as the first clear description of the camera obscura in Hockney, p. 240.

"Truth is sought for its own sake. And those who are engaged upon the quest for anything for its own sake are not interested in other things. Finding the truth is difficult, and the road to it is rough." – Alhazen (Ibn Al-Haytham 965 – c. 1040) *Critique of Ptolemy*, translated by S. Pines, *Actes X Congrès internationale d'histoire des sciences*, Vol I Ithaca 1962, as quoted in Sambursky 1974, p. 139. (This quotation is from Alhazen's critique of Ptolemy's books *Almagest, Planetary Hypotheses*, and *Optics* as translated into English by A. Mark Smith.)

[27] Galilei, Galileo (1638), *Discorsi e Dimonstrazioni Matematiche, intorno a due nuove scienze*, Leida: Apresso gli Elsevirri, ISBN 0-486-60099-8, Dover reprint of the 1914 Macmillan translation by Henry Crew and Alfonso de Salvio of *Two New Sciences*. Galileo Galilei Linceo (1638). Additional publication information is from the collection of first editions of the Library of Congress surveyed by Bruno 1989, pp. 261–64.

[28] Godfrey-Smith 2003 p. 236.

[29] Gauch 2003, p. 3

[30] Schuster and Powers (2005), Translational and Experimental Clinical Research. Ch. 1. Link. This chapter also discusses the different types of research questions and how they are produced.

[31] This phrasing is attributed to Marshall Nirenberg.

[32] Note: for a discussion of multiple hypotheses, see Bayesian inference#Informal

[33] McCarty1985

[34] October 1951, as noted in McElheny 2004, p. 40:"That's what a helix should look like!" Crick exclaimed in delight (This is the Cochran-Crick-Vand-Stokes theory of the transform of a helix).

[35] June 1952, as noted in McElheny 2004, p. 43: Watson had succeeded in getting X-ray pictures of TMV showing a diffraction pattern consistent with the transform of a helix.

[36] Watson did enough work on Tobacco mosaic virus to produce the diffraction pattern for a helix, per Crick's work on the transform of a helix. pp. 137–38, Horace Freeland Judson (1979) *The Eighth Day of Creation* ISBN 0-671-22540-5

[37] – Cochran W, Crick FHC and Vand V. (1952) "The Structure of Synthetic Polypeptides. I. The Transform of Atoms on a Helix", *Acta Crystallogr.*, **5**, 581–86.

[38] Friday, January 30, 1953. Tea time, as noted in McElheny 2004, p. 52: Franklin confronts Watson and his paper – "Of course it [Pauling's pre-print] is wrong. DNA is not a helix." However, Watson then visits Wilkins' office, sees photo 51, and immediately recognizes the diffraction pattern of a helical structure. But additional questions remained, requiring additional iterations of their research. For example, the number of strands in the backbone of the helix (Crick suspected 2 strands, but cautioned Watson to examine that more critically), the location of the base pairs (inside the backbone or outside the backbone), etc. One key point was that they realized that the quickest way to reach a result was not to continue a mathematical analysis, but to build a physical model.

[39] "The instant I saw the picture my mouth fell open and my pulse began to race." – Watson 1968, p. 167 Page 168 shows the X-shaped pattern of the B-form of DNA, clearly indicating crucial details of its helical structure to Watson and Crick.

- McElheny 2004 p. 52 dates the Franklin-Watson confrontation as Friday, January 30, 1953. Later that evening, Watson urges Wilkins to begin model-building immediately. But Wilkins agrees to do so only after Franklin's departure.

[40] Saturday, February 28, 1953, as noted in McElheny 2004, pp. 57–59: Watson found the base pairing mechanism which explained Chargaff's rules using his cardboard models.

[41] Galileo Galilei (1638) *Two new sciences*

[42] Reconstruction of Galileo Galilei's experiment – the inclined plane

[43] In *Two new sciences*, there are three 'reviewers': Simplicio, Sagredo, and Salviati, who serve as foil, antagonist, and protagonist. Galileo speaks for himself only briefly. But note that Einstein's 1905 papers were not peer reviewed before their publication.

[44] Fleck 1979, pp. xxvii–xxviii

[45] "NIH Data Sharing Policy."

[46] Stanovich, Keith E. (2007), *How to Think Straight About Psychology*. Boston: Pearson Education. p. 123

[47] Tow, David Hunter (2010). *The Future of Life: A Unified Theory of Evolution*. Future of Life Series. Future of Life Media. p. 262. Retrieved 2016-12-11. On further examination however, the scientific method bears a striking similarity to the larger process of evolution itself. [...] Of great significance is the evolutionary algorithm, which uses a simplified subset of the process of natural evolution applied to find the solution to problems that are too complex to solve by traditional analytic methods. In essence it is a process of accelerated and rigorous trial and error building on previous knowledge to refine an existing hypothesis, or discarding it altogether to find a better model. [...] The evolutionary algorithm is a technique derived from the evolution of knowledge processing applied within the context of science and technology, itself an outcome of evolution. The scientific method continues to evolve through adaptive reward, trial and error and application of the method to itself.

[48] Brody 1993, pp. 44–45

[49] Hall, B. K.; Hallgrímsson, B., eds. (2008). *Strickberger's Evolution* (4th ed.). Jones & Bartlett. p. 762. ISBN 0-7637-0066-5.

[50] Cracraft, J.; Donoghue, M. J., eds. (2005). *Assembling the tree of life*. Oxford University Press. p. 592. ISBN 0-19-517234-5.

[51] Needham & Wang 1954 p. 166 shows how the 'flying gallop' image propagated from China to the West.

[52] "A myth is a belief given uncritical acceptance by members of a group ..." – Weiss, *Business Ethics* p. 15, as cited by Ronald R. Sims (2003) *Ethics and corporate social responsibility: why giants fall* p. 21

[53] Imre Lakatos (1976), *Proofs and Refutations*. Taleb 2007, p. 72 lists ways to avoid the narrative fallacy and confirmation bias.

[54] For more on the narrative fallacy, see also Fleck 1979, p. 27: "Words and ideas are originally phonetic and mental equivalences of the experiences coinciding with them. ... Such proto-ideas are at first always too broad and insufficiently specialized. ... Once a structurally complete and closed system of opinions consisting of many details and relations has been formed, it offers enduring resistance to anything that contradicts it."

[55] The scientific method requires testing and validation *a posteriori* before ideas are accepted. "Invariably one came up against fundamental physical limits to the accuracy of measurement. ... The art of physical measurement seemed to be a matter of compromise, of choosing between reciprocally related uncertainties. ... Multiplying together the conjugate pairs of uncertainty limits mentioned, however, I found that they formed invariant products of not one but two distinct kinds. ... The first group of limits were calculable *a priori* from a specification of the instrument. The second group could be calculated only *a posteriori* from a specification of what was *done* with the instrument. ... In the first case each unit [of information] would add one additional *dimension* (conceptual category), whereas in the second each unit would add one additional *atomic fact*.", pp. 1–4: MacKay, Donald M. (1969), *Information, Mechanism, and Meaning*. Cambridge, MA: MIT Press, ISBN 0-262-63-032-X

[56] See the hypothethico-deductive method, for example, Godfrey-Smith 2003, p. 236.

[57] Jevons 1874, pp. 265–66.

[58] pp. 65, 73, 92, 398 – Andrew J. Galambos, *Sic Itur ad Astra* ISBN 0-88078-004-5(AJG learned scientific method from Felix Ehrenhaft

[59] Galileo 1638, pp. v–xii, 1–300

[60] Brody 1993, pp. 10–24 calls this the "epistemic cycle": "The epistemic cycle starts from an initial model; iterations of the cycle then improve the model until an adequate fit is achieved."

[61] Iteration example: Chaldean astronomers such as Kidinnu compiled astronomical data. Hipparchus was to use this data to calculate the precession of the Earth's axis. Fifteen hundred years after Kidinnu, Al-Batani, born in what is now Turkey, would use the collected data and improve Hipparchus' value for the precession of the Earth's axis. Al-Batani's value, 54.5 arc-seconds per year, compares well to the current value of 49.8 arc-seconds per year (26,000 years for Earth's axis to complete a circle around the Ecliptic pole).

[62] Recursion example: the Earth is itself a magnet, with its own North and South Poles William Gilbert (in Latin 1600) *De Magnete*, or *On Magnetism and Magnetic Bodies*. Translated from Latin to English, selection by Moulton & Schifferes 1960, pp. 113–17. Gilbert created a *terrella*, a lodestone ground into a spherical shape, which served as Gilbert's model for the Earth itself, as noted in Bruno 1989, p. 277.

[63] "The foundation of general physics ... is experience. These ... everyday experiences we do not discover without deliberately directing our attention to them. Collecting information about these is *observation*." – Hans Christian Ørsted("First Introduction to General Physics" ¶13, part of a series of public lectures at the University of Copenhagen. Copenhagen 1811, in Danish, printed by Johan Frederik Schulz. In Kirstine Meyer's 1920 edition of Ørsted's works, vol.**III** pp. 151–90.) "First Introduction to Physics: the Spirit, Meaning, and Goal of Natural Science". Reprinted in German in 1822, Schweigger's *Journal für Chemie und Physik* **36**, pp. 458–88, as translated in Ørsted 1997, p. 292

[64] "When it is not clear under which law of nature an effect or class of effect belongs, we try to fill this gap by means of a guess. Such guesses have been given the name *conjectures* or *hypotheses*." – Hans Christian Ørsted(1811) "First Introduction to General Physics" as translated in Ørsted 1997, p. 297.

[65] "In general we look for a new law by the following process. First we guess it. ...", – Feynman 1965, p. 156

[66] "... the statement of a law – A depends on B – always transcends experience." – Born 1949, p. 6

[67] "The student of nature ... regards as his property the experiences which the mathematician can only borrow. This is why he deduces theorems directly from the nature of an effect while the mathematician only arrives at them circuitously." – Hans Christian Ørsted(1811) "First Introduction to General Physics" ¶17. as translated in Ørsted 1997, p. 297.

[68] Salviati speaks: "I greatly doubt that Aristotle ever tested by experiment whether it be true that two stones, one weighing ten times as much as the other, if allowed to fall, at the same instant, from a height of, say, 100 cubits, would so differ in speed that when the heavier had reached the ground, the other would not have fallen more than 10 cubits." Two New Sciences (1638) – Galileo 1638, pp. 61–62. A more extended quotation is referenced by Moulton & Schifferes 1960, pp. 80–81.

[69] In the inquiry-based education paradigm, the stage of "characterization, observation, definition, ..." is more briefly summed up under the rubric of a Question

[70] "To raise new questions, new possibilities, to regard old problems from a new angle, requires creative imagination and marks real advance in science." – Einstein & Infeld 1938, p. 92.

[71] Crawford S, Stucki L (1990), "Peer review and the changing research record", "J Am Soc Info Science", vol. 41, pp. 223–28

[72] *See, e.g.*, Gauch 2003, esp. chapters 5–8

[73] Andreas Vesalius, *Epistola, Rationem, Modumque Propinandi Radicis Chynae Decocti* (1546), 141. Quoted and translated in C.D. O'Malley, *Andreas Vesalius of Brussels*, (1964), 116. As quoted by Bynum & Porter 2005, p. 597: Andreas Vesalius, 597#1.

[74] Crick, Francis (1994), *The Astonishing Hypothesis* ISBN 0-684-19431-7 p. 20

[75] McElheny 2004 p. 34

[76] Glen 1994, pp. 37–38.

[77] John R. Platt (16 October 1964) Strong Inference *Science* vol **146** (3642) p.347 doi:10.1126/science.146.3642.347

[78] "The structure that we propose is a three-chain structure, each chain being a helix" – Linus Pauling, as quoted on p. 157 by Horace Freeland Judson (1979), *The Eighth Day of Creation* ISBN 0-671-22540-5

[79] McElheny 2004, pp. 49–50: January 28, 1953 – Watson read Pauling's pre-print, and realized that in Pauling's model, DNA's phosphate groups had to be un-ionized. But DNA is an acid, which contradicts Pauling's model.

[80] June 1952, as noted in McElheny 2004, p. 43: Watson had succeeded in getting X-ray pictures of TMV showing a diffraction pattern consistent with the transform of a helix.

[81] McElheny 2004 p. 68: *Nature* April 25, 1953.

[82] In March 1917, the Royal Astronomical Society announced that on May 29, 1919, the occasion of a total eclipse of the sun would afford favorable conditions for testing Einstein's General theory of relativity. One expedition, to Sobral, Ceará, Brazil, and Eddington's expedition to the island of Principe yielded a set of photographs, which, when compared to photographs taken at Sobral and at Greenwich Observatory showed that the deviation of light was measured to be 1.69 arc-seconds, as compared to Einstein's desk prediction of 1.75 arc-seconds. – Antonina Vallentin (1954), *Einstein*, as quoted by Samuel Rapport and Helen Wright (1965), *Physics*, New York: Washington Square Press, pp. 294–95.

[83] Mill, John Stuart, "A System of Logic", University Press of the Pacific, Honolulu, 2002, ISBN 1-4102-0252-6.

[84] al-Battani, *De Motu Stellarum* translation from Arabic to Latin in 1116, as cited by "Battani, al-" (c. 858–929) *Encyclopædia Britannica*, 15th. ed. Al-Battani is known for his accurate observations at al-Raqqah in Syria, beginning in 877. His work includes measurement of the annual precession of the equinoxes.

[85] McElheny 2004 p. 53: The weekend (January 31 – February 1) after seeing photo 51, Watson informed Bragg of the X-ray diffraction image of DNA in B form. Bragg gave them permission to restart their research on DNA (that is, model building).

[86] McElheny 2004 p. 54: On Sunday February 8, 1953, Maurice Wilkes gave Watson and Crick permission to work on models, as Wilkes would not be building models until Franklin left DNA research.

[87] McElheny 2004 p. 56: Jerry Donohue, on sabbatical from Pauling's lab and visiting Cambridge, advises Watson that textbook form of the base pairs was incorrect for DNA base pairs; rather, the keto form of the base pairs should be used instead. This form allowed the bases' hydrogen bonds to pair 'unlike' with 'unlike', rather than to pair 'like' with 'like', as Watson was inclined to model, on the basis of the textbook statements. On February 27, 1953, Watson was convinced enough to make cardboard models of the nucleotides in their keto form.

[88] "Suddenly I became aware that an adenine-thymine pair held together by two hydrogen bonds was identical in shape to a guanine-cytosine pair held together by at least two hydrogen bonds. ..." – Watson 1968, pp. 194–97.

- McElheny 2004 p. 57 Saturday, February 28, 1953, Watson tried 'like with like' and admitted these base pairs didn't have hydrogen bonds that line up. But after trying 'unlike with unlike', and getting Jerry Donohue's approval, the base pairs turned out to be identical in shape (as Watson stated above in his 1968 *Double Helix* memoir quoted above). Watson now felt confident enough to inform Crick. (Of course, 'unlike with unlike' increases the number of possible codons, if this scheme were a genetic code.)

[89] See, e.g., *Physics Today*, **59**(1), p. 42. Richmann electrocuted in St. Petersburg (1753)

[90] Aristotle, "Prior Analytics", Hugh Tredennick (trans.), pp. 181–531 in *Aristotle, Volume 1*, Loeb Classical Library, William Heinemann, London, UK, 1938.

[91] "What one does not in the least doubt one should not pretend to doubt; but a man should train himself to doubt," said Peirce in a brief intellectual autobiography; see Ketner, Kenneth Laine (2009) "Charles Sanders Peirce: Interdisciplinary Scientist" in *The Logic of Interdisciplinarity*). Peirce held that actual, genuine doubt originates externally, usually in surprise, but also that it is to be sought and cultivated, "provided only that it be the weighty and noble metal itself, and no counterfeit nor paper substitute"; in "Issues of Pragmaticism", *The Monist*, v. XV, n. 4, pp. 481–99, see p. 484, and p. 491. (Reprinted in *Collected Papers* v. 5, paragraphs 438–63, see 443 and 451).

[92] But see Scientific method and religion.

[93] Peirce (1898), "Philosophy and the Conduct of Life", Lecture 1 of the Cambridge (MA) Conferences Lectures, published in *Collected Papers* v. 1, paragraphs 616–48 in part and in *Reasoning and the Logic of Things*, Ketner (ed., intro.) and Putnam (intro., comm.), pp. 105–22, reprinted in *Essential Peirce* v. 2, pp. 27–41.

[94] " ... in order to learn, one must desire to learn ..." – Peirce (1899), "F.R.L." [First Rule of Logic], *Collected Papers* v. 1, paragraphs 135–40. "Eprint". Archived from the original on January 6, 2012. Retrieved 2012-01-06.

[95] Peirce, Charles Sanders (1877). *How to Make Our Ideas Clear*. **12**. Wikisource. pp. 286–302 wslink==How to Make Our Ideas Clear.

[96] Peirce (1868), "Some Consequences of Four Incapacities", *Journal of Speculative Philosophy* v. 2, n. 3, pp. 140–57. Reprinted *Collected Papers* v. 5, paragraphs 264–317, *The Essential Peirce* v. 1, pp. 28–55, and elsewhere. *Arisbe* Eprint

[97] Peirce (1878), "The Doctrine of Chances", *Popular Science Monthly* v. 12, pp. 604–15, see pp. 610–11 via *Internet Archive*. Reprinted *Collected Papers* v. 2, paragraphs 645–68, *Essential Peirce* v. 1, pp. 142–54. "...death makes the number of our risks, the number of our inferences, finite, and so makes their mean result uncertain. The very idea of probability and of reasoning rests on the assumption that this number is indefinitely great. ....logicality inexorably requires that our interests shall not be limited. .... Logic is rooted in the social principle."

[98] Peirce (c. 1906). "PAP (Prolegomena for an Apology to Pragmatism)" (Manuscript 293, not the like-named article), *The New Elements of Mathematics* (NEM) 4:319–20, see first quote under "Abduction" at *Commens Dictionary of Peirce's Terms*.

[99] Peirce, Carnegie application (L75, 1902), *New Elements of Mathematics* v. 4, pp. 37–38:

> For it is not sufficient that a hypothesis should be a justifiable one. Any hypothesis which explains the facts is justified critically. But among justifiable hypotheses we have to select that one which is suitable for being tested by experiment.

[100] Peirce (1902), Carnegie application, see MS L75.329330, from Draft D of Memoir 27:

> Consequently, to discover is simply to expedite an event that would occur sooner or later, if we had not troubled ourselves to make the discovery. Consequently, the art of discovery is purely a question of economics. The economics of research is, so far as logic is concerned, the leading doctrine with reference to the art of discovery. Consequently, the conduct of abduction, which is chiefly a question of heuretic and is the first question of heuretic, is to be governed by economical considerations.

[101] Peirce (1903), "Pragmatism – The Logic of Abduction", *Collected Papers* v. 5, paragraphs 195–205, especially 196. Eprint.

[102] Peirce, "On the Logic of Drawing Ancient History from Documents", *Essential Peirce* v. 2, see pp. 107–09. On Twenty Questions, p. 109:

> Thus, twenty skillful hypotheses will ascertain what 200,000 stupid ones might fail to do.

[103] Peirce (1878), "The Probability of Induction", *Popular Science Monthly*, v. 12, pp. 705–18, see 718 *Google Books*; 718 via *Internet Archive*. Reprinted often, including (*Collected Papers* v. 2, paragraphs 669–93), (*The Essential Peirce* v. 1, pp. 155–69).

[104] Peirce (1905 draft "G" of "A Neglected Argument"), "Crude, Quantitative, and Qualitative Induction", *Collected Papers* v. 2, paragraphs 755–60, see 759. Find under "Induction" at *Commens Dictionary of Peirce's Terms*.

[105] . Brown, C. (2005) Overcoming Barriers to Use of Promising Research Among Elite Middle East Policy Groups, Journal of Social Behaviour and Personality, Select Press.

[106] Smaldino, PE; McElreath, R. (2016-09-21). "The natural selection of bad science". R. Soc. open sci. 3: 160384. **3**. Bibcode:2016RSOS....360384S. arXiv:1605.09511. doi:10.1098/rsos.160384.

[107] Einstein, Albert (1936, 1956) One may say "the eternal mystery of the world is its comprehensibility." From the article "Physics and Reality" (1936), reprinted in *Out of My Later Years* (1956). 'It is one of the great realizations of Immanuel Kant that the setting up of a real external world would be senseless without this comprehensibility.'

[108] Thurs, Daniel P. (2015), "That the scientific method accurately reflects what scientists actually do", in Numbers, Ronald L.; Kampourakis, Kostas, *Newton's Apple and Other Myths about Science*, Harvard University Press, pp. 210–18

[109] Hanson, Norwood (1958), *Patterns of Discovery*, Cambridge University Press. ISBN 0-521-05197-5

[110] Kuhn 1962, p. 113 ISBN 978-1-4432-5544-8

[111] Feyerabend, Paul K (1960) "Patterns of Discovery" The Philosophical Review (1960) vol. 69 (2) pp. 247–52

[112] Kuhn, Thomas S., "The Function of Measurement in Modern Physical Science", *ISIS* 52(2), 161–93, 1961.

[113] Feyerabend, Paul K., *Against Method, Outline of an Anarchistic Theory of Knowledge*, 1st published, 1975. Reprinted, Verso, London, UK, 1978.

[114]
- *Higher Superstition: The Academic Left and Its Quarrels with Science*, The Johns Hopkins University Press, 1997
- *Fashionable Nonsense: Postmodern Intellectuals' Abuse of Science*, Picador; 1st Picador USA Pbk. Ed edition, 1999
- *The Sokal Hoax: The Sham That Shook the Academy*, University of Nebraska Press, 2000 ISBN 0-8032-7995-7
- *A House Built on Sand: Exposing Postmodernist Myths About Science*, Oxford University Press, 2000
- *Intellectual Impostures*, Economist Books, 2003

[115] Dunbar, K., & Fugelsang, J. (2005). Causal thinking in science: How scientists and students interpret the unexpected. In M. E. Gorman, R. D. Tweney, D. Gooding & A. Kincannon (Eds.), Scientific and Technical Thinking (pp. 57–79). Mahwah, NJ: Lawrence Erlbaum Associates.

[116] Oliver, J.E. (1991) Ch2. of The incomplete guide to the art of discovery. New York:NY, Columbia University Press.

[117] Riccardo Pozzo (2004) *The impact of Aristotelianism on modern philosophy*. CUA Press. p. 41. ISBN 0-8132-1347-9

[118] Achinstein, Peter (2004). *General Introduction. Science Rules: A Historical Introduction to Scientific Methods*. Johns Hopkins University Press. pp. 1–5. ISBN 0-8018-7943-4.

[119] "When we are working intensively, we feel keenly the progress of our work; we are elated when our progress is rapid, we are depressed when it is slow." – the mathematician Pólya 1957, p. 131 in the section on 'Modern heuristic'.

[120] "Philosophy [i.e., physics] is written in this grand book – I mean the universe – which stands continually open to our gaze, but it cannot be understood unless one first learns to comprehend the language and interpret the characters in which it is written. It is written in the language of mathematics, and its characters are triangles, circles, and other geometrical figures, without which it is humanly impossible to understand a single word of it; without these, one is wandering around in a dark labyrinth." – Galileo Galilei, *Il Saggiatore* (*The Assayer*, 1623), as translated by Stillman Drake (1957), *Discoveries and Opinions of Galileo* pp. 237–38, as quoted by di Francia 1981, p. 10.

[121] Pólya 1957 2nd ed.

[122] George Pólya (1954), *Mathematics and Plausible Reasoning Volume I: Induction and Analogy in Mathematics*.

[123] George Pólya (1954), *Mathematics and Plausible Reasoning Volume II: Patterns of Plausible Reasoning*.

[124] Pólya 1957, p. 142

[125] Pólya 1957, p. 144

[126] Mackay 1991 p. 100

[127] See the development, by generations of mathematicians, of Euler's formula for polyhedra as documented by Lakatos, Imre (1976), *Proofs and refutations*, Cambridge: Cambridge University Press. ISBN 0-521-29038-4

[128] Lakatos, Imre (Worrall & Zahar, eds. 1976) *Proofs and Refutations*, p. 55

[129] Rosenberg, Nathan; Luther Earle Birdzell; Mitchell, Glenn William. *How the West grew Rich*. Popular Prakashan, 1986.

[130] Ioannidis, John P. A. (2005-08-01). "Why Most Published Research Findings Are False". *PLoS Medicine*. **2** (8). ISSN 1549-1277. PMC 1182327. PMID 16060722. doi:10.1371/journal.pmed.0020124.

18.12 References

- Born, Max (1949), *Natural Philosophy of Cause and Chance*, Peter Smith, also published by Dover, 1964. From the Waynflete Lectures, 1948. On the web. N.B.: the web version does not have the 3 addenda by Born, 1950, 1964, in which he notes that all knowledge is subjective. Born then proposes a solution in Appendix 3 (1964)

- Brody, Thomas A. (1993), *The Philosophy Behind Physics*, Springer Verlag, ISBN 0-387-55914-0. (Luis de la Peña and Peter E. Hodgson, eds.)

- Bruno, Leonard C. (1989), *The Landmarks of Science*, ISBN 0-8160-2137-6

- Bynum, W.F.; Porter, Roy (2005), *Oxford Dictionary of Scientific Quotations*, Oxford. ISBN 0-19-858409-1.

18.12. REFERENCES

- Dales, Richard C. (1973), *The Scientific Achievement of the Middle Ages (The Middle Ages Series)*, University of Pennsylvania Press. ISBN 9780812210576

- di Francia, G. Toraldo (1981), *The Investigation of the Physical World*, Cambridge University Press. ISBN 0-521-29925-X.

- Einstein, Albert; Infeld, Leopold (1938), *The Evolution of Physics: from early concepts to relativity and quanta*, New York: Simon and Schuster, ISBN 0-671-20156-5

- Feynman, Richard (1965), *The Character of Physical Law*, Cambridge: M.I.T. Press, ISBN 0-262-56003-8.

- Fleck, Ludwik (1979), *Genesis and Development of a Scientific Fact*, Univ. of Chicago, ISBN 0-226-25325-2. (written in German, 1935, *Entstehung und Entwickelung einer wissenschaftlichen Tatsache: Einführung in die Lehre vom Denkstil und Denkkollectiv*) English translation, 1979

- Galileo (1638), *Two New Sciences*, Leiden: Lodewijk Elzevir, ISBN 0-486-60099-8 Translated from Italian to English in 1914 by Henry Crew and Alfonso de Salvio. Introduction by Antonio Favaro. xxv+300 pages, index. New York: Macmillan, with later reprintings by Dover.

- Gauch, Hugh G., Jr. (2003), *Scientific Method in Practice*, Cambridge University Press, ISBN 0-521-01708-4 435 pages

- Glen, William (ed.) (1994), *The Mass-Extinction Debates: How Science Works in a Crisis*, Stanford, CA: Stanford University Press, ISBN 0-8047-2285-4.

- Godfrey-Smith, Peter (2003), *Theory and Reality: An introduction to the philosophy of science*, University of Chicago Press, ISBN 0-226-30063-3.

- Goldhaber, Alfred Scharff; Nieto, Michael Martin (January–March 2010), "Photon and graviton mass limits", *Rev. Mod. Phys.*, American Physical Society, **82**: 939, Bibcode:2010RvMP...82..939G, arXiv:0809.1003, doi:10.1103/RevModPhys.82.939, pp. 939–79.

- Jevons, William Stanley (1874), *The Principles of Science: A Treatise on Logic and Scientific Method*, Dover Publications, ISBN 1-4304-8775-5. 1877, 1879. Reprinted with a foreword by Ernst Nagel, New York, NY, 1958.

- Kuhn, Thomas S. (1962), *The Structure of Scientific Revolutions*, Chicago, IL: University of Chicago Press. 2nd edition 1970. 3rd edition 1996.

- Lindberg, David C. (2007), *The Beginnings of Western Science*, University of Chicago Press 2nd edition 2007.

- Mackay, Alan L. (ed.) (1991), *Dictionary of Scientific Quotations*, London: IOP Publishing Ltd, ISBN 0-7503-0106-6

- McElheny, Victor K. (2004), *Watson & DNA: Making a scientific revolution*, Basic Books, ISBN 0-7382-0866-3.

- Moulton, Forest Ray; Schifferes, Justus J. (eds., Second Edition) (1960), *The Autobiography of Science*, Doubleday.

- Needham, Joseph; Wang, Ling (🕮) (1954), *Science and Civilisation in China*, 1 *Introductory Orientations*, Cambridge University Press

- Newton, Isaac (1999) [1687, 1713, 1726], *Philosophiae Naturalis Principia Mathematica*, University of California Press, ISBN 0-520-08817-4, Third edition. From I. Bernard Cohen and Anne Whitman's 1999 translation, 974 pages.

- Ørsted, Hans Christian (1997), *Selected Scientific Works of Hans Christian Ørsted*, Princeton, ISBN 0-691-04334-5. Translated to English by Karen Jelved, Andrew D. Jackson, and Ole Knudsen, (translators 1997).

- Peirce, C. S. – see Charles Sanders Peirce bibliography.

- Poincaré, Henri (1905), *Science and Hypothesis* Eprint

- Pólya, George (1957), *How to Solve It*, Princeton University Press, ISBN 978-4871878302, OCLC 706968824 (reprinted 2009)

- Popper, Karl R. (1959), *The Logic of Scientific Discovery* 1934, 1959.

- Sambursky, Shmuel (ed.) (1974), *Physical Thought from the Presocratics to the Quantum Physicists*, Pica Press, ISBN 0-87663-712-8.

- Sanches, Francisco; Limbrick, Elaine. Introduction, Notes, and Bibliography; Thomson, Douglas F.S. Latin text established, annotated, and translated. (1988), *That Nothing is Known*, Cambridge: Cambridge University Press, ISBN 0-521-35077-8 Critical edition.

- Taleb, Nassim Nicholas (2007), *The Black Swan*, Random House, ISBN 978-1-4000-6351-2

- Watson, James D. (1968), *The Double Helix*, New York: Atheneum, Library of Congress card number 68-16217.

18.13 Further reading

- Bauer, Henry H., *Scientific Literacy and the Myth of the Scientific Method*, University of Illinois Press, Champaign, IL, 1992
- Beveridge, William I. B., *The Art of Scientific Investigation*, Heinemann, Melbourne, Australia, 1950.
- Bernstein, Richard J., *Beyond Objectivism and Relativism: Science, Hermeneutics, and Praxis*, University of Pennsylvania Press, Philadelphia, PA, 1983.
- Brody, Baruch A. and Capaldi, Nicholas, *Science: Men, Methods, Goals: A Reader: Methods of Physical Science*, W. A. Benjamin, 1968
- Brody, Baruch A., and Grandy, Richard E., *Readings in the Philosophy of Science*, 2nd edition, Prentice Hall, Englewood Cliffs, NJ, 1989.
- Burks, Arthur W., *Chance, Cause, Reason – An Inquiry into the Nature of Scientific Evidence*, University of Chicago Press, Chicago, IL, 1977.
- Alan Chalmers. *What is this thing called science?*, Queensland University Press and Open University Press, 1976.
- Crick, Francis (1988), *What Mad Pursuit: A Personal View of Scientific Discovery*, New York: Basic Books. ISBN 0-465-09137-7.
- Crombie, A. C. (1953), *Robert Grosseteste and the Origins of Experimental Science 1100–1700*, Oxford
- Dewey, John, *How We Think*, D.C. Heath, Lexington, MA, 1910. Reprinted, Prometheus Books, Buffalo, NY, 1991.
- Earman, John (ed.), *Inference, Explanation, and Other Frustrations: Essays in the Philosophy of Science*, University of California Press, Berkeley & Los Angeles, CA, 1992.
- Fraassen, Bas C. van, *The Scientific Image*, Oxford University Press, Oxford, UK, 1980.
- Franklin, James (2009), *What Science Knows: And How It Knows It*, New York: Encounter Books, ISBN 1-59403-207-6.
- Gadamer, Hans-Georg, *Reason in the Age of Science*, Frederick G. Lawrence (trans.), MIT Press, Cambridge, MA, 1981.
- Giere, Ronald N. (ed.), *Cognitive Models of Science*, vol. 15 in 'Minnesota Studies in the Philosophy of Science', University of Minnesota Press, Minneapolis, MN, 1992.
- Hacking, Ian, *Representing and Intervening. Introductory Topics in the Philosophy of Natural Science*, Cambridge University Press, Cambridge, UK, 1983.
- Heisenberg, Werner, *Physics and Beyond, Encounters and Conversations*, A.J. Pomerans (trans.), Harper and Row, New York, NY 1971, pp. 63–64.
- Holton, Gerald, *Thematic Origins of Scientific Thought, Kepler to Einstein*, 1st edition 1973, revised edition, Harvard University Press, Cambridge, MA, 1988.
- Kuhn, Thomas S., *The Essential Tension, Selected Studies in Scientific Tradition and Change*, University of Chicago Press, Chicago, IL, 1977.
- Latour, Bruno, *Science in Action, How to Follow Scientists and Engineers through Society*, Harvard University Press, Cambridge, MA, 1987.
- Losee, John, *A Historical Introduction to the Philosophy of Science*, Oxford University Press, Oxford, UK, 1972. 2nd edition, 1980.
- Maxwell, Nicholas, *The Comprehensibility of the Universe: A New Conception of Science*, Oxford University Press, Oxford, 1998. Paperback 2003.
- McCarty, Maclyn (1985), *The Transforming Principle: Discovering that genes are made of DNA*, New York: W. W. Norton. p. 252. ISBN 0-393-30450-7. Memoir of a researcher in the Avery–MacLeod–McCarty experiment.
- McComas, William F., ed. "The Principal Elements of the Nature of Science: Dispelling the Myths" (PDF). (189 KB), from *The Nature of Science in Science Education*, pp. 53–70. Kluwer Academic Publishers, Netherlands 1998.
- Misak, Cheryl J., *Truth and the End of Inquiry, A Peircean Account of Truth*, Oxford University Press, Oxford, UK, 1991.
- Piattelli-Palmarini, Massimo (ed.), *Language and Learning, The Debate between Jean Piaget and Noam Chomsky*, Harvard University Press, Cambridge, MA, 1980.
- Popper, Karl R., *Unended Quest, An Intellectual Autobiography*, Open Court, La Salle, IL, 1982.
- Putnam, Hilary, *Renewing Philosophy*, Harvard University Press, Cambridge, MA, 1992.
- Rorty, Richard, *Philosophy and the Mirror of Nature*, Princeton University Press, Princeton, NJ, 1979.

- Salmon, Wesley C., *Four Decades of Scientific Explanation*, University of Minnesota Press, Minneapolis, MN, 1990.

- Shimony, Abner, *Search for a Naturalistic World View: Vol. 1, Scientific Method and Epistemology, Vol. 2, Natural Science and Metaphysics*, Cambridge University Press, Cambridge, UK, 1993.

- Thagard, Paul, *Conceptual Revolutions*, Princeton University Press, Princeton, NJ, 1992.

- Ziman, John (2000). *Real Science: what it is, and what it means*. Cambridge, UK: Cambridge University Press.

18.14 External links

- Andersen, Anne; Hepburn, Brian. "Scientific Method". *Stanford Encyclopedia of Philosophy*.

- "Confirmation and Induction". *Internet Encyclopedia of Philosophy*.

- Scientific method at PhilPapers

- Scientific method at the Indiana Philosophy Ontology Project

- An Introduction to Science: Scientific Thinking and a scientific method by Steven D. Schafersman.

- Introduction to the scientific method at the University of Rochester

- Theory-ladenness by Paul Newall at The Galilean Library

- Lecture on Scientific Method by Greg Anderson

- Using the scientific method for designing science fair projects

- *Scientific Methods* an online book by Richard D. Jarrard

- Richard Feynman on the Key to Science (one minute, three seconds), from the Cornell Lectures.

- Lectures on the Scientific Method by Nick Josh Karean, Kevin Padian, Michael Shermer and Richard Dawkins

Chapter 19

Outline of scientific method

The following outline is provided as an overview of and topical guide to scientific method:

Scientific method – body of techniques for investigating phenomena and acquiring new knowledge, as well as for correcting and integrating previous knowledge. It is based on observable, empirical, reproducible, measurable evidence, and subject to the laws of reasoning.

19.1 Nature of scientific method

Scientific method

- Science
- Philosophy of science
- Sociology of knowledge
- Process of science
- Knowledge

19.2 Elements of scientific method

Research

19.2.1 Observation

Observation

- Scientific method
- Causation
- Investigation
- Measurement

19.2.2 Hypothesis

Hypothesis

- pro:Karl Popper
    - Falsifiability
- con:Paul Feyerabend
- Statistical hypothesis testing

19.2.3 Experiment

Experiment

- Laboratory
- Laboratory techniques
- Design of experiments
- Scientific control
- Natural experiment
- Observational study
- Field experiment
- Self-experimentation
    - Self-experimentation in medicine
- Placebo effect

19.2.4 Theory

- Scientific theory

Prediction

- Prediction
    - Bayesian inference – subjective use of statistical reasoning
    - Deductive reasoning
    - Retrodiction

19.2.5 Evaluation by scientific community

- Peer review
- Medical peer review

19.3 Scientific method concepts

19.3.1 Empirical methods

Empirical methods

- Empiricism
- Robert Grosseteste
- Peter Parker
- Bitoy's method
- Empirical validation
- Operationalization

19.3.2 Use of statistics

- *Uncomfortable science* — Inference from a limited sample of data
- Exploratory data analysis
- Confirmatory data analysis

19.3.3 Paradigm change

- Thomas Kuhn
    - The Structure of Scientific Revolutions
    - Paradigm
    - Paradigm shift

19.3.4 Problem of induction

The problem of induction questions the logical basis of scientific statements.

- Inductive reasoning appears to lie at the core of scientific method, yet also appears to be invalid.
- David Hume was the person who first pointed out the problem of induction.
- Karl Popper offered one solution, Falsifiability

19.3.5 Scientific creativity

- Tacit knowledge

19.3.6 Deviations from the scientific method

- Bad science
- Junk science
- Pseudoscience
- Pathological science

19.3.7 Critique of scientific method

- Paul Feyerabend argued that the search for a definitive scientific method was misplaced, and even counterproductive.
- Imre Lakatos attempted to bridge the gap between Popper and Kuhn.
- Sociology of scientific knowledge
- Scientism

19.3.8 Relationship of scientific method to technology

- Science and technology studies
- Theories of technology

19.3.9 Aesthetics in the scientific method

- Elegance
- Occam's razor

19.4 History of scientific method

Main articles: History of scientific method, Timeline of the history of scientific method, and History of science

19.4.1 Publications

- Ibn al-Haytham's *Book of Optics*
- Avicenna's *The Canon of Medicine*
- Roger Bacon's *Opus Majus*
- Francis Bacon's *Novum Organum*

19.4.2 Persons influential in the development of scientific method

- Alhazen
- Francis Bacon
- Galileo Galilei
- René Descartes
- Charles Sanders Peirce

19.4.3 Why didn't the scientific method arise elsewhere?

- China
- Greece
- India
- Korean Peninsula
- Malay Archipelago
- Mesoamerica
- Sub-Saharan Africa

19.5 See also

- Bayesian probability
    - Quasi-empirical methods
    - Foundation ontology
    - Ontology
    - Philosophy of mathematics
    - Mathematics
- Epistemology
    - Post-processual archaeology is a methodological curiosity from Archaeology.
    - Structuralism
        - Post-structuralism
        - Deconstruction
        - Postmodernism
        - Latour, Bruno
        - Secularism-
- Physical law
    - Science policy
    - Scientific Revolution
    - Sociology of knowledge
    - Science studies

19.6 External links

Chapter 20

Empirical research

Empirical research is research using empirical evidence. It is a way of gaining knowledge by means of direct and indirect observation or experience. Empiricism values such research more than other kinds. Empirical evidence (the record of one's direct observations or experiences) can be analyzed quantitatively or qualitatively. Quantifying the evidence or making sense of it in qualitative form, a researcher can answer empirical questions, which should be clearly defined and answerable with the evidence collected (usually called data). Research design varies by field and by the question being investigated. Many researchers combine qualitative and quantitative forms of analysis to better answer questions which cannot be studied in laboratory settings, particularly in the social sciences and in education.

In some fields, quantitative research may begin with a research question (e.g., "Does listening to vocal music during the learning of a word list have an effect on later memory for these words?") which is tested through experimentation. Usually, a researcher has a certain theory regarding the topic under investigation. Based on this theory statements, or hypotheses, will be proposed (e.g., "Listening to vocal voice has a negative effect on learning a word list."). From these hypotheses predictions about specific events are derived (e.g., "People who study a word list while listening to vocal music will remember fewer words on a later memory test than people who study a word list in silence."). These predictions can then be tested with a suitable experiment. Depending on the outcomes of the experiment, the theory on which the hypotheses and predictions were based will be supported or not,[1] or may need to be modified and then subjected to further testing.

20.1 Terminology

The term empirical was originally used to refer to certain ancient Greek practitioners of medicine who rejected adherence to the dogmatic doctrines of the day, preferring instead to rely on the observation of phenomena as perceived in experience. Later empiricism referred to a theory of knowledge in philosophy which adheres to the principle that knowledge arises from experience and evidence gathered specifically using the senses. In scientific use the term empirical refers to the gathering of data using only evidence that is observable by the senses or in some cases using calibrated scientific instruments. What early philosophers described as empiricist and empirical research have in common is the dependence on observable data to formulate and test theories and come to conclusions.

20.2 Usage

The researcher attempts to describe accurately the interaction between the instrument (or the human senses) and the entity being observed. If instrumentation is involved, the researcher is expected to calibrate his/her instrument by applying it to known standard objects and documenting the results before applying it to unknown objects. In other words, it describes the research that has not taken place before and their results.

In practice, the accumulation of evidence for or against any particular theory involves planned research designs for the collection of empirical data, and academic rigor plays a large part of judging the merits of research design. Several typologies for such designs have been suggested, one of the most popular of which comes from Campbell and Stanley.[2] They are responsible for popularizing the widely cited distinction among pre-experimental, experimental, and quasi-experimental designs and are staunch advocates of the central role of randomized experiments in educational research.

20.2.1 Scientific research

Accurate analysis of data using standardized statistical methods in scientific studies is critical to determining the validity of empirical research. Statistical formulas such as regression, uncertainty coefficient, t-test, chi square, and various types of ANOVA (analyses of variance) are fundamental to forming logical, valid conclusions. If empirical data reach significance under the appropriate statistical formula, the research hypothesis is supported. If not, the null hypothesis is supported (or, more accurately, not rejected), meaning no effect of the independent variable(s) was observed on the dependent variable(s).

It is important to understand that the outcome of empirical research using statistical hypothesis testing is never *proof*. It can only *support* a hypothesis, *reject* it, or do neither. These methods yield only probabilities.

Among scientific researchers, empirical *evidence* (as distinct from empirical *research*) refers to objective evidence that appears the same regardless of the observer. For example, a thermometer will not display different temperatures for each individual who observes it. Temperature, as measured by an accurate, well calibrated thermometer, is empirical evidence. By contrast, non-empirical evidence is subjective, depending on the observer. Following the previous example, observer A might truthfully report that a room is warm, while observer B might truthfully report that the same room is cool, though both observe the same reading on the thermometer. The use of empirical evidence negates this effect of personal (i.e., subjective) experience or time.

The varying perception of empiricism and rationalism shows concern with the limit to which there is dependency on experience of sense as an effort of gaining knowledge. According to rationalism, there are a number of different ways in which sense experience is gained independently for the knowledge and concepts. According to empiricism, sense experience is considered as the main source of every piece of knowledge and the concepts. In reference with a specific piece of knowledge, this paper will focus on differentiating between rationalism and empiricism or rational views and empirical views. In general, rationalists are known for the development of their own views following two different way. First, the key argument can be placed that there are cases in which the content of knowledge or concepts end up outstripping the information. This outstripped information is provided by the sense experience (Hjørland, 2010, 2). Second, there is construction of accounts as to how reasoning helps in the provision of addition knowledge about a specific or broader scope. Empiricists are known to be presenting complementary senses related to thought. First there is development of accounts of how there is provision of information by experience that is cited by rationalists. This is insofar for having it in the initial place. At times, empiricists tend to be opting skepticism as an option of rationalism. If experience is not helpful in the provision of knowledge or concept cited by rationalists, then they do not exist (Pearce, 2010, 35). Second, empiricists hold the tendency of attacking the accounts of rationalists while considering reasoning to be an important source of knowledge or concepts. The overall disagreement between empiricists and rationalists show primary concerns in how there is gaining of knowledge with respect to the sources of knowledge and concept. In some of the cases, disagreement at the point of gaining knowledge results in the provision of conflicting responses to other aspects as well. There might be a disagreement in the overall feature of warrant, while limiting the knowledge and thought. Empiricists are known for sharing the view that there is no existence of innate knowledge and rather that is derivation of knowledge out of experience. These experiences are either reasoned using the mind or sensed through the five senses human possess (Bernard, 2011, 5). On the other hand, rationalists are known to be sharing the view that there is existence of innate knowledge and this is different for the objects of innate knowledge being chosen. In order to follow rationalism, there must be adoption of one of the three claims related to the theory that are Deduction or Intuition, Innate Knowledge, and Innate Concept. The more there is removal of concept from mental operations and experience, there can be performance over experience with increased plausibility in being innate. Further ahead, empiricism in context with a specific subject provides a rejection of corresponding version related to innate knowledge and deduction or intuition (Weiskopf, 2008, 16). Insofar as there is acknowledgement of concepts and knowledge within the area of subject, the knowledge has major dependence on experience through human senses.

20.3 Empirical cycle

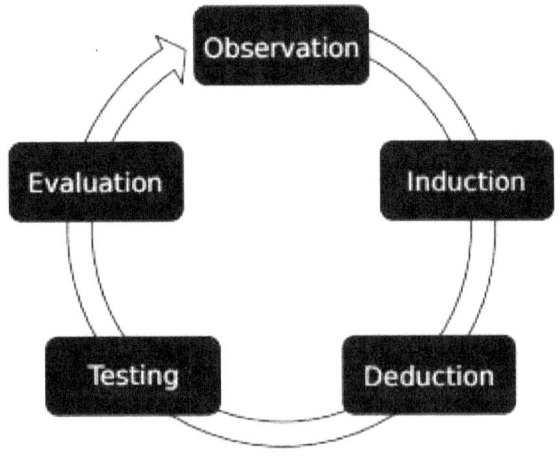

Empirical cycle according to A.D. de Groot

A.D. de Groot's empirical cycle:[3]

1. Observation: The observation of a phenomenon and

inquiry concerning its causes.

2. Induction: The formulation of hypotheses - generalized explanations for the phenomenon.

3. Deduction: The formulation of experiments that will test the hypotheses (i.e. confirm them if true, refute them if false).

4. Testing: The procedures by which the hypotheses are tested and data are collected.

5. Evaluation: The interpretation of the data and the formulation of a theory - an abductive argument that presents the results of the experiment as the most reasonable explanation for the phenomenon.

20.4 See also

- Case study
- Fact

20.5 References

[1] Goodwin, C. J. (2005). *Research in Psychology: Methods and Design*. USA: John Wiley & Sons, Inc.

[2] Campbell, D. & Stanley, J. (1963). *Experimental and quasi-experimental designs for research*. Boston: Houghton Mifflin Company.

[3] Heitink, G. (1999). *Practical Theology: History, Theory, Action Domains: Manual for Practical Theology*. Grand Rapids, MI: Wm. B. Eerdmans Publishing. p. 233. ISBN 9780802842947

20.6 External links

- The dictionary definition of empirical research at Wiktionary
- Some Key Concepts for the Design and Review of Empirical Research

Chapter 21

Deductive-nomological model

The **deductive-nomological model** (**DN model**), also known as **Hempel's model**, the **Hempel–Oppenheim model**, the **Popper–Hempel model**, or the **covering law model**, is a formal view of scientifically answering questions asking, "Why...?". The DN model poses scientific explanation as a deductive structure—that is, one where truth of its premises entails truth of its conclusion—hinged on accurate prediction or postdiction of the phenomenon to be explained.

Because of problems concerning humans' ability to define, discover, and know causality, it was omitted in initial formulations of the DN model. Causality was thought to be incidentally approximated by realistic selection of premises that *derive* the phenomenon of interest from observed starting conditions plus general laws. Still, DN model formally permitted causally irrelevant factors. Also, derivability from observations and laws sometimes yielded absurd answers.

Upon the fall of logical empiricism's in the 1960s, the DN model was widely seen as a flawed or greatly incomplete model of scientific explanation. Nonetheless, it remained an idealized version of scientific explanation, and one that was rather accurate when applied to modern physics. In the early 1980s, revision to DN model emphasized *maximal specificity* for relevance of the conditions and axioms stated. Together with Hempel's inductive-statistical model, the DN model forms scientific explanation's **covering law model**, which is also termed, from critical angle, **subsumption theory**.

21.1 Form

The term *deductive* distinguishes the DN model's intended determinism from the probabilism of inductive inferences.[1] The term *nomological* is derived from the Greek word νόμος or *nomos*, meaning "law".[1] The DN model holds to a view of scientific explanation whose *conditions of adequacy* (CA)—semiformal but stated classically—are *derivability* (CA1), *lawlikeness* (CA2), *empirical content* (CA3), and *truth* (CA4).[2]

In the DN model, a law axiomatizes an unrestricted generalization from antecedent A to consequent B by conditional proposition—*If A, then B*—and has empirical content testable.[3] A law differs from mere true regularity—for instance, *George always carries only $1 bills in his wallet*—by supporting counterfactual claims and thus suggesting what *must* be true,[4] while following from a scientific theory's axiomatic structure.[5]

The phenomenon to be explained is the *explanandum*—an event, law, or theory—whereas the premises to explain it are *explanans*, true or highly confirmed, containing at least one universal law, and entailing the explanandum.[6][7] Thus, given the explanans as initial, specific conditions C_1, $C_2 \ldots C_n$ plus general laws $L_1, L_2 \ldots L_n$, the phenomenon E as explanandum is a deductive consequence, thereby scientifically explained.[6]

21.2 Roots

Aristotle's scientific explanation in *Physics* resemble the DN model, an idealized form of scientific explanation.[7] The framework of Aristotelian physics—Aristotelian metaphysics—reflected the perspective of this principally biologist, who, amid living entities' undeniable purposiveness, formalized vitalism and teleology, an intrinsic morality in nature.[8] With emergence of Copernicanism, however, Descartes introduced mechanical philosophy, then Newton rigorously posed lawlike explanation, both Descartes and especially Newton shunning teleology within natural philosophy.[9] At 1740, David Hume[10] staked Hume's fork,[11] highlighted the problem of induction,[12] and found humans ignorant of either necessary or sufficient causality.[13][14] Hume also highlighted the fact/value gap, as what *is* does not itself reveal what *ought*.[15]

Near 1780, countering Hume's ostensibly radical empiricism, Immanuel Kant highlighted extreme rationalism—as by Descartes or Spinoza—and sought middle ground. Inferring the mind to arrange experience of the world into *substance, space,* and *time,* Kant placed the mind as part of the causal constellation of experience and thereby found Newton's theory of motion universally true,[16] yet knowledge of things in themselves impossible.[14] Safeguarding science, then, Kant paradoxically stripped it of scientific realism.[14][17][18] Aborting Francis Bacon's inductivist mission to dissolve the veil of appearance to uncover the *noumena*—metaphysical view of nature's ultimate truths—Kant's transcendental idealism tasked science with simply modeling patterns of *phenomena.* Safeguarding metaphysics, too, it found the mind's constants holding also universal moral truths,[19] and launched German idealism, increasingly speculative.

Auguste Comte found the problem of induction rather irrelevant since enumerative induction is grounded on the empiricism available, while science's point is not metaphysical truth. Comte found human knowledge had evolved from theological to metaphysical to scientific—the ultimate stage—rejecting both theology and metaphysics as asking questions unanswerable and posing answers unverifiable. Comte in the 1830s expounded positivism—the first modern philosophy of science and simultaneously a political philosophy[20]—rejecting conjectures about unobservables, thus rejecting search for *causes.*[21] Positivism predicts observations, confirms the predictions, and states a *law,* thereupon applied to benefit human society.[22] From late 19th century into the early 20th century, the influence of positivism spanned the globe.[20] Meanwhile, evolutionary theory's natural selection brought the Copernican Revolution into biology and eventuated in the first conceptual alternative to vitalism and teleology.[8]

21.3 Growth

Whereas Comtean positivism posed science as *description,* logical positivism emerged in the late 1920s and posed science as *explanation,* perhaps to better unify empirical sciences by covering not only fundamental science—that is, fundamental physics—but special sciences, too, such as biology, psychology, economics, and anthropology.[23] After defeat of National Socialism with World War II's close in 1945, logical positivism shifted to a milder variant, *logical empiricism.*[24] All variants of the movement, which lasted until 1965, are neopositivism,[25] sharing the quest of verificationism.[26]

Neopositivists led emergence of the philosophy subdiscipline philosophy of science, researching such questions and aspects of scientific theory and knowledge.[24] Scientific realism takes scientific theory's statements at face value, thus accorded either falsity or truth—probable or approximate or actual.[17] Neopositivists held scientific antirealism as instrumentalism, holding scientific theory as simply a device to predict observations and their course, while statements on nature's unobservable aspects are elliptical at or metaphorical of its observable aspects, rather.[27]

DN model received its most detailed, influential statement by Carl G Hempel, first in his 1942 article "The function of general laws in history", and more explicitly with Paul Oppenheim in their 1948 article "Studies in the logic of explanation".[28][29] Leading logical empiricist, Hempel embraced the Humean empiricist view that humans observe sequence of sensory events, not cause and effect,[23] as causal relations and casual mechanisms are unobservables.[30] DN model bypasses causality beyond mere constant conjunction: first an event like *A,* then always an event like *B.*[23]

Hempel held natural laws—empirically confirmed regularities—as satisfactory, and if included realistically to approximate causality.[6] In later articles, Hempel defended DN model and proposed probabilistic explanation by *inductive-statistical model* (IS model).[6] DN model and IS model—whereby the probability must be high, such as at least 50%[31]—together form *covering law model,*[6] as named by a critic, William Dray.[32] Derivation of statistical laws from other statistical laws goes to the *deductive-statistical model* (DS model).[31][33] Georg Henrik von Wright, another critic, named the totality *subsumption theory.*[34]

21.4 Decline

Amid failure of neopositivism's fundamental tenets,[35] Hempel in 1965 abandoned verificationism, signaling neopositivism's demise.[36] From 1930 onward, Karl Popper had refuted any positivism by asserting falsificationism, which Popper claimed had killed positivism, although, paradoxically, Popper was commonly mistaken for a positivist.[37][38] Even Popper's 1934 book[39] embraces DN model,[7][28] widely accepted as the model of scientific explanation for as long as physics remained the model of science examined by philosophers of science.[30][40]

In the 1940s, filling the vast observational gap between cytology[41] and biochemistry,[42] cell biology arose[43] and established existence of cell organelles besides the nucleus. Launched in the late 1930s, the molecular biology research program cracked a genetic code in the early 1960s and then converged with cell biology as *cell and molecular biology,* its breakthroughs and discoveries defying DN model by arriving in quest not of lawlike explanation but of causal mechanisms.[30] Biology became a new model of science, while special sciences were no longer thought defective by lacking universal laws, as borne by physics.[40]

In 1948, when explicating DN model and stating scientific explanation's semiformal *conditions of adequacy*, Hempel and Oppenheim acknowledged redundancy of the third, *empirical content*, implied by the other three—derivability, lawlikeness, and truth.[2] In the early 1980s, upon widespread view that causality ensures the explanans' relevance, Wesley Salmon called for returning *cause* to *because*,[44] and along with James Fetzer helped replace CA3 *empirical content* with CA3' *strict maximal specificity*.[45]

Salmon introduced *causal mechanical* explanation, never clarifying how it proceeds, yet reviving philosophers' interest in such.[30] Via shortcomings of Hempel's inductive-statistical model (IS model), Salmon introduced *statistical-relevance model* (SR model).[7] Although DN model remained an idealized form of scientific explanation, especially in applied sciences,[7] most philosophers of science consider DN model flawed by excluding many types of explanations generally accepted as scientific.[33]

21.5 Strengths

As theory of knowledge, epistemology differs from ontology, which is a subbranch of metaphysics, theory of reality.[46] Ontology poses which categories of being—what sorts of things exist—and so, although a scientific theory's ontological commitment can be modified in light of experience, an ontological commitment inevitably precedes empirical inquiry.[46]

Natural laws, so called, are statements of humans' observations, thus are epistemological—concerning human knowledge—the *epistemic*. Causal mechanisms and structures existing putatively independently of minds exist, or would exist, in the natural world's structure itself, and thus are ontological, the *ontic*. Blurring epistemic with ontic—as by incautiously presuming a natural law to refer to a causal mechanism, or to trace structures realistically during unobserved transitions, or to be true regularities always unvarying—tends to generate a *category mistake*.[47][48]

Discarding ontic commitments, including causality *per se*, DN model permits a theory's laws to be reduced to—that is, subsumed by—a more fundamental theory's laws. The higher theory's laws are explained in DN model by the lower theory's laws.[5][6] Thus, the epistemic success of Newtonian theory's law of universal gravitation is reduced to—thus explained by—Einstein's general theory of relativity, although Einstein's discards Newton's ontic claim that universal gravitation's epistemic success predicting Kepler's laws of planetary motion[49] is through a causal mechanism of a straightly attractive force instantly traversing absolute space despite absolute time.

Covering law model reflects neopositivism's vision of empirical science, a vision interpreting or presuming unity of science, whereby all empirical sciences are either fundamental science—that is, fundamental physics—or are special sciences, whether astrophysics, chemistry, biology, geology, psychology, economics, and so on.[40][50][51] All special sciences would network via covering law model.[52] And by stating *boundary conditions* while supplying *bridge laws*, any special law would reduce to a lower special law, ultimately reducing—theoretically although generally not practically—to fundamental science.[53][54] (*Boundary conditions* are specified conditions whereby the phenomena of interest occur. *Bridge laws* translate terms in one science to terms in another science.)[53][54]

21.6 Weaknesses

By DN model, if one asks, "Why is that shadow 20 feet long?", another can answer, "Because that flagpole is 15 feet tall, the Sun is at x angle, and laws of electromagnetism".[6] Yet by problem of symmetry, if one instead asked, "Why is that flagpole 15 feet tall?", another could answer, "Because that shadow is 20 feet long, the Sun is at x angle, and laws of electromagnetism", likewise a deduction from observed conditions and scientific laws, but an answer clearly incorrect.[6] By the problem of irrelevance, if one asks, "Why did that man not get pregnant?", one could in part answer, among the explanans, "Because he took birth control pills"—if he factually took them, and the law of their preventing pregnancy—as covering law model poses no restriction to bar that observation from the explanans.

Many philosophers have concluded that causality is integral to scientific explanation.[55] DN model offers a necessary condition of a causal explanation—successful prediction—but not sufficient conditions of causal explanation, as a universal regularity can include spurious relations or simple correlations, for instance Z always following Y, but not Z because of Y, instead Y and then Z as an effect of X.[55] By relating temperature, pressure, and volume of gas within a container, Boyle's law permits prediction of an unknown variable—volume, pressure, or temperature—but does not explain *why* to expect that unless one adds, perhaps, the kinetic theory of gases.[55][56]

Scientific explanations increasingly pose not determinism's universal laws, but probabilism's chance,[57] *ceteris paribus* laws.[40] Smoking's contribution to lung cancer fails even the inductive-statistical model (IS model), requiring probability over 0.5 (50%).[58] (Probability standardly ranges from 0 (0%) to 1 (100%).) An applied science that applies statistics seeking associations between events, epidemiology cannot show causality, but consistently found higher incidence of lung cancer in smokers versus otherwise similar nonsmokers, although the proportion of smokers who develop lung cancer is modest.[59] Versus nonsmokers, however, smokers as a group showed over 20 times the risk of lung cancer, and in conjunction with basic research, consensus followed that smoking had been scientifically explained as *a* cause of lung cancer,[60] responsible for some cases that without smoking would not have occurred,[59] a probabilistic counterfactual causality.[61][62]

21.7 Covering action

Through lawlike explanation, fundamental physics—often perceived as fundamental science—has proceeded through intertheory relation and theory reduction, thereby resolving experimental paradoxes to great historical success,[63] resembling covering law model.[64] In early 20th century, Ernst Mach as well as Wilhelm Ostwald had resisted Ludwig Boltzmann's reduction of thermodynamics—and thereby Boyle's law[65]—to statistical mechanics partly *because* it rested on kinetic theory of gas,[56] hinging on atomic/molecular theory of matter.[66] Mach as well as Ostwald viewed matter as a variant of energy, and molecules as mathematical illusions,[66] as even Boltzmann thought possible.[67]

In 1905, via statistical mechanics, Albert Einstein predicted the phenomenon Brownian motion—unexplained since reported in 1827 by botanist Robert Brown.[66] Soon, most physicists accepted that atoms and molecules were unobservable yet real.[66] Also in 1905, Einstein explained the electromagnetic field's energy as distributed in *particles*, doubted until this helped resolve atomic theory in the 1910s and 1920s.[68] Meanwhile, all known physical phenomena were gravitational or electromagnetic,[69] whose two theories misaligned.[70] Yet belief in aether as the source of all physical phenomena was virtually unanimous.[71][72][73][74] At experimental paradoxes,[75] physicists modified the aether's hypothetical properties.[76]

Finding the luminiferous aether a useless hypothesis,[77] Einstein in 1905 *a priori* unified all inertial reference frames to state special *principle* of relativity,[78] which, by omitting aether,[79] converted space and time into *relative* phenomena whose relativity aligned electrodynamics with the Newtonian principle Galilean relativity or invariance.[63][80] Originally epistemic or instrumental, this was interpreted as ontic or realist—that is, a causal mechanical explanation—and the *principle* became a *theory*,[81] refuting Newtonian gravitation.[79][82] By predictive success in 1919, general relativity apparently overthrew Newton's theory, a revolution in science[83] resisted by many yet fulfilled around 1930.[84]

In 1925, Werner Heisenberg as well as Erwin Schrödinger independently formalized quantum mechanics (QM).[85][86] Despite clashing explanations,[86][87] the two theories made identical predictions.[85] Paul Dirac's 1928 model of the electron was set to special relativity, launching QM into the first quantum field theory (QFT), quantum electrodynamics (QED).[88] From it, Dirac interpreted and predicted the electron's antiparticle, soon discovered and termed *positron*,[89] but the QED failed electrodynamics at high energies.[90] Elsewhere and otherwise, strong nuclear force and weak nuclear force were discovered.[91]

In 1941, Richard Feynman introduced QM's path integral formalism, which if taken toward *interpretation* as a causal mechanical model clashes with Heisenberg's matrix formalism and with Schrödinger's wave formalism,[87] although all three are empirically identical, sharing predictions.[85] Next, working on QED, Feynman sought to model particles without fields and find the vacuum truly empty.[92] As each known fundamental force[93] is apparently an effect of a field, Feynman failed.[92] Louis de Broglie's waveparticle duality had rendered atomism—indivisible particles in a void—untenable, and highlighted the very notion of discontinuous particles as selfcontradictory.[94]

Meeting in 1947, Freeman Dyson, Richard Feynman, Julian Schwinger, and Sin-Itiro Tomonaga soon introduced *renormalization*, a procedure converting QED to physics' most predictively precise theory,[90][95] subsuming chemistry, optics, and statistical mechanics.[63][96] QED thus won physicists' general acceptance.[97] Paul Dirac criticized its need for renormalization as showing its unnaturalness,[97] and called for an aether.[98] In 1947, Willis Lamb had found unexpected motion of electron orbitals, shifted since the vacuum is not truly empty.[99] Yet *emptiness* was catchy, abolishing aether conceptually, and physics proceeded ostensibly without it,[92] even suppressing it.[98] Meanwhile, "sickened by untidy math, most philosophers of physics tend to neglect QED".[97]

Physicists have feared even mentioning *aether*,[100] renamed *vacuum*,[98][101] which—as such—is nonexistent.[98][102] General philosophers of science commonly believe that aether, rather, is fictitious,[103] "relegated to the dustbin of scientific history ever since" 1905 brought special relativity.[104] Einstein was noncommittal to aether's nonexistence,[77] simply said it superfluous.[79] Abolishing Newtonian motion for electrodynamic primacy, however, Einstein inadvertently reinforced aether,[105] and to explain motion was led back to aether in general relativity.[106][107][108] Yet resistance to relativity theory[109] became associated with earlier theories of aether, whose word and concept became taboo.[110] Einstein explained special relativity's compatibility with an aether,[107] but Einstein aether, too, was opposed.[100] Objects became conceived as pinned directly on space and time[111] by abstract geometric relations lacking ghostly or fluid medium.[100][112]

By 1970, QED along with weak nuclear field was reduced to electroweak theory (EWT), and the strong nuclear field was modeled as quantum chromodynamics (QCD).[90] Comprised by EWT, QCD, and Higgs field, this Standard Model of particle physics is an "effective theory",[113] not truly fundamental.[114][115] As QCD's particles are considered nonexistent in the everyday world,[92] QCD especially suggests an aether,[116] routinely found by physics experiments to exist and to exhibit relativistic symmetry.[110] Confirmation of the Higgs particle, modeled as a condensation within the Higgs field, corroborates aether,[100][115] although physics need not state or even include aether.[100] Organizing regularities of *observations*—as in the covering law model—physicists find superfluous the quest to discover *aether*.[64]

In 1905, from special relativity, Einstein deduced mass-energy equivalence,[117] particles being variant forms of distributed energy,[118] how particles colliding at vast speed experience that energy's transformation into mass, producing heavier particles,[119] although physicists' talk promotes confusion.[120] As "the contemporary locus of metaphysical research", QFTs pose particles not as existing individually, yet as *excitation modes* of fields,[114][121] the particles and their masses being states of aether,[92] apparently unifying all physical phenomena as the more fundamental causal reality,[101][115][116] as long ago foreseen.[73] Yet a *quantum* field is an intricate abstraction—a *mathematical* field—virtually inconceivable as a *classical* field's physical properties.[121] Nature's deeper aspects, still unknown, might elude any possible field theory.[114][121]

Though discovery of causality is popularly thought science's aim, search for it was shunned by the Newtonian research program,[14] even more Newtonian than was Isaac Newton.[92][122] By now, most theoretical physicists infer that the four, known fundamental interactions would reduce to superstring theory, whereby atoms and molecules, after all, are energy vibrations holding mathematical, geometric forms.[63] Given uncertainties of scientific realism,[18] some conclude that the concept *causality* raises comprehensibility of scientific explanation and thus is key folk science, but compromises precision of scientific explanation and is dropped as a science matures.[123] Even epidemiology is maturing to heed the severe difficulties with presumptions about causality.[14][57][59] Covering law model is among Carl G Hempel's admired contributions to philosophy of science.[124]

21.8   See also

Types of inference

- Deductive reasoning
- Inductive reasoning
- Abductive reasoning

Related subjects

- Explanandum and explanans
- Hypothetico-deductive model
- Models of scientific inquiry
- Philosophy of science
- Scientific method

21.9   Notes

[1] Woodward, "Scientific explanation", §2 "The DN model", in *SEP*, 2011.

[2] James Fetzer, ch 3 "The paradoxes of Hempelian explanation", in Fetzer, ed, *Science, Explanation, and Rationality* (Oxford U P, 2000), p 113.

[3] Montuschi, *Objects in Social Science* (Continuum, 2003), pp 61–62.

[4] Bechtel, *Philosophy of Science* (Lawrence Erlbaum, 1988), ch 2, subch "DN model of explanation and HD model of theory development", pp 25–26.

[5] Bechtel, *Philosophy of Science* (Lawrence Erlbaum, 1988), ch 2, subch "Axiomatic account of theories", pp 27–29.

[6] Suppe, "Afterword—1977", "Introduction", §1 "Swan song for positivism", §1A "Explanation and intertheoretical reduction", pp 619–24, in Suppe, ed, *Structure of Scientific Theories*, 2nd edn (U Illinois P, 1977).

[7] Kenneth F Schaffner, "Explanation and causation in biomedical sciences", pp 79–125, in Laudan, ed, *Mind and Medicine* (U California P, 1983), p 81.

[8] G Montalenti, ch 2 "From Aristotle to Democritus via Darwin", in Ayala & Dobzhansky, eds, *Studies in the Philosophy of Biology* (U California P, 1974).

[9] In the 17th century, Descartes as well as Isaac Newton firmly believed in God as nature's designer and thereby firmly believed in natural purposiveness, yet found teleology to be outside science's inquiry (Bolotin, *Approach to Aristotle's Physics*, pp 31–33). By 1650, formalizing heliocentrism and launching mechanical philosophy, Cartesian physics overthrew geocentrism as well as Aristotelian physics. In the 1660s, Robert Boyle sought to lift chemistry as a new discipline from alchemy. Newton more especially sought the laws of nature—simply the regularities of phenomena—whereby Newtonian physics, reducing celestial science to terrestrial science, ejected from physics the vestige of

Aristotelian metaphysics, thus disconnecting physics and alchemy/chemistry, which then followed its own course, yielding chemistry around 1800.

[10] Nicknames for principles attributed to Hume—Hume's fork, problem of induction, Hume's law—were not created by Hume but by later philosophers labeling them for ease of reference.

[11] By Hume's fork, the truths of mathematics and logic as formal sciences are universal through "relations of ideas"—simply abstract truths—thus knowable without experience. On the other hand, the claimed truths of empirical sciences are contingent on "fact and real existence", knowable only upon experience. By Hume's fork, the two categories never cross. Any treatises containing neither can contain only "sophistry and illusion". (Flew, *Dictionary*, "Hume's fork", p 156).

[12] Not privy to the world's either necessities or impossibilities, but by force of habit or mental nature, humans experience sequence of sensory events, find seeming constant conjunction, make the unrestricted generalization of an enumerative induction, and justify it by presuming uniformity of nature. Humans thus attempt to justify a minor induction by adding a major induction, both logically invalid and unverified by experience—the problem of induction—how humans irrationally presume discovery of causality. (Chakraborti, *Logic*, p 381; Flew, *Dictionary*, "Hume", p 156.

[13] For more discursive discussions of types of causality—necessary, sufficient, necessary and sufficient, component, sufficient component, counterfactual—see Rothman & Greenland, Parascandola & Weed, as well as Kundi. Following is more direct elucidation:

A *necessary cause* is a causal condition *required* for an event to occur. A *sufficient cause* is a causal condition *complete* to produce an event. Necessary is not always sufficient, however, since other casual factors—that is, other *component causes*—might be required to produce the event. Conversely, a sufficient cause is not always a necessary cause, since differing sufficient causes might likewise produce the event. Strictly speaking, a sufficient cause cannot be a single factor, as any causal factor must act casually through many other factors. And although a necessary cause might exist, humans cannot verify one, since humans cannot check every possible state of affairs. (Language can state necessary causality as a tautology—a statement whose terms' arrangement and meanings render it logically true by mere definition—which, as an *analytic* statement, is uninformative about the actual world. A statement referring to and contingent on the world's actualities is a *synthetic* statement, rather.)

Sufficient causality is more actually *sufficient component causality*—a complete set of component causes interacting within a causal constellation—which, however, is beyond humans' capacity to fully discover. Yet humans tend intuitively to conceive of causality as *necessary and sufficient*—a single factor both required and complete—the one and only cause, *the* cause. One may so view flipping a light switch.

The switch's flip was not sufficient cause, however, but contingent on countless factors—intact bulb, intact wiring, circuit box, bill payment, utility company, neighborhood infrastructure, engineering of technology by Thomas Edison and Nikola Tesla, explanation of electricity by James Clerk Maxwell, harnessing of electricity by Benjamin Franklin, metal refining, metal mining, and on and on—while, whatever the tally of events, nature's causal mechanical structure remains a mystery.

From a Humean perspective, the light's putative inability to come on without the switch's flip is neither a logical necessity nor an empirical finding, since no experience ever reveals that the world either is or will remain universally uniform as to the aspects appearing to bind the switch's flip as the necessary event for the light's coming on. If the light comes on without switch flip, surprise will affect one's *mind*, but one's mind cannot know that the event violated *nature*. As just a mundane possibility, an activity within the wall could have connected the wires and completed the circuit without the switch's flip.

Though apparently enjoying the scandals that trailed his own explanations, Hume was very practical and his skepticism was quite uneven (Flew p 156). Although Hume rejected orthodox theism and sought to reject metaphysics, Hume supposedly extended Newtonian method to the human mind, which Hume, in a sort of antiCopernican move, placed as the pivot of human knowledge (Flew p 154). Hume thus placed his own theory of knowledge on par with Newton's theory of motion (Buckle pp 70–71, Redman pp 182–83, Schliesser § abstract). Hume found enumerative induction an unavoidable custom required for one to live (Gattei pp 28–29). Hume found constant conjunction to reveal a modest causality type: *counterfactual causality*. Silent as to causal role—whether necessity, sufficiency, component strength, or mechanism—counterfactual causality is simply that alteration of a factor prevents or produces the event of interest.

[14] Kundi M (2006). "Causality and the interpretation of epidemiologic evidence". *Environmental Health Perspectives*. **114** (7): 969–974. PMC 1513293. PMID 16835045. doi:10.1289/ehp.8297.

[15] Hume noted that authors ubiquitously continue for some time stating facts and then suddenly switch to stating norms—supposedly what should be—with barely explanation. Yet such values, as in ethics or aesthetics or political philosophy, are not found true merely by stating facts: *is* does not itself reveal *ought*. Hume's law is the principle that the fact/value gap is unbridgeable—that no statements of facts can ever justify norms—although Hume himself did not state that. Rather, some later philosophers found Hume to merely stop short of stating it, but to have communicated it. Anyway, Hume found that humans acquired morality through experience by communal reinforcement. (Flew, *Dictionary*, "Hume's law", p 157 & "Naturalistic fallacy", pp 240–41; Wootton, *Modern Political Thought*, p 306.)

[16] Kant inferred that the mind's constants arrange space holding Euclidean geometry—like Newton's absolute space—

while objects interact temporally as modeled in Newton's theory of motion, whose law of universal gravitation is a truth *synthetic a priori*, that is, contingent on experience, indeed, but known universally true without universal experience. Thus, the mind's innate constants cross the tongs of Hume's fork and lay Newton's universal gravitation as *a priori* truth.

[17] Chakravartty, "Scientific realism", §1.2 "The three dimensions of realist commitment", in *SEP*, 2013: "Semantically, realism is committed to a literal interpretation of scientific claims about the world. In common parlance, realists take theoretical statements at 'face value'. According to realism, claims about scientific entities, processes, properties, and relations, whether they be observable or unobservable, should be construed literally as having truth values, whether true or false. This semantic commitment contrasts primarily with those of so-called instrumentalist epistemologies of science, which interpret descriptions of unobservables simply as instruments for the prediction of observable phenomena, or for systematizing observation reports. Traditionally, instrumentalism holds that claims about unobservable things have no literal meaning at all (though the term is often used more liberally in connection with some antirealist positions today). Some antirealists contend that claims involving unobservables should not be interpreted literally, but as elliptical for corresponding claims about observables".

[18] Challenges to scientific realism are captured succinctly by Bolotin, *Approach to Aristotle's Physics* (SUNY P, 1998), pp 33–34, commenting about modern science, "But it has not succeeded, of course, in encompassing all phenomena, at least not yet. For it laws are mathematical idealizations, idealizations, moreover, with no immediate basis in experience and with no evident connection to the ultimate causes of the natural world. For instance, Newton's first law of motion (the law of inertia) requires us to imagine a body that is always at rest or else moving aimlessly in a straight line at a constant speed, even though we never see such a body, and even though according to his own theory of universal gravitation, it is impossible that there can be one. This fundamental law, then, which begins with a claim about what would happen in a situation that never exists, carries no conviction except insofar as it helps to predict observable events. Thus, despite the amazing success of Newton's laws in predicting the observed positions of the planets and other bodies, Einstein and Infeld are correct to say, in *The Evolution of Physics*, that 'we can well imagine another system, based on different assumptions, might work just as well'. Einstein and Infeld go on to assert that 'physical concepts are free creations of the human mind, and are not, however it may seem, uniquely determined by the external world'. To illustrate what they mean by this assertion, they compare the modern scientist to a man trying to understand the mechanism of a closed watch. If he is ingenious, they acknowledge, this man 'may form some picture of a mechanism which would be responsible for all the things he observes'. But they add that he 'may never quite be sure his picture is the only one which could explain his observations. He will never be able to compare his picture with the real mechanism and he cannot even imagine the possibility or the meaning of such a comparison'. In other words, modern science cannot claim, and it will never be able to claim, that it has the definite understanding of any natural phenomenon".

[19] Whereas a hypothetical imperative is practical, simply what one ought to do if one seeks a particular outcome, the categorical imperative is morally universal, what everyone always ought to do.

[20] Bourdeau, "Auguste Comte", §§ "Abstract" & "Introduction", in Zalta, ed, *SEP*, 2013.

[21] Comte, *A General View of Positivism* (Trübner, 1865), pp 49–50, including the following passage: "As long as men persist in attempting to answer the insoluble questions which occupied the attention of the childhood of our race, by far the more rational plan is to do as was done then, that is, simply to give free play to the imagination. These spontaneous beliefs have gradually fallen into disuse, not because they have been disproved, but because humankind has become more enlightened as to its wants and the scope of its powers, and has gradually given an entirely new direction to its speculative efforts".

[22] Flew, *Dictionary* (St Martin's, 1984), "Positivism", p 283.

[23] Woodward, "Scientific explanation", §1 "Background and introduction", in *SEP*, 2011.

[24] Friedman, *Reconsidering Logical Positivism* (Cambridge U P, 1999), p xii.

[25] Any *positivism* placed in the 20th century is generally *neo*, although there was Ernst Mach's positivism nearing 1900, and a general positivistic approach to science—traceable to the inductivist trend from Bacon at 1620, the Newtonian research program at 1687, and Comptean positivism at 1830—that continues in a vague but usually disavowed sense within popular culture and some sciences.

[26] Neopositivists are sometimes called "verificationists".

[27] • Chakravartty, "Scientific realism", §4 "Antirealism: Foils for scientific realism", §4.1 "Empiricism", in *SEP*, 2013: "Traditionally, instrumentalists maintain that terms for unobservables, by themselves, have no meaning; construed literally, statements involving them are not even candidates for truth or falsity. The most influential advocates of instrumentalism were the logical empiricists (or logical positivists), including Carnap and Hempel, famously associated with the Vienna Circle group of philosophers and scientists as well as important contributors elsewhere. In order to rationalize the ubiquitous use of terms which might otherwise be taken to refer to unobservables in scientific discourse, they adopted a non-literal semantics according to which these terms acquire meaning by being associated with terms for observables (for example, 'electron' might mean 'white

streak in a cloud chamber'), or with demonstrable laboratory procedures (a view called 'operationalism'). Insuperable difficulties with this semantics led ultimately (in large measure) to the demise of logical empiricism and the growth of realism. The contrast here is not merely in semantics and epistemology: a number of logical empiricists also held the neo-Kantian view that ontological questions 'external' to the frameworks for knowledge represented by theories are also meaningless (the choice of a framework is made solely on pragmatic grounds), thereby rejecting the metaphysical dimension of realism (as in Carnap 1950)".

- Okasha, *Philosophy of Science* (Oxford U P, 2002), p 62: "Strictly we should distinguish two sorts of anti-realism. According to the first sort, talk of unobservable entities is not to be understood literally at all. So when a scientist puts forward a theory about electrons, for example, we should not take him to be asserting the existence of entities called 'electrons'. Rather, his talk of electrons is metaphorical. This form of anti-realism was popular in the first half of the 20th century, but few people advocate it today. It was motivated largely by a doctrine in the philosophy of language, according to which it is not possible to make meaningful assertions about things that cannot in principle be observed, a doctrine that few contemporary philosophers accept. The second sort of anti-realism accepts that talk of unobservable entities should be taken at face value: if a theory says that electrons are negatively charged, it is true if electrons do exist and are negatively charged, but false otherwise. But we will never know which, says the anti-realist. So the correct attitude towards the claims that scientists make about unobservable reality is one of total agnosticism. They are either true or false, but we are incapable of finding out which. Most modern anti-realism is of this second sort".

[28] Woodward, "Scientific explanation", in Zalta, ed, *SEP*, 2011, abstract.

[29] Carl G Hempel & Paul Oppenheim, "Studies in the logic of explanation", *Philosophy of Science*, 1948 Apr; 15(2):135–175.

[30] Bechtel, *Discovering Cell Mechanisms* (Cambridge U P, 2006), esp pp 24–25.

[31] Woodward, "Scientific explanation", §2 "The DN model", §2.3 "Inductive statistical explanation", in Zalta, ed, *SEP*, 2011.

[32] von Wright, *Explanation and Understanding* (Cornell U P, 1971), p 11.

[33] Stuart Glennan, "Explanation", § "Covering-law model of explanation", in Sarkar & Pfeifer, eds, *Philosophy of Science* (Routledge, 2006), p 276.

[34] Manfred Riedel, "Causal and historical explanation", in Manninen & Tuomela, eds, *Essays on Explanation and Understanding* (D Reidel, 1976), pp 3–4.

[35] Neopositivism's fundamental tenets were the verifiability criterion of *cognitive meaningfulness*, the analytic/synthetic gap, and the observation/theory gap. From 1950 to 1951, Carl Gustav Hempel renounced the verifiability criterion. In 1951 Willard Van Orman Quine attacked the analytic/synthetic gap. In 1958, Norwood Russell Hanson blurred the observational/theoretical gap. In 1959, Karl Raimund Popper attacked all of verificationism—he attacked, actually, any type of positivism—by asserting falsificationism. In 1962, Thomas Samuel Kuhn overthrew foundationalism, which was erroneously presumed to be a fundamental tenet of neopositivism.

[36] Fetzer, "Carl Hempel", §3 "Scientific reasoning", in *SEP*, 2013: "The need to dismantle the verifiability criterion of meaningfulness together with the demise of the observational/theoretical distinction meant that logical positivism no longer represented a rationally defensible position. At least two of its defining tenets had been shown to be without merit. Since most philosophers believed that Quine had shown the analytic/synthetic distinction was also untenable, moreover, many concluded that the enterprise had been a total failure. Among the important benefits of Hempel's critique, however, was the production of more general and flexible criteria of *cognitive significance* in Hempel (1965b), included in a famous collection of his studies, *Aspects of Scientific Explanation* (1965d). There he proposed that *cognitive significance* could not be adequately captured by means of principles of verification or falsification, whose defects were parallel, but instead required a far more subtle and nuanced approach. Hempel suggested multiple criteria for assessing the *cognitive significance* of different theoretical systems, where significance is not categorical but rather a matter of degree: 'Significant systems range from those whose entire extralogical vocabulary consists of observation terms, through theories whose formulation relies heavily on theoretical constructs, on to systems with hardly any bearing on potential empirical findings' (Hempel 1965b: 117). The criteria Hempel offered for evaluating the 'degrees of significance' of theoretical systems (as conjunctions of hypotheses, definitions, and auxiliary claims) were (a) the clarity and precision with which they are formulated, including explicit connections to observational language; (b) the systematic—explanatory and predictive—power of such a system, in relation to observable phenomena; (c) the formal simplicity of the systems with which a certain degree of systematic power is attained; and (d) the extent to which those systems have been confirmed by experimental evidence (Hempel 1965b). The elegance of Hempel's study laid to rest any lingering aspirations for simple criteria of 'cognitive significance' and signaled the demise of logical positivism as a philosophical movement".

[37] Popper, "Against big words", *In Search of a Better World* (Routledge, 1996), pp 89-90.

[38] Hacohen, *Karl Popper: The Formative Years* (Cambridge U P, 2000), pp 212-13.

[39] *Logik der Forschung*, published in Austria in 1934, was translated by Popper from German to English, *The Logic*

of Scientific Discovery, and arrived in the English-speaking world in 1959.

[40] Reutlinger, Schurz & Hüttemann, "Ceteris paribus", § 1.1 "Systematic introduction", in Zalta, ed, *SEP*, 2011.

[41] As scientific study of cells, cytology emerged in the 19th century, yet its technology and methods were insufficient to clearly visualize and establish existence of any cell organelles beyond the nucleus.

[42] The first famed biochemistry experiment was Edward Buchner's in 1897 (Morange, *A History*, p 11). The biochemistry discipline soon emerged, initially investigating colloids in biological systems, a "biocolloidology" (Morange p 12; Bechtel, *Discovering*, p 94). This yielded to macromolecular theory, the term *macromolecule* introduced by German chemist Hermann Staudinger in 1922 (Morange p 12).

[43] Cell biology emerged principally at Rockefeller Institute through new technology (electron microscope and ultracentrifuge) and new techniques (cell fractionation and advancements in staining and fixation).

[44] James Fetzer, ch 3 "The paradoxes of Hempelian explanation", in Fetzer J, ed, *Science, Explanation, and Rationality* (Oxford U P, 2000), pp 121–122.

[45] Fetzer, ch 3 in Fetzer, ed, *Science, Explanation, and Rationality* (Oxford U P, 2000), p 129.

[46] Bechtel, *Philosophy of Science* (Lawrence Erlbaum, 1988), ch 1, subch "Areas of philosophy that bear on philosophy of science", § "Metaphysics", pp 8–9, § "Epistemology", p 11.

[47] H Atmanspacher, R C Bishop & A Amann, "Extrinsic and intrinsic irreversibility in probabilistic dynamical laws", in Khrennikov, ed, *Proceedings* (World Scientific, 2001), pp 51–52.

[48] Fetzer, ch 3, in Fetzer, ed, *Science, Explanation, and Rationality* (Oxford U P, 2000), p 118, poses some possible ways that natural laws, so called, when epistemic can fail as ontic: "The underlying conception is that of bringing order to our *knowledge* of the universe. Yet there are at least three reasons why even complete knowledge of every empirical regularity that obtains during the world's history might not afford an adequate inferential foundation for discovery of the world's laws. First, some laws might remain uninstantiated and therefore not be displayed by any regularity. Second, some regularities may be accidental and therefore not display any law of nature. And, third, in the case of probabilistic laws, some frequencies might deviate from their generating nomic probabilities 'by chance' and therefore display natural laws in ways that are unrepresentative or biased".

[49] This theory reduction occurs if, and apparently only if, the Sun and one planet are modeled as a two-body system, excluding all other planets (Torretti, *Philosophy of Physics*, pp 60–62).

[50] Spohn, *Laws of Belief* (Oxford U P, 2012), p 305.

[51] Whereas fundamental physics has sought laws of universal regularity, special sciences normally include *ceteris paribus* laws, which are predictively accurate to high probability in "normal conditions" or with "all else equal", but have exceptions [Reutlinger *et al* § 1.1]. Chemistry's laws seem exceptionless in their domains, yet were in principle reduced to fundamental physics [Feynman p 5, Schwarz Fig 1, and so are special sciences.

[52] Bechtel, *Philosophy of Science* (Lawrence Erlbaum, 1988), ch 5, subch "Introduction: Relating disciplines by relating theories" pp 71–72.

[53] Bechtel, *Philosophy of Science* (Lawrence Erlbaum, 1988), ch 5, subch "Theory reduction model and the unity of science program" pp 72–76.

[54] Bem & de Jong, *Theoretical Issues* (Sage, 2006), pp 45–47.

[55] O'Shaughnessy, *Explaining Buyer Behavior* (Oxford U P, 1992), pp 17–19.

[56] Spohn, *Laws of Belief* (Oxford U P, 2012), p 306.

[57] Karhausen, L. R. (2000). "Causation: The elusive grail of epidemiology". *Medicine, health care, and philosophy*. 3 (1): 59–67. PMID 11080970. doi:10.1023/A:1009970730507.

[58] Bechtel, *Philosophy of Science* (Lawrence Erlbaum, 1988), ch 3, subch "Repudiation of DN model of explanation", pp 38–39.

[59] Rothman, K. J., Greenland, S. (2005). "Causation and Causal Inference in Epidemiology". *American Journal of Public Health*. 95: S144–S150. PMID 16030331. doi:10.2105/AJPH.2004.059204.

[60] Boffetta, "Causation in the presence of weak associations", *Crit Rev Food Sci Nutr*, 2010; **50**(S1):13–16.

[61] Making no commitment as to the particular causal *role*—such as necessity, or sufficiency, or component strength, or mechanism—*counterfactual causality* is simply that alteration of a factor from its factual state prevents or produces by any which way the event of interest.

[62] In epidemiology, the counterfactual causality is not deterministic, but probabilistic (Parascandola & Weed, "Causation in epidemiology", *J Epidemiol Community Health*, 2001; **55**:905–12) PMID 11707485.

[63] Schwarz, "Recent developments in string theory", *Proc Natl Acad Sci U S A*, 1998; **95**:2750–7, esp Fig 1.

[64] Ben-Menahem, *Conventionalism* (Cambridge U P, 2006), p 71.

[65] Instances of falsity limited Boyle's law to special cases, thus ideal gas law.

[66] Newburgh *et al*, "Einstein, Perrin, and the reality of atoms", *Am J Phys*, 2006, p 478.

[67] For brief review of Boltmann's view, see ch 3 "Philipp Frank", § 1 "T S Kuhn's interview", in Blackmore *et al*, eds, *Ernst Mach's Vienna 1895–1930* (Kluwer, 2001), p 63, as Frank was a student of Boltzmann soon after Mach's retirement. See "Notes", pp 79–80, #12 for views of Mach and of Ostwald, #13 for views of contemporary physicists generally, and #14 for views of Einstein. The more relevant here is #12: "Mach seems to have had several closely related opinions concerning atomism. First, he often thought the theory might be useful in physics as long as one did not believe in the reality of atoms. Second, he believed it was difficult to apply the atomic theory to both psychology and physics. Third, his own theory of elements is often called an 'atomistic theory' in psychology in contrast with both gestalt theory and a continuum theory of experience. Fourth, when critical of the reality of atoms, he normally meant the Greek sense of 'indivisible substance' and thought Boltzmann was being evasive by advocating divisible atoms or 'corpuscles' such as would become normal after J J Thomson and the distinction between electrons and nuclei. Fifth, he normally called physical atoms 'things of thought' and was very happy when Ostwald seemed to refute the reality of atoms in 1905. And sixth, after Ostwald returned to atomism in 1908, Mach continued to defend Ostwald's 'energeticist' alternative to atomism".

[68] Physicists had explained the electromagnetic field's energy as *mechanical* energy, like an ocean wave's bodily impact, not water droplets individually showered (Grandy, *Everyday Quantum Reality*, pp 22–23). In the 1890s, the problem of blackbody radiation was paradoxical until Max Planck theorized *quantum* exhibiting Planck's constant—a minimum unit of energy. The quanta were mysterious, not viewed as *particles*, yet simply as units of *energy*. Another paradox, however, was the photoelectric effect.

As shorter wavelength yields more waves per unit distance, lower wavelength is higher wave frequency. Within the electromagnetic spectrum's visible portion, frequency sets the color. Light's intensity, however, is the wave's amplitude as the wave's height. In a strictly wave explanation, a greater intensity—higher wave amplitude—raises the mechanical energy delivered, namely, the wave's impact, and thereby yields greater physical effect. And yet in the photoelectric effect, only a certain color and beyond—a certain frequency and higher—was found to knock electrons off a metal surface. Below that frequency or color, raising the intensity of the light still knocked no electrons off.

Einstein modeled Planck's quanta as each a particle whose individual energy was Planck's constant multiplied by the light's wave's frequency: at only a certain frequency and beyond would each particle be energetic enough to eject an electron from its orbital. Although elevating the intensity of light would deliver more energy—more total particles— each individual particle would still lack sufficient energy to dislodge an electron. Einstein's model, far more intricate, used probability theory to explain rates of electrons ejections as rates of collisions with electromagnetic particles. This revival of the particle hypothesis of light—generally attributed to Newton—was widely doubted. By 1920, however, the explanation helped solve problems in atomic theory, and thus quantum mechanics emerged. In 1926, Gilbert N Lewis termed the particles *photons*. QED models them as the electromagnetic field's messenger particles or force carriers, emitted and absorbed by electrons and by other particles undergoing transitions.

[69] Wolfson, *Simply Einstein* (W W Norton & Co, 2003), p 67.

[70] Newton's gravitational theory at 1687 had postulated absolute space and absolute time. To fit Young's transverse wave theory of light at 1804, space was theoretically filled with Fresnel's luminiferous aether at 1814. By Maxwell's electromagnetic field theory of 1865, light always holds a constant speed, which, however, must be relative to something, apparently to aether. Yet if light's speed is constant relative to aether, then a body's motion through aether would be relative to—thus vary in relation to—light's speed. Even Earth's vast speed, multiplied by experimental ingenuity with an interferometer by Michelson & Morley at 1887, revealed no apparent *aether drift*—light speed apparently constant, an absolute. Thus, both Newton's gravitational theory and Maxwell's electromagnetic theory each had its own relativity principle, yet the two were incompatible. For brief summary, see Wilczek, *Lightness of Being* (Basic Books, 2008), pp 78–80.

[71] Cordero, *EPSA Philosophy of Science* (Springer, 2012), pp 26–28.

[72] Hooper, *Aether and Gravitation* (Chapman & Hall, 1903), pp 122–23.

[73] Lodge, "The ether of space", *Sci Am Suppl*, 1909; **67**:202–03.

[74] Even Mach, who shunned all hypotheses beyond direct sensory experience, presumed an aether, required for motion to not violate mechanical philosophy's founding principle, *No instant interaction at a distance* (Einstein, "Ether", *Sidelights* (Methuen, 1922), pp 15–18).

[75] Rowlands, *Oliver Lodge* (Liverpool U P, 1990), pp 159–60: "Lodge's ether experiments have become part of the historical background leading up to the establishment of special relativity and their significance is usually seen in this context. Special relativity, it is stated, eliminated both the ether and the concept of absolute motion from physics. Two experiments were involved: that of Michelson and Morley, which showed that bodies do not move with respect to a stationary ether, and that of Lodge, which showed that moving bodies do not drag ether with them. With the emphasis on relativity, the Michelson–Morley experiment has come to be seen as the more significant of the two, and Lodge's experiment becomes something of a detail, a matter of eliminating the final, and less likely, possibility of a nonstationary, viscous, all-pervading medium. It could be argued that almost the exact opposite may have been the case. The Michelson–Morley experiment did not prove that there was no absolute motion, and it did not prove that there was no stationary ether. Its results—and the FitzGerald–Lorentz contraction—could

have been predicted on Heaviside's, or even Maxwell's, theory, even if no experiment had ever taken place. The significance of the experiment, though considerable, is purely historical, and in no way factual. Lodge's experiment, on the other hand, showed that, if an ether existed, then its properties must be quite different from those imagined by mechanistic theorists. The ether which he always believed existed had to acquire entirely new properties as a result of this work".

[76] Mainly Hendrik Lorentz as well as Henri Poincaré modified electrodynamic theory and, more or less, developed special theory of relativity before Einstein did (Ohanian, *Einstein's Mistakes*, pp 281–85). Yet Einstein, free a thinker, took the next step and stated it, more elegantly, without aether (Torretti, *Philosophy of Physics*, p 180).

[77] Tavel, *Contemporary Physics* (Rutgers U P, 2001), pp , 66.

[78] Introduced soon after Einstein explained Brownian motion, special relativity holds only in cases of inertial motion, that is, unaccelerated motion. Inertia is the state of a body experiencing no acceleration, whether by change in speed—either quickening or slowing—or by change in direction, and thus exhibits constant velocity, which is speed plus direction.

[79] Cordero, *EPSA Philosophy of Science* (Springer, 2012), pp 29–30.

[80] To explain absolute light speed without aether, Einstein modeled that a body at motion in an electromagnetic field experiences length contraction and time dilation, which Lorentz and Poincaré had already modeled as Lorentz-FitzGerald contraction and Lorentz transformation but by hypothesizing dynamic states of the aether, whereas Einstein's special relativity was simply kinematic, that is, positing no causal mechanical explanation, simply describing positions, thus showing how to align measuring devices, namely, clocks and rods. (Ohanian, *Einstein's Mistakes*, pp 281–85).

[81] Ohanian, *Einstein's Mistakes* (W W Norton, 2008), pp 281–85.

[82] Newton's theory required absolute space and time.

[83] Buchen, "May 29, 1919", *Wired*, 2009.
Moyer, "Revolution", in *Studies in the Natural Sciences* (Springer, 1979), p 55.
Melia, *Black Hole* (Princeton U P, 2003), pp 83–87.

[84] Crelinsten, *Einstein's Jury* (Princeton U P, 2006), p 28.

[85] From 1925 to 1926, independently but nearly simultaneously, Werner Heisenberg as well as Erwin Schrödinger developed quantum mechanics (Zee in Feynman, *QED*, p xiv). Schrödinger introduced wave mechanics, whose wave function is discerned by a partial differential equation, now termed *Schrödinger equation* (p xiv). Heisenberg, who also stated the uncertainty principle, along with Max Born and Pascual Jordan introduced matrix mechanics, which rather confusingly talked of *operators* acting on *quantum states* (p xiv). If taken as causal mechanically explanatory, the two formalisms vividly disagree, and yet are indiscernible empirically, that is, when not used for *interpretation*, and taken as simply *formalism* (p xv).
In 1941, at a party in a tavern in Princeton, New Jersey, visiting physicist Herbert Jehle mentioned to Richard Feynman a different formalism suggested by Paul Dirac, who developed bra–ket notation, in 1932 (p xv). The next day, Feynman completed Dirac's suggested approach as *sum over histories* or *sum over paths* or *path integrals* (p xv). Feynman would joke that this approach—which sums all possible paths that a particle could take, as though the particle actually takes them all, canceling themselves out except for one pathway, the particle's most efficient—abolishes the uncertainty principle (p xvi). All empirically equivalent, Schrödinger's wave formalism, Heisenberg's matrix formalism, and Feynman's path integral formalism all incorporate the uncertain principle (p xvi).
There is no particular barrier to additional formalisms, which could be, simply have not been, developed and widely disseminated (p xvii). In a particular physical discipline, however, and on a particular problem, one of the three formalisms might be easier than others to operate (pp xvi–xvii). By the 1960s, path integral formalism virtually vanished from use, while matrix formalism was the "canonical" (p xvii). In the 1970s, path integral formalism made a "roaring comeback", became the predominant means to make predictions from QFT, and impelled Feynman to an aura of mystique (p xviii).

[86] Cushing, *Quantum Mechanics* (U Chicago P, 1994), pp 113–18.

[87] Schrödinger's wave mechanics posed an electron's charge smeared across space as a waveform, later reinterpreted as the electron manifesting across space probabilistically but nowhere definitely while eventually building up that deterministic waveform. Heisenberg's matrix mechanics confusingly talked of *operators* acting on *quantum states*. Richard Feynman introduced QM's path integral formalism—interpretable as a particle traveling all paths imaginable, canceling themselves, leaving just one, the most efficient—predictively identical with Heisenberg's matrix formalism and with Schrödinger's wave formalism.

[88] Torretti, *Philosophy of Physics* (Cambridge U P, 1999), pp 393–95.

[89] Torretti, *Philosophy of Physics* (Cambridge U P, 1999), p 394.

[90] Torretti, *Philosophy of Physics* (Cambridge U P, 1999), p 395.

[91] Recognition of strong force permitted Manhattan Project to engineer Little Boy and Fat Man, dropped on Japan, whereas effects of weak force were seen in its aftermath—radioactive fallout—of diverse health consequences.

[92] Wilczek, "The persistence of ether", *Phys Today*, 1999; **52**:11,13, p 13.

[93] The four, known fundamental interactions are gravitational, electromagnetic, weak nuclear, and strong nuclear.

[94] Grandy, *Everyday Quantum Reality* (Indiana U P, 2010), pp 24–25.

[95] Schweber, *QED and the Men who Made it* (Princeton U P, 1994).

[96] Feynman, *QED* (Princeton U P, 2006), p 5.

[97] Torretti, *Philosophy of Physics*. (Cambridge U P, 1999), pp 395–96.

[98] Cushing, *Quantum Mechanics* (U Chicago P, 1994), pp 158–59.

[99] Close, "Much ado about nothing", *Nova*, PBS/WGBH, 2012: "This new quantum mechanical view of nothing began to emerge in 1947, when Willis Lamb measured spectrum of hydrogen. The electron in a hydrogen atom cannot move wherever it pleases but instead is restricted to specific paths. This is analogous to climbing a ladder: You cannot end up at arbitrary heights above ground, only those where there are rungs to stand on. Quantum mechanics explains the spacing of the rungs on the atomic ladder and predicts the frequencies of radiation that are emitted or absorbed when an electron switches from one to another. According to the state of the art in 1947, which assumed the hydrogen atom to consist of just an electron, a proton, and an electric field, two of these rungs have identical energy. However, Lamb's measurements showed that these two rungs differ in energy by about one part in a million. What could be causing this tiny but significant difference? "When physicists drew up their simple picture of the atom, they had forgotten something: Nothing. Lamb had become the first person to observe experimentally that the vacuum is not empty, but is instead seething with ephemeral electrons and their anti-matter analogues, positrons. These electrons and positrons disappear almost instantaneously, but in their brief mayfly moment of existence they alter the shape of the atom's electromagnetic field slightly. This momentary interaction with the electron inside the hydrogen atom kicks one of the rungs of the ladder just a bit higher than it would be otherwise.
"This is all possible because, in quantum mechanics, energy is not conserved on very short timescales, or for very short distances. Stranger still, the more precisely you attempt to look at something—or at nothing—the more dramatic these energy fluctuations become. Combine that with Einstein's $E=mc^2$, which implies that energy can congeal in material form, and you have a recipe for particles that bubble in and out of existence even in the void. This effect allowed Lamb to literally measure something from nothing".

[100] • Vongehr "Higgs discovery rehabilitating despised Einstein Aether", *Science 2.0*, 2011.
• Vongehr, "Supporting abstract relational space-time as fundamental without doctrinism against emergence", arXiv:0912.3069, 2011.

[101] Riesselmann "Concept of ether in explaining forces", *Inquiring Minds*, Fermilab, 2008.

[102] Close, "Much ado about nothing", *Nova*, PBS/WGBH, 2012.

[103] On "historical examples of empirically successful theories that later turn out to be false", Okasha, *Philosophy of Science* (Oxford U P, 2002), p 65, concludes, "One that remains is the wave theory of light, first put forward by Christian Huygens in 1690. According to this theory, light consists of wave-like vibrations in an invisible medium called the ether, which was supposed to permeate the whole universe. (The rival to the wave theory was the particle theory of light, favoured by Newton, which held that light consists of very small particles emitted by the light source.) The wave theory was not widely accepted until the French physicist Auguste Fresnel formulated a mathematical version of the theory in 1815, and used it to predict some surprising new optical phenomena. Optical experiments confirmed Fresnel's predictions, convincing many 19th-century scientists that the wave theory of light must be true. But modern physics tells us that the theory is not true: there is no such thing as the ether, so light doesn't consist of vibrations in it. Again, we have an example of a false but empirically successful theory".

[104] Pigliucci, *Answers for Aristotle* (Basic Books, 2012), p 119: "But the antirealist will quickly point out that plenty of times in the past scientists have posited the existence of unobservables that were apparently necessary to explain a phenomenon, only to discover later on that such unobservables did not in fact exist. A classic case is the aether, a substance that was supposed by nineteenth-century physicists to permeate all space and make it possible for electromagnetic radiation (like light) to propagate. It was Einstein's special theory of relativity, proposed in 1905, that did away with the necessity of aether, and the concept has been relegated to the dustbin of scientific history ever since. The antirealists will relish pointing out that modern physics features a number of similarly unobservable entities, from quantum mechanical 'foam' to dark energy, and that the current crop of scientists seems just as confident about the latter two as their nineteenth-century counterparts were about aether".

[105] Wilczek, *Lightness of Being* (Basic Books, 2008), pp 78–80.

[106] Laughlin, *A Different Universe* (Basic Books, 2005), pp 120–21.

[107] Einstein, "Ether", *Sidelights* (Methuen, 1922), pp 14–18.

[108] Lorentz aether was at absolute rest—acting *on* matter but not acted on *by* matter. Replacing it and resembling Ernst Mach's aether, Einstein aether is spacetime itself—which is the gravitational field—receiving motion from a body and transmitting it to other bodies while propagating at light speed, waving. An unobservable, however, Einstein aether is not a privileged reference frame—is not to be assigned a state of absolute motion or absolute rest.

[109] Relativity theory comprises both special relativity (SR) and general relativity (GR). Holding for inertial reference frames, SR is as a limited case of GR, which holds for all reference frames, both inertial and accelerated. In GR,

all motion—inertial, accelerated, or gravitational—is consequent of the geometry of 3D space stretched onto the 1D axis of time. By GR, no force distinguishes acceleration from inertia. Inertial motion is consequence simply of *uniform* geometry of spacetime, acceleration is consequence simply of *nonuniform* geometry of spacetime, and gravitation is simply acceleration.

[110] Laughlin, *A Different Universe*, (Basic Books, 2005), pp 120–21: "The word 'ether' has extremely negative connotations in theoretical physics because of its past association with opposition to relativity. This is unfortunate because, stripped of these connotations, it rather nicely captures the way most physicists actually think about the vacuum. ... Relativity actually says nothing about the existence or nonexistence of matter pervading the universe, only that any such matter must have relativistic symmetry. It turns out that such matter exists. About the time that relativity was becoming accepted, studies of radioactivity began showing that the empty vacuum of space had spectroscopic structure similar to that of ordinary quantum solids and fluids. Subsequent studies with large particle accelerators have now led us to understand that space is more like a piece of window glass than ideal Newtonian emptiness. It is filled with 'stuff' that is normally transparent but can be made visible by hitting it sufficiently hard to knock out a part. The modern concept of the vacuum of space, confirmed every day by experiment, is a relativistic ether. But we do not call it this because it is taboo".

[111] In Einstein's 4D spacetime, 3D space is stretched onto the 1D axis of time flow, which slows while space additionally contracts in the vicinity of mass or energy.

[112] Torretti, *Philosophy of Physics* (Cambridge U P, 1999), p 180.

[113] As an effective field theory, once adjusted to particular domains, Standard Model is predictively accurate until a certain, vast energy scale that is a cutoff, whereupon more fundamental phenomena—regulating the effective theory's modeled phenomena—would emerge. (Burgess & Moore, *Standard Model*, p xi; Wells, *Effective Theories*, pp 55–56).

[114] Torretti, *Philosophy of Physics* (Cambridge U P, 1999), p 396.

[115] Jegerlehner, "The Standard Model as a low-energy effective theory", arXiv:1304.7813: "We understand the SM as a low energy effective emergence of some unknown physical system—we may call it 'ether'—which is located at the Planck scale with the Planck length as a 'microscopic' length scale. Note that the cutoff, though very large, in any case is finite".

[116] Wilczek, *Lightness of Being* (Basic Books, 2008), ch 8 "The grid (persistence of ether)", p 73: "For natural philosophy, the most important lesson we learn from QCD is that what we perceive as empty space is in reality a powerful medium whose activity molds the world. Other developments in modern physics reinforce and enrich that lesson. Later, as we explore the current frontiers, we'll see how the concept of 'empty' space as a rich, dynamic medium empowers our best thinking about how to achieve the unification of forces".

[117] Mass–energy equivalence is formalized in the equation $E=mc^2$.

[118] Einstein, "Ether", *Sidelights* (Methuen, 1922), p 13: "[A]ccording to the special theory of relativity, both matter and radiation are but special forms of distributed energy, ponderable mass losing its isolation and appearing as a special form of energy".

[119] Braibant, Giacomelli & Spurio, *Particles and Fundamental Interactions* (Springer, 2012), p 2: "Any particle can be created in collisions between two high energy particles thanks to a process of transformation of energy in mass".

[120] Brian Greene explained, "People often have the wrong image of what happens inside the LHC, and I am just as guilty as anyone of perpetuating it. The machine does not smash together particles to pulverise them and see what is inside. Rather, it collides them at extremely high energy. Since, by dint of Einstein's famous equation, $E=mc^2$, energy and mass are one and the same, the combined energy of the collision can be converted into a mass, in other words, a particle, that is heavier than either of the colliding protons. The more energy is involved in the collision, the heavier the particles that might come into being" [Avent, "The Q&A", *Economist*, 2012].

[121] Kuhlmann, "Physicists debate", *Sci Am*, 2013.

[122] Whereas Newton's *Principia* inferred absolute space and absolute time, omitted an aether, and, by Newton's law of universal gravitation, formalized action at a distance—a supposed force of gravitation spanning the entire universe instantly—Newton's later work *Optiks* introduced an aether binding bodies' matter, yet denser outside bodies, and, not uniformly distributed across all space, in some locations condensed, whereby "aethereal spirits" mediate electricity, magnetism, and gravitation. (Whittaker, *A History of Theories of Aether* (Longmans, Green & Co; 1910), pp 17–18)

[123] Norton, "Causation as folk science", in Price & Corry, eds, *Mature Causation, Physics, and the Constitution of Reality* (Oxford U P, 2007), esp p 12.

[124] Fetzer, ch 3, in Fetzer, ed, *Science, Explanation, and Rationality* (Oxford U P, 2000), p 111.

21.10 Sources

- Avent, Ryan, "The Q&A: Brian Greene—life after the Higgs", *The Economist* blog: *Babbage*, 19 Jul 2012.

- Ayala, Francisco J & Theodosius G Dobzhansky, eds, *Studies in the Philosophy of Biology: Reduction and Related Problems* (Berkeley & Los Angeles: University of California Press, 1974).

- Bechtel, William, *Discovering Cell Mechanisms: The Creation of Modern Cell Biology* (New York: Cambridge University Press, 2006).

- Bechtel, William, *Philosophy of Science: An Overview for Cognitive Science* (Hillsdale, NJ: Lawrence Erlbaum Associates, 1988).

- Bem, Sacha & Huib L de Jong, *Theoretical Issues in Psychology: An Introduction*, 2nd edn (London: Sage Publications, 2006).

- Ben-Menahem, Yemima, *Conventionalism: From Poincaré to Quine* (Cambridge: Cambridge University Press, 2006).

- Blackmore, J T & R Itagaki, S Tanaka, eds, *Ernst Mach's Vienna 1895–1930: Or Phenomenalism as Philosophy of Science* (Dordrecht: Kluwer Academic Publishers, 2001).

- Boffetta, Paolo, "Causation in the presence of weak associations", *Critical Reviews in Food Science and Nutrition*, 2010 Dec; **50**(s1):13–16.

- Bolotin, David, *An Approach to Aristotle's Physics: With Particular Attention to the Role of His Manner of Writing* (Albany: State University of New York Press, 1998).

- Bourdeau, Michel, "Auguste Comte", in Edward N Zalta, ed, *The Stanford Encyclopedia of Philosophy*, Winter 2013 edn.

- Braibant, Sylvie & Giorgio Giacomelli, Maurizio Spurio, *Particles and Fundamental Interactions: An Introduction to Particle Physics* (Dordrecht, Heidelberg, London, New York: Springer, 2012).

- Buchen, Lizzie, "May 29, 1919: A major eclipse, relatively speaking", *Wired*, 29 May 2009.

- Buckle, Stephen, *Hume's Enlightenment Tract: The Unity and Purpose of An Enquiry Concerning Human Understanding* (New York: Oxford University Press, 2001).

- Burgess, Cliff & Guy Moore, *The Standard Model: A Primer* (New York: Cambridge University Press, 2007).

- Chakraborti, Chhanda, *Logic: Informal, Symbolic and Inductive* (New Delhi: Prentice-Hall of India, 2007).

- Chakravartty, Anjan, "Scientific realism", in Edward N Zalta, ed, *The Stanford Encyclopedia of Philosophy*, Summer 2013 edn.

- Close, Frank, "Much ado about nothing", *Nova: The Nature of Reality*, PBS Online / WGBH Educational Foundation, 13 Jan 2012.

- Comte, Auguste, auth, J. H. Bridges, trans, *A General View of Positivism* (London: Trübner and Co, 1865) [English translation from French as Comte's 2nd edn in 1851, after the 1st edn in 1848].

- Cordero, Alberto, ch 3 "Rejected posits, realism, and the history of science", pp 23–32, in Henk W de Regt, Stephan Hartmann & Samir Okasha, eds, *EPSA Philosophy of Science: Amsterdam 2009* (New York: Springer, 2012).

- Crelinsten, Jeffrey, *Einstein's Jury: The Race to Test Relativity* (Princeton: Princeton University Press, 2006).

- Cushing, James T, *Quantum Mechanics: Historical Contingency and the Copenhagen Hegemony* (Chicago: University of Chicago Press, 1994).

- Einstein, Albert, "Ether and the theory of relativity", pp 3–24, *Sidelights on Relativity* (London: Methuen, 1922), the English trans of Einstein, "Äther und Relativitätstheorie" (Berlin: Verlag Julius, 1920), based on Einstein's 5 May 1920 address at University of Leyden, and collected in Jürgen Renn, ed, *The Genesis of General Relativity*, Volume 3 (Dordrecht: Springer, 2007).

- Fetzer, James H, "Carl Hempel", in Edward N Zalta, ed, *The Stanford Encyclopedia of Philosophy*, Spring 2013 edn.

- Fetzer, James H., ed, *Science, Explanation, and Rationality: Aspects of the Philosophy of Carl G Hempel* (New York: Oxford University Press, 2000).

- Feynman, Richard P., *QED: The Strange Theory of Light and Matter*, w/new intro by A Zee (Princeton: Princeton University Press, 2006).

- Flew, Antony G, *A Dictionary of Philosophy*, 2nd edn (New York: St Martin's Press, 1984), "Positivism", p 283.

- Friedman, Michael, *Reconsidering Logical Positivism* (New York: Cambridge University Press, 1999).

- Gattei, Stefano, *Karl Popper's Philosophy of Science: Rationality without Foundations* (New York: Routledge, 2009), ch 2 "Science and philosophy".

- Grandy, David A., *Everyday Quantum Reality* (Bloomington, Indiana : Indiana University Press, 2010).

- Hacohen, Malachi H, *Karl Popper—the Formative Years, 1902–1945: Politics and Philosophy in Interwar Vienna* (Cambridge: Cambridge University Press, 2000).

- Jegerlehner, Fred, "The Standard Model as a low-energy effective theory: What is triggering the Higgs mechanism?", *arXiv* (High Energy Physics—Phenomenology):1304.7813, 11 May 2013 (last revised).

- Karhausen, Lucien R, "Causation: The elusive grail of epidemiology", *Medicine, Health Care, and Philosophy*, 2000; **3**(1):59–67. PMID 11080970

- Kay, Lily E, *Molecular Vision of Life: Caltech, the Rockefeller Foundation, and the Rise of the New Biology* (New York: Oxford University Press, 1993).

- Khrennikov, K, ed, *Proceedings of the Conference: Foundations of Probability and Physics* (Singapore: World Scientific Publishing, 2001).

- Kuhlmann, Meinard, "Physicists debate whether the world is made of particles or fields—or something else entirely", *Scientific American*, 24 July 2013.

- Kundi, Michael, "Causality and the interpretation of epidemiologic evidence", *Environmental Health Perspectives*, 2006 Jul; **114**(7):969–74. PMID 16835045

- Laudan, Larry, ed, *Mind and Medicine: Problems of Explanation and Evaluation in Psychiatry and the Biomedical Sciences* (Berkeley, Los Angeles, London: University of California Press, 1983).

- Laughlin, Robert B, *A Different Universe: Reinventing Physics from the Bottom Down* (New York: Basic Books, 2005).

- Lodge, Oliver, "The ether of space: A physical conception", *Scientific American Supplement*, 1909 Mar 27; **67**(1734):202–03.

- Manninen, Juha & Raimo Tuomela, eds, *Essays on Explanation and Understanding: Studies in the Foundation of Humanities and Social Sciences* (Dordrecht: D Reidel, 1976).

- Melia, Fulvio, *The Black Hole at the Center of Our Galaxy* (Princeton: Princeton University Press, 2003).

- Montuschi, Eleonora, *Objects in Social Science* (London & New York: Continuum Books, 2003).

- Morange, Michel, trans by Michael Cobb, *A History of Molecular Biology* (Cambridge MA: Harvard University Press, 2000).

- Moyer, Donald F, "Revolution in science: The 1919 eclipse test of general relativity", in Arnold Perlmutter & Linda F Scott, eds, *Studies in the Natural Sciences: On the Path of Einstein* (New York: Springer, 1979).

- Newburgh, Ronald & Joseph Peidle, Wolfgang Rueckner, "Einstein, Perrin, and the reality of atoms: 1905 revisited", *American Journal of Physics*, 2006 June; **74**(6):478–481.

- Norton, John D, "Causation as folk science", *Philosopher's Imprint*, 2003; **3**(4), collected as ch 2 in Price & Corry, eds, *Causation, Physics, and the Constitution of Reality* (Oxford U P, 2007).

- Ohanian, Hans C, *Einstein's Mistakes: The Human Failings of Genius* (New York: W W Norton & Company, 2008).

- Okasha, Samir, *Philosophy of Science: A Very Short Introduction* (New York: Oxford University Press, 2002).

- O'Shaughnessy, John, *Explaining Buyer Behavior: Central Concepts and Philosophy of Science Issues* (New York: Oxford University Press, 1992).

- Parascandola, M & D L Weed, "Causation in epidemiology", *Journal of Epidemiology and Community Health*, 2001 Dec; **55**(12):905–12. PMID 11707485

- Pigliucci, Massimo, *Answers for Aristotle: How Science and Philosophy Can Lead Us to a More Meaningful Life* (New York: Basic Books, 2012).

- Popper, Karl, "Against big words", *In Search of a Better World: Lectures and Essays from Thirty Years* (New York: Routledge, 1996).

- Price, Huw & Richard Corry, eds, *Causation, Physics, and the Constitution of Reality: Russell's Republic Revisited* (New York: Oxford University Press, 2007).

- Redman, Deborah A, *The Rise of Political Economy as a Science: Methodology and the Classical Economists* (Cambridge MA: MIT Press, 1997).

- Reutlinger, Alexander & Gerhard Schurz, Andreas Hüttemann, "Ceteris paribus laws", in Edward N Zalta, ed, *The Stanford Encyclopedia of Philosophy*, Spring 2011 edn.

- Riesselmann, Kurt, "Concept of ether in explaining forces", *Inquiring Minds: Questions About Physics*, US Department of Energy: Fermilab, 28 Nov 2008.

- Rothman, Kenneth J & Sander Greenland, "Causation and causal inference in epidemiology", *American Journal of Public Health*, 2005; **95**(Suppl 1):S144–50. PMID 16030331

- Rowlands, Peter, *Oliver Lodge and the Liverpool Physical Society* (Liverpool: Liverpool University Press, 1990).

- Sarkar, Sahotra & Jessica Pfeifer, eds. *The Philosophy of Science: An Encyclopedia, Volume 1: A–M* (New York: Routledge, 2006).

- Schwarz, John H, "Recent developments in superstring theory", *Proceedings of the National Academy of Sciences of the United States of America*, 1998 Mar 17; 95(6):2750–7. PMID 9501161

- Schweber, Silvan S, *QED and the Men who Made it: Dyson, Feynman, Schwinger, and Tomonaga* (Princeton: Princeton University Press, 1994).

- Schliesser, Eric, "Hume's Newtonianism and anti-Newtonianism", in Edward N Zalta, ed, *The Stanford Encyclopedia of Philosophy*, Winter 2008 edn.

- Spohn, Wolfgang, *The Laws of Belief: Ranking Theory and Its Philosophical Applications* (Oxford: Oxford University Press, 2012).

- Suppe, Frederick, ed. *The Structure of Scientific Theories*, 2nd edn (Urbana, Illinois: University of Illinois Press, 1977).

- Tavel, Morton, *Contemporary Physics and the Limits of Knowledge* (Piscataway, NJ: Rutgers University Press, 2002).

- Torretti, Roberto, *The Philosophy of Physics* (New York: Cambridge University Press, 1999).

- Vongehr, Sascha, "Higgs discovery rehabilitating despised Einstein Ether", *Science 2.0: Alpha Meme* website, 13 Dec 2011.

- Vongehr, Sascha, "Supporting abstract relational space-time as fundamental without doctrinism against emergence, *arXiv* (History and Philosophy of Physics):0912.3069, 2 Oct 2011 (last revised).

- von Wright, Georg Henrik, *Explanation and Understanding* (Ithaca, NY: Cornell University Press, 1971/2004).

- Wells, James D, *Effective Theories in Physics: From Planetary Orbits to Elementary Particle Masses* (Heidelberg, New York, Dordrecht, London: Springer, 2012).

- Wilczek, Frank, *The Lightness of Being: Mass, Ether, and the Unification of Forces* (New York: Basic Books, 2008).

- Whittaker, Edmund T, *A History of the Theories of Aether and Electricity: From the Age of Descartes to the Close of the Nineteenth Century* (London, New York, Bombay, Calcutta: Longmans, Green, and Co, 1910 / Dublin: Hodges, Figgis, & Co, 1910).

- Wilczek, Frank, "The persistence of ether", *Physics Today*, 1999 Jan; 52:11,13.

- Wolfson, Richard, *Simply Einstein: Relativity Demystified* (New York: W W Norton & Co, 2003).

- Woodward, James, "Scientific explanation", in Edward N Zalta, ed, *The Stanford Encyclopedia of Philosophy*, Winter 2011 edn.

- Wootton, David, ed, *Modern Political Thought: Readings from Machiavelli to Nietzsche* (Indianapolis: Hackett Publishing, 1996).

21.11 Further reading

- Carl G. Hempel, *Aspects of Scientific Explanation and other Essays in the Philosophy of Science* (New York: Free Press, 1965).

- Randolph G. Mayes, "Theories of explanation", in Fieser Dowden, ed, *Internet Encyclopedia of Philosophy*, 2006.

- Ilkka Niiniluoto, "Covering law model", in Robert Audi, ed., *The Cambridge Dictionary of Philosophy*, 2nd edn (New York: Cambridge University Press, 1996).

- Wesley C. Salmon, *Four Decades of Scientific Explanation* (Minneapolis: University of Minnesota Press, 1990 / Pittsburgh: University of Pittsburgh Press, 2006).

Chapter 22

Scientific modelling

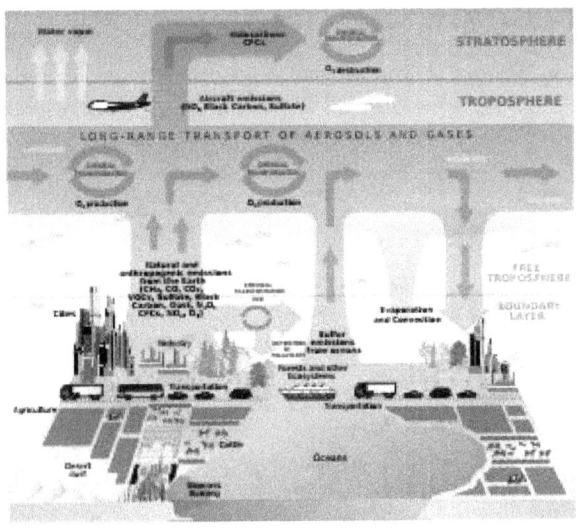

Example of scientific modelling. A schematic of chemical and transport processes related to atmospheric composition.

Scientific modelling is a scientific activity, the aim of which is to make a particular part or feature of the world easier to understand, define, quantify, visualize, or simulate by referencing it to existing and usually commonly accepted knowledge. It requires selecting and identifying relevant aspects of a situation in the real world and then using different types of models for different aims, such as conceptual models to better understand, operational models to operationalize, mathematical models to quantify, and graphical models to visualize the subject. Modelling is an essential and inseparable part of many scientific disciplines, each of which have their own ideas about specific types of modelling.[1][2]

There is also an increasing attention to scientific modelling[3] in fields such as science education, philosophy of science, systems theory, and knowledge visualization. There is growing collection of methods, techniques and meta-theory about all kinds of specialized scientific modelling.

22.1 Overview

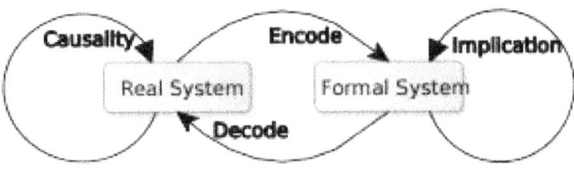

A scientific model seeks to represent empirical objects, phenomena, and physical processes in a logical and objective way. All models are *in simulacra*, that is, simplified reflections of reality that, despite being approximations, can be extremely useful.[4] Building and disputing models is fundamental to the scientific enterprise. Complete and true representation may be impossible, but scientific debate often concerns which is the better model for a given task, e.g., which is the more accurate climate model for seasonal forecasting.[5]

Attempts to formalize the principles of the empirical sciences use an interpretation to model reality, in the same way logicians axiomatize the principles of logic. The aim of these attempts is to construct a formal system that will not produce theoretical consequences that are contrary to what is found in reality. Predictions or other statements drawn from such a formal system mirror or map the real world only insofar as these scientific models are true.[6][7]

For the scientist, a model is also a way in which the human thought processes can be amplified.[8] For instance, models that are rendered in software allow scientists to leverage computational power to simulate, visualize, manipulate and gain intuition about the entity, phenomenon, or process being represented. Such computer models are *in silico*. Other types of scientific models are *in vivo* (living models, such as laboratory rats) and *in vitro* (in glassware, such as tissue culture).[9]

22.2 Basics of scientific modelling

22.2.1 Modelling as a substitute for direct measurement and experimentation

Models are typically used when it is either impossible or impractical to create experimental conditions in which scientists can directly measure outcomes. Direct measurement of outcomes under controlled conditions (see Scientific method) will always be more reliable than modelled estimates of outcomes.

Within modelling and simulation, a model is a task-driven, purposeful simplification and abstraction of a perception of reality, shaped by physical, legal, and cognitive constraints.[10] It is task-driven, because a model is captured with a certain question or task in mind. Simplifications leave all the known and observed entities and their relation out that are not important for the task. Abstraction aggregates information that is important, but not needed in the same detail as the object of interest. Both activities, simplification and abstraction, are done purposefully. However, they are done based on a perception of reality. This perception is already a *model* in itself, as it comes with a physical constraint. There are also constraints on what we are able to legally observe with our current tools and methods, and cognitive constraints which limit what we are able to explain with our current theories. This model comprises the concepts, their behavior, and their relations in formal form and is often referred to as a conceptual model. In order to execute the model, it needs to be implemented as a computer simulation. This requires more choices, such as numerical approximations or the use of heuristics.[11] Despite all these epistemological and computational constraints, simulation has been recognized as the third pillar of scientific methods: theory building, simulation, and experimentation.[12]

22.2.2 Simulation

A simulation is the implementation of a model. A steady state simulation provides information about the system at a specific instant in time (usually at equilibrium, if such a state exists). A dynamic simulation provides information over time. A simulation brings a model to life and shows how a particular object or phenomenon will behave. Such a simulation can be useful for testing, analysis, or training in those cases where real-world systems or concepts can be represented by models.[13]

22.2.3 Structure

Structure is a fundamental and sometimes intangible notion covering the recognition, observation, nature, and stability of patterns and relationships of entities. From a child's verbal description of a snowflake, to the detailed scientific analysis of the properties of magnetic fields, the concept of structure is an essential foundation of nearly every mode of inquiry and discovery in science, philosophy, and art.[14]

22.2.4 Systems

A system is a set of interacting or interdependent entities, real or abstract, forming an integrated whole. In general, a system is a construct or collection of different elements that together can produce results not obtainable by the elements alone.[15] The concept of an 'integrated whole' can also be stated in terms of a system embodying a set of relationships which are differentiated from relationships of the set to other elements, and from relationships between an element of the set and elements not a part of the relational regime. There are two types of system models: 1) discrete in which the variables change instantaneously at separate points in time and, 2) continuous where the state variables change continuously with respect to time.[16]

22.2.5 Generating a model

Modelling is the process of generating a model as a conceptual representation of some phenomenon. Typically a model will deal with only some aspects of the phenomenon in question, and two models of the same phenomenon may be essentially different—that is to say, that the differences between them comprise more than just a simple renaming of components.

Such differences may be due to differing requirements of the model's end users, or to conceptual or aesthetic differences among the modellers and to contingent decisions made during the modelling process. Considerations that may influence the structure of a model might be the modeller's preference for a reduced ontology, preferences regarding statistical models versus deterministic models, discrete versus continuous time, etc. In any case, users of a model need to understand the assumptions made that are pertinent to its validity for a given use.

Building a model requires abstraction. Assumptions are used in modelling in order to specify the domain of application of the model. For example, the special theory of relativity assumes an inertial frame of reference. This assumption was contextualized and further explained by the general theory of relativity. A model makes accurate predictions when its assumptions are valid, and might well not make accurate predictions when its assumptions do not hold. Such assumptions are often the point with which older theories are succeeded by new ones (the general theory of relativity works in non-inertial reference frames as well).

The term "assumption" is actually broader than its standard use, etymologically speaking. The Oxford English Dictionary (OED) and online Wiktionary indicate its Latin source as *assumere* ("accept, to take to oneself, adopt, usurp"), which is a conjunction of *ad-* ("to, towards, at") and *sumere* (to take). The root survives, with shifted meanings, in the Italian *sumere* and Spanish *sumir*. In the OED, "assume" has the senses of (i) "investing oneself with (an attribute)," (ii) "to undertake" (especially in Law), (iii) "to take to oneself in appearance only, to pretend to possess," and (iv) "to suppose a thing to be." Thus, "assumption" connotes other associations than the contemporary standard sense of "that which is assumed or taken for granted; a supposition, postulate," and deserves a broader analysis in the philosophy of science.

22.2.6 Evaluating a model

See also: Models of scientific inquiry § Choice of a theory

A model is evaluated first and foremost by its consistency to empirical data; any model inconsistent with reproducible observations must be modified or rejected. One way to modify the model is by restricting the domain over which it is credited with having high validity. A case in point is Newtonian physics, which is highly useful except for the very small, the very fast, and the very massive phenomena of the universe. However, a fit to empirical data alone is not sufficient for a model to be accepted as valid. Other factors important in evaluating a model include:

- Ability to explain past observations
- Ability to predict future observations
- Cost of use, especially in combination with other models
- Refutability, enabling estimation of the degree of confidence in the model
- Simplicity, or even aesthetic appeal

People may attempt to quantify the evaluation of a model using a utility function.

22.2.7 Visualization

Visualization is any technique for creating images, diagrams, or animations to communicate a message. Visualization through visual imagery has been an effective way to communicate both abstract and concrete ideas since the dawn of man. Examples from history include cave paintings, Egyptian hieroglyphs, Greek geometry, and Leonardo da Vinci's revolutionary methods of technical drawing for engineering and scientific purposes.

22.2.8 Space mapping

Space mapping refers to a methodology that employs a "quasi-global" modeling formulation to link companion "coarse" (ideal or low-fidelity) with "fine" (practical or high-fidelity) models of different complexities. In engineering optimization, space mapping aligns (maps) a very fast coarse model with its related expensive-to-compute fine model so as to avoid direct expensive optimization of the fine model. The alignment process iteratively refines a "mapped" coarse model (surrogate model).

22.3 Types of scientific modelling

22.4 Applications

22.4.1 Modelling and simulation

One application of scientific modelling is the field of modelling and simulation, generally referred to as "M&S". M&S has a spectrum of applications which range from concept development and analysis, through experimentation, measurement and verification, to disposal analysis. Projects and programs may use hundreds of different simulations, simulators and model analysis tools.

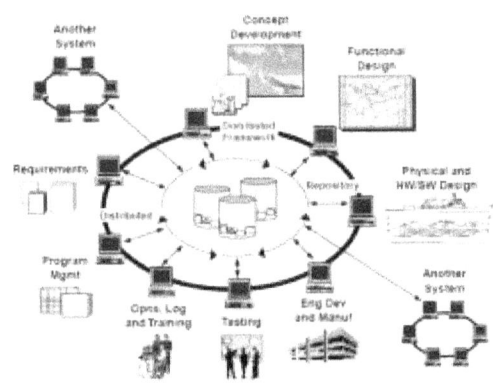

Example of the integrated use of Modelling and Simulation in Defence life cycle management. The modelling and simulation in this image is represented in the center of the image with the three containers.[13]

The figure shows how Modelling and Simulation is used as a central part of an integrated program in a Defence capability development process.[13]

22.4.2 Model-based learning in education

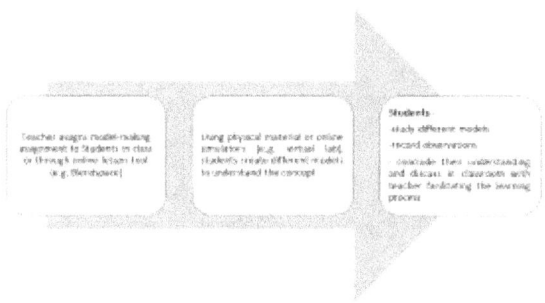

Flowchart Describing One Style of Model-based Learning

Model–based learning in education, particularly in relation to learning science involves students creating models for scientific concepts in order to:[17]

- Gain insight of the scientific idea(s)
- Acquire deeper understanding of the subject through visualization of the model
- Improve student engagement in the course

Different types of model based learning techniques include:[17]

- Physical macrocosms
- Representational systems
- Syntactic models
- Emergent models

Model–making in education is an iterative exercise with students refining, developing and evaluating their models over time. This shifts learning from the rigidity and monotony of traditional curriculum to an exercise of students' creativity and curiosity. This approach utilizes the constructive strategy of social collaboration and learning scaffold theory. Model based learning includes cognitive reasoning skills where existing models can be improved upon by construction of newer models using the old models as a basis.[18]

"Model–based learning entails determining target models and a learning pathway that provide realistic chances of understanding." [19] Model making can also incorporate blended learning strategies by using web based tools and simulators, thereby allowing students to:

- Familiarize themselves with on-line or digital resources
- Create different models with various virtual materials at little or no cost
- Practice model making activity any time and any place
- Refine existing models

"A well-designed simulation simplifies a real world system while heightening awareness of the complexity of the system. Students can participate in the simplified system and learn how the real system operates without spending days, weeks or years it would take to undergo this experience in the real world." [20]

The teacher's role in the overall teaching and learning process is primarily that of a facilitator and arranger of the learning experience. He or she would assign the students, a model making activity for a particular concept and provide relevant information or support for the activity. For virtual model making activities, the teacher can also provide information on the usage of the digital tool and render troubleshooting support in case of glitches while using the same. The teacher can also arrange the group discussion activity between the students and provide the platform necessary for students to share their observations and knowledge extracted from the model making activity.

Model–based learning evaluation could include the use of rubrics that assess the ingenuity and creativity of the student in the model construction and also the overall classroom participation of the student vis-a-vis the knowledge constructed through the activity.

It is important, however, to give due consideration to the following for successful model–based learning to occur:

- Use of the right tool at the right time for a particular concept
- Provision within the educational setup for model-making activity: e.g., computer room with internet facility or software installed to access simulator or digital tool

22.5 See also

- Heuristic
- Grey box completion and validation

- Scientific visualization
- Simulation
- Statistical model
- Systems engineering
- Toy model

22.6 References

[1] Cartwright, Nancy. 1983. *How the Laws of Physics Lie*. Oxford University Press

[2] Hacking, Ian. 1983. *Representing and Intervening. Introductory Topics in the Philosophy of Natural Science*. Cambridge University Press

[3] Frigg and Hartmann (2009) state: "Philosophers are acknowledging the importance of models with increasing attention and are probing the assorted roles that models play in scientific practice". Source: Frigg, Roman and Hartmann, Stephan, "Models in Science", *The Stanford Encyclopedia of Philosophy* (Summer 2009 Edition), Edward N. Zalta (ed.). (source)

[4] Box, George E.P. & Draper, N.R. (1987). [Empirical Model-Building and Response Surfaces.] *Wiley*. p. 424

[5] Hagedorn, R. et al. (2005) http://www.ecmwf.int/staff/paco_doblas/abstr/tellus05_1.pdf *Tellus* 57A:219-33

[6] Leo Apostel (1961). "Formal study of models". In: *The Concept and the Role of the Model in Mathematics and Natural and Social*. Edited by Hans Freudenthal. Springer. pp. 8–9 (Source)].

[7] Ritchey, T. (2012) Outline for a Morphology of Modelling Methods: Contribution to a General Theory of Modelling

[8] C. West Churchman, *The Systems Approach*, New York: Dell publishing. 1968. p. 61

[9] Griffiths, E. C. (2010) What is a model?

[10] Tolk, A. (2015). Learning something right from models that are wrong – Epistemology of Simulation. In Yilmaz, L. (Ed.) *Concepts and Methodologies in Modeling and Simulation*. Springer-Verlag. pp. 87–106

[11] Oberkampf, W. L., DeLand, S. M., Rutherford, B. M., Diegert, K. V., & Alvin, K. F. (2002). Error and uncertainty in modeling and simulation. *Reliability Engineering & System Safety* 75(3): 333–57.

[12] Ihrig, M. (2012). A New Research Architecture For The Simulation Era. In *European Council on Modelling and Simulation*. pp. 715–20).

[13] *Systems Engineering Fundamentals*. Defense Acquisition University Press. 2003.

[14] Pullan, Wendy (2000). *Structure*. Cambridge: Cambridge University Press. ISBN 0-521-78258-9.

[15] Fishwick PA. (1995). Simulation Model Design and Execution: Building Digital Worlds. Upper Saddle River, NJ: Prentice Hall.

[16] Sokolowski, J.A., Banks, C.M.(2009). Principles of Modelling and Simulation. Hoboken, NJ: John Wiley and Sons.

[17] Lehrer, Richard; Schauble, Leona (2006). *The Cambridge Handbook of Learning Sciences*. Cambridge, UK: Cambridge University Press. p. 371. ISBN 978-0-521-84554-0.

[18] Nersessian, Nancy J (2002). *The Cognitive Basis of Science*. Cambridge, UK: Cambridge University Press. p. 133. ISBN 0-521-01177-9.

[19] Clement, JJ; Rea-Ramirez, Mary Anne (2008). *Model Based Learning and Instruction in Science* (2 ed.). Springer Science & Business Media. p. 45. ISBN 978-1-4020-6493-7.

[20] Blumschein, Patrick; Hung, Woei; Jonassen, David; Strobel, Johannes (2009). *Model-Based Approaches to Learning* (PDF). Netherlands: Sense Publishers. ISBN 978-90-8790-711-2.

22.7 Further reading

Nowadays there are some 40 magazines about scientific modelling which offer all kinds of international forums. Since the 1960s there is a strong growing number of books and magazines about specific forms of scientific modelling. There is also a lot of discussion about scientific modelling in the philosophy-of-science literature. A selection:

- Rainer Hegselmann, Ulrich Müller and Klaus Troitzsch (eds.) (1996). *Modelling and Simulation in the Social Sciences from the Philosophy of Science Point of View*. Theory and Decision Library. Dordrecht: Kluwer.

- Paul Humphreys (2004). *Extending Ourselves: Computational Science, Empiricism, and Scientific Method*. Oxford: Oxford University Press.

- Johannes Lenhard, Günter Küppers and Terry Shinn (Eds.) (2006) "Simulation: Pragmatic Constructions of Reality", Springer Berlin.

- Tom Ritchey (2012). "Outline for a Morphology of Modelling Methods: Contribution to a General Theory of Modelling". In: *Acta Morphologica Generalis*, Vol 1. No 1. pp. 1–20.

- Fritz Rohrlich (1990). "Computer Simulations in the Physical Sciences". In: *Proceedings of the Philosophy of Science Association, Vol. 2*, edited by Arthur Fine et al., 507-518. East Lansing: The Philosophy of Science Association.

- Rainer Schnell (1990). "Computersimulation und Theoriebildung in den Sozialwissenschaften". In: *Kölner Zeitschrift für Soziologie und Sozialpsychologie* 1, 109-128.

- Sergio Sismondo and Snait Gissis (eds.) (1999). *Modeling and Simulation. Special Issue of Science in Context* 12.

- Eric Winsberg (2001). "Simulations, Models and Theories: Complex Physical Systems and their Representations". In: *Philosophy of Science* 68 (Proceedings): 442-454.

- Eric Winsberg (2003). "Simulated Experiments: Methodology for a Virtual World". In: *Philosophy of Science* 70: 105–125.

- Eric Winsberg (2010) Science in the Age of Computer Simulation Chicago: University of Chicago Press

- Tomáš Helikar, Jim A Rogers (2009). "ChemChains: a platform for simulation and analysis of biochemical networks aimed to laboratory scientists". BioMed Central.

22.8 External links

- Models. Entry in the *Internet Encyclopedia of Philosophy*

- Models in Science. Entry in the *Stanford Encyclopedia of Philosophy*

- The World as a Process: Simulations in the Natural and Social Sciences, in: R. Hegselmann et al. (eds.), Modelling and Simulation in the Social Sciences from the Philosophy of Science Point of View, Theory and Decision Library. Dordrecht: Kluwer 1996, 77-100.

- Research in simulation and modelling of various physical systems

- Modelling Water Quality Information Center, U.S. Department of Agriculture

- Ecotoxicology & Models

- A Morphology of Modelling Methods. Acta Morphologica Generalis, Vol 1, No 1, pp. 1–20.

Template:Computer modelling

Chapter 23

Hypothetico-deductive model

The **hypothetico-deductive model** or **method** is a proposed description of scientific method. According to it, scientific inquiry proceeds by formulating a hypothesis in a form that could conceivably be falsified by a test on observable data. A test that could and does run contrary to predictions of the hypothesis is taken as a falsification of the hypothesis. A test that could but does not run contrary to the hypothesis corroborates the theory. It is then proposed to compare the explanatory value of competing hypotheses by testing how stringently they are corroborated by their predictions.

23.1 Example

Main article: Scientific method

One example of an algorithmic statement of the hypothetico-deductive method is as follows:[1]

> *1.* Use your experience: Consider the problem and try to make sense of it. Gather data and look for previous explanations. If this is a new problem to you, then move to step *2*.
>
> *2.* Form a conjecture (hypothesis): When nothing else is yet known, try to state an explanation, to someone else, or to your notebook.
>
> *3.* Deduce predictions from the hypothesis: if you assume *2* is true, what consequences follow?
>
> *4.* Test (or experiment): Look for evidence (observations) that conflict with these predictions in order to disprove *2*. It is a logical error to seek *3* directly as proof of *2*. This formal fallacy is called *affirming the consequent*.[2]

One possible sequence in this model would be *1, 2, 3, 4*. If the outcome of *4* holds, and *3* is not yet disproven, you may continue with *3, 4, 1*, and so forth; but if the outcome of *4* shows *3* to be false, you will have to go back to *2* and try to invent a *new 2*, deduce a *new 3*, look for *4*, and so forth.

Note that this method can never absolutely **verify** (prove the truth of) *2*. It can only **falsify** *2*.[3] (This is what Einstein meant when he said, "No amount of experimentation can ever prove me right; a single experiment can prove me wrong."[4])

23.2 Discussion

Additionally, as pointed out by Carl Hempel (1905–1997), this simple view of the scientific method is incomplete; a conjecture can also incorporate probabilities, e.g., the drug is effective about 70% of the time.[5] Tests, in this case, must be repeated to substantiate the conjecture (in particular, the probabilities). In this and other cases, we can quantify a probability for our confidence in the conjecture itself and then apply a Bayesian analysis, with each experimental result shifting the probability either up or down. Bayes' theorem shows that the probability will never reach exactly 0 or 100% (no absolute certainty in either direction), but it can still get very close to either extreme. See also confirmation holism.

Qualification of corroborating evidence is sometimes raised as philosophically problematic. The raven paradox is a famous example. The hypothesis that 'all ravens are black' would appear to be corroborated by observations of only black ravens. However, 'all ravens are black' is logically equivalent to 'all non-black things are non-ravens' (this is the contraposition form of the original implication). 'This is a green tree' is an observation of a non-black thing that is a non-raven and therefore corroborates 'all non-black things are non-ravens'. It appears to follow that the observation 'this is a green tree' is corroborating evidence for the hypothesis 'all ravens are black'. Attempted resolutions may distinguish:

- non-falsifying observations as to strong, moderate, or

weak corroborations

- investigations that do or do not provide a potentially falsifying test of the hypothesis.[6]

Evidence contrary to a hypothesis is itself philosophically problematic. Such evidence is called a falsification of the hypothesis. However, under the theory of confirmation holism it is always possible to save a given hypothesis from falsification. This is so because any falsifying observation is embedded in a theoretical background, which can be modified in order to save the hypothesis. Popper acknowledged this but maintained that a critical approach respecting methodological rules that avoided such *immunizing stratagems* is conducive to the progress of science.[7]

Physicist Sean Carroll claims the model ignores underdetermination.[8]

The hypothetico-deductive model (or approach) versus other research models

The hypothetico-deductive approach contrasts with other research models such as the inductive approach or grounded theory. In the data percolation methodology, the hypothetico-deductive approach is included in a paradigm of pragmatism by which four types of relations between the variables can exist: descriptive, of influence, longitudinal or causal. The variables are classified in two groups, structural and functional, a classification that drives the formulation of hypotheses and the statistical tests to be performed on the data so as to increase the efficiency of the research. [9]

23.3 See also

- Confirmation bias
- Deductive-nomological
- Explanandum and explanans
- Inquiry
- Models of scientific inquiry
- Philosophy of science
- Pragmatism
- Scientific method
- Verifiability theory of meaning
- Will to Believe Doctrine

23.3.1 Types of inference

- Strong inference
- Abductive reasoning
- Deductive reasoning
- Inductive reasoning
- Analogy

23.4 Citations

[1] Peter Godfrey-Smith (2003) *Theory and Reality*, p. 236.

[2] Taleb 2007 e.g., p. 58. devotes his chapter 5 to *the error of confirmation*.

[3] "I believe that we do not know anything for certain, but everything probably." —Christiaan Huygens, Letter to Pierre Perrault, 'Sur la préface de M. Perrault de son traité del'Origine des fontaines' [1763], *Oeuvres Complétes de Christiaan Huygens* (1897), Vol. 7, 298. Quoted in Jacques Roger, *The Life Sciences in Eighteenth-Century French Thought*, ed. Keith R. Benson and trans. Robert Ellrich (1997), 163. Quotation selected by Bynum & Porter 2005, p. 317 Huygens 317#4.

[4] As noted by Alice Calaprice (ed. 2005) *The New Quotable Einstein* Princeton University Press and Hebrew University of Jerusalem, ISBN 0-691-12074-9 p. 291. Calaprice denotes this not as an exact quotation, but as a paraphrase of a translation of A. Einstein's "Induction and Deduction". *Collected Papers of Albert Einstein* 7 Document 28. Volume 7 is *The Berlin Years: Writings, 1918-1921*. A. Einstein; M. Janssen, R. Schulmann, et al., eds.

[5] Murzi, Mauro (2001, 2008), "Carl Gustav Hempel (1905—1997)", *Internet Encyclopedia of Philosophy*. Murzi used the term relative frequency rather than probability.

[6] John N.W. Watkins (1984), *Science and Skepticism*, p. 319.

[7] Karl R. Popper (1979, Rev. ed.), *Objective Knowledge*, pp. 30, 360.

[8] Sean Carroll. "What is Science?".

[9] Mesly, Olivier (2015), *Creating Models in Psychological Research*, United States: Springer Psychology, p. 126, ISBN 978-3-319-15752-8

23.5 References

- Brody, Thomas A. (1993), *The Philosophy Behind Physics*, Springer Verlag, ISBN 0-387-55914-0. (Luis de la Peña and Peter E. Hodgson, eds.)

- Bynum, W.F.; Porter, Roy (2005), *Oxford Dictionary of Scientific Quotations*, Oxford, ISBN 0-19-858409-1.

- Godfrey-Smith, Peter (2003), *Theory and Reality: An introduction to the philosophy of science*, University of Chicago Press, ISBN 0-226-30063-3

- Taleb, Nassim Nicholas (2007), *The Black Swan*, Random House, ISBN 978-1-4000-6351-2

Chapter 24

Branches of science

Further information: Outline of science § Branches of science

The **branches of science** (also referred as "sciences",

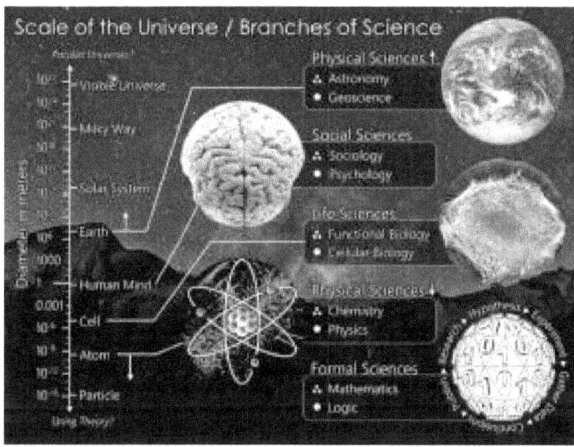

The scale of the universe mapped to the branches of science, with formal sciences as the foundation

"scientific fields", or "scientific disciplines") are commonly divided into three major groups:

- **Natural sciences**: the study of natural phenomena (including cosmological, geological, chemical, and biological factors of the universe)
- **Formal sciences**: the study of mathematics and logic, which use an *a priori*, as opposed to factual, methodology.
- **Social sciences**: the study of human behavior and societies.

Natural and social sciences are empirical sciences, meaning that the knowledge must be based on observable phenomena and must be capable of being verified by other researchers working under the same conditions. [1]

Natural, social, and formal science make up the fundamental sciences, which form the basis of interdisciplinary and applied sciences such as engineering and medicine. Specialized scientific disciplines that exist in multiple categories may include parts of other scientific disciplines but often possess their own terminologies and expertises.[2]

24.1 Natural/Pure Science

Main articles: Natural science and Outline of natural science

Natural science is a branch of science that seeks to elucidate the rules that govern the natural world by applying an empirical and scientific method to the study of the universe. The term natural sciences is used to distinguish it from the social sciences, which apply the scientific method to study human behavior and social patterns; the humanities, which use a critical, or analytical approach to the study of the human condition; and the formal sciences.

24.1.1 Physical science

Main articles: Physical science and Outline of physical science

Physical science is an encompassing term for the branches of natural science and science that study non-living systems, in contrast to the life sciences. However, the term "physical" creates an unintended, somewhat arbitrary distinction, since many branches of physical science also study biological phenomena. There is a difference between physical science and physics.

Physics

Main articles: Physics and Outline of physics

Physics(from Ancient Greek: φύσις *physis* "nature") is a natural science that involves the study of matter[3] and its motion through spacetime, along with related concepts such as energy and force.[4] More broadly, it is the general analysis of nature, conducted in order to understand how the universe behaves.[5][6][7]

Physics is one of the oldest academic disciplines, perhaps the oldest through its inclusion of astronomy.[8] Over the last two millennia, physics was a part of natural philosophy along with chemistry, certain branches of mathematics, and biology, but during the Scientific Revolution in the 16th century, the natural sciences emerged as unique research programs in their own right.[9] Certain research areas are interdisciplinary, such as biophysics and quantum chemistry, which means that the boundaries of physics are not rigidly defined. In the nineteenth and twentieth centuries physicalism emerged as a major unifying feature of the philosophy of science as physics provides fundamental explanations for every observed natural phenomenon. New ideas in physics often explain the fundamental mechanisms of other sciences, while opening to new research areas in mathematics and philosophy.

Chemistry

Main articles: Chemistry and Outline of chemistry

Chemistry (the etymology of the word has been much disputed)[10] is the science of matter and the changes it undergoes. The science of matter is also addressed by physics, but while physics takes a more general and fundamental approach, chemistry is more specialized, being concerned by the composition, behavior (or reaction), structure, and properties of matter, as well as the changes it undergoes during chemical reactions.[11] It is a physical science which studies various substances, atoms, molecules, and matter (especially carbon based); biochemistry, the study of substances found in biological organisms; physical chemistry, the study of chemical processes using physical concepts such as thermodynamics and quantum mechanics; and analytical chemistry, the analysis of material samples to gain an understanding of their chemical composition and structure. Many more specialized disciplines have emerged in recent years, e.g. neurochemistry the chemical study of the nervous system (see subdisciplines).

Earth science

Main articles: Earth science and Outline of earth science

Earth science (also known as *geoscience, the geosciences* or *the Earth sciences*) is an all-embracing term for the sciences related to the planet Earth.[12] It is arguably a special case in planetary science, the Earth being the only known life-bearing planet. There are both reductionist and holistic approaches to Earth sciences. The formal discipline of Earth sciences may include the study of the atmosphere, hydrosphere, oceans and biosphere, as well as the solid earth. Typically Earth scientists will use tools from physics, chemistry, biology, geography, chronology and mathematics to build a quantitative understanding of how the Earth system works, and how it evolved to its current state.

Ecology Main articles: Ecology and Outline of ecology

Ecology (from Greek: οἶκος, "house"; -λογία, "study of") is the scientific study of the relationships that living organisms have with each other and with their abiotic environment. Topics of interest to ecologists include the composition, distribution, amount (biomass), number, and changing states of organisms within and among ecosystems.

Oceanography Main article: Oceanography

Oceanography, or marine science, is the branch of Earth science that studies the ocean. It covers a wide range of topics, including marine organisms and ecosystem dynamics; ocean currents, waves, and geophysical fluid dynamics; plate tectonics and the geology of the sea floor; and fluxes of various chemical substances and physical properties within the ocean and across its boundaries. These diverse topics reflect multiple disciplines that oceanographers blend to further knowledge of the world ocean and understanding of processes within it: biology, chemistry, geology, meteorology, and physics as well as geography.

Geology Main articles: Geology and Outline of geology

Geology (from the Greek γῆ, gê, "earth" and λόγος, logos, "study") is the science comprising the study of solid Earth, the rocks of which it is composed, and the processes by which they change. Geology can also refer generally to the study of the solid features of any celestial body (such as the geology of the Moon or geology of Mars).

Geology gives insight into the history of the Earth, as it provides the primary evidence for plate tectonics, the evolutionary history of life, and past climates. In modern times, geology is commercially important for mineral and hydrocarbon exploration and exploitation and for evaluating water resources. It is publicly important for the prediction and understanding of natural hazards, the remediation of environmental problems, and for providing insights into past climate change. Geology plays a role in geotechnical engineering and is a major academic discipline.

Meteorology Main articles: Meteorology and Outline of meteorology

Meteorology is the interdisciplinary scientific study of the atmosphere<especially>. Studies in the field stretch back millennia, though significant progress in meteorology did not occur until the 17th century. The 19th century saw breakthroughs occur after observing networks developed across several countries. After the development of the computer in the latter half of the 20th century, breakthroughs in weather forecasting were achieved.

Space Science or Astronomy

Main articles: Space Science and Outline of space science

Space science or Astronomy is the study of everything in outer space.[13] This has sometimes been called astronomy, but recently astronomy has come to be regarded as a division of broader space science, which has grown to include other related fields,[14] such as studying issues related to space travel and space exploration (including space medicine), space archaeology[15] and science performed in outer space (see space research).

24.1.2 Life science

Main articles: Life science and Outline of life sciences

Life science comprises the branches of science[16] that involve the scientific study of living organisms, like plants, animals, and human beings. However, the study of behavior of organisms, such as practiced in ethology and psychology, is only included in as much as it involves a clearly biological aspect. While biology remains the centerpiece of life science, technological advances in molecular biology and biotechnology have led to a burgeoning of specializations and new, often interdisciplinary, fields.

Biology

Main articles: Biology and Outline of biology

Biology is the branch of natural science concerned with the study of life and living organisms, including their structure, function, growth, origin, evolution, distribution, and taxonomy.[17] Biology is a vast subject containing many subdivisions, topics, and disciplines.

Zoology Main articles: Zoology and Outline of zoology

Zoology /zoʊˈɒlədʒi/, occasionally spelled zoölogy, is the branch of science that relates to the animal kingdom, including the structure, embryology, evolution, classification, habits, and distribution of all animals, both living and extinct. The term is derived from Ancient Greek ζῷον (zōon, "animal") + λόγος (logos, "knowledge").

Human biology Main articles: Human biology and Outline of human biology

Human biology is an interdisciplinary academic field of biology, biological anthropology, nutrition and medicine which focuses on humans; it is closely related to primate biology, and a number of other fields.

Some branches of biology include: microbiology, anatomy, neurology and neuroscience, immunology, genetics, physiology, pathology, biophysics, biolinguistics, and ophthalmology.

Botany Main articles: Botany and Outline of botany

Botany, plant science, or plant biology is a branch of biology that involves the scientific study of plant life. Botany covers a wide range of scientific disciplines including structure, growth, reproduction, metabolism, development, diseases, chemical properties, and evolutionary relationships among taxonomic groups. Botany began with early human efforts to identify edible, medicinal and poisonous plants, making it one of the oldest sciences. Today botanists study over 550,000 species of living organisms. The term "botany" comes from Greek βοτάνη, meaning "pasture, grass, fodder", perhaps via the idea of a livestock keeper needing to know which plants are safe for livestock to eat.

24.2 Social sciences

Main articles: Social sciences and Outline of social science

The *social sciences* are the fields of scholarship that study society. "Social science" is commonly used as an umbrella term to refer to a plurality of fields outside of the natural sciences. These include: anthropology, archaeology, business administration, communication, criminology, economics, education, government, linguistics, international relations, political science, psychology (especially social psychology), theology, sociology and, in some contexts, geography, history and law.[18][19]

24.3 Formal sciences

Main articles: Formal sciences and Outline of formal science

The *formal sciences* are the branches of science that are concerned with formal systems, such as logic, mathematics, theoretical computer science, information theory, systems theory, decision theory, statistics, and theoretical linguistics.

Unlike other sciences, the formal sciences are not concerned with the validity of theories based on observations in the real world (empirical knowledge), but rather with the properties of formal systems based on definitions and rules. Methods of the formal sciences are, however, essential to the construction and testing of scientific models dealing with observable reality,[20] and major advances in formal sciences have often enabled major advances in the empirical sciences.

24.3.1 Decision theory

Main article: Decision theory

Decision theory in economics, psychology, philosophy, mathematics, and statistics is concerned with identifying the values, uncertainties and other issues relevant in a given decision, its rationality, and the resulting optimal decision. It is very closely related to the field of game law.

24.3.2 Logic

Main articles: Logic and Outline of logic

Logic (from the Greek λογική logikē)[21] is the formal systematic study of the principles of valid inference and correct reasoning. Logic is used in most intellectual activities, but is studied primarily in the disciplines of philosophy, mathematics, semantics, and computer science. Logic examines general forms which arguments may take, which forms are valid, and which are fallacies. In philosophy, the study of logic figures in most major areas: epistemology, ethics, metaphysics. In mathematics and computer science, it is the study of valid inferences within some formal language.[22] Logic is also studied in argumentation theory.[23]

24.3.3 Mathematics

Main articles: Mathematics and Outline of mathematics

Mathematics, first of all known as The Science of numbers which is classified in Arithmetic and Algebra, is classified as a formal science,[24][25] has both similarities and differences with the empirical sciences (the natural and social sciences). It is similar to empirical sciences in that it involves an objective, careful and systematic study of an area of knowledge; it is different because of its method of verifying its knowledge, using *a priori* rather than empirical methods.[26]

24.3.4 Statistics

Main articles: Statistics and Outline of statistics

Statistics is the study of the collection, organization, and interpretation of data.[27][28] It deals with all aspects of this, including the planning of data collection in terms of the design of surveys and experiments.[27]

A statistician is someone who is particularly well versed in the ways of thinking necessary for the successful application of statistical analysis. Such people have often gained this experience through working in any of a wide number of fields. There is also a discipline called *mathematical statistics*, which is concerned with the theoretical basis of the subject.

The word *statistics*, when referring to the scientific discipline, is singular, as in "Statistics is an art."[29] This should not be confused with the word *statistic*, referring to a quantity (such as mean or median) calculated from a set of data,[30] whose plural is *statistics* ("this statistic seems wrong" or "these statistics are misleading").

24.3.5 Systems theory

Main article: Systems theory

Systems theory is the transdisciplinary study of systems in general, with the goal of elucidating principles that can be applied to all types of systems in all fields of research. The term does not yet have a well-established, precise meaning, but systems theory can reasonably be considered a specialization of systems thinking and a generalization of systems science. The term originates from Bertalanffy's General System Theory (GST) and is used in later efforts in other fields, such as the action theory of Talcott Parsons and the system-theory of Niklas Luhmann.

In this context the word *systems* is used to refer specifically to self-regulating systems, i.e. that are self-correcting through feedback. Self-regulating systems are found in nature, including the physiological systems of our body, in local and global ecosystems, and in climate.

24.3.6 Theoretical computer science

Main article: Theoretical computer science

Theoretical computer science (TCS) is a division or subset of general computer science and focuses on more abstract or mathematical aspects of computing.

These divisions and subsets include analysis of algorithms and formal semantics of programming languages. Technically, there are hundreds of divisions and subsets besides these two. Each of the multiple parts have their own individual personal leaders (of popularity) and there are many associations and professional social groups and publications of distinction.

24.4 Applied sciences

Main articles: Applied science and Outline of applied sciences

Applied science is the application of scientific knowledge transferred into a physical environment. Examples include testing a theoretical model through the use of formal science or solving a practical problem through the use of natural science.

Applied science differs from fundamental science, which seeks to describe the most basic objects and forces, having less emphasis on practical applications. Applied science can be like biological science and physical science.

Example fields of applied science include

- Engineering
- Applied mathematics
- Applied physics
- Medicine
- Computer science

Fields of engineering are closely related to applied sciences. Applied science is important for technology development. Its use in industrial settings is usually referred to as research and development (R&D).

24.5 See also

- Index of branches of science
- Outline of science
  - Exact science
  - Fundamental science
  - Hard and soft science
- Branches of philosophy
  - Philosophy of science
- Engineering Science
  - Moral Science

24.6 Notes

[1] Popper 2002, p. 20.

[2] see: Editorial Staff (March 7, 2008). "Scientific Method: Relationships among Scientific Paradigms". Seed magazine. Retrieved 2007-09-12.

[3] Richard Feynman begins his *Lectures* with the atomic hypothesis, as his most compact statement of all scientific knowledge: "If, in some cataclysm, all of scientific knowledge were to be destroyed, and only one sentence passed on to the next generations ..., what statement would contain the most information in the fewest words? I believe it is ... that *all things are made up of atoms – little particles that move around in perpetual motion, attracting each other when they are a little distance apart, but repelling upon being squeezed into one another. ...*" R.P. Feynman; R.B. Leighton; Matthew Sands (1963). *The Feynman Lectures on Physics*. 1. p. I-2. ISBN 0-201-02116-1.

[4] J.C. Maxwell (1878). *Matter and Motion*. D. Van Nostrand. p. 9. ISBN 0-486-66895-9. Physical science is that department of knowledge which relates to the order of nature, or, in other words, to the regular succession of events.

[5] H.D. Young; R.A. Freedman (2004). *University Physics with Modern Physics* (11th ed.). Addison Wesley. p. 2. Physics is an *experimental* science. Physicists observe the phenomena of nature and try to find patterns and principles that relate these phenomena. These patterns are called physical theories or, when they are very well established and of broad use, physical laws or principles.

[6] S. Holzner (2006). *Physics for Dummies*. Wiley. p. 7. ISBN 0-470-61841-8. Physics is the study of your world and the world and universe around you.

[7] Note: The term 'universe' is defined as everything that physically exists: the entirety of space and time, all forms of matter, energy and momentum, and the physical laws and constants that govern them. However, the term 'universe' may also be used in slightly different contextual senses, denoting concepts such as the cosmos or the philosophical world.

[8] Evidence exists that the earliest civilizations dating back to beyond 3000 BCE, such as the Sumerians, Ancient Egyptians, and the Indus Valley Civilization, all had a predictive knowledge and a very basic understanding of the motions of the Sun, Moon, and stars.

[9] Francis Bacon's 1620 *Novum Organum* was critical in the development of scientific method.

[10] **See:** Chemistry (etymology) for possible origins of this word.

[11] Chemistry. (n.d.). Merriam-Webster's Medical Dictionary. Retrieved August 19, 2007.

[12] Wordnet Search: Earth science

[13] space science – definition of space science by the Free Online Dictionary, Thesaurus and Encyclopedia

[14] National Space Science Data Center (NSSDC) – NASA Science

[15] Space science | Define Space science at Dictionary.com

[16] Branches of Science

[17] Based on definition from Aquarena Wetlands Project glossary of terms. Archived June 8, 2004, at the Wayback Machine.

[18] Verheggen; et al. (1999). "From shared representations to consensually coordinated actions". In Morrs, John; et al. *Theoretical Issues in Psychology*. International Society for Theoretical Psychology.

[19] Garai, L.; Kocski, M. (1995). "Another crisis in the psychology: A possible motive for the Vygotsky-boom". *Journal of Russian and East-European Psychology*. **33** (1): 82–94. doi:10.2753/RPO1061-0405330182.

[20] Popper 2002, pp. 79–82.

[21] "possessed of reason, intellectual, dialectical, argumentative", also related to λόγος (logos), "word, thought, idea, argument, account, reason, or principle" (Liddell & Scott 1999; Online Etymology Dictionary 2001).

[22] Hofweber, T. (2004). "Logic and Ontology". In Zalta, Edward N. *Stanford Encyclopedia of Philosophy*.

[23] Cox, J. Robert; Willard, Charles Arthur, eds. (1983). *Advances in Argumentation Theory and Research*. Southern Illinois University Press. ISBN 978-0-8093-1050-0.

[24] Marcus Tomalin (2006) *Linguistics and the Formal Sciences*

[25] Benedikt Löwe (2002) "The Formal Sciences: Their Scope, Their Foundations, and Their Unity"

[26] Popper 2002, pp. 10–11.

[27] Dodge, Y. (2003) *The Oxford Dictionary of Statistical Terms*. OUP. ISBN 0-19-920613-9

[28] The Free Online Dictionary

[29] "Statistics". *Merriam-Webster Online Dictionary*.

[30] "Statistic". *Merriam-Webster Online Dictionary*.

24.7 References

- Popper, Karl R. (2002) [1959]. *The Logic of Scientific Discovery*. New York, NY: Routledge Classics. ISBN 0-415-27844-9. OCLC 59377149.

Chapter 25

Exact sciences

Ulugh Beg's meridian arc for precise astronomical measurements (15th c.)

The **exact sciences**, sometimes called the **exact mathematical sciences**[1] are those sciences "which admit of absolute precision in their results"; especially the mathematical sciences.[2] Examples of the exact sciences are mathematics, optics, astronomy, and physics, which many philosophers from Descartes, Leibniz, and Kant to the logical positivists took as paradigms of rational and objective knowledge.[3] These sciences have been practiced in many cultures from Antiquity[4][5] to modern times.[6][7] Given their ties to mathematics, the exact sciences are characterized by accurate quantitative expression, precise predictions and/or rigorous methods of testing hypotheses involving quantifiable predictions and measurements.

The distinction between the quantitative exact sciences and those sciences which deal with the causes of things is due to Aristotle, who distinguished mathematics from natural philosophy and considered the exact sciences to be the "more natural of the branches of mathematics."[8] Thomas Aquinas employed this distinction when he pointed out that astronomy explains the spherical shape of the Earth by mathematical reasoning while physics explains it by material causes.[9] This distinction was widely, but not universally, accepted until the scientific revolution of the Seventeenth Century.[10] Edward Grant has proposed that a fundamental change leading to the new sciences was the unification of the exact sciences and physics by Kepler, Newton, and others, which resulted in a quantitative investigation of the physical causes of natural phenomena.[11]

25.1 See also

- Hard and soft science
- Fundamental science
- Demarcation problem

25.2 References

[1] Grant, Edward (2007), *A History of Natural Philosophy: From the Ancient World to the Nineteenth Century*, Cambridge: Cambridge University Press, p. 43, ISBN 9781139461092

[2] "Exact, *adj.*¹", *Oxford English Dictionary*, Online version (2nd ed.), Oxford: Oxford University Press, June 2016

[3] Friedman, Michael (1992), "Philosophy and the Exact Sciences: Logical Positivism as a Case Study", in Earman, John, *Inference, Explanation, and Other Frustrations: Essays in the*

Philosophy of Science, Pittsburgh series in philosophy and history of science, **14**, Berkeley and Los Angeles: University of California Press, p. 84, ISBN 9780520075771

[4] Neugebauer, Otto (1962), *The Exact Sciences in Antiquity*, The Science Library (2nd. reprint ed.), New York: Harper & Bros.

[5] Sarkar, Benoy Kumar (1918), *Hindu Achievements in Exact Science: A Study in the History of Scientific Development*, London / New York: Longmans, Green and Company

[6] Harman, Peter M.; Shapiro, Alan E. (2002), *The Investigation of Difficult Things: Essays on Newton and the History of the Exact Sciences in Honour of D. T. Whiteside*, Cambridge: Cambridge University Press, ISBN 9780521892667

[7] Pyenson, Lewis (1993), "Cultural Imperialism and Exact Sciences Revisited", *Isis*: 103–108, JSTOR 235556, doi:10.1086/356376, [M]any of the exact sciences... between Claudius Ptolemy and Tycho Brahe were in a common register, whether studied in the diverse parts of the Islamic world, in India, in Christian Europe, in China, or apparently in Mesoamerica.

[8] Grant, Edward (2007), *A History of Natural Philosophy: From the Ancient World to the Nineteenth Century*, Cambridge: Cambridge University Press, pp. 42–43, ISBN 9781139461092

[9] Aquinas, Thomas, *Summa Theologica*, Part I, Q. 1, Art. 1, Reply 2, retrieved 3 September 2016, For the astronomer and the physicist both may prove the same conclusion: that the earth, for instance, is round: the astronomer by means of mathematics (i.e. abstracting from matter), but the physicist by means of matter itself.

[10] Grant, Edward (2007), *A History of Natural Philosophy: From the Ancient World to the Nineteenth Century*, Cambridge: Cambridge University Press, pp. 303–305, ISBN 9781139461092

[11] Grant, Edward (2007), *A History of Natural Philosophy: From the Ancient World to the Nineteenth Century*, Cambridge: Cambridge University Press, pp. 303, 312–313, ISBN 9781139461092

Chapter 26

History of scientific method

The **history of scientific method** considers changes in the methodology of scientific inquiry, as distinct from the history of science itself. The development of rules for scientific reasoning has not been straightforward; scientific method has been the subject of intense and recurring debate throughout the history of science, and eminent natural philosophers and scientists have argued for the primacy of one or another approach to establishing scientific knowledge. Despite the disagreements about approaches, scientific method has advanced in definite steps. Rationalist explanations of nature, including atomism, appeared both in ancient Greece in the thought of Leucippus and Democritus, and in ancient India, in the Nyaya, Vaisesika and Buddhist schools, while Charvaka materialism rejected inference as a source of knowledge in favour of an empiricism that was always subject to doubt. Aristotle pioneered scientific method in ancient Greece alongside his empirical biology and his work on logic, rejecting a purely deductive framework in favour of generalisations made from observations of nature.

Some of the most important debates in the history of scientific method center on: rationalism, especially as advocated by René Descartes; inductivism, which rose to particular prominence with Isaac Newton and his followers; and hypothetico-deductivism, which came to the fore in the early 19th century. In the late 19th and early 20th centuries, a debate over realism vs. antirealism was central to discussions of scientific method as powerful scientific theories extended beyond the realm of the observable, while in the mid-20th century some prominent philosophers argued against any universal rules of science at all.[1]

26.1 Early methodology

There are few explicit discussions of scientific methodologies in surviving records from early cultures. The most that can be inferred about the approaches to undertaking science in this period stems from descriptions of early investigations into nature, in the surviving records. An Egyptian medical

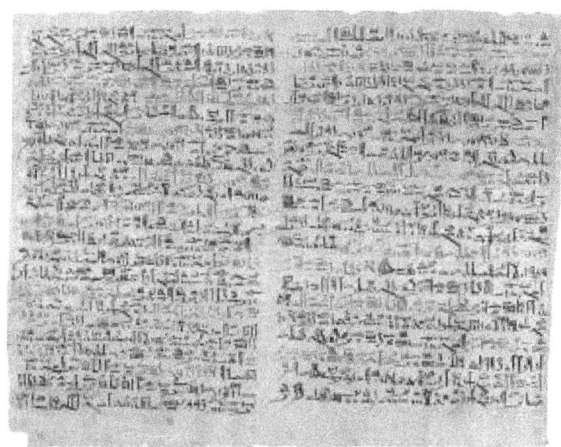

Edwin Smith papyrus

textbook, the Edwin Smith papyrus, (c. 1600 BCE), applies the following components: examination, diagnosis, treatment and prognosis, to the treatment of disease,[2] which display strong parallels to the basic empirical method of science and according to G. E. R. Lloyd[3] played a significant role in the development of this methodology. The Ebers papyrus (c. 1550 BCE) also contains evidence of traditional empiricism.

By the middle of the 1st millennium BCE in Mesopotamia, Babylonian astronomy had evolved into the earliest example of a scientific astronomy, as it was "the first and highly successful attempt at giving a refined mathematical description of astronomical phenomena." According to the historian Asger Aaboe, "all subsequent varieties of scientific astronomy, in the Hellenistic world, in India, in Islam, and in the West – if not indeed all subsequent endeavour in the exact sciences – depend upon Babylonian astronomy in decisive and fundamental ways."[4]

The early Babylonians and Egyptians developed much technical knowledge, crafts, and mathematics[5] used in practical tasks of divination, as well as a knowledge of medicine,[6] and made lists of various kinds. While the

Babylonians in particular had engaged in the earliest forms of an empirical mathematical science, with their early attempts at mathematically describing natural phenomena, they generally lacked underlying rational theories of nature.[4][7][8] It was the ancient Greeks who engaged in the earliest forms of what is today recognized as a rational theoretical science,[7][9] with the move towards a more rational understanding of nature which began at least since the Archaic Period (650 – 480 BCE) with the Presocratic school. Thales was the first to use natural explanations, proclaiming that every event had a natural cause, even though he is known for saying "all things are full of gods" and sacrificed an ox when he discovered his theorem.[10] Leucippus, went on to develop the theory of atomism – the idea that everything is composed entirely of various imperishable, indivisible elements called atoms. This was elaborated in great detail by Democritus.

Similar atomist ideas emerged independently among ancient Indian philosophers of the Nyaya, Vaisesika and Buddhist schools.[11] In particular, like the Nyaya, Vaisesika, and Buddhist schools, the Cārvāka epistemology was materialist, and skeptical enough to admit perception as the basis for unconditionally true knowledge, while cautioning that if one could only infer a truth, then one must also harbor a doubt about that truth; an inferred truth could not be unconditional.[12]

Towards the middle of the 5th century BCE, some of the components of a scientific tradition were already heavily established, even before Plato, who was an important contributor to this emerging tradition, thanks to the development of deductive reasoning, as propounded by his student, Aristotle. In *Protagoras* (318d-f), Plato mentioned the teaching of arithmetic, astronomy and geometry in schools. The philosophical ideas of this time were mostly freed from the constraints of everyday phenomena and common sense. This denial of reality as we experience it reached an extreme in Parmenides who argued that the world is one and that change and subdivision do not exist.

In the 3rd and 4th centuries BCE, the Greek physicians Herophilos (335–280 BCE) and Erasistratus of Chios employed experiments to further their medical research; Erasistratus at one time repeatedly weighing a caged bird, and noting its weight loss between feeding times.[13]

26.1.1 Aristotle

Further information: Aristotle's biology

Aristotle's inductive-deductive method used inductions from observations to infer general principles, deductions from those principles to check against further observations,

Aristotle's philosophy involved both inductive and deductive reasoning.

and more cycles of induction and deduction to continue the advance of knowledge.[14]

The *Organon* (Greek: Ὄργανον, meaning "instrument, tool, organ") is the standard collection of Aristotle's six works on logic. The name *Organon* was given by Aristotle's followers, the Peripatetics. The order of the works is not chronological (the chronology is now difficult to determine) but was deliberately chosen by Theophrastus to constitute a well-structured system. Indeed, parts of them seem to be a scheme of a lecture on logic. The arrangement of the works was made by Andronicus of Rhodes around 40 BCE.[15]

The *Organon* comprises the following six works:

1. The *Categories* (Latin: *Categoriae*) introduces Aristotle's 10-fold classification of that which exists: substance, quantity, quality, relation, place, time, situation, condition, action, and passion.

2. *On Interpretation* (Latin:*De Interpretatione*, Greek *Perihermenias*) introduces Aristotle's conception of proposition and judgment, and the various relations between affirmative, negative, universal, and particular propositions. Aristotle discusses the square of opposition or square of Apuleius in Chapter 7 and its ap-

26.1. EARLY METHODOLOGY

pendix Chapter 8. Chapter 9 deals with the problem of future contingents.

3. The *Prior Analytics* (Latin: *Analytica Priora*) introduces Aristotle's syllogistic method (see term logic), argues for its correctness, and discusses inductive inference.

4. The *Posterior Analytics* (Latin: *Analytica Posteriora*) deals with demonstration, definition, and scientific knowledge.

5. The *Topics* (Latin: *Topica*) treats of issues in constructing valid arguments, and of inference that is probable, rather than certain. It is in this treatise that Aristotle mentions the predicables, later discussed by Porphyry and by the scholastic logicians.

6. The *Sophistical Refutations* (Latin: *De Sophisticis Elenchis*) gives a treatment of logical fallacies, and provides a key link to Aristotle's work on rhetoric.

Aristotle's *Metaphysics* has some points of overlap with the works making up the *Organon* but is not traditionally considered part of it; additionally there are works on logic attributed, with varying degrees of plausibility, to Aristotle that were not known to the Peripatetics.

Aristotle introduced what may be called a scientific method.[16] His demonstration method is found in *Posterior Analytics*. He provided another of the ingredients of scientific tradition: empiricism. For Aristotle, universal truths can be known from particular things via induction. To some extent then, Aristotle reconciles abstract thought with observation, although it would be a mistake to imply that Aristotelian science is empirical in form. Indeed, Aristotle did not accept that knowledge acquired by induction could rightly be counted as scientific knowledge. Nevertheless, induction was for him a necessary preliminary to the main business of scientific enquiry, providing the primary premises required for scientific demonstrations.

Aristotle largely ignored inductive reasoning in his treatment of scientific enquiry. To make it clear why this is so, consider this statement in the *Posterior Analytics*:

> We suppose ourselves to possess unqualified scientific knowledge of a thing, as opposed to knowing it in the accidental way in which the sophist knows, when we think that we know the cause on which the fact depends, as the cause of that fact and of no other, and, further, that the fact could not be other than it is.

It was therefore the work of the philosopher to demonstrate universal truths and to discover their causes.[17] While induction was sufficient for discovering universals by generalization, it did not succeed in identifying causes. For this task Aristotle used the tool of deductive reasoning in the form of syllogisms. Using the syllogism, scientists could infer new universal truths from those already established.

Aristotle developed a complete normative approach to scientific inquiry involving the syllogism, which he discusses at length in his *Posterior Analytics*. A difficulty with this scheme lay in showing that derived truths have solid primary premises. Aristotle would not allow that demonstrations could be circular (supporting the conclusion by the premises, and the premises by the conclusion). Nor would he allow an infinite number of middle terms between the primary premises and the conclusion. This leads to the question of how the primary premises are found or developed, and as mentioned above, Aristotle allowed that induction would be required for this task.

Towards the end of the *Posterior Analytics*, Aristotle discusses knowledge imparted by induction.

> Thus it is clear that we must get to know the primary premises by induction; for the method by which even sense-perception implants the universal is inductive. [...] it follows that there will be no scientific knowledge of the primary premises, and since except intuition nothing can be truer than scientific knowledge, it will be intuition that apprehends the primary premises. [...] If, therefore, it is the only other kind of true thinking except scientific knowing, intuition will be the originative source of scientific knowledge.

The account leaves room for doubt regarding the nature and extent of Aristotle's empiricism. In particular, it seems that Aristotle considers sense-perception only as a vehicle for knowledge through intuition. He restricted his investigations in natural history to their natural settings,[18] such as at the Pyrrha lagoon,[19] now called Kalloni, at Lesbos. Aristotle and Theophrastus together formulated the new science of biology,[20] inductively, case by case, for two years before Aristotle was called to tutor Alexander. Aristotle performed no modern-style experiments in the form in which they appear in today's physics and chemistry laboratories.[21] Induction is not afforded the status of scientific reasoning, and so it is left to intuition to provide a solid foundation for Aristotle's science. With that said, Aristotle brings us somewhat closer an empirical science than his predecessors.

26.1.2 Epicurus

In his work Κανών ('canon', a straight edge or ruler, thus any type of measure or standard, referred to as 'canonic'), Epicurus laid out his first rule for inquiry in physics: 'that

CHAPTER 26. HISTORY OF SCIENTIFIC METHOD

Some philosophers held that there are only atoms and void; others that the atoms are divine fire, others only wind, others only water, others only earth.

26.2 Emergence of inductive experimental method

During the Middle Ages issues of what is now termed science began to be addressed. There was greater emphasis on combining theory with practice in the Islamic world than there had been in Classical times, and it was common for those studying the sciences to be artisans as well, something that had been "considered an aberration in the ancient world." Islamic experts in the sciences were often expert instrument makers who enhanced their powers of observation and calculation with them.[24] Muslim scientists used experiment and quantification to distinguish between competing scientific theories, set within a generically empirical orientation, as can be seen in the works of Jābir ibn Hayyān (721–815)[25] and Alkindus (801–873)[26] as early examples. Several scientific methods thus emerged from the medieval Muslim world by the early 11th century, all of which emphasized experimentation as well as quantification to varying degrees.

26.2.1 Ibn al-Haytham

"How does light travel through transparent bodies? Light travels through transparent bodies in straight lines only.... We have explained this exhaustively in our Book of Optics.*"[27]* —Alhazen

the *first concepts be seen*,[22]:p.20 and that they *not require demonstration* '.[22]:pp.35–47

His second rule for inquiry was that prior to an investigation, *we are to have self-evident concepts*,[22]:pp.61–80 so that we might infer [ἔχωμεν οἷς σημειωσόμεθα] both what is expected [τὸ προσμένον], and also what is non-apparent [τὸ ἄδηλον].[22]:pp.83–103

Epicurus applies his method of inference (the use of observations as signs, Asmis' summary, p. 333: *the method of using the phenomena as signs (σημεῖα) of what is unobserved*)[22]:pp.175–196 immediately to the atomic theory of Democritus. In Aristotle's *Prior Analytics*, Aristotle himself employs the use of signs.[22]:pp.212–224[23] But Epicurus presented his 'canonic' as rival to Aristotle's logic.[22]:pp.19–34 See: Lucretius (c. 99 BCE – c. 55 BCE) *De rerum natura* (On the nature of things) a didactic poem explaining Epicurus' philosophy and physics.

The Arab physicist Ibn al-Haytham (Alhazen) used experimentation to obtain the results in his *Book of Optics* (1021). He combined observations, experiments and rational arguments to support his intromission theory of vision, in which rays of light are emitted from objects rather than from the eyes. He used similar arguments to show that the ancient emission theory of vision supported by Ptolemy and Euclid (in which the eyes emit the rays of light used for seeing), and the ancient intromission theory supported by Aristotle (where objects emit physical particles to the eyes), were both wrong.[28]

Experimental evidence supported most of the propositions in his *Book of Optics* and grounded his theories of vision, light and colour, as well as his research in catoptrics and dioptrics. His legacy was elaborated through the 'reforming' of his *Optics* by Kamal al-Din al-Farisi (d. c. 1320) in the latter's *Kitab Tanqih al-Manazir* (*The Revision of* [Ibn al-Haytham's] *Optics*).[29][30]

Alhazen viewed his scientific studies as a search for truth: "Truth is sought for its own sake. And those who are engaged upon the quest for anything for its own sake are not interested in other things. Finding the truth is difficult, and the road to it is rough. ..."[31]

Alhazen's work included the conjecture that "Light travels through transparent bodies in straight lines only", which he was able to corroborate only after years of effort. He stated, "[This] is clearly observed in the lights which enter into dark rooms through holes. ... the entering light will be clearly observable in the dust which fills the air."[27] He also demonstrated the conjecture by placing a straight stick or a taut thread next to the light beam.[32]

Ibn al-Haytham also employed scientific skepticism and emphasized the role of empiricism. He also explained the role of induction in syllogism, and criticized Aristotle for his lack of contribution to the method of induction, which Ibn al-Haytham regarded as superior to syllogism, and he considered induction to be the basic requirement for true scientific research.[33]

Something like Occam's razor is also present in the *Book of Optics*. For example, after demonstrating that light is generated by luminous objects and emitted or reflected into the eyes, he states that therefore "the extramission of [visual] rays is superfluous and useless."[34] He may also have been the first scientist to adopt a form of positivism in his approach. He wrote that "we do not go beyond experience, and we cannot be content to use pure concepts in investigating natural phenomena", and that the understanding of these cannot be acquired without mathematics. After assuming that light is a material substance, he does not further discuss its nature but confines his investigations to the diffusion and propagation of light. The only properties of light he takes into account are those treatable by geometry and verifiable by experiment.[35]

26.2.2 Al-Biruni

The Persian scientist Abū Rayhān al-Bīrūnī introduced early scientific methods for several different fields of inquiry during the 1020s and 1030s. For example, in his treatise on mineralogy, *Kitab al-Jawahir* (*Book of Precious Stones*), al-Biruni is "the most exact of experimental scientists", while in the introduction to his study of India, he declares that "to execute our project, it has not been possible to follow the geometric method" and thus became one of the pioneers of comparative sociology in insisting on field experience and information.[36] He also developed an early experimental method for mechanics.[37]

Al-Biruni's methods resembled the modern scientific method, particularly in his emphasis on repeated experimentation. Biruni was concerned with how to conceptualize and prevent both systematic errors and observational biases, such as "errors caused by the use of small instruments and errors made by human observers." He argued that if instruments produce errors because of their imperfections or idiosyncratic qualities, then multiple observations must be taken, analyzed qualitatively, and on this basis, arrive at a "common-sense single value for the constant sought", whether an arithmetic mean or a "reliable estimate."[38] In his scientific method, "universals came out of practical, experimental work" and "theories are formulated after discoveries", as with inductivism.[36]

26.2.3 Ibn Sina (Avicenna)

In the *On Demonstration* section of *The Book of Healing* (1027), the Persian philosopher and scientist Avicenna (Ibn Sina) discussed philosophy of science and described an early scientific method of inquiry. He discussed Aristotle's *Posterior Analytics* and significantly diverged from it on several points. Avicenna discussed the issue of a proper procedure for scientific inquiry and the question of "How does one acquire the first principles of a science?" He asked how a scientist might find "the initial axioms or hypotheses of a deductive science without inferring them from some more basic premises?" He explained that the ideal situation is when one grasps that a "relation holds between the terms, which would allow for absolute, universal certainty." Avicenna added two further methods for finding a first principle: the ancient Aristotelian method of induction (*istiqra*), and the more recent method of examination and experimentation (*tajriba*). Avicenna criticized Aristotelian induction, arguing that "it does not lead to the absolute, universal, and certain premises that it purports to provide." In its place, he advocated "a method of experimentation as a means for scientific inquiry."[39]

Earlier, in *The Canon of Medicine* (1025), Avicenna

was also the first to describe what is essentially methods of agreement, difference and concomitant variation which are critical to inductive logic and the scientific method.[40][41][42] However, unlike his contemporary al-Biruni's scientific method, in which "universals came out of practical, experimental work" and "theories are formulated after discoveries", Avicenna developed a scientific procedure in which "general and universal questions came first and led to experimental work."[36] Due to the differences between their methods, al-Biruni referred to himself as a mathematical scientist and to Avicenna as a philosopher, during a debate between the two scholars.[43]

26.2.4 Robert Grosseteste

During the European Renaissance of the 12th century, ideas on scientific methodology, including Aristotle's empiricism and the experimental approaches of Alhazen and Avicenna, were introduced to medieval Europe via Latin translations of Arabic and Greek texts and commentaries. Robert Grosseteste's commentary on the *Posterior Analytics* places Grosseteste among the first scholastic thinkers in Europe to understand Aristotle's vision of the dual nature of scientific reasoning. Concluding from particular observations into a universal law, and then back again, from universal laws to prediction of particulars. Grosseteste called this "resolution and composition". Further, Grosseteste said that both paths should be verified through experimentation to verify the principles.[44]

26.2.5 Roger Bacon

Roger Bacon was inspired by the writings of Grosseteste. In his account of a method, Bacon described a repeating cycle of *observation*, *hypothesis*, *experimentation*, and the need for independent *verification*. He recorded the way he had conducted his experiments in precise detail, perhaps with the idea that others could reproduce and independently test his results.

About 1256 he joined the Franciscan Order and became subject to the Franciscan statute forbidding Friars from publishing books or pamphlets without specific approval. After the accession of Pope Clement IV in 1265, the Pope granted Bacon a special commission to write to him on scientific matters. In eighteen months he completed three large treatises, the *Opus Majus*, *Opus Minus*, and *Opus Tertium* which he sent to the Pope.[45] William Whewell has called *Opus Majus* at once the Encyclopaedia and Organon of the 13th century.[46]

- Part I (pp. 1–22) treats of the four causes of error: authority, custom, the opinion of the unskilled many, and the concealment of real ignorance by a pretense of knowledge.

- Part VI (pp. 445–477) treats of experimental science, *domina omnium scientiarum*. There are two methods of knowledge: the one by argument, the other by experience. Mere argument is never sufficient; it may decide a question, but gives no satisfaction or certainty to the mind, which can only be convinced by immediate inspection or intuition, which is what experience gives.

- Experimental science, which in the *Opus Tertium* (p. 46) is distinguished from the speculative sciences and the operative arts, is said to have three great prerogatives over all sciences:

    1. It verifies their conclusions by direct experiment;
    2. It discovers truths which they could never reach;
    3. It investigates the secrets of nature, and opens to us a knowledge of past and future.

- Roger Bacon illustrated his method by an investigation into the nature and cause of the rainbow, as a specimen of inductive research.[47]

26.2.6 Renaissance humanism and medicine

Aristotle's ideas became a framework for critical debate beginning with absorption of the Aristotelian texts into the university curriculum in the first half of the 13th century.[48] Contributing to this was the success of medieval theologians in reconciling Aristotelian philosophy with Christian theology. Within the sciences, medieval philosophers were not afraid of disagreeing with Aristotle on many specific issues, although their disagreements were stated within the language of Aristotelian philosophy. All medieval natural philosophers were Aristotelians, but "Aristotelianism" had become a somewhat broad and flexible concept. With the end of Middle Ages, the Renaissance rejection of medieval traditions coupled with an extreme reverence for classical sources led to a recovery of other ancient philosophical traditions, especially the teachings of Plato.[49] By the 17th century, those who clung dogmatically to Aristotle's teachings were faced with several competing approaches to nature.[50]

The discovery of the Americas at the close of the 15th century showed the scholars of Europe that new discoveries could be found outside of the authoritative works of Aristotle, Pliny, Galen, and other ancient writers.

Galen of Pergamon (129 – c. 200 AD) had studied with four schools in antiquity — Platonists, Aristotelians, Stoics, and Epicureans, and at Alexandria, the center of medicine

26.2. EMERGENCE OF INDUCTIVE EXPERIMENTAL METHOD

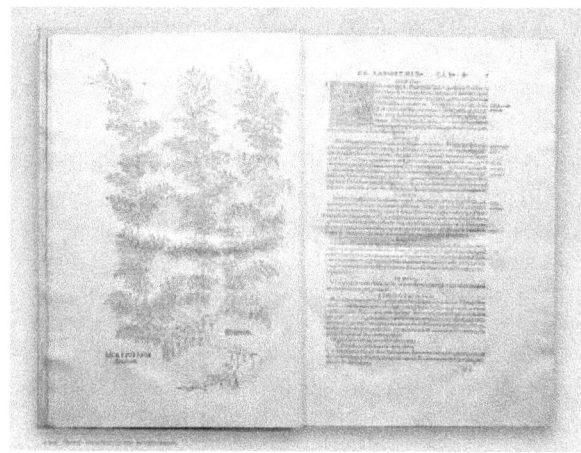

Leonhart Fuchs' drawing of absinthe plant, De Historia Stirpium. Basle 1542

at the time. In his *Methodus Medendi*, Galen had synthesized the empirical and dogmatic schools of medicine into his own method, which was preserved by Arab scholars. After the translations from Arabic were critically scrutinized, a backlash occurred and demand arose in Europe for translations of Galen's medical text from the original Greek. Galen's method became very popular in Europe. Thomas Linacre, the teacher of Erasmus, thereupon translated *Methodus Medendi* from Greek into Latin for a larger audience in 1519.[51] Limbrick 1988 notes that 630 editions, translations, and commentaries on Galen were produced in Europe in the 16th century, eventually eclipsing Arabic medicine there, and peaking in 1560, at the time of the scientific revolution.[52]

By the late 15th century, the physician-scholar Niccolò Leoniceno was finding errors in Pliny's *Natural History*. As a physician, Leoniceno was concerned about these botanical errors propagating to the materia medica on which medicines were based.[53] To counter this, a botanical garden was established at Orto botanico di Padova, University of Padua (in use for teaching by 1546), in order that medical students might have empirical access to the plants of a pharmacopia. Other Renaissance teaching gardens were established, notably by the physician Leonhart Fuchs, one of the founders of botany.[54]

The first published work devoted to the concept of method is Jodocus Willichius, *De methodo omnium artium et disciplinarum informanda opusculum* (1550).

26.2.7 Skepticism as a basis for understanding

In 1562 "Outlines of Pyrrhonism" by Sextus Empiricus (c. 160-210 AD) appeared in print and in Latin, quickly placing the arguments of classical skepticism in the European mainstream. Skepticism either denies or strongly doubts (depending on the school) the possibility of certain knowledge. Descartes' famous "Cogito" argument is an attempt to overcome skepticism and reestablish a foundation for certainty but other thinkers responded by revising what the search for knowledge, particularly physical knowledge, might be.

The first of these, philosopher and physician Francisco Sanches, was led by his medical training at Rome, 1571–73, to search for a true method of knowing (*modus sciendi*), as nothing clear can be known by the methods of Aristotle and his followers[55] — for example, 1) syllogism fails upon circular reasoning; 2) Aristotle's modal logic was not stated clearly enough for use in medieval times, and remains a research problem to this day.[56] Following the physician Galen's *method of medicine*, Sanches lists the methods of judgement and experience, which are faulty in the wrong hands,[57] and we are left with the bleak statement *That Nothing is Known* (1581, in Latin *Quod Nihil Scitur*). This challenge was taken up by René Descartes in the next generation (1637), but at the least, Sanches warns us that we ought to refrain from the methods, summaries, and commentaries on Aristotle, if we seek scientific knowledge. In this, he is echoed by Francis Bacon who was influenced by another prominent exponent of skepticism, Montaigne: Sanches cites the humanist Juan Luis Vives who sought a better educational system, as well as a statement of human rights as a pathway for improvement of the lot of the poor.

"Sanches develops his scepticism by means of an intellectual critique of Aristotelianism, rather than by an appeal to the history of human stupidity and the variety and contrariety of previous theories." —Popkin 1979, p. 37, as cited by Sanches, Limbrick & Thomson 1988, pp. 24–5

"To work, then; and if you know something, then teach me; I shall be extremely grateful to you. In the meantime, as I prepare to examine *Things*, I shall raise the question anything is *known*, and if so, how, in the introductory passages of another book,[58] a book in which I will expound, as far as human frailty allows,[59] the *method of knowing*. Farewell.

WHAT IS TAUGHT HAS NO MORE STRENGTH THAN IT DERIVES FROM HIM WHO IS TAUGHT.

WHAT?" —Francisco Sanches (1581) *Quod Nihil Scitur* p. 100[60]

26.2.8 Francis Bacon's eliminative induction

Main article: Baconian method

> "If a man will begin with certainties, he shall end in doubts; but if he will be content to begin with doubts, he shall end in certainties."
> —Francis Bacon (1605) *The Advancement of Learning*, Book 1, v. 8

Francis Bacon (1561–1626) entered Trinity College, Cambridge in April 1573, where he applied himself diligently to the several sciences as then taught, and came to the conclusion that the methods employed and the results attained were alike erroneous; he learned to despise the current Aristotelian philosophy. He believed philosophy must be taught its true purpose, and for this purpose a new method must be devised. With this conception in his mind, Bacon left the university.[61]

Bacon attempted to describe a rational procedure for establishing causation between phenomena based on induction. Bacon's induction was, however, radically different than that employed by the Aristotelians. As Bacon put it,

> [A]nother form of induction must be devised than has hitherto been employed, and it must be used for proving and discovering not first principles (as they are called) only, but also the lesser axioms, and the middle, and indeed all. For the induction which proceeds by simple enumeration is childish. —*Novum Organum* section CV

Bacon's method relied on experimental *histories* to eliminate alternative theories.[62] Bacon explains how his method is applied in his *Novum Organum* (published 1620). In an example he gives on the examination of the nature of heat, Bacon creates two tables, the first of which he names "Table of Essence and Presence", enumerating the many various circumstances under which we find heat. In the other table, labelled "Table of Deviation, or of Absence in Proximity", he lists circumstances which bear resemblance to those of the first table except for the absence of heat. From an analysis of what he calls the *natures* (light emitting, heavy, colored, etc.) of the items in these lists we are brought to conclusions about the *form nature*, or cause, of heat. Those natures which are always present in the first table, but never in the second are deemed to be the cause of heat.

The role experimentation played in this process was twofold. The most laborious job of the scientist would be to gather the facts, or 'histories', required to create the tables of presence and absence. Such histories would document a mixture of common knowledge and experimental results. Secondly, *experiments of light*, or, as we might say, crucial experiments would be needed to resolve any remaining ambiguities over causes.

Bacon showed an uncompromising commitment to experimentation. Despite this, he did not make any great scientific discoveries during his lifetime. This may be because he was not the most able experimenter.[63] It may also be because hypothesising plays only a small role in Bacon's method compared to modern science.[64] Hypotheses, in Bacon's method, are supposed to emerge during the process of investigation, with the help of mathematics and logic. Bacon gave a substantial but secondary role to mathematics "*which ought only to give definiteness to natural philosophy, not to generate or give it birth*" (*Novum Organum* XCVI). An over-emphasis on axiomatic reasoning had rendered previous non-empirical philosophy impotent, in Bacon's view, which was expressed in his *Novum Organum*:

> XIX. There are and can be only two ways of searching into and discovering truth. The one flies from the senses and particulars to the most general axioms, and from these principles, the truth of which it takes for settled and immoveable, proceeds to judgment and to the discovery of middle axioms. And this way is now in fashion. The other derives axioms from the senses and particulars, rising by a gradual and unbroken ascent, so that it arrives at the most general axioms last of all. This is the true way, but as yet untried.

In Bacon's utopian novel, *The New Atlantis*, the ultimate role is given for inductive reasoning:

> Lastly, we have three that raise the former discoveries by experiments into greater observations, axioms, and aphorisms. These we call interpreters of nature.

26.2.9 Descartes

Main article: Cartesianism

In 1619, René Descartes began writing his first major treatise on proper scientific and philosophical thinking, the unfinished *Rules for the Direction of the Mind*. His aim was to create a complete science that he hoped would overthrow the Aristotelian system and establish himself as the sole architect[65] of a new system of guiding principles for scientific research.

This work was continued and clarified in his 1637 treatise, *Discourse on Method*, and in his 1641 *Meditations*. Descartes describes the intriguing and disciplined thought experiments he used to arrive at the idea we instantly associate with him: *I think therefore I am*.

From this foundational thought, Descartes finds proof of the existence of a God who, possessing all possible perfections, will not deceive him provided he resolves "[...] never to accept anything for true which I did not clearly know to be such; that is to say, carefully to avoid precipitancy and prejudice, and to comprise nothing more in my judgment than what was presented to my mind so clearly and distinctly as to exclude all ground of methodic doubt."[66]

This rule allowed Descartes to progress beyond his own thoughts and judge that there exist extended bodies outside of his own thoughts. Descartes published seven sets of objections to the *Meditations* from various sources[67] along with his replies to them. Despite his apparent departure from the Aristotelian system, a number of his critics felt that Descartes had done little more than replace the primary premises of Aristotle with those of his own. Descartes says as much himself in a letter written in 1647 to the translator of Principles of Philosophy,

> a perfect knowledge [...] must necessarily be deduced from first causes [...] we must try to deduce from these principles knowledge of the things which depend on them, that there be nothing in the whole chain of deductions deriving from them that is not perfectly manifest.[68]

And again, some years earlier, speaking of Galileo's physics in a letter to his friend and critic Mersenne from 1638,

> without having considered the first causes of nature, [Galileo] has merely looked for the explanations of a few particular effects, and he has thereby built without foundations.[69]

Whereas Aristotle purported to arrive at his first principles by induction, Descartes believed he could obtain them using reason only. In this sense, he was a Platonist, as he believed in the innate ideas, as opposed to Aristotle's blank slate (*tabula rasa*), and stated that the seeds of science are inside us.[70]

Unlike Bacon, Descartes successfully applied his own ideas in practice. He made significant contributions to science, in particular in aberration-corrected optics. His work in analytic geometry was a necessary precedent to differential calculus and instrumental in bringing mathematical analysis to bear on scientific matters.

26.2.10 Galileo Galilei

Galileo Galilei, 1564–1642, a father of scientific method

During the period of religious conservatism brought about by the Reformation and Counter-Reformation, Galileo Galilei unveiled his new science of motion. Neither the contents of Galileo's science, nor the methods of study he selected were in keeping with Aristotelian teachings. Whereas Aristotle thought that a science should be demonstrated from first principles, Galileo had used experiments as a research tool. Galileo nevertheless presented his treatise in the form of mathematical demonstrations without reference to experimental results. It is important to understand that this in itself was a bold and innovative step in terms of scientific method. The usefulness of mathematics in obtaining scientific results was far from obvious.[71] This is because mathematics did not lend itself to the primary pursuit of Aristotelian science: the discovery of causes.

Whether it is because Galileo was realistic about the acceptability of presenting experimental results as evidence or because he himself had doubts about the epistemological status of experimental findings is not known. Nevertheless, it is not in his Latin treatise on motion that we find reference to experiments, but in his supplementary dialogues written in the Italian vernacular. In these dialogues experimental results are given, although Galileo may have found them inadequate for persuading his audience. Thought experiments showing logical contradictions in Aristotelian thinking, presented in the skilled rhetoric of Galileo's dialogue were further enticements for the reader.

Modern replica of Galileo's inclined plane experiment: The distance covered by a uniformly accelerated body is proportional to the square of the time elapsed.

As an example, in the dramatic dialogue titled *Third Day* from his *Two New Sciences*, Galileo has the characters of the dialogue discuss an experiment involving two free falling objects of differing weight. An outline of the Aristotelian view is offered by the character Simplicio. For this experiment he expects that "a body which is ten times as heavy as another will move ten times as rapidly as the other". The character Salviati, representing Galileo's persona in the dialogue, replies by voicing his doubt that Aristotle ever attempted the experiment. Salviati then asks the two other characters of the dialogue to consider a thought experiment whereby two stones of differing weights are tied together before being released. Following Aristotle, Salviati reasons that "the more rapid one will be partly retarded by the slower, and the slower will be somewhat hastened by the swifter". But this leads to a contradiction, since the two stones together make a heavier object than either stone apart, the heavier object should in fact fall with a speed greater than that of either stone. From this contradiction, Salviati concludes that Aristotle must, in fact, be wrong and the objects will fall at the same speed regardless of their weight, a conclusion that is borne out by experiment.

In his 1991 survey of developments in the modern accumulation of knowledge such as this Charles Van Doren[72] considers that the Copernican Revolution really is the Galilean Cartesian (René Descartes) or simply the Galilean revolution on account of the courage and depth of change brought about by the work of Galileo.

26.2.11 Isaac Newton

Sir Isaac Newton, the discoverer of universal gravitation and one of the most influential scientists in history

Main article: Newton's Rules for Science

Both Bacon and Descartes wanted to provide a firm foundation for scientific thought that avoided the deceptions of the mind and senses. Bacon envisaged that foundation as essentially empirical, whereas Descartes provides a metaphysical foundation for knowledge. If there were any doubts about the direction in which scientific method would develop, they were set to rest by the success of Isaac Newton. Implicitly rejecting Descartes' emphasis on rationalism in favor of Bacon's empirical approach, he outlines his four "rules of reasoning" in the *Principia*,

1. We are to admit no more causes of natural things than such as are both true and sufficient to explain their appearances.
2. Therefore to the same natural effects we must, as far as possible, assign the same causes.
3. The qualities of bodies, which admit neither intension nor remission of degrees, and which are found to belong to all bodies

within the reach of our experiments, are to be esteemed the universal qualities of all bodies whatsoever.

4. In experimental philosophy we are to look upon propositions collected by general induction from phænomena as accurately or very nearly true, notwithstanding any contrary hypotheses that may be imagined, until such time as other phænomena occur, by which they may either be made more accurate, or liable to exceptions.[73]

But Newton also left an admonition about a theory of everything:

To explain all nature is too difficult a task for any one man or even for any one age. 'Tis much better to do a little with certainty, and leave the rest for others that come after you, than to explain all things.[74]

Newton's work became a model that other sciences sought to emulate, and his inductive approach formed the basis for much of natural philosophy through the 18th and early 19th centuries. Some methods of reasoning were later systematized by Mill's Methods (or Mill's canon), which are five explicit statements of what can be discarded and what can be kept while building a hypothesis. George Boole and William Stanley Jevons also wrote on the principles of reasoning.

26.3 Integrating deductive and inductive method

Attempts to systematize a scientific method were confronted in the mid-18th century by the problem of induction, a positivist logic formulation which, in short, asserts that nothing can be known with certainty except what is actually observed. David Hume took empiricism to the skeptical extreme; among his positions was that there is no logical necessity that the future should resemble the past, thus we are unable to justify inductive reasoning itself by appealing to its past success. Hume's arguments, of course, came on the heels of many, many centuries of excessive speculation upon excessive speculation not grounded in empirical observation and testing. Many of Hume's radically skeptical arguments were argued against, but not resolutely refuted, by Immanuel Kant's *Critique of Pure Reason* in the late 18th century.[75] Hume's arguments continue to hold a strong lingering influence and certainly on the consciousness of the educated classes for the better part of the 19th century when the argument at the time became the focus on whether or not the inductive method was valid.

Hans Christian Ørsted, (Ørsted is the Danish spelling; *Oersted* in other languages) (1777–1851) was heavily influenced by Kant, in particular, Kant's *Metaphysische Anfangsgründe der Naturwissenschaft* (*Metaphysical Foundations of Natural Science*).[76] The following sections on Ørsted encapsulate our current, common view of scientific method. His work appeared in Danish, most accessibly in public lectures, which he translated into German, French, English, and occasionally Latin. But some of his views go beyond Kant:

Ørsted observed the deflection of a compass from a voltaic circuit in 1820

"In order to achieve completeness in our knowledge of nature, we must start from two extremes, from experience and from the intellect itself. ... The former method must conclude with natural laws, which it has abstracted from experience, while the latter must begin with principles, and gradually, as it develops more and more, it becomes ever more detailed. Of course, I speak here about the method as manifested in the process of the human intellect itself, not as found in textbooks, where the laws of nature which have been abstracted from the consequent experiences are placed first because they are required to explain the experiences. When the empiricist in his regression towards general laws of nature meets the metaphysician in his progression, science will reach its perfection."[77]

Ørsted's "First Introduction to General Physics" (1811) exemplified the steps of observation,[78] hypothesis,[79] deduction[80] and experiment. In 1805, based on his researches on electromagnetism Ørsted came to believe that electricity is propagated by undulatory action (i.e., fluctuation). By 1820, he felt confident enough in his beliefs that he resolved to demonstrate them in a public lecture, and in fact observed a small magnetic effect from a galvanic circuit (i.e., voltaic circuit), *without rehearsal*;[81][82]

In 1831 John Herschel (1792–1871) published *A Preliminary Discourse on the study of Natural Philosophy*, setting out the principles of science. Measuring and comparing observations was to be used to find generalisations in "empirical laws", which described regularities in phenomena, then natural philosophers were to work towards the higher aim of finding a universal "law of nature" which explained the causes and effects producing such regularities. An explanatory hypothesis was to be found by evaluating true causes (Newton's "vera causae") derived from experience, for example evidence of past climate change could be due to changes in the shape of continents, or to changes in Earth's orbit. Possible causes could be inferred by analogy to known causes of similar phenomena.[83][84] It was essential to evaluate the importance of a hypothesis; "our next step in the verification of an induction must, therefore, consist in extending its application to cases not originally contemplated; in studiously varying the circumstances under which our causes act, with a view to ascertain whether their effect is general; and in pushing the application of our laws to extreme cases."[85]

William Whewell (1794–1866) regarded his *History of the Inductive Sciences, from the Earliest to the Present Time* (1837) to be an introduction to the *Philosophy of the Inductive Sciences* (1840) which analyzes the method exemplified in the formation of ideas. Whewell attempts to follow Bacon's plan for discovery of an effectual art of discovery. He named the hypothetico-deductive method (which *Encyclopædia Britannica* credits to Newton[86]); Whewell also coined the term *scientist*. Whewell examines ideas and attempts to construct science by uniting ideas to facts. He analyses induction into three steps:

1. the selection of the fundamental idea, such as space, number, cause, or likeness
2. a more special modification of those ideas, such as a circle, a uniform force, etc.
3. the determination of magnitudes

Upon these follow special techniques applicable for quantity, such as the method of least squares, curves, means, and special methods depending on resemblance (such as pattern matching, the method of gradation, and the method of natural classification (such as cladistics). But no art of discovery, such as Bacon anticipated, follows, for "invention, sagacity, genius" are needed at every step.[87] Whewell's sophisticated concept of science had similarities to that shown by Herschel, and he considered that a good hypothesis should connect fields that had previously been thought unrelated, a process he called consilience. However, where Herschel held that the origin of new biological species would be found in a natural rather than a miraculous process, Whewell opposed this and considered that no natural cause had been shown for adaptation so an unknown divine cause was appropriate.[83]

John Stuart Mill (1806–1873) was stimulated to publish *A System of Logic* (1843) upon reading Whewell's *History of the Inductive Sciences*. Mill may be regarded as the final exponent of the empirical school of philosophy begun by John Locke, whose fundamental characteristic is the duty incumbent upon all thinkers to investigate for themselves rather than to accept the authority of others. Knowledge must be based on experience.[88]

In the mid-19th century Claude Bernard was also influential, especially in bringing the scientific method to medicine. In his discourse on scientific method, *An Introduction to the Study of Experimental Medicine* (1865), he described what makes a scientific theory good and what makes a scientist a true discoverer. Unlike many scientific writers of his time, Bernard wrote about his own experiments and thoughts, and used the first person.[89]

William Stanley Jevons' *The Principles of Science: a treatise on logic and scientific method* (1873, 1877) Chapter XII "The Inductive or Inverse Method", Summary of the Theory of Inductive Inference, states "Thus there are but three steps in the process of induction :-

1. Framing some hypothesis as to the character of the general law.
2. Deducing some consequences of that law.
3. Observing whether the consequences agree with the particular tasks under consideration."

Jevons then frames those steps in terms of probability, which he then applied to economic laws. Ernest Nagel notes that Jevons and Whewell were not the first writers to argue for the centrality of the hypothetico-deductive method in the logic of science.[90]

26.3.1 Charles Sanders Peirce

In the late 19th century, Charles Sanders Peirce proposed a schema that would turn out to have considerable influence in the further development of scientific method generally. Peirce's work quickly accelerated the progress on several fronts. Firstly, speaking in broader context in "How to Make Our Ideas Clear" (1878),[91] Peirce outlined an objectively verifiable method to test the truth of putative knowledge on a way that goes beyond mere foundational alternatives, focusing upon both Deduction and Induction. He thus placed induction and deduction in a complementary rather than competitive context (the latter of which had been the primary trend at least since David Hume a century before). Secondly, and of more direct importance to scientific method, Peirce put forth the basic schema for hypothesis-testing that continues to prevail today. Extracting the theory of inquiry from its raw materials in classical logic, he refined it in parallel with the early development of symbolic logic to address the then-current problems in scientific reasoning. Peirce examined and articulated the three fundamental modes of reasoning that play a role in scientific inquiry today, the processes that are currently known as abductive, deductive, and inductive inference. Thirdly, he played a major role in the progress of symbolic logic itself – indeed this was his primary specialty.

Charles S. Peirce was also a pioneer in statistics. Peirce held that science achieves statistical probabilities, not certainties, and that chance, a veering from law, is very real. He assigned probability to an argument's conclusion rather than to a proposition, event, etc., as such. Most of his statistical writings promote the frequency interpretation of probability (objective ratios of cases), and many of his writings express skepticism about (and criticize the use of) probability when such models are not based on objective randomization.[92] Though Peirce was largely a frequentist, his possible world semantics introduced the "propensity" theory of probability. Peirce (sometimes with Jastrow) investigated the probability judgments of experimental subjects, pioneering decision analysis.

Peirce was one of the founders of statistics. He formulated modern statistics in "Illustrations of the Logic of Science" (1877–1878) and "A Theory of Probable Inference" (1883). With a repeated measures design, he introduced blinded, controlled randomized experiments (before Fisher). He invented an optimal design for experiments on gravity, in which he "corrected the means". He used logistic regression, correlation, and smoothing, and improved the treatment of outliers. He introduced terms "confidence" and "likelihood" (before Neyman and Fisher). (See the historical books of Stephen Stigler.) Many of Peirce's ideas were later popularized and developed by Ronald A. Fisher, Jerzy Neyman, Frank P. Ramsey, Bruno de Finetti, and Karl Popper.

26.3.2 Popper and Kuhn

Karl Popper (1902–1994) is generally credited with providing major improvements in the understanding of the scientific method in the mid-to-late 20th century. In 1934 Popper published *The Logic of Scientific Discovery*, which repudiated the by then traditional observationalist-inductivist account of the scientific method. He advocated empirical falsifiability as the criterion for distinguishing scientific work from non-science. According to Popper, scientific theory should make predictions (preferably predictions not made by a competing theory) which can be tested and the theory rejected if these predictions are shown not to be correct. Following Peirce and others, he argued that science would best progress using deductive reasoning as its primary emphasis, known as critical rationalism. His astute formulations of logical procedure helped to rein in the excessive use of inductive speculation upon inductive speculation, and also helped to strengthen the conceptual foundations for today's peer review procedures.

Critics of Popper, chiefly Thomas Kuhn, Paul Feyerabend and Imre Lakatos, rejected the idea that there exists a *single* method that applies to all science and could account for its progress. In 1962 Kuhn published the influential book *The Structure of Scientific Revolutions* which suggested that scientists worked within a series of paradigms, and argued there was little evidence of scientists actually following a falsificationist methodology. Kuhn quoted Max Planck who had said in his autobiography, "a new scientific truth does not triumph by convincing its opponents and making them see the light, but rather because its opponents eventually die, and a new generation grows up that is familiar with it."[93]

These debates clearly show that there is no universal agreement as to what constitutes *the* "scientific method".[94] There remain, nonetheless, certain core principles that are the foundation of scientific inquiry today.[95]

26.4 Mention of the topic

In *Quod Nihil Scitur* (1581), Francisco Sanches refers to another book title, *De modo sciendi* (on the method of knowing). This work appeared in Spanish as *Método universal de las ciencias*.[59]

In 1833 Robert and William Chambers published their 'Chambers's information for the people'. Under the rubric 'Logic' we find a description of investigation that is familiar as scientific method,

> Investigation, or the art of inquiring into the nature of causes and their operation, is a leading characteristic of reason [...] Investigation implies three things – Observation, Hypothesis, and Experiment [...] The first step in the process, it will be perceived, is to observe...[96]

In 1885, the words "Scientific method" appear together with

a description of the method in Francis Ellingwood Abbot's 'Scientific Theism'.

> Now all the established truths which are formulated in the multifarious propositions of science have been won by the use of Scientific Method. This method consists in essentially three distinct steps (1) observation and experiment, (2) hypothesis, (3) verification by fresh observation and experiment.[97]

The Eleventh Edition of *Encyclopædia Britannica* did not include an article on scientific method; the Thirteenth Edition listed scientific management, but not method. By the Fifteenth Edition, a 1-inch article in the *Micropædia* of Britannica was part of the 1975 printing, while a fuller treatment (extending across multiple articles, and accessible mostly via the index volumes of Britannica) was available in later printings.[98]

26.5 Current issues

In the past few centuries, some statistical methods have been developed, for reasoning in the face of uncertainty, as an outgrowth of methods for eliminating error. This was an echo of the program of Francis Bacon's *Novum Organum* of 1620. Bayesian inference acknowledges one's ability to alter one's beliefs in the face of evidence. This has been called belief revision, or defeasible reasoning: the models in play during the phases of scientific method can be reviewed, revisited and revised, in the light of further evidence. This arose from the work of Frank P. Ramsey[99] (1903–1930), of John Maynard Keynes[100] (1883–1946), and earlier, of William Stanley Jevons[101][102] (1835–1882) in economics.

Later in the 20th century, methodological naturalism was emphasized by Robert T. Pennock as central to scientific method, partly in response to rise of creation science.

26.6 Science and pseudoscience

The question of how science operates and therefore how to distinguish genuine science from pseudoscience has importance well beyond scientific circles or the academic community. In the judicial system and in public policy controversies, for example, a study's deviation from *accepted scientific practice* is grounds for rejecting it as junk science or pseudoscience. However, the high public perception of science means that pseudoscience is widespread. An advertisement in which an actor wears a white coat and product ingredients are given Greek or Latin sounding names is intended to give the impression of scientific endorsement. Richard Feynman has likened pseudoscience to cargo cults in which many of the external forms are followed, but the underlying basis is missing. Fringe or alternative theories often present themselves with a pseudoscientific appearance.

26.7 See also

- Timeline of the history of scientific method

26.8 Notes and references

[1] Peter Achinstein, "General Introduction" (pp. 1–5) to *Science Rules: A Historical Introduction to Scientific Methods*. Johns Hopkins University Press. 2004. ISBN 0-8018-7943-4

[2] http://www.britannica.com/eb/article?tocId=9032043&query=Edwin%20Smith%20papyrus&ct=

[3] Lloyd, G. E. R. "The development of empirical research", in his *Magic, Reason and Experience: Studies in the Origin and Development of Greek Science*.

[4] A. Aaboe (2 May 1974). "Scientific Astronomy in Antiquity". *Philosophical Transactions of the Royal Society*. **276** (1257): 21–42. Bibcode:1974RSPTA.276...21A. JSTOR 74272. doi:10.1098/rsta.1974.0007.

[5] "The cradle of mathematics is in Egypt." – Aristotle, *Metaphysics*, as cited on page 1 of Olaf Pedersen (1993) *Early physics and astronomy: a historical introduction* Cambridge: Cambridge University Press, revised edition

[6] "There each man is a leech skilled beyond all human kind; yea, for they are of the race of Paeeon." – Homer, *Odyssey* book IV, acknowledges the skill of the ancient Egyptians in medicine.

[7] Pingree, David (December 1992). "Hellenophilia versus the History of Science". *Isis*. University of Chicago Press. **83** (4): 554–563. Bibcode:1992Isis...83..554P. JSTOR 234257. doi:10.1086/356288.

[8] Rochberg, Francesca (October–December 1999). "Empiricism in Babylonian Omen Texts and the Classification of Mesopotamian Divination as Science". *Journal of the American Oriental Society*. American Oriental Society. **119** (4): 559–569. JSTOR 604834. doi:10.2307/604834.

[9] Yves Gingras, Peter Keating, and Camille Limoges, *Du scribe au savant: Les porteurs du savoir de l'antiquité à la révolution industrielle*, Presses universitaires de France. 1998.

26.8. NOTES AND REFERENCES

[10] Harrison, Peter (2015). *The Territories of Science and Religion*. University of Chicago Press. p. 24. ISBN 9780226184487.

[11] Oliver Leaman, *Key Concepts in Eastern Philosophy*. Routledge, 1999, page 269.

[12] Kamal, M.M. (1998). "The Epistemology of the Carvaka Philosophy", *Journal of Indian and Buddhist Studies*, 46(2): pp.13–16

[13] Barnes, *Hellenistic Philosophy and Science*, pp. 383–384

[14] Gauch, Hugh G. (2003). *Scientific Method in Practice*. Cambridge University Press. p. 45. ISBN 978-0-521-01708-4. Retrieved 10 February 2015.

[15] Hammond, p. 64, "Andronicus Rhodus"

[16] "In the days when the Arabs inherited the culture of ancient Greece, Greek thought was chiefly interested in science, Athens was replaced by Alexandria, and Hellenism had an entirely "modern "outlook. This was an attitude with which Alexandria and its scholars were directly connected, but it was by no means confined to Alexandria. It was a logical outcome of the influence of Aristotle who before all else was a patient observer of nature, and was, in fact, the founder of modern science." Ch.1, Introduction —De Lacy O'Leary (1949), *How Greek Science Passed to the Arabs*, London: Routledge & Kegan Paul Ltd., ISBN 0-7100-1903-3

[17] See Nominalism#The problem of universals for several approaches to this goal.

[18] Aristotle (fl. 4th c. BCE, d. 322 BCE), *History of Animals*, including vivisection of the tortoise and chameleon. His theory of spontaneous generation was not experimentally disproved until Francesco Redi (1668).

[19] Armand Leroi, Aristotle's Lagoon - Lesvos island - Greece name of Pyrrha lagoon, now called Kalloni, minute 5:06/57:55. His spontaneous generation disproved, minute 50:00/57:55. His lack of experiment, minute 51:00/57:55

[20] Armand Leroi, following in Aristotle's footsteps, projects that Aristotle interviewed the fishermen of Lesbos to learn empirical details about the animals. (Leroi, *Aristotle's Lagoon*)

[21] See: Ancient Greek medicine#Aristotle, which cites Annas, Julia *Classical Greek Philosophy*. In Boardman, John; Griffin, Jasper; Murray, Oswyn (ed.) *The Oxford History of the Classical World*. Oxford University Press: New York, 1986. ISBN 0-19-872112-9

[22] Asmis 1984

[23] Madden, Edward H. (Apr., 1957) "Aristotle's Treatment of Probability and Signs" *Philosophy of Science* 24(2), pp. 167-172 via JSTOR discusses Aristotle's enthymeme (70a, 5ff.) in *Prior Analytics*

[24] David C. Lindberg (1980). *Science in the Middle Ages*, University of Chicago Press, p. 21, ISBN 0-226-48233-2

[25] Holmyard, E. J. (1931), *Makers of Chemistry*, Oxford: Clarendon Press, p. 56

[26] Plinio Prioreschi, "Al-Kindi, A Precursor Of The Scientific Revolution", *Journal of the International Society for the History of Islamic Medicine*, 2002 (2): 17–19 [17].

[27] Alhazen, translated into English from German by M. Schwarz, from "Abhandlung über das Licht", J. Baarmann (ed. 1882) *Zeitschrift der Deutschen Morgenländischen Gesellschaft* Vol **36** as referenced on p.136 by Shmuel Sambursky (1974) *Physical thought from the Presocratics to the Quantum Physicists* ISBN 0-87663-712-8

[28] D. C. Lindberg, *Theories of Vision from al-Kindi to Kepler*, (Chicago, Univ. of Chicago Pr., 1976), pp. 60–7.

[29] Nader El-Bizri, "A Philosophical Perspective on Alhazen's Optics," *Arabic Sciences and Philosophy*, Vol. 15, Issue 2 (2005), pp. 189–218 (Cambridge University Press)

[30] Nader El-Bizri, "Ibn al-Haytham," in *Medieval Science, Technology, and Medicine: An Encyclopedia*, eds. Thomas F. Glick, Steven J. Livesey, and Faith Wallis (New York – London: Routledge, 2005), pp. 237–240.

[31] Alhazen (Ibn Al-Haytham) *Critique of Ptolemy*, translated by S. Pines, *Actes X Congrès internationale d'histoire des sciences*, Vol I Ithaca 1962, as referenced on p.139 of Shmuel Sambursky (ed. 1974) *Physical Thought from the Presocratics to the Quantum Physicists* ISBN 0-87663-712-8

[32] p.136, as quoted by Shmuel Sambursky (1974) *Physical thought from the Presocratics to the Quantum Physicists* ISBN 0-87663-712-8

[33] Plott, C. (2000), *Global History of Philosophy: The Period of Scholasticism*, Motilal Banarsidass, p. 462, ISBN 81-208-0551-8

[34] Alhazen; Smith, A. Mark (2001). *Alhacen's Theory of Visual Perception: A Critical Edition, with English Translation and Commentary of the First Three Books of Alhacen's De Aspectibus, the Medieval Latin Version of Ibn al-Haytham's Kitab al-Manazir*, DIANE Publishing, pp. 372 & 408, ISBN 0-87169-914-1

[35] Rashed, Roshdi (2007), "The Celestial Kinematics of Ibn al-Haytham", *Arabic Sciences and Philosophy*, Cambridge University Press, **17**: 7–55 [19]. doi:10.1017/S0957423907000355:

> "In reforming optics he, as it were, adopted "positivism" (before the term was invented): we do not go beyond experience, and we cannot be content to use pure concepts in investigating natural phenomena. Understanding of these cannot be acquired without mathematics. Thus, once he has assumed light is a material substance, Ibn al-Haytham does not discuss its nature further, but confines himself to considering its propagation and diffusion. In his optics "the

smallest parts of light", as he calls them, retain only properties that can be treated by geometry and verified by experiment; they lack all sensible qualities except energy."

[36] Sardar, Ziauddin (1998), "Science in Islamic philosophy", *Islamic Philosophy*, Routledge Encyclopedia of Philosophy, retrieved 2008-02-03

[37] Mariam Rozhanskaya and I. S. Levinova (1996), "Statics", p. 642, in (Morelon & Rashed 1996, pp. 614–642):

> "Using a whole body of mathematical methods (not only those inherited from the antique theory of ratios and infinitesimal techniques, but also the methods of the contemporary algebra and fine calculation techniques), Arabic scientists raised statics to a new, higher level. The classical results of Archimedes in the theory of the centre of gravity were generalized and applied to three-dimensional bodies, the theory of ponderable lever was founded and the 'science of gravity' was created and later further developed in medieval Europe. The phenomena of statics were studied by using the dynamic approach so that two trends – statics and dynamics – turned out to be inter-related within a single science, mechanics. The combination of the dynamic approach with Archimedean hydrostatics gave birth to a direction in science which may be called medieval hydrodynamics. [...] Numerous fine experimental methods were developed for determining the specific weight, which were based, in particular, on the theory of balances and weighing. The classical works of al-Biruni and al-Khazini can by right be considered as the beginning of the application of experimental methods in medieval science."

[38] Glick, Thomas F.; Livesey, Steven John; Wallis, Faith (2005), *Medieval Science, Technology, and Medicine: An Encyclopedia*, Routledge, pp. 89–90, ISBN 0-415-96930-1

[39] McGinnis, Jon (July 2003), "Scientific Methodologies in Medieval Islam", *Journal of the History of Philosophy*, **41** (3): 307–327, doi:10.1353/hph.2003.0033

[40] Lenn Evan Goodman (2003), *Islamic Humanism*, p. 155, Oxford University Press, ISBN 0-19-513580-6.

[41] Lenn Evan Goodman (1992), *Avicenna*, p. 33, Routledge, ISBN 0-415-01929-X.

[42] James Franklin (2001), *The Science of Conjecture: Evidence and Probability Before Pascal*, pp. 177–8, Johns Hopkins University Press, ISBN 0-8018-6569-7.

[43] Dallal, Ahmad (2001–2002), *The Interplay of Science and Theology in the Fourteenth-century Kalam*, From Medieval to Modern in the Islamic World, Sawyer Seminar at the University of Chicago, retrieved 2008-02-02

[44] A. C. Crombie, *Robert Grosseteste and the Origins of Experimental Science, 1100–1700*, (Oxford: Clarendon Press, 1971), pp. 52–60.

[45] Jeremiah Hackett, "Roger Bacon: His Life, Career, and Works," in Hackett, *Roger Bacon and the Sciences*, pp. 13–17.

[46] "Roger Bacon", *Encyclopædia Britannica*, Eleventh Edition

[47] "Roger Bacon", *Encyclopædia Britannica Eleventh Edition* (R[obert] Ad[amson]; X.) This article incorporates text from a publication now in the public domain: Chisholm, Hugh, ed. (1911), "article name needed", *Encyclopædia Britannica* (11th ed.), Cambridge University Press.

[48] Instead of reading Aristotle directly from Greek texts, students of these texts would rely on summaries and translations of Aristotle's work, coupled with commentary by the translators, according to Elaine Limbrick, who cites Michel Reulos, "L'Enseignement d'Aristote dans les collèges au XVIe siècle" in *Platon et Aristote à la Renaissance* ed. J.-C. Margolin (Paris: Vrin, 1976) pp147-154:Sanches, Limbrick & Thomson 1988, p. 26

[49] Edward Grant, *The Foundations of Modern Science in the Middle Ages: Their Religious, Institutional, and Intellectual Contexts*, (Cambridge: Cambridge Univ. Pr., 1996, pp. 164–7.

[50] "Even Aristotle would have laughed at the stupidity of his commentators." — Vives 1531 attacks obscurity in Aristotle's works, as cited by Sanches, Limbrick & Thomson 1988, pp. 28–9

[51] Galenus, Claudius (1519) *Galenus methodus medendi, vel de morbis curandis. T. Linacro ... interprete, libri quatuordecim* Lutetiae, as cited by Sanches, Limbrick & Thomson 1988, p. 301

[52] Richard J. Durling (1961) "A Chronological Census of Renaissance Editions and Translations of Galen", in *Journal of the Warburg and Courtald Institutes* **24** pp.242-3 as cited on p. 300 of Sanches, Limbrick & Thomson 1988

[53] Niccolò Leoniceno (1509), *De Plinii et aliorum erroribus liber* apud Ferrara, as cited by Sanches, Limbrick & Thomson 1988, p. 13

[54] Fuchs's book on the methods of Galen and Hippocrates became a standard medical text of 809 pages: Leonhart Fuchs (1560) *Institutionum medicinae, sive methodi ad Hippocratis, Galeni, aliorumque veterum scripta recte intelligenda mire utiles libri quinque ... Editio secunda*. Lugduni. As cited in Sanches, Limbrick & Thomson 1988, pp. 61 and 301.

[55] 'I have sometimes seen a verbose quibbler attempting to persuade some ignorant person that white was black; to which the latter replied, "I do not understand your reasoning, since I have not studied as much as you have; yet I honestly believe that white differs from black. But pray go on refuting me for just as long as you like." ' — Sanches, Limbrick & Thomson 1988, p. 276

[56] Susanne Bobzien, "Aristotle's modal logic" *Stanford Encyclopedia of Philosophy*

[57] Sanches, Limbrick & Thomson 1988, p. 278.

[58] "Since, as he had shown, nothing can be known, Sanches put forward a procedure, not to gain knowledge but to deal constructively with human experience. This procedure, for which he introduced the term (for the first time) scientific method, "Metodo universal de las ciencias," consists in patient, careful empirical research and cautious judgment and evaluation of the data we observe. This would not lead, as his contemporary Francis Bacon thought, to a key to knowledge of the world. But it would allow us to obtain the best information available. ...In advancing this limited or constructive view of science, Sanches was the first Renaissance sceptic to conceive of science in its modern form, as the fruitful activity about the study of nature that remained after one had given up the search for absolutely certain knowledge of the nature of things. Popkin 2003, p. 41"

[59] Sanches, Limbrick & Thomson 1988, p. 292 lists *De modo sciendi* under the Unpublished, Lost, or Projected Works [of Francisco Sanches]. This work appeared in Spanish as *Metodo universal de las ciencias*, as cited by Guy Patin (1701) *Naudeana et Patiniana* pp.72-3

[60] Sanches, Limbrick & Thomson 1988, p. 290

[61] "Francis Bacon", *Encyclopædia Britannica Eleventh Edition* (R[obert]. Ad[amson]; J[ohn] M[alcolm] M[itchell]) This article incorporates text from a publication now in the public domain: Chisholm, Hugh, ed. (1911). "article name needed". *Encyclopædia Britannica* (11th ed.). Cambridge University Press.

[62] In this sense, it has been seen as a precursor to falsificationism of Charles Sanders Peirce and Karl Popper. However, Bacon believed his method would produce certain knowledge, similar to Peirce's view of scientific methods as ultimately approaching the truth; with the goal of attaining knowledge of the truth, Bacon's philosophy is less sceptical than Popper's philosophy.

- Bacon precedes Peirce in another sense – his reliance on doubt: "If a man will begin with certainties, he shall end in doubts; but if he will be content to begin with doubts he shall end in certainties." – Francis Bacon, *The Advancement of Learning* (1605), Book I, v, 8.

[63] B. Gower, *Scientific Method, An Historical and Philosophical Introduction*, (Routledge, 1997), pp. 48–2.

[64] B. Russell, *History of Western Philosophy*, (Routledge, 2000), pp. 529–3.

[65] Descartes compares his work to that of an architect: "there is less perfection in works composed of several separate pieces and by difference masters, than those in which only one person has worked.", *Discourse on Method and The Meditations*, (Penguin, 1968), pp. 35. (see too his letter to Mersenne (28. January 1641 [AT III, 297–8]).

[66] This is the first of four rules Descartes resolved "never once to fail to observe", *Discourse on Method and The Meditations*, (Penguin, 1968), pp. 41.

[67] René Descartes, *Meditations on First Philosophy: With Selections from the Objections and Replies*, (Cambridge: Cambridge Univ. Pr., 2nd ed., 1996), pp. 63–107.

[68] René Descartes, *The Philosophical Writings of Descartes: Principles of Philosophy, Preface to French Edition*, translated by J. Cottingham, R. Stoothoff, D. Murdoch (Cambridge: Cambridge Univ. Pr., 1985), vol. 1, pp. 179–189.

[69] René Descartes, *Oeuvres De Descartes*, edited by Charles Adam and Paul Tannery (Paris: Librairie Philosophique J. Vrin, 1983), vol. 2, pp. 380.

[70] Koyré, Alexandre: Introduction a la Lecture de Platon, suivi de Entretiens sur Descartes, Gallimard, p. 203

[71] For more about the role of mathematics in science around the time of Galileo see R. Feldhay, *The Cambridge Companion to Galileo: The use and abuse of mathematical entities*, (Cambridge: Cambridge Univ. Pr., 1998), pp. 80–133.

[72] Van Doren, Charles. A History of Knowledge. (New York, Ballantine, 1991)

[73] Rule IV, Philosophiae Naturalis Principia Mathematica#Rules of Reasoning in Philosophy:

- Newton states "This rule we must follow that the argument of induction may not be evaded by hypotheses", in the Motte translation (p. 400 in the Cajori revision, volume 2)
- Newton's comment is also rendered as "This rule should be followed so that arguments based on induction may not be nullified by hypotheses" on p. 796 of Newton, Isaac (1999), *Philosophiae Naturalis Principia Mathematica*, University of California Press, ISBN 0-520-08817-4, Third edition: 1687, 1713, 1726. From I. Bernard Cohen and Anne Whitman's 1999 translation, 974 pages.

[74] Statement from unpublished notes for the Preface to *Opticks* (1704) quoted in *Never at Rest: A Biography of Isaac Newton* (1983) by Richard S. Westfall, p. 643

[75] "Hume awakened Kant from his dogmatic slumbers"

[76] Karen Jelved, Andrew D. Jackson, and Ole Knudsen, (1997) translators for *Selected Scientific Works of Hans Christian Ørsted*, ISBN 0-691-04334-5, p. x. The succeeding Ørsted references are contained in this book.

[77] "Fundamentals of the Metaphysics of Nature Partly According to a New Plan", a special reprint of Hans Christian Ørsted (1799), *Philosophisk Repertorium*, printed by Boas Brünnich, Copenhagen, in Danish. Kirstine Meyer's 1920 edition of Ørsted's works, vol.I, pp. 33–78. English translation by Karen Jelved, Andrew D. Jackson, and Ole Knudsen, (1997) ISBN 0-691-04334-5 pp. 46–47.

[78] "The foundation of general physics ... is experience. These ... everyday experiences we do not discover without deliberately directing our attention to them. Collecting information about these is *observation*." – Hans Christian Ørsted("First Introduction to General Physics" ¶13, part of a series of public lectures at the University of Copenhagen. Copenhagen 1811, in Danish, printed by Johan Frederik Schulz. In Kirstine Meyer's 1920 edition of Ørsted's works, vol.**III** pp. 151–190.) "First Introduction to Physics: the Spirit, Meaning, and Goal of Natural Science". Reprinted in German in 1822, Schweigger's *Journal für Chemie und Physik* **36**, pp. 458–488. ISBN 0-691-04334-5 p. 292

[79] "When it is not clear under which law of nature an effect or class of effect belongs, we try to fill this gap by means of a guess. Such guesses have been given the name *conjectures* or *hypotheses*." – Hans Christian Ørsted(1811) "First Introduction to General Physics" ¶18. *Selected Scientific Works of Hans Christian Ørsted*, ISBN 0-691-04334-5 p. 297

[80] "The student of nature ... regards as his property the experiences which the mathematician can only borrow. This is why he deduces theorems directly from the nature of an effect while the mathematician only arrives at them circuitously." – Hans Christian Ørsted(1811) "First Introduction to General Physics" ¶17. *Selected Scientific Works of Hans Christian Ørsted*, ISBN 0-691-04334-5 p. 297

[81] Hans Christian Ørsted(1820) ISBN 0-691-04334-5 preface, p.xvii

[82] Hans Christian Ørsted(1820) ISBN 0-691-04334-5, 1820 and other public experiments, pp. 421–445

[83] Young, David (2007). *The discovery of evolution*. Cambridge ; New York: Cambridge University Press. pp. 105–106, 113. ISBN 0-521-68746-2.

[84] Herschel, John Frederick William (1840), *A Preliminary Discourse on the study of Natural Philosophy*, Dionysius Lardner's Cabinet Cyclopædia, London: Longman, Rees, Orme, Brown & Green; John Taylor, retrieved 5 March 2013

[85] Armstrong, Patrick (1992), *Darwin's desolate islands: A naturalist in the Falklands, 1833 and 1834*, Chippenham: Picton Publishing, retrieved 5 March 2013

[86] "Science, Philosophy of", *Encyclopædia Britannica* Fifteenth Ed. (1979) ISBN 0-85229-297-X pp. 378–9

[87] "William Whewell", *Encyclopædia Britannica Eleventh Edition* This article incorporates text from a publication now in the public domain: Chisholm, Hugh, ed. (1911). "article name needed". *Encyclopædia Britannica* (11th ed.). Cambridge University Press.

[88] "John Stuart Mill", *Encyclopædia Britannica Eleventh Edition* This article incorporates text from a publication now in the public domain: Chisholm, Hugh, ed. (1911). "article name needed". *Encyclopædia Britannica* (11th ed.). Cambridge University Press.

[89] All page references refer to the Dover edition of 1957.

- Bernard, Claude. *An Introduction to the Study of Experimental Medicine*, 1865. First English translation by Henry Copley Greene, published by Macmillan & Co., Ltd., 1927; reprinted in 1949. The Dover Edition of 1957 is a reprint of the original translation with a new Foreword by I. Bernard Cohen of Harvard University.

[90] William Stanley Jevons (1873, 1877) *The Principles of Science: a treatise on logic and scientific method* Dover edition p.li with a new preface by Ernest Nagel (1958)

[91] Charles S. Peirce How to Make Our Ideas Clear, Popular Science Monthly 12 (January 1878), pp. 286–302

[92] Peirce condemned the use of "certain likelihoods" even more strongly than he criticized Bayesian methods. Indeed Peirce used Bayesian inference in criticizing parapsychology.

[93] Max Planck (1949) *Scientific Autobiography and Other Papers*, pp. 33–34 ISBN 0-8371-0194-8, as cited by Kuhn, Thomas (1997), *The Structure of Scientific Revolutions* (3rd ed.), University of Chicago Press, p. 151

[94] Jerry Wellington, *Secondary Science: Contemporary Issues and Practical Approaches* (Routlege, 1994, p. 41)

[95] Gauch, Hugh G. (2003). *Scientific Method in Practice* (Reprint ed.). Cambridge University Press. p. 3. ISBN 9780521017084. The scientific method 'is often misrepresented as a fixed sequence of steps,' rather than being seen for what it truly is, 'a highly variable and creative process' (AAAS 2000:18). The claim here is that science has general principles that must be mastered to increase productivity and enhance perspective, not that these principles provide a simple and automated sequence of steps to follow.

[96] William Chambers, Robert Chambers, *Chambers's information for the people: a popular encyclopaedia*, Volume 1, pp. 363–4

[97] Francis Ellingwood Abbot, *Scientific Theism* p. 60

[98] *Encyclopædia Britannica*, Fifteenth Edition ISBN 0-85229-493-X Index L-Z "scientific method" pp. 588–9

[99] A review and defense of Frank P.Ramsey's formulation can be found in Alan Hájek, "Scotching Dutch Books?" *Philosophical Perspectives* **19**

[100] John Maynard Keynes(1921) *Treatise on Probability*

[101] William Stanley Jevons(1888) *The Theory of Political Economy*

[102] William Stanley Jevons(1874), *The Principles of Science*, p. 267, reprinted by Dover in 1958

26.9 Sources

- Asmis, Elizabeth (January 1984), *Epicurus' Scientific method*, **42**, Cornell University Press, p. 386, ISBN 978-0-8014-6682-3, JSTOR 10.7591/j.cttq45z9

- Debus, Allen G. (1978), *Man and Nature in the Renaissance*, Cambridge: Cambridge University Press, ISBN 0-521-29328-6

- Popkin, Richard H. (1979), *The History of Scepticism from Erasmus to Spinoza*, University of California Press, ISBN 0-520-03876-2

- Popkin, Richard H. (2003), *The History of Scepticism from Savonarola to Bayle*, Oxford University Press, ISBN 0-19-510768-3. Third enlarged edition.

- Sanches, Francisco (1636), *Opera medica. His iuncti sunt tratus quidam philosophici non insubtiles*, Toulosae tectosagum as cited by Sanches, Limbrick & Thomson 1988

- Sanches, Francisco (1649), *Tractatus philosophici. Quod Nihil Scitur. De divinatione per somnum, ad Aristotlem. In lib. Aristoteles Physionomicon commentarius. De longitudine et brevitate vitae.*, Roterodami: ex officina Arnoldi Leers as cited by Sanches, Limbrick & Thomson 1988

- Sanches, Francisco; Limbrick, Elaine. Introduction, Notes, and Bibliography; Thomson, Douglas F.S. Latin text established, annotated, and translated. (1988), *That Nothing is Known*, Cambridge: Cambridge University Press, ISBN 0-521-35077-8 Critical edition of Sanches' *Quod Nihil Scitur* Latin:(1581, 1618, 1649, 1665), Portuguese:(1948, 1955, 1957), Spanish:(1944, 1972), French:(1976, 1984), German:(2007)

- Vives, Ioannes Lodovicus (1531), *De Disciplinis libri XX*, Antwerpiae: exudebat M. Hillenius English translation: *On Discipline*.

    - Part 1: De causis corruptarum artium.
    - Part 2: De tradendis disciplinis
    - Part 3: De artibus

26.10 Text and image sources, contributors, and licenses

26.10.1 Text

- **Science** *Source:* https://en.wikipedia.org/wiki/Science?oldid=801484064 *Contributors:* AxelBoldt, Eloquence, Zundark, The Anome, Stephen Gilbert, Malcolm Farmer, Ed Poor, RK, Andre Engels, Eclecticology, Vignaux, XJaM, Fredbauder, PierreAbbat, Fubar Obfusco, William Avery, SimonP, Anthere, KF, Hephaestos, JDG, ChrisSteinbach, Twilsonb, Stevertigo, Edward, Lir, Infrogmation, Michael Hardy, Fred Bauder, Lexor, Grizzly, BoNoMoJo (old), Tannin, Mic, Ixfd64, Lquilter, Tango, Deljr, GTBacchus, Pagingmrherman, Ahoerstemeier, Arwel Parry, Snoyes, Angela, Den fjättrade ankan~enwiki, JWSchmidt, Glenn, RadRafe, Cyan, Poor Yorick, Big iron, Rotem Dan, Andres, Kaihsu, Evercat, Sethmahoney, Mxn, Hemmer, Ec5618, Charles Matthews, Guaka, Pingchen, Wikiborg, Ed Cormany, Reddi, Terse, Fuzheado, Greenrd, Lord Kenneth, DJ Clayworth, Markhurd, Tpbradbury, Ksolway, Maximus Rex, Morwen, Saltine, Martinphi, SEWilco, Paul-L~enwiki, Omegatron, Buridan, Fvw, Stormie, Wetman, Gakrivas, Secretlondon, Pilaf~enwiki, Flockmeal, Banno, Francs2000, Owen, PuzzletChung, Phil Boswell, Gentgeen, Robbot, Chrism, Fredrik, Gwrede, Zandperl, R3m0t, RedWolf, Goethean, Altenmann, Modulatum, Lowellian, Gkochanowsky, Stewartadcock, Rholton, Rursus, Texture, Davodd, Hadal, UtherSRG, Robinh, HaeB, Guy Peters, SpellBott, Dina, Alan Liefting, Enochlau, Leighxucl, Vaoverland, Ancheta Wis, Matthew Stannard, Centrx, Giftlite, Christopher Parham, Pretzelpaws, Inter, Lee J Haywood, Tom harrison, Lysis~enwiki, Brian Kendig, Fastfission, Aphaia, Hokanomono, Everyking, Curps, Bensaccount, Wikibob, Jorend, Jfdwolff, Avsa, Ezhiki, Mboverload, Prosfilaes, AlistairMcMillan, Solipsist, Costyn, SWAdair, AdamJacobMuller, Tagishsimon, Wmahan, OldakQuill, James Crippen, Gadfium, Andycjp, R. fiend, Quadell, Stephan Leclercq, Antandrus, The Singing Badger, Beland, Estel~enwiki, OverlordQ, PDH, Jossi, Karol Langner, Rdsmith4, APH, Maximaximax, Bodnotbod, Huwr, Pethan, DanielDemaret, Sam Hocevar, Rlcantwell, Popadopolis, Darksun, Iantresman, Neutrality, Joyous!, Ukexpat, Jh51681, Frau Holle, Karl Dickman, Eduardoporcher, Grm wnr, Deglr6328, Gleet, Zondor, Adashiel, Grunt, Bluemask, Mike Rosoft, Brianjd, SimonEast, Reinthal, Juan Ponderas, Haiduc, EugeneZelenko, Discospinster, ElTyrant, Rich Farmbrough, KillerChihuahua, Rhobite, Guanabot, FT2, Schuetzm, Vsmith, HeikoEvermann, Smyth, Dave souza, Cagliost, 1pezguy, Paul August, SpookyMulder, BBB~enwiki, Bender235, ESkog, Melamed, Eric Forste, Brian0918, TFK~enwiki, RJHall, El C, Cap'n Refsmmat, Lycurgus, Mwanner, QuartierLatin1968, Skeppy, Aude, Shanes, Art LaPella, Riyehn, Adambro, Guettarda, Shoujun, Bobo192, Flxmghvgvk, Shenme, Jung dalglish, Maurreen, Ziggurat, Guiltyspark, Greenleaf~enwiki, Jeffreyn, Dzou, VBGFscJUn3, Malcolm rowe, Vanished user 19794758563875, John Fader, MPerel, Pharos, Pearle, Nsaa, Mdd, Ekhalom, HasharBot~enwiki, ADM, Jumbuck, Beyondthislife, Poweroid, Alansohn, JYolkowski, Eleland, Polarscribe, Arthena, Atlant, Paleorthid, Plumbago, JoaoRicardo, Logologist, Riana, AzaToth, Lightdarkness, Eukesh, Malo, Titanium Dragon, Avenue, Caesura, Cortonin, Cugel~enwiki, Velella, SidP, Tycho, Rick Sidwell, Gaussianzz, Knowledge Seeker, Suruena, Garzo, Evil Monkey, Dinoguy2, Omphaloscope, Dmccabe, Harej, Tony Sidaway, Amorymeltzer, Grenavitar, CloudNine, Sciurinæ, Mikeo, W7KyzmJt, Redvers, Kenyon, Mullet, FrancisTyers, The JPS, Simetrical, Mel Etitis, OwenX, Woohookitty, Mindmatrix, N1r4v, Pmberry, Georgia guy, Consequencefree, Swamp Ig, Daniel Case, Brunnock, Before My Ken, Ruud Koot, MONGO, Eleassar777, Tygar, Friarslantern, WikiкIrsc, Ledouche, I64s, Striver, Sengkang, GregorB, Kralizec!, Noetica, Wayward, Joke137, Gimboid13, MarcoTolo, Phlebas, Allen3, LexCorp, GSlicer, DavidParfitt, Raguks, Graham87, Alienus, Magister Mathematicae, BD2412, Qwertyus, Kbdank71, FreplySpang, Malangthon, Island, Icey, Sjö, Drbogdan, Sjakkalle, Rjwilmsi, Mayumashu, Nightscream, Dpark, Jake Wartenberg, Commander, Vary, Bob A. Mll, MarSch, Quiddity, Jiohdi, Xosé, Rschen7754, HolyApocalypse, MZMcBride, Mork the delayer, Tawker, Plotinus~enwiki, Mm35173, Bhadani, Dar-Ape, GregAsche, Jesus Is Love, Cassowary, Tommy Kronkvist, Falphin, Titoxd, Ian Pitchford, RobertG, Airumel, Nihiltres, MethodicEvolution, SouthernNights, Nivix, Chanting Fox, Hottentot, Andy85719, Pathoschild, RexNL, Alexjohnc3, AndriuZ, Agesilaus II, Diza, Malhonen, Dayed, Snailwalker, Imnotminkus, King of Hearts, Chobot, Mhking, VolatileChemical, Skraz, Gwernol, Roboto de Ajvol, Summalogicae, YurikBot, Wavelength, TexasAndroid, Sceptre, Sarranduin, WhatPotato?, Jlc46, Mark Ironie, Cswrye, Markus Schmaus, Netscott, SpuriousQ, Ansell, Matt Fitzpatrick, Akamad, Stephenb, Grubber, Cate, Gaius Cornelius, CambridgeBayWeather, Alex Bakharev, KSchutte, Ergzay, Ugur Basak, MosheA, Shanel, NawlinWiki, Rick Norwood, Nowa, Wiki alf, Hwasungmars, Deskana, Jaxl, Johann Wolfgang, InformationalAnarchist, Ino5hiro, Jfsaiya, Dureo, Nick, Ragesoss, Brythain, Banes, Daniel Mietchen, Rmky87, Raven4x4x, Stevenwmccrary58, Alex43223, Nate1481, RonCram, PrimeCupEevee, Mysid, Gadget850, DeadEyeArrow, Tachs, ThreePD, Haemo, Xpclient, Werdna, Efbrazil, Wknight94, Pooryorick~enwiki, FF2010, Enormousdude, Rolf-Peter Wille, Zzuuzz, Snotface, Andrew Lancaster, Mike Dillon, Closedmouth, The Son of Oink, Arthur Rubin, KGasso, Reyk, Dspradau, Jake Spooky, BorgQueen, GraemeL, JoanneB, CWenger, Cjwright79, HereToHelp, ArielGold, CKelly, Ilmari Karonen, Kungfuadam, Junglecat, Pfistermeister, Aeosynth, Meegs, Bsod2, Benandorsqueaks, Paul Erik, Asterion, MansonP, DVD R W, Algae, Jdcompguy, Luk, Sardanaphalus, Xygtshadow, A bit iffy, BonsaiViking, SmackBot, NSLE, Rtc, Zazaban, Brianyoumans, Prodego, KnowledgeOfSelf, TestPilot, Hydrogen Iodide, Od Mishehu, Vald, Rokfaith, Bomac, Jagged 85, Gabi bart, Anastrophe, Delldot, Alan McBeth, Hardyplants, RobotJcb, Canthusus, Chauncey27, Shamalyguy, Jpvinall, Edgar181, Commander Keane bot, M stone, Sloman, Portillo, Ohnoitsjamie, Hmains, Skizzik, Ppntori, Jwgraham, Kmarinas86, Lindosland, Anwar saadat, Wigren, Bluebot, SlimJim, Samosa Poderosa, Bartimaeus, Persian Poet Gal, NCurse, Tito4000, Bduke, Cattus, Stubblyhead, MartinPoulter, Lddnhan, Fplay, Silly rabbit, Papa November, Ryan Paddy, SchfiftyThree, Deli nk, Sadads, Mark7-2, J. Spencer, Go for it!, CMacMillan, DHN-bot~enwiki, Sbharris, Hallenrm, A. B., Reaper X, D-Rock, Can't sleep, clown will eat me, MisterHand, Nick Levine, Милан Јелисавчић, Danielkueh, Frap, Sommers, Aelsi, Darthgriz98, Voyajer, Xiner, Run!, Addshore, Kcordina, Meepster, SundarBot, Zophar1, Barkman34, Fuhghettaboutit, Cybercobra, Wapp~enwiki, Bowlhover, Nakon, Theodore7, Jiddisch~enwiki, Nick125, Rezecib, Dacoutts, Salt Yeung, BullRangifer, Clean Copy, Polonium, Jon Awbrey, Adrigon, Hammer1980, Jklin, Wizardman, Just plain Bill, Metamagician3000, Mystaker1, Sadi Carnot, Ck lostsword, Pilotguy, Kensor, Kukini, Dkusic~enwiki, Ged UK, Byelf2007, Chwech, Lambiam, Mchavez, ArglebargleIV, Rory096, Orbicle, Giovanni33, Paaerduag, Zahid Abdassabur, Kuru, UberCryxic, Vgy7ujm, J 1982, Nzgabriel, JoshuaZ, Chodorkovskiy, JorisvS, Dumelow, Mgiganteus1, CaptainVindaloo, Stefan2, Singh.vish, Runningfridgesrule, IKR1, AdAdAdAd, 16@r, Cjackb, Stwalkerster, Martinp23, Laogeodritt, Wstomv, Mr Stephen, Bendzh, Z E U S, Waggers, Icez, Dhp1080, Tuspm, Anonymous anonymous, RichardF, Jose77, RHB, Osame, Travia21, Snezzy, ShakingSpirit, Phuzion, Hu12, Stephen B Streater, Tawkerbot, Ginkgo100, Levineps, Kevlar992, K. Dekaels~enwiki, Paul venter, David Little, TWIS, J Di, Mrdthree, Andrew Hampe, Shoreranger, Lenoxus, Secretpizaparty, AGK, Az1568, Courcelles, Audiosmurf, Peteweez, Nkayesmith, Secos5, Tawkerbot2, MarylandArtLover, Blueracer6, Kurtan~enwiki, Lahiru k, Trubkozub, MightyWarrior, Patrickwoodridge, Firefly111, Farny1, Efrum, JForget, Wolfdog, Ken McRitchie, Phillip J, CmdrObot, Wafulz, Insanephantom, Van helsing, Makeemlighter, Enselic, Page Up, Taylorhewitt, Dan2119, GHe, Dark24spn, Dgw, Yarnalgo, Ballista, Ezrakilty, ButterApple, Neelix, GregW, Longshot.222, Andrew Delong, Myasuda, Murderd2death, Funnyfarmofdoom, TJDay, Slazenger, Bkessler23, Don.atreides, Mualphachi, Mato, Mortus Est, Michaelas10, Gogo Dodo, Red Director, JFreeman, Chasingsol, Eu.stefan, Dancter, He Who Is, Tawkerbot4, Shirulashem, DumbBOT, Teratornis, FastLizard4, NaLalina, Crana, IComputerSaysNo,

26.10. TEXT AND IMAGE SOURCES, CONTRIBUTORS, AND LICENSES

Joe11miles, Omicronpersei8, Dyanega, Wexcan, Trev M, Casliber, FrancoGG, BetacommandBot, Smellyk, Alquri, Epbr123, Barticus88, Wikid77, Btball, Qwyrxian, Goods21, N5iln, Headbomb, Marek69, Joeprempeh, John254, Tapir Terrific, Second Quantization, Peter Gulutzan, Tellyaddict, Sturm55, BlytheG, Random Tree, SusanLesch, Beezle1999, DblGkid, AlefZet, Eleuther, K12worker, Porqin, WikiSlasher, AntiVandalBot, RobotG, Luna Santin, Guy Macon, Why My Fleece?, Wenisboy111, Emeraldcityserendipity, TimVickers, Vicè, Geogeogeo, Vanjagenije, Pheoinixflame, Danger, Lperez2029, Gdo01, Spencer, Zidane tribal, Myanw, ClassicSC, Ioeth, Mikenorton, JAnDbot, Narssarssuaq, Husond, Poga, JenLouise, MER-C, Plantsurfer, The Transhumanist, Gtorell, Hello32020, Arturo 7, Plm209, Andonic, Dcooper, Martinkunev, Hut 8.5, Lirter, Tstrobaugh, Maias, Acroterion, Magioladitis, Pedro, Bennybp, Bongwarrior, VoABot II, Dekimasu, Wikidudeman, Yandman, JamesBWatson, CattleGirl, A10brown, Genedoug, Avicennasis, Wolfram.Tungsten, Cardamon, Rachita Sephiroth, D-rew, Viconpr, Dinohunter, User86654, Spacegoat, Bloodredrover, Ciaccona, Biokinetica, Allstarecho, Chivista~enwiki, Cpl Syx, Gomm, MCG, Vssun, Sabedon, Glen, DerHexer, LW77, WLU, Calltech, Akhil999in, Lightnin Boltz, 0612, Weiojranwie v5a, MartinBot, BetBot~enwiki, It334~enwiki, Arjun01, Tvoz, Jessoupe, Rettetast, Alsee, Sm8900, Ottantotto, R'n'B, Snozzer, Boston, Erkan Yilmaz, Artaxiad, Gizmo321, J.delanoy, Pharaoh of the Wizards, Trusilver, Bogey97, Psycho Kirby, Maurice Carbonaro, Nigholith, Ginsengbomb, Eliz81, Jason-rule, Kimhyunwoo, Taop, G. Campbell, Josisb, Heat023.robes, Jkaplan, SharkD, Dontrustme, Nnamdinwokoro, Smeira, Jeepday, Olithal, Gabe mayne, Pyrospirit, Rossenglish, The Transhumanist (AWB), Arms & Hearts, NewEnglandYankee, Antony-22, Philmacrackin, DadaNeem, Luctor IV, Mrfriedchicken, Daerg, Unknownguy123456789, Jjdukejj403, Olegwiki, Shoessss, 2help, Miaferron, Juliancolton, Cometstyles, WJBscribe, Remember the dot, Zara1709, C-word, Tae Guk Gi, Diego, Khargas, Andy Marchbanks, The Fat Guy, Useight, Axle12693, Markguitar333, Nigger1234567, Nwanda, Xiahou, CardinalDan, Cromoser, Idioma-bot, Wikieditor06, DaDawg22, Vranak, Deor, VolkovBot, TreasuryTag, Swfcowls, ABF, Somebodyreallycool, Arialboundaries123, Alexandria, AlnoktaBOT, Bacchus87, Mugander, Fences and windows, Miguelzinho, Voronwae, WOSlinker, QuackGuru, Soap Poisoning, Sześćsetsześćdziesiątsześć, Tzetzes, Eedo Bee, Philip Trueman, Nerm12, TXiKiBoT, Rollo44, KateBerry, Cosmic Latte, The Original Wildbear, Red Act, A4bot, Hqb, Scilit, Applerw, Joel Kincaid, Jazzwick, GDonato, Gerrish, Ask123, Figgisfiggisfiggis, Qxz, Taimaster, PolarBearoughey, Forrest1966, Indy 900, Retiono Virginian, Sciencegrl101, Littlealien182, H2ono2, Seth103, Kiwi1234, Explosiv, Dendodge, Gjgarrett, The Tetrast, Martin451, Blacktriangle10k, RedAndr, Rexeken, Abdullais4u, Fbs. 13, LeaveSleaves, Supernerd 10, Themcman1, PDFbot, 1yesfan, StillTrill, Koolkatie, Colin stuart, Songrit, Aphilo, Mwilso24, Lethalraptor, Blurpeace, Simonwerner, Lerdthenerd, SQL, Farkas János, Synthebot, Lainer21193, Falcon8765, Enviroboy, Sam1993, SeizeThe Dayy, Thanatos666, Insanity Incarnate, Sebastjanmm, Dmcq, Pjoef, AlleborgoBot, Roadcreature, S4ndp4perm4n, TheXenocide, Riverwaste, TimProof, Randula~enwiki, Roberdor, SieBot, Tiddly Tom, Nihil novi, Graham Beards, ElphabaThropp95, Scarian, Lemonflash, Elmllama, Parhamr, Dawn Bard, Mammamiamania, Whiteghost.ink, Breakyunit, GlassCobra, Keilana, Bobcrankins, Ujjwol, Elvissenthil, Tiptoety, Radon210, Michaelgerety, Nopetro, Oiws, Copperwing0, JSpung, Elmacenderesi, Oxymoron83, Faradayplank, Yoshimad123, Lightmouse, LaidOff, Jimmycleveland, RyanParis, AMackenzie, Sunrise, DancingPhilosopher, Yoda of Borg, Mojoworker, Mygerardromance, Fox red star, Markster2000, Vig vimarsh, Yotex9, ProductofSociety, Neurophysics, Amahoney, Verdadero, JL-Bot, TracySurya, Onemado, Myrvin, Asher196, Ainlina, TheCatalyst31, Afiya27, Kleinhev, Martarius, Tanvir Ahmmed, Elassint, ClueBot, PipepBot, Jackollie, Jncc0, Panoptik, The Thing That Should Not Be, Chocoforfriends, Rjd0060, Papa Smurf11, Tractorboy60, Pwitham, Ukabia, Drmies, VQuakr, Polyamorph, J8079s, Boing! said Zebedee, CounterVandalismBot, Themully, Ryan1182, The mullisk, Briankohl, Turbo566, Neverquick, Jasonssmith94, Puchiko, MindstormsKid, DragonBot, Excirial, Anne Prouse, Keithbowden, SkE, Vanisheduser12345, Jjvikingsfan, Lingo pen, Estirabot, Mit027♦, Carau, Cenarium, JoelDick, Jotterbot, Medos2, Tppp7, Tnxman307, Singhalawap, B-man79210, Laughitup2, Dekisugi, Banime, SchreiberBike, Spykodemon, Casualpsycho, Xme, Polly, Dpthurs, GFHandel, Invisibill, Thingg, Aitias, JDPhD, Ertemplin, Versus22, LieAfterLie, Johnuniq, Macderv15h, Relly Komaruzaman, JKeck, Dakrismeno, Against the current, XLinkBot, Nicholiser, PseudoOne, Jovianeye, UESParules, Saeed.Veradi, Dsgdfshfdshdsfh, Little Mountain 5, Rreagan007, Whisky it Up!, SilvonenBot, Pogipogi, Mifter, Sikig, Tannerthegreat, Badgernet, HarlandQPitt, Navy Blue, The Rationalist, Branrile09, HexaChord, Rmiddl, Pamejudd, Loueiler, Addbot, Cxz111, Willking1979, Fireheart7397, Nickenge, Bobafett29, DOI bot, Jojhutton, Captain-tucker, Binary TSO, DougsTech, Fgnievinski, EliteAthlete, Chris19910, Ronhjones, TutterMouse, Screwdis, Wikiwizzard123, Camarinha, Scient~enwiki, Ilya-108, Leszek Jańczuk, Kapaleev, Devrit, Looie496, MrOllie, Download, Chamal N, CarsracBot, Glane23, Dizzle13, Debresser, NittyG, Dr. Universe, Favonian, LemmeyBOT, AtheWeatherman, Azurefox, Connect1, Numbo3-bot, Tide rolls, Lightbot, Jan eissfeldt, Krano, Tenth Plague, Luckas Blade, Gail, Greyhood, Trotter, Quantumobserver, Archeologo, Luckas-bot, ZX81, Finbob83, Yobot, 2D, ALL OF YOU ST, Ajh16, THEN WHO WAS PHONE?, Runinbraces12, Nallimbot, Thehappymoustaches, Jimmysevolution, KamikazeBot, Sumail, 2008CM, Alexkin, Tempodivalse, Rlogan2, Licor, AnomieBOT, Tryptofish, Archon 2488, Kristen Eriksen, Rubinbot, VX, Jim1138, Shock Brigade Harvester Boris, Piano non troppo, PhaseChanger, AdjustShift, Quispiam, Ulric1313, Flewis, Materialscientist, Limideen, The High Fin Sperm Whale, Citation bot, Alkhowarizmi, François Pichette, GB fan, ArthurBot, Xqbot, Timir2, Marko Grobelnik, Intelati, Cureden, JimVC3, Raphyortanez, Capricorn42, Dsoconno, A455bcd9, TechBot, Jeffrey Mall, Stsang, Prettygirlswebshow, Fancy steve, Grim23, Edward Luva, Jmundo, Jakwra, Locos epraix, The Evil IP address, FlightTime, Billybob131, Aa77zz, Miguel in Portugal, Ewindward, Bakaw69, Srich32977, NEDM4EVER, Nicholas.a.chambers, ScreaminXD, Almabot, IntellectToday, ClareZeBearZe, J04n, Shuzo123456789, Elitefart505, Flaucinausihilipilifacation, Corruptcopper, Biggedawg, Rockmasterdan, XXIsuelXx, Yuhh, KEEHAM, Redpanda900, Omnipaedista, Robert froste, Tclgb, Yugolervan, Earlypsychosis, Gott wisst, Voheezy, Thogan3, Robert leon, Beatson121, Jamierobb893, 14albeev, Sexyz, Sports36, Doulos Christos, XxHolyDiverxx, Persontgssbdx, Nanana180, Sahebco, RFC posting script, Hamamelis, AlimanRuna, SchnitzelMannGreek, Doodoocacapeepee, Sicronet, Legobot III, Hugetim, GliderMaven, Nagualdesign, FrescoBot, Liridon, Paine Ellsworth, Tobby72, Sky Attacker, Alberttruong, VS6507, Strot, Machine Elf 1735, Drew R. Smith, Xhaoz, ClickRick, Citation bot 1, Killian441, Gravityguy, Tnt uncontested, Skunen1, Pinethicket, Kiefer.Wolfowitz, Per Ardua, Therustinator132, MJ94, Calmer Waters, MTDinoHunter, Shan3coley, Fancynancy1244521, Elnene15, Georgiaham, RedBot, Lalalllala, Tensil, SpaceFlight89, Meaghan, I am a ginger bread man, EdoDodo, Gilbeysjame, Holybassist, Saayiit, Anticent, Dude1818, Max Duchess, FoxBot, TobeBot, DixonDBot, Kellymaj, Sznax, WebEdHC, Jovenmae, Gfdfgshdhdhdfhfghgfhfh, Mrgarcia94, Jdavie, DragonofFire, Toniiiix, NicholasCarlough, Suffusion of Yellow, Tbhotch, Deanmullen09, Bricaniwi, RjwilmsiBot, TjBot, DHooke1973, Tesseract2, DASHBot, Ficz44, Ejamtiger, EmausBot, Proud Liberal 6, Dasher246, WikitanvirBot, Sciencenerdphd, Gfoley4, Dominus Vobisdu, Gcastellanos, Chloeey97, Farpre, GoingBatty, Outriggr, Chricho, Oceans and oceans, Hhhippo, JSquish, ZéroBot, Josve05a, MithrandirAgain, Andyman1125, The Nut, AshforkAZ, Aeonx, H3llBot, Mburdis, GrindtXX, Henesuri, Staszek Lem, L Kensington, Lilgas52, Phronetic, Neuberliner, KyleAraujo, Imagoofygooberyeah, Freecie1, Kiesewetter, RockMagnetist, ClamDip, Sharonmil, U3964057, Teapeat, WordDiver, Helpsome, ClueBot NG, W.Kaleem, Another n00b, Movses-bot, Nahiyan8, Schicagos, Frietjes, Delusion23, Fbarrera09, IvoryMeerkat, Skiles1611, Helpful Pixie Bot, Calabe1992, Bibcode Bot, Plantdrew, BG19bot, Ryker-Smith, Hallows AG, ElphiBot, AvocatoBot, Wingroras, Solomon7968, Ramos1990, Air-tractor sledge, Mthoodhood, Mbotee, I am huge liar listen, Thestickman91, Jakakla, Soerfm, CitationCleanerBot, Samyriup, Hairandfashion142, Seanpkenny, Smsagro, In11Chaudri, BattyBot, Bart49, Cyberbot II, Warheadpat, Layzeeboi, Kenixkil, Karanhbhatt, Tow, IjonTichyIjonTichy, Linkman22, Dexbot, Renamed user qh37rbwki62h19772b, Mogism, AndiPersti, Numbermaniac, वागेश्वर मंत्र, Sowlos,

Josophie, Hillbillyholiday, Alexis1812w, MarchOrDie, Zjrong, Thepooj9, Theo's Little Bot, I am One of Many, Jp4gs, Globalcooling400, Melonkelon, Biomedicinal, Rakomwolveshane, Praemonitus, AKYF, Arnlodg, The French Rat, Sssssss340, Sol1, Mrm7171, Hansmuller, Finnusertop, Jackmcbarn, Frogger48, VeryCrocker, Tudor1999, Man of Steel 85, Avelez00, Loganate123, TheG3NERAL, John 3:16, Monkbot, Cityrailsaints, Teaksmitty, ShawntheGod, Filedelinkerbot, Teemome, Hkeyser, UglowT, 115ash, VanishedUser sdu9aya9fs654654, Loraof, Cynulliad, Randomeditor1000, Batsgasps, Jack Pepa, Yprpyqp, DavidJac, BeUtkarsh, Knightplex, ???, Serten II, LadyLeodia, Moorrests, Isambard Kingdom, Rctillinghast, Vonbergh, Angelababy00, Jeunine, KasparBot, JorisEnter, Eulalefty, Continentaleurope, Sir Cumference, BD2412bot, The name, BU Rob13, Feminist, Primetime637, Twitteristhebest, Zenedits, Ermahgerd9, MartinZ, Charlotte135, Chrisvacc, InternetArchiveBot, TssRnapII, Motivação, Mynameisparatha, Bender the Bot, Metaphysicswar, 001blondjamie, Reason is Immortal, 000meow, Catboy7006, Nixinova, Heididoerr061, Laurdecl, Wittgenstein123, Sheila Ki Jawani, Magic links bot, ENamel5, KolbertBot and Anonymous: 1277

- **Formal science** *Source:* https://en.wikipedia.org/wiki/Formal_science?oldid=799868047 *Contributors:* Michael Hardy, Deljr, Ojs, Ancheta Wis, Mdd, Trylks, Cenobite, Woohookitty, Ruud Koot, Tomy108~enwiki, Rjwilmsi, Chobot, RussBot, Igiffin, SmackBot, RDBury, Jagged 85, Alsandro, Powo, Nbarth, Baronnet, Jajhill, JudahH, Salt Yeung, Jon Awbrey, Lambiam, OcarinaOfTime, RichardF, Rlinfinity, K, Shoreranger, Thomasmeeks, Gregbard, Kilva, Vanjagenije, Magioladitis, Damuna, R'n'B, Vvitor, Rjclaudio, Squids and Chips, VolkovBot, Jimmaths, Dendodge, Jackbars, SieBot, Tim Thomason, ImageRemovalBot, MenoBot, Eric Wester, Hans Adler, Roger491127, Tdslk, Addbot, SpBot, J05HYYY, Legobot, Luckas-bot, AnomieBOT, JackieBot, ???, Xqbot, Dsoconno, NOrbeck, Sophivorus, J04n, GrouchoBot, Aaron Kauppi, CES1596, FrescoBot, HRoestBot, Kiefer.Wolfowitz, Galoa2804~enwiki, Gamewizard71, Duoduoduo, EmausBot, WikitanvirBot, Slightsmile, ZéroBot, Erianna, Architectchao, Tijfo098, ChuispastonBot, RockMagnetist, Scortchi, ClueBot NG, Widr, Muhammad Shuaib Nadwi, BattyBot, Jibun. bukiyou desu kara, Jemee012, Comp.arch, Iº½, VanishedUser sdu9aya9fs654654, Isambard Kingdom, Squiver, CyberWarfare, Anareth, Hofhof, AndoniyaK, Magic links bot and Anonymous: 78

- **Logic** *Source:* https://en.wikipedia.org/wiki/Logic?oldid=802787893 *Contributors:* AxelBoldt, Vicki Rosenzweig, The Anome, Toby Bartels, Ryguasu, Hirzel, Dwheeler, Stevertigo, Edward, Patrick, Chas zzz brown, Michael Hardy, Lexor, TakuyaMurata, Bagpuss, Looxix~enwiki, Ahoerstemeier, Notheruser, BigFatBuddha, Александър, Glenn, Marco Krohn, Rossami, Tim Retout, Rotem Dan, Evercat, EdH, DesertSteve, Caffelice~enwiki, Mxn, Michael Voytinsky, Ehn, Peter Damian (original account), Rzach, Charles Matthews, Dcoetzee, Paul Stansifer, Dysprosia, Jitse Niesen, Xiaodai~enwiki, Markhurd, MikeS, Carol Fenijn, SEWilco, Samsara, J D, Shizhao, Power~enwiki, Olathe, Jusjih, Ldo, Banno, Chuunen Baka, Robbot, Iwpg, Fredrik, R3m0t, Altenmann, MathMartin, Rorro, Rholton, Saforrest, Borislav, Robertoalencar, Michael Snow, Raeky, Guy Peters, Jooler, Tea2min, Filemon, Ancheta Wis, Exploding Boy, Giftlite, Recentchanges, Inter, Wolfkeeper, Lee J Haywood, COMPATT, Everyking, Rookkey, Malyctenar, Andris, Bovlb, Jason Quinn, Sundar, Siroxo, Deus Ex, Rheun, LiDaobing, Roachgod, Quadell, Starbane, Piotrus, Ludimer~enwiki, Karol Langner, CSTAR, Rdsmith4, APH, JimWae, OwenBlacker, Kntg, Mysidia, Pmanderson, TiMike, Eduardoporcher, Eliazar, Grunt, Guppyfinsoup, Mike Rosoft, Freakofnurture, Ultratomio, Lorenzo Martelli, Discospinster, Rich Farmbrough, KillerChihuahua, Rhobite, Guanabot, Leibniz, Hippojazz, Vsmith, Raistlinjones, Slipstream, ChadMiller, Paul August, Bender235, El C, Chalst, Mwanner, Tverbeek, Bobo192, Cretog8, Johnkarp, Shenme, Amerindianarts, Passw0rd, Knucmo2, Storm Rider, Red Winged Duck, Alansohn, Anthony Appleyard, Shadikka, Rh~enwiki, Chira, ABCD, Kurt Shaped Box, SlimVirgin, Batmanand, Denniss, Yummifruitbat, Shinjiman, Velella, Sciurinæ, MIT Trekkie, Alai, CranialNerves, Velho, Mel Etitis, Mindmatrix, Camw, Kokoriko, Kzollman, Ruud Koot, Orz, MONGO, Apokrif, Jok2000, WikikIrsc, CharlesC, MarcoTolo, DRHansen, Gerbrant, Tslocum, Graham87, Alienus, BD2412, Porcher, Sjö, Rjwilmsi, Mayumashu, Саша Стефановић, GOD, Bruce1ee, Salix alba, Crazynas, Ligulem, Baryonic Being, Titoxd, FlaBot, Kwhittingham, Latka, Mathbot, Twipley, Nihiltres, SportsMaster, RexNL, AndriuZ, Quuxplusone, Celendin, Influence, R Lee E, JegaPRIME, Malhonen, Spencerk, Chobot, DVdm, Bgwhite, EamonnPKeane, Roboto de Ajvol, Wavelength, Deeptrivia, KSmrq, Raquel Baranow, Endgame~enwiki, Polyvios, CambridgeBayWeather, KSchutte, NawlinWiki, Rick Norwood, SEWilcoBot, Mipadi, Brimstone~enwiki, LaszloWalrus, AJHalliwell, Trovatore, Pontifexmaximus, Chunky Rice, Cleared as filed, Nick, Darkfred, Wjwma, Googl, Mendicott, StuRat, Open2universe, ChrisGriswold, Nikkimaria, OEMCUST, David Biddulph, Nahaj, Extreme Unction, Allens, KNHaw, Sardanaphalus, Johndc, SmackBot, Lestrade, InverseHypercube, Pschelden, Jim62sch, Jagged 85, WookieInHeat, Josephprymak, Timotheus Canens, Srnec, LonesomeDrifter, Yamaguchi???, Collingsworth, Gilliam, Skizzik, RichardClarke, Heliostellar, Chris the speller, Jaymay, Da nuke, Unbreakable MJ, MK8, Andrew Parodi, Kevin Hanse, MalafayaBot, Clconway, Sciyoshi~enwiki, Go for it!, Mikker, Zsinj, Can't sleep, clown will eat me, Misgnomer, Grover cleveland, Fuhghettaboutit, Cybercobra, Nakon, Jiddisch~enwiki, Richard001, MEJ119, Kabain52, Lacatosias, Jon Awbrey, DMacks, Henning Makholm, Ged UK, Ceoil, Byelf2007, SashatoBot, Lambiam, Dbtfz, Deaconse, UberCryxic, FrozenMan, Heimstern, Shlomke, Shadowlynk, Joshua Scott, F15 sanitizing eagle, Prince153, WithstyleCMC, Hvn0413, Meco, RichardF, Novangelis, Vagary, Pamplmoose, KJS77, Hu12, Levineps, BananaFiend, K, Lottamiata, Catherineyronwode, Mrdthree, Igoldste, Themanofnines, Adambiswanger1, Satarnion, Tawkerbot2, Galex, SkyWalker, CRGreathouse, CBM, Editorius, Rubberchix, Gregbard, Kpossin, Cydebot, Pce3@ij.net, Jasperdoomen, Samuell, Quinnculver, Peterdjones, Asgrim, Travelbird, Pv2b, Drksl, JamesLucas, Julian Mendez, Dancter, Tawkerbot4, Shirulashem, Doug Weller, DumbBOT, Garik, Progicnet, Mattisse, Letranova, Thijs!bot, Epbr123, Kredal, Smee, Marek69, AgentPeppermint, OrenBochman, Dawnseeker2000, Escarbot, Eleuther, Mentifisto, Vafthrudnir, AntiVandalBot, Peoppenheimer, Majorly, Gioto, Hidayat ullah, GeePriest, Dougher, Dhrm77, Sluzzelin, JAnDbot, Narssarssuaq, MER-C, The Transhumanist, Avaya1, Zizon, Frankie816, Savant13, Dr mindbender, LittleOldMe, Bongwarrior, VoABot II, SDas, JNW, Arno Matthias, Appraiser, Gammy, Smihael, Caesarjbsquitti, Midgrid, Bubba hotep, Moopiefoof, GeorgeFThomson, Virtlink, David Eppstein, Epsilon0, DerHexer, Waninge, Exbuzz, MartinBot, Wylve, CommonsDelinker, EdBever, C.R.Selvakumar, J.delanoy, Trusilver, Jbessie, Fictionpuss, Cpiral, RJMalko, McSly, Lightest~enwiki, Classicalsubjects, Mrg3105, Daniel5Ko, The Transhumanist (AWB), Policron, MetsFan76, Kenneth M Burke, Steel1943, Idioma-bot, Spellcast, WraithM, VolkovBot, Cireshoe, Rucha58, Macedonian, Hotfeba, Indubitably, Fundamental metric tensor, Jimmaths, Station1, Djhmoore, Aesopos, Oshwah, Rei-bot, Llamabr, Ontoraul, Philogo, Leafyplant, Sanfranman59, Abdullais4u, Jackfork, Cullowheean, Wiae, Maxim, Myscience, LIBLAHLIBLAHTIMMAH, Synthebot, Rurik3, Koolo, Nagy, Symane, PGWG, W4chris, Prom2008, FlyingLeopard2014, Radagast3, Demmy, JonnyJD, Newbyguesses, Linguist1, SieBot, StAnselm, Maurauth, Paradoctor, Gerakibot, RJaguar3, Yintan, Whiteghost.ink, Bjrslogii, Soler97, Til Eulenspiegel, Flyer22 Reborn, DanEdmonds, Undead Herle King, Crowstar, Redmarkviolinist, Spinethetic, Thelogicthinker, DancingPhilosopher, Svick, Valeria.depaiva, Adhawk, Sginc, Tognopop, CBM2, Yotex9, 3rdAlcove, PsyberS, Francvs, Classicalecon, Khirurg, Mx. Granger, Atif.t2, Martarius, ClueBot, Andrew Nutter, Snigbrook, The Thing That Should Not Be, Taroaldo, Ukabia, Mild Bill Hiccup, TheOldJacobite, Boing! said Zebedee, Niceguyedc, Blancbardb, DragonBot, Jessieslame, Excirial, Alexbot, Jusdafax, Watchduck, AENAON, NuclearWarfare, Arjayay, SchreiberBike, Thingg, JDPhD, Wirjadisastra, Scalhotrod, Budelberger, Skunkboy74, Gerhardvalentin, Duncan, Saeed.Veradi, Mcgauley08, NellieBly, Noctibus, Aunt Entropy, Jjfuller123, Spidz, Addbot, Rdanneskjold, Proofreader77, Atethnekos, Sully111, Logicist, Vitruvius3, Rchard2scout, Glane23, Uber WoMensch!, Chzz, Favonian, LinkFA-Bot, AgadaUrbanit, Numbo3-bot, Ehrenkater, Tide rolls, Lightbot, Macro Shell, Zorrobot, Jarble, JEN9841, Aarsalankhalid,

GorgeUbuasha, Yobot, Arcvirgos 08, Jammie101, Francos22, Azcolvin429, MassimoAr, AnomieBOT, Hairhorn, Jim1138, IRP, AdjustShift, Melune, NickK, Materialscientist, Neurolysis, ArthurBot, Gemtpm, LilHelpa, Blueeyedbombshell, Junho7391, Xqbot, RJGray, Gilo1969, The Land Surveyor, Tyrol5, A157247, Petropoxy (Lithoderm Proxy), Uarrin, GrouchoBot, Peter Damian, Hifcelik, Omnipaedista, ירמ"ש78,26, مهدی چمن‌بدی, Tales23, GhalyBot, Aaron Kauppi, GliderMaven, FrescoBot, Liridon, D'Artagnol, Tobby72, D'ohBot, Mewulwe, Itisnotme, Cannolis, Rhalah, Citation bot 1, Chenopodiaceous, AstaBOTh15, Gus the mouse, Pinethicket, Vicenarian, A8UDI, Ninjasaves, Seryred123, Serols, Wikiain, PlyrStar93, Maokart444, Jandalhandler, Gamewizard71, FoxBot, TobeBot, Burritoburritoburrito, Mysticcooperfox, Lotje, GregKaye, Vistascan, Vrenator, Duoduoduo, נעשית ב, Merlinsorca, Literateur, Jarpup, Whisky drinker, Mean as custard, Rlnewma, TjBot, Walkinxyz, EmausBot, Orphan Wiki, Nick Moyes, The Kytan Apprentice, Pologic, Faolin42, Jedstamas, Wham Bam Rock II, Solarra, ZéroBot, PBS-AWB, Leminh91, Josve05a, Shuipzv3, Mar4d, Wayne Slam, Frigotoni, Staszek Lem, Resprinter123, FrankFlanagan, L Kensington, Danmuz, Eranderson, Donner60, Chewings72, Puffin, GKaczinsky, ChuispastonBot, NTox, Poopnubblet, Xanchester, Rememberway, ClueBot NG, W.Kaleem, Jack Greenmaven, Satellizer, WikiTatik, Quantamflux, Validlessness, Bastianperrot, Wdchk, Snotbot, Masssly, Widr, Lawsonstu, ESL75, Helpful Pixie Bot, Anav2221, BG19bot, Daniel Zsenits, Norma Romm, PTJoshua, Northamerica1000, Graham11, Geegeeg, JohnChrysostom, Frze, Chjohnson39, Marcocapelle, Alex.Ramek, CJMacalister, CitationCleanerBot, Jilliandivine, Flosfa, Chrisct1993, Brad7777, Lrq3000, Mewhho18, A.coolmcfly, Compulogger, Cyberbot II, Roger Smalling, The Illusive Man, NanishaOpaenyak, Rhlozier, EagerToddler39, Dexbot, Marius siuram, Табалдыев Ысламбек, Omanchandy007, RideLightning, Jochen Burghardt, Wieldthespade, Hippocamp, Wickid123, Matticusmadness, JMCF125, NIXONDIXON, CsDix, I am One of Many, Biogeographist, רלבי, Tentinator, EvergreenFir, Babitaarora, Ugog Nizdast, Melody Lavender, JustBerry, Skansi.sandro, Ginsuloft, Robf00f1235, Calvinator8, The Annoyed Logician, Liz, GreyWinterOwl, ByDash, Jbob13, Henniepenny, Matthew Derick B Cruz, Filedelinkerbot, Sherlock502, Equilibrium103, Fvdedphill, Norwo037, Karnaoui, IagoQnsi, Claireney, Pat132, The Expedia, Sbcdave, Muneeb Masoud, A Great Catholic Person, Jacksplay, Asdklf;, Esicam, ChamithN, WillemienH, Ntuser123, PsychopathicAssassin, Cthulhu is love cthulhu is life, Jiten Dhandha, Loraof, Julietdeltalima, Adamrobson28, Josmust222, Rubbish computer, Layfi, KcBessy, SamiLayfi, Lanzdsey, SoSivr, Human3015, MeshCollider, ZanderEdmunds, Amccann421, KasparBot, Bestusername-ign, Sparky Macgillicuddy, BjörnF, Mithisharma, MindForgedManacle, Citation requested but not required, Nyetoson, CLCStudent, SaundersLane, TBNRGiazo, Baking Soda, Harmon758, InternetArchiveBot, Isuredid, Entranced98, Anareth, Rachel Benedict, Fluttershy (totally not SonicFan007FTW), JILOmed, Robot psychiatrist, Gulumeemee, Hanlucky, OZiefOx, Sora 190, Apollo The Logician, Mr EVERYTHING, Sense1024, What cat?, Kjhajkfa, L8 ManeValidus, WolfGargan, Vhbbbhbhbbbhfcfg, Mai Onee-sama, Sophia emmi, Astatzcyn, NickTheTurtle and Anonymous: 833

- **Mathematics** *Source:* https://en.wikipedia.org/wiki/Mathematics?oldid=800892143 *Contributors:* AxelBoldt, Magnus Manske, LC-enwiki, Brion VIBBER, Eloquence, Mav, Bryan Derksen, Zundark, The Anome, Tarquin, Koyaanis Qatsi, Ap, Gareth Owen, --April, RK, Iwnbap, LA2, Youssefsan, XJaM, Arvindn, Christian List, Matusz, Toby Bartels, PierreAbbat, Little guru, Miguel-enwiki, Rade Kutil, DavidLevinson, FvdP, Daniel C. Boyer, David spector, Camembert, Netesq, Zippy, Olivier, Ram-Man, Stevertigo, Spiff-enwiki, Edward, Quintessent, Ghyll-enwiki, D, Chas zzz brown, JohnOwens, Michael Hardy, Chris-martin, JakeVortex, Lexor, Isomorphic, Dominus, Nixdorf, Grizzly, Kku, Mic, Ixfd64, Firebirth, Alireza Hashemi, Deljr, Sannse, TakuyaMurata, Karada, Minesweeper, Alho, Tregoweth, Dgrant, CesarB, Ahoerstemeier, Cyp, Ronz, Muriel Gottrop-enwiki, Snoyes, Notheruser, Angela, Den fjättrade ankan-enwiki, Kingturtle, LittleDan, Kevin Baas, Salsa Shark, Glenn, Jschwa1, Bogdangiusca, BenKovitz, Poor Yorick, Rossami, Tim Retout, Rotem Dan, Evercat, Rl, Jonik, Madir, Mxn, Smack, Silverfish, Vargenau, Pizza Puzzle, Nikola Smolenski, Charles Matthews, Guaka, Timwi, Spacemonkey-enwiki, Nohat, Ralesk, MarcusVox, Dysprosia, Jitse Niesen, Fuzheado, Gutza, Piolinfax, Selket, DJ Clayworth, Markhurd, Vancouverguy, Tpbradbury, Maximus Rex, Hyacinth, Saltine, AndrewKepert, Fibonacci, Zero0000, Phys, Ed g2s, Wakka, Samsara, Bevo, McKay, Traroth, Fvw, Power-enwiki, Babalouloù, Secretlondon, Jusjih, Cvaneg, Flockmeal, Guppy, Francs2000, Dmytro, Lumos3, Jni, PuzzletChung, Donarreiskoffer, Robbot, Fredrik, RedWolf, Peak, Romanm, Lowellian, Gandalf61, Georg Muntingh, Merovingian, HeadCase, Sverdrup, Henrygb, Academic Challenger, IIR, Thesilverbail, Hadal, Mark Krueger, Wereon, Robinh, Borislav, GarnetRChaney, Ilya (usurped), Michael Snow, Fuelbottle, ElBenevolente, Lupo, PrimeFan, Zhymkus-enwiki, Dmn, Cutler, Dina, Mlk, Alan Liefting, Rock69-enwiki, Cedars, Ancheta Wis, Fabiform, Centrx, Giftlite, Dbenbenn, Christopher Parham, Fennec, Markus Krötzsch, Mikez, Inter, Wolfkeeper, Ævar Arnfjörð Bjarmason, Netoholic, Lethe, Tom harrison, Lupin, MathKnight, Bfinn, Ayman, Everyking, No Guru, Curps, Jorend, Ssd, Niteowlneils, Gareth Wyn, Andris, Guanaco, Sundar, Daniel Brockman, Siroxo, Node ue, Eequor, Arne List, Matt Crypto, Python eggs, Avala, Jackol, Marlonbraga, Bobblewik, Deus Ex, Golbez, Gubbubu, Kennethduncan, Cap601, Geoffspear, Utcursch, Andycjp, CryptoDerk, LucasVB, Quadell, Frogjim-enwiki, Antandrus, BozMo, Rajasekaran Deepak, Beland, WhiteDragon, Beameron54, Kaldari, PDH, Profvk, Jossi, Alexturse, Adamsan, CSTAR, Rdsmith4, APH, John Foley, Elektron, Pethan, Mysidia, Pmanderson, Elroch, Sam Hocevar, Arcturus, Gseshoyru, Stephen j omalley, Jew69, Ukexpat, Eduardoporcher, Qef, Random account 47, Zondor, Adashiel, Trevor MacInnis, Grunt, Kate, Bluemask, PhotoBox, Mike Rosoft, Vesta-enwiki, Shahab, Oskar Sigvardsson, Brianjd, D6, CALR, DanielCD, Olga Raskolnikova, EugeneZelenko, Discospinster, Rich Farmbrough, Guanabot, FiP, Clawed, Inkypaws, Spundun, Andrewferrier, ArnoldReinhold, HeikoEvermann, Smyth, Notinasnaid, AlanBarrett, Paul August, MarkS, DcoetzeeBot-enwiki, Bender235, ESkog, Geoking66, Ben Standeven, Tompw, GabrielAPetrie, RJHall, MisterSheik, Mr. Billion, El C, Chalst, Shanes, Haxwell, Briséis-enwiki, Art LaPella, RoyBoy, Lyght, Jpgordon, JRM, Porton, Bobo192, Ntmatter, Fir0002, Mike Schwartz, Wood Thrush, Func, Teorth, Flxmghvgvk, Archfalhwyl, Jung dalglish, Maurreen, Man vyi, Alphax, Rje, Sam Korn, Krellis, Sean Kelly, Jonathunder, Mdd, Tsirel, Passw0rd, Lawpjc, Vesal, Storm Rider, Stephen G. Brown, Danski14, Msh210, Poweroid, Alansohn, Gary, JYolkowski, Anthony Appleyard, Blackmail-enwiki, Mo0, Polarscribe, ChristopherWillis, Lordthees, Rgclegg, Jet57, Muffin-enwiki, Mmmready, Riana, AzaToth, Lectonar, Lightdarkness, Giant toaster, Cjnm, Mysdaao, Hu, Malo, Avenue, Blobglob, LavosBacons, Schapel, Orionix, BanyanTree, Saga City, Knowledge Seeker, ReyBrujo, Danhash, Garzo, Huerlisi, Jon Cates, RainbowOfLight, CloudNine, TenOfAllTrades, Mcmillin24, Bsadowski1, Itsmine, Blaxthos, HenryLi, Bookandcoffee, Kz8, Oleg Alexandrov, Ashujo, Stemonitis, Novacatz, Angr, DealPete, Kelly Martin, Wikiworkerindividual***, TSP, OwenX, Woohookitty, Linas, Masterjamie, Yansa, Brunnock, Carcharoth, BillC, Ruud Koot, WadeSimMiser, Orz, Hdante, MONGO, Mpatel, Abhilaa, Al E., WikiklrSc, Bbatsell, Damicatz, I64s, MFH, Sengkang, Zzyzx11, Noetica, , Xiong Chiamiov, Gimboid13, Liface, Asdfdsa, PeregrineAY, Thirty-seven, Graham87, Magister Mathematicae, BD2412, Chun-hian, FreplySpang, JIP, Island, Zoz, Icey, BorgHunter, Josh Parris, Paul13-enwiki, Rjwilmsi, Mayumashu, MJSkia1, Prateekrr, Vary, MarSch, Amire80, Tangotango, Staecker, Omnieiunium, Salix alba, Tawker, Zhurovai, Crazynas, Ligulem, Juan Marquez, Slac, R.e.b., The wub, Sango123, Yamamoto Ichiro, Kasparov, Staples, Titoxd, Pruneau, RobertG, Latka, Mathbot, Harmil, Narxysus, Andy85719, RexNL, Gurch, Short Verses, Quuxplusone, Celendin, Ichudov, Jagginess, Alphachimp, Malhonen, David H Braun (1964), Snailwalker, Mongreilf, Chobot, Jersey Devil, DONZOR, DVdm, Cactus.man, John-Haggerty, Gwernol, Elfguy, Buggi22, Roboto de Ajvol, Raelx, JPD, YurikBot, Wavelength, Karlscherer3, Jeremybub, Doug Alford, Grifter84, RobotE, Elapsed, Dmharvey, Gmackematix, 4C-enwiki, RussBot, Michael Slone, Geologician, Red Slash, Jtkiefer, Muchness, Anonymous editor, Albert Einsteins pipe, Nobs01, Soltras, Bhny, Pi Delport, CanadianCaesar, Polyvios, Akamad, Stephenb, Yakuzai, Sacre, Bovineone, Tungsten, Ugur

Basak, David R. Ingham, NawlinWiki, Vanished user kjdioejh329io3rksdkj, Rick Norwood, Misos, SEWilcoBot, Wiki alf, Mipadi, Lacipac, Armindo, Deskana, Johann Wolfgang, Trovatore, Joel7687, GrumpyTroll, LMSchmitt, Schlafly, Eighty–enwiki, Herve661, JocK, Mccready, Tearlach, Apokryltaros, JDoorjam, Abb3w, Misza13, My Cat inn, Vikvik, Mvsmith, Brucevdk, DryaUnda, SFC9394, Font, Tachyon01, Mgnbar, Jemebius, Nlu, Mike92591, Dna-webmaster, Tonywalton, Joshurtree, Wknight94, Pooryorick–enwiki, Avraham, Mkns, Googl, Noosfractal, SimonMorgan, Tigershrike, FF2010, Cursive, Scheinwerfermann, Enormousdude, TheKoG, Donald Albury, Zsynopsis, Skullfission, Claygate, MaNeMeBasat, GraemeL, JoanneB, Shawnc, Bentong Isles, Donhalcon, JLaTondre, RenamedUser jaskldjslak904, Spliffy, Flowersofnight, 158-152-12-77, RunOrDie, Kungfuadam, Canadianism, Ben D., Greatal386, JDspeeder1, NeilN, Saboteur–enwiki, Asterion, Shmm70, Pentasyllabic, Lunch, DVD R W, Finell, Capitalist, Sardanaphalus, Crystallina, JJL, SmackBot, RDBury, YellowMonkey, Selfworm, Smitz, Bobet, Diggyba, Warhawkhalo101, Estoy Aquí, Reedy, Tarret, KnowledgeOfSelf, Royalguard11, Melchoir, McGeddon, Falustra77, Masparasol, Pgk, C.Fred, AndyZ, Kilo-Lima, Jagged 85, PizzaMargherita, CapitalSasha, Antibubbles, AnOddName, Canthusus, BiT, Nscheffey, Amystreet, Ekilfeather, Papep, Jaichander, Ohnoitsjamie, Hmains, Skizzik, Richfife, ERcheck, Hopper5, Squiddy, Armenia, Durova, Qtoktok, Wigren, Keegan, Woofboy, Rmt2m, Fplay, Christopher denman, Miquonranger03, MalafayaBot, Silly rabbit, Alink, Dlohcierekim's sock, Richard Woods, Kungming2, Go for it!, Baa, RDT, Spellchecker, Baronnet, Colonies Chris, Ulises Sarry–enwiki, Nevada, Zachorious, Chendy, J•A•K, Can't sleep, clown will eat me, RyanEberhart, Timothy Clemans, Милан Јелисавчић, TheGerm, HoodedMan, Chlewbot, Vanished User 0001, Joshua Boniface, TheKMan, Rrburke, Addshore, Mr.Z-man, SundarBot, AndySimpson, Emre D., Iapetus, Jwy, CraigDesjardins, Daqu, Nakon, VegaDark, Jiddisch–enwiki, Maxwahrhaftig, Salt Yeung, Danielkwalsh, Diocles, Pg2114, Jon Awbrey, Ruwanraj, Jklin, Xen 1986, Just plain Bill, Knuckles sonic8, Where, Bart v M, ScWizard, Pilotguy, Nov ialiste, TenPoundHammer, JoeTrumpet, Math hater, Lambiam, Nishkid64, TachyonP, ArglebargleIV, Doug Bell, Harryboyles, Srikeit, Dbtfz, Kuru, JackLumber, Simonkoldyk, Vgy7ujm, Nat2, J 1982, Cronholm144, Heimstern, Gobonobo, Mfishergt, Coastergeekperson04, Sir Nicholas de Mimsy-Porpington, Dumelow, Jazriel, Gnevin, Unterdenlinden, Ckatz, Loadmaster, Special-T, Dozing, Mr Stephen, Mudcower, AxG, Optakeover, SandyGeorgia, Mets501, Funnybunny, Markjdb, Ryulong, Gff–enwiki, RichardF, Limaner, Jose77, Asyndeton, Stephen B Streater, Politepunk, DabMachine, Levineps, Hetar, BranStark, Roland Deschain, Kevlar992, Iridescent, K, Kencf0618, Zootsuits, Onestone, Nilamdoc, C. Lee, CzarB, Polymerbringer, Joseph Solis in Australia, Newone, White wolf753, Muéro, David Little, Igoldste, Amakuru, Marysunshine, Maelor, Masshaj, Jatrius, Experiment123, Tawkerbot2, Daniel5127, Joshuagross, Emote, Pikminiman, Heyheyhey99, JForget, Smkumar0, Sakowski, Wolfdog, Sleeping123, CRGreathouse, Wafulz, Sir Vicious, Triage, Iced Kola, CBM, Page Up, Jester-Tester, Taylorhewitt, Nczempin, GHe, Green caterpillar, Phanu9000, Yarnalgo, Thomasmeeks, McVities, Requestion, FlyingToaster, MarsRover, Tac-Tics, Some P. Erson, Tim1988, Tuluat, Alaymehta, MrFish, Oo7565, Gregbard, Captmog, El3m3nt09, Antiwiki–enwiki, Cydebot, Meznaric, Cantras, Funwithbig, MC10, Meno25, Gogo Dodo, DVokes, ST47, Srinath555, Pascal.Tesson, Goldencako, Benjiboi, Andrewm1986, Michael C Price, Tawkerbot4, Dragomiloff, Juansempere, M a s, Chrislk02, Brotown3, Mamounjo, 5300abc, Roccorossi, Abtract, Daven200520, Omicronpersei8, Vanished User jdksfajlasd, Daniel Olsen, Ventifact, TAU710, Aditya Kabir, BetacommandBot, Thijs!bot, Epbr123, Bezking, Jpark3591, Daemen, TheEmaciatedStilson, MCrawford, Opabinia regalis, Mattyboy500, Kilva, Daniel, Loudsox, Ucanlookitup, Hazmat2, Wootwootwoot, Brian G. Wilson, Timo3, Mojo Hand, Djfeldman, Pjvpjv, West Brom 4ever, John254, Alientraveller, Mnemeson, Ollyrobotham, BadKarma14, Sethdoe92, Dfrg.msc, RobHar, CharlotteWebb, Dawnseeker2000, RoboServien, Escarbot, Itsfrankie1221, Thomaswgc, Thadius856, Sidasta, AntiVandalBot, Ais523, RobotG, Gioto, Luna Santin, Dark Load, DarkAudit, Ringleader1489, Dylan Lake, Doktor Who, Chill doubt, AxiomShell, Abc30, Matheor, Archmagusrm, Falconleaf, Labongo, Spacefarer, Chocolatepizza, JAnDbot, Kaobear, MyNamesLogan, MER-C, The Transhumanist, Db099221, AussieOzborn au, Thenub314, Mosesroses, Hut 8.5, Kipholbeck, Xact, Twospoonfuls, .anacondabot, Yahel Guhan, Bencherlite, Yurei-eggtart, Bongwarrior, VoABot II, JamesBWatson, Swpb, Redaktor, EdwardLockhart, SineWave, Charlielee111, Cic, Ryeterrell, Caesarjbsquitti, Wikiwhat?, Bubba hotep, KConWiki, Meb43, Faustnh, Hiplibrarianship, Johnbibby, Seberle, MetsBox, Pawl Kennedy, 28421u2232nfenfcenc, David Eppstein, Systemlover, Bmeguru, Hotmedal, Just James, EstebanF, Glen, Rajpaj, Memorymentor, TheRanger, Calltech, Gun Powder Ma, Welshleprechaun, Robin S, Seba5618, SquidSK, 0612, J0equ1nn, Riccardobot, Jtir, Hdt83, MartinBot, Vladimir m, Arjun01, Quanticle, Nocklas, Rettetast, Fuzzyhair2, R'n'B, Pbroks13, Cmurphy au, Snozzer, Ben2then, PrestonH, Crazybobson, Thefutureschannel, RockMFR, Hrishikesh.24889, J.delanoy, Nev1, Unlockitall, Phoenix1177, Numbo3, Sp3000, Maurice Carbonaro, Nigholith, Hellonicole, -jmac-, Boris Allen, 2boobies, Jerry, TheSeven, NerdyNSK, Syphertext, Yadar677, Taop, G. Campbell, Wayp123, Keesiewonder, Matt1314, Ksucemfof, Gzkn, Ivelnaps, Smeira, DarkFalls, Renamed user 5417514488, Vishi-vie, Washington8785, Xyzaxis, Arkuski, JDQuimby, Batmanfan77, Alphapeta, Trd89, HiLo48, The Transhumanist (AWB), NewEnglandYankee, RANDP, MKoltnow, MhordeXsnipa, Milogardner, Nacrha, Balaam42, Mviergujerghs89fhsdifds, Cfrehr, Elvisfan2095, Tiyoringo, Juliancolton, Cometstyles, DavidCBryant, SlightlyMad, Jamesontai, Remember the dot, Ilya Voyager, Huzefahamid, Dandy mandy, Andreas2001, Ishap, Sarregouset, CANUTELOOL2, CANUTELOOL3, Devonboy69, Jeyarathan, Death blaze, Emo kid you?, Thedudester, Samlyn.josfyn, Mother69, Vinsfan368, Cartiod, Helldude99, Sternkampf, Steel1943, CardinalDan, RJASE1, Idiomabot, Remi0o, Lights, Tamillimat, Bandaidboy, C.lettingaAV, VolkovBot, Somebodyreallycool, Pleasantville, Jeff G., JohnBlackburne, Hhjk, The Catcher in The Rye D:, Alexandria, AlnoktaBOT, Dboerstl, NikolaiLobachevsky, Bangvang, 62 (number), Tseay11, Soliloquial, Headforaheadeyeforaneye, Barneca, Sześćsetsześćdziesiątsześć, Zeuron, Yoyoyo9, Trehansiddharth, TXiKiBoT, Katoa, Jacob Lundberg, Candy-Panda, Chickenclucker, Antoni Barau, Walor, Nxavar, Anonymous Dissident, Qxz, Nukemason4, Retiono Virginian, Ocolon, Savagepine, Denny-Colt, Digby Tantrum, JhsBot, Leafyplant, Beanai, 20em89.01, Cremepuff222, Geometry guy, Canyonsupreme, Natural Philosopher, Teller33, Mathsmad, Unknown 987, Tarten5, Nickmuller, Robomonster, Wolfrock, Jacob501, Kreemy, Synthebot, Tomaxer, Careercornerstone, Enviroboy, Rurik3, Sardonicone, Evanbrown326, Alliashax, Sylent, Rubentimothy, SMIE SMIE, Gamahucher, Braindamage3, Animalalley12895, Moohahaha, Thanatos666, Dillydumdum, AlleborgoBot, Voicework, Symane, Katzmik, Monkeynuts27, Demmy, Cam275, GoonerDP, SieBot, Mikemoral, LovelyLillith, James Banogon, BotMultichill, Timgregg96, Triwbe, 5150pacer, Soler97, Andersmusician, Anubhav29, Keilana, Tiptoety, Arbor to SJ, Undead Herle King, Richardcraig, Paolo.dL, Boogster, Oxymoron83, Henry Delforn (old), Avnjay, MiNombreDeGuerra, RW Marloe, SH84, Deejaye6, Musse-kloge, Jorgen W, Kumioko, Correogsk, MadmanBot, Nomoneynotime, Nickm4c, Darkmyst932, Anchor Link Bot, Jacob.jose, Randomblue, Melcombe, Yotex9, CaptainIron555, Yhkhoo, Dabomb87, Jat99, Pinkadelica, Francvs, Khirurg, Ooswesthoesbes, ClueBot, Volcom5347, Gladysamuel, GPdB, Bwfrank, DFRussia, PipepBot, Fox, Dobermanji, C1932, Remus John Lupin, Chocoforfriends, Smithpith, ArdClose, IceUnshattered, Plastikspork, Lawrence Cohen, Gawaxay, Nnemo, Ukabia, Michael.Urban, Niceguyedc, Xenon54, Mspraveen, DragonBot, Isaac25, 4pario, Donkeyboya, Excirial, CBOrgatrope, Bedsandbellies, Soccermaster3112, Alexbot, Tony-Ballioni, Pjb14, Ona01der, Andy pyro, Wikibobspider, BrentLeah, Eeekster, Anonymous13243546576879808978675645342311, Mycatiscool, Greenjuice, Chance Jeong, Arunta007, Greenjuice3.0, Greenjuice4, AnimeFan7, MacedonianBoy, ZuluPapa5, NuclearWarfare, JoelDick, Honeyspots3121, Blondeychek7, Faty148, Jotterbot, RC-0722, Wulfric1, Thingg, Franklin.vp, Aitias, DerBorg, Versus22, Hwalee76, SoxBot III, Apparition11, Mofeed.sawan, Slayerteez, XLinkBot, Marc van Leeuwen, Moocow444, Joejill67–enwiki, Little Mountain 5, Drumbeatsofeden, SilvonenBot, Planb 89, Alexius08, Vianello, MystBot, Zodon, RyanCross, Aetherealize, Zoltan808, Northdevonian, Aceleo, Blanche of King's

Lynn, Jetsboy101, Willking1979, Mattguzy, 3Nigma, DOI bot, Cdt laurence, Fgnievinski, Yobmod, Aaronthegr8, CanadianLinuxUser, Potatoscrub, MrOllie, Download, Protonk, Chamal N, CarsracBot, Favonian, LinkFA-Bot, ViskonBot, Barak Sh, Aldermalhir, Jubeidono, PRL42, Lightbot, Ann Logsdon, Floccinocin123, Matěj Grabovský, Fivexthethird, TeH nOmInAtOr, Jarble, Herve1729, Sitehut, Ptbotgourou, Senator Palpatine, TaBOT-zerem, Legobot II, Kan8eDie, Nirvana888, Gugtup, Washburnmav, Mikeedla, THEN WHO WAS PHONE?, Skyeliam, MeatJustice, Wierdox, AnomieBOT, Nastor, ThaddeusB, Connectonline, Taskualads, Themantheman, Galoubet, Neko85, Noahschultz, JackieBot, Commander Shepard, Chingchangriceball, Piano non troppo, Supersmashballs123, Agroose, Pm11189, Riekuh, Hamleto, Deverenn, Frank2710, Chief Heath, Easton12, Codycash33, Archaeopteryx, Citation bot, Merlissimo, ArthurBot, Tatarian, MauritsBot, Xqbot, TinucherianBot II, Sketchmoose, Timir2, Capricorn42, Johnferrer, Jmundo, Locos epraix, Br77rino, Isheden, Inferno, Lord of Penguins, Srich32977, Ragityman, Uarrin, LevenBoy, Quixotex, GrouchoBot, Resident Mario, ProtectionTaggingBot, Omnipaedista, Point-set topologist, Gott wisst, RibotBOT, Charvest, KrazyKosbyKidz, MarilynCP, Gingerninja12, Caleb7693, Deathiscomin90919, VictorPorton, Grg222, Daryl7569, Petes2176, GhalyBot, ThibautLienart, Prozo3190, Family400005, Bupsiij, Aaron Kauppi, Har56, Dr. Klim, Velblod, CES1596, GliderMaven, Thomascjackson, FrescoBot, RTFVerterra, Triwikanto, Tobby72, Mark Renier, Onetive15, VS6507, Alpboyraz, ParaDoxus, Sławomir Biały, Xefer, Zhentmdfan, Tzurvah MeRabannan, Citation bot 1, Amplitude101, Tkuvho, Rotje66, Kiefer.Wolfowitz, AwesomeHersh, ElNuevoEinstein, Cnwilliams, Gamewizard71, FoxBot, TobeBot, DixonDBot, Burritoburritoburrito, Fama Clamosa, Lotje, Dinamik-bot, Raiden09, Mrjames99, DJTrickyM, Stephen MUFC, Tbhotch, RjwilmsiBot, TjBot, Ripchip Bot, Galois fu, Alphanumeric Sheep Pig, BertSeghers, Mr magnolias, DarkLightA, LibertyDodzo, EmausBot, PrisonerOfIce, Nima1024, WikitanvirBot, Surlyduff50, AThornyKoanz, Mehdiirfani, Legajoe, Wham Bam Rock II, Dcirovic, Bethnim, ZéroBot, John Cline, Josve05a, Leafiest of Futures, Battoe19, Anmol9999, Scythia, Brandmeister, Vanished user fijtji34toksdcknqrjn54yoimascj, Prototypehumanoid, Ain92, Agatecat2700, Herk1955, Teapeat, Mjbmrbot, Anita5192, Liuthar, ClueBot NG, Incompetence, Wcherowi, Movses-bot, Kindyin, LJosil, SilentResident, Braincricket, Rbellini, Zackaback, MillingMachine, Helpful Pixie Bot, Thisthat2011, Curb Chain, AnandVivekSatpathi, Nashhinton, EmilyREditor, Ariel C.M.K., Fraqtive42, Graham11, AvocatoBot, Davidiad, Ropestring, Edward Gordon Gey, EliteforceMMA, Karthickraj007, VirusKA, MYustin, Brad7777, Idresjafary, Nbrothers, IkamusumeFan, Kavy32, Toploftical, Cyberbot II, Sklange, Blevintron, BlevintronBot, Sulphuric Glue, Dexbot, Rezonansowy, Mudcap, Sglooney316, AndiPersti, CuriousMind01, Augustus Leonhardus Cartesius, Pankaj Jyoti Mahanta, Ybidzian, Praemonitus, Comp.arch, TycoonSaad, Finnusertop, Jarash, George8211, Chern038, FireflySixtySeven, Kind Tennis Fan, Justin86789, 12visakhva, Dodi 8238, Suelru, Rcehy, Cosmia Nebula, Leegrc, Vanisheduser00348374562342, 115ash, AdditionSubtraction, Holypod, Mario Castelán Castro, Loraof, Arvind asia, GoldCoastPrior, Atvica, Retillinghast, User000name, KasparBot, Kafishabbir, The name, Aviartm, Ermahgerd9, MarkYabloko, MartinZ, Baking Soda, InternetArchiveBot, Jrheller1, Ttt74, GreenC bot, Dominic3203, Thephilosopher6, Hawkeye75, Apollo The Logician, Deacon Vorbis, Jon Kolbert, King Prithviraj II, Fluffyyyy, Wikigenstein, KolbertBot, Thechinesekid and Anonymous: 1222

- **Mathematical logic** *Source:* https://en.wikipedia.org/wiki/Mathematical_logic?oldid=799090642 *Contributors:* AxelBoldt, Michael Hardy, Modster, Dominus, TakuyaMurata, Tregoweth, Peter Damian (original account), Charles Matthews, Dcoetzee, Nohat, Mwoolf, Dysprosia, OkPerson, Piolinfax, Hyacinth, David Shay, Aleph4, Robbot, Romanm, Gandalf61, MathMartin, Ojigiri~enwiki, Tea2min, Giftlite, Recentchanges, Lethe, Fleminra, Jason Quinn, Matt Crypto, Edcolins, Beefalo, Kntg, Eduardoporcher, Barnaby dawson, HedgeHog, Smimram, EugeneZelenko, Guanabot, Leibniz, Wclark, Ivan Bajlo, Paul August, Spayrard, Lycurgus, Chalst, Kwamikagami, Irr*ti*nal, Art LaPella, Atomique~enwiki, Nicke Lilltroll~enwiki, Jojit fb, Obradovic Goran, Mdd, Tsirel, Msh210, Chira, CuriousOne, Sligocki, Samohyl Jan, Lebob (renamed), Jak86, Mindmatrix, Guardian of Light, Ruud Koot, Graham87, BD2412, Rjwilmsi, Salix alba, R.e.b., Reinis, Nigosh, Mathbot, Tillmo, Jersey Devil, FrankTobia, Roboto de Ajvol, Wavelength, Hairy Dude, RussBot, Icarus3, SpuriousQ, Polyvios, KSchutte, Rick Norwood, Meloman, Trovatore, Musteval, Tony1, Bota47, Robertbyrne, Reyk, Claygate, MaNeMeBasat, Aeosynth, Otto ter Haar, Sardanaphalus, JJL, SmackBot, SaxTeacher, Bomac, Jagged 85, RockRockOn, Logic2go, Ohnoitsjamie, Chris the speller, Jnorden, Persian Poet Gal, MK8, MalafayaBot, Sholto Maud, Nixeagle, Maksim-bot, Allan McInnes, Stevenmitchell, Jon Awbrey, DMacks, Sammy1339, Bidabadi~enwiki, Tkos, Byelf2007, The undertow, Lambiam, Rsimmonds01, Bjankuloski06en-enwiki, Snem, Nabeth, Dan Gluck, Iridescent, Stotr~enwiki, Francl, Mrdthree, JRSpriggs, Atomobot, CRGreathouse, CBM, Thomasmeeks, Myasuda, Gregbard, Gogo Dodo, Julian Mendez, Thijs!bot, Brian G. Wilson, Blah3, Dgies, Malcolm, Alphachimpbot, VictorAnyakin, JAnDbot, Avaya1, Meeples, Ling.Nut, Robin S, Metamusing, Maurice Carbonaro, NewEnglandYankee, Policron, Uhai, DavidCBryant, Treisijs, Alan U. Kennington, VolkovBot, JohnBlackburne, Am Fiosaigear~enwiki, TXiKiBoT, Ontoraul, The Tetrast, Martin451, Kowsari, Davin, Cremepuff222, Popopp, Dmcq, Palaeovia, Ohiostandard, Flyer22 Reborn, Hxhbot, Blueclaw, OKBot, Valeria.depaiva, IsleLaMotte, CBM2, Butane Goddess, LarRan, Francvs, ClueBot, LAX, The Thing That Should Not Be, Smithpith, WikiSBTR, Compellingelegance, Razimantv, ScNewcastle, Alexbot, Kanguole, Hans Adler, PergolesiCoffee, Certes, Skolemizer, Good Olfactory, Addbot, NjardarBot, MrOllie, Ozob, TeH nOmInAtOr, Legobot, Math Champion, Ddzhafar, Pcap, AnomieBOT, Götz, Citation bot, Xqbot, RJGray, GrouchoBot, Omnipaedista, The Wiki ghost, Sophus Bie, Tales23, Mark Renier, VS6507, Orhanghazi, Qiemem, Zhentmdfan, Citation bot 1, Tkuvho, I dream of horses, Jonesey95, Sa'y, TheIndianWikiEditor, Rover6891, Steve2011, BostX, Ham and bacon, No One of Consequence, Onel5969, Jmencisom, Anirudh Emani, Chharvey, Future ahead, Karthikndr, Philippe BINANT, 28bot, Maxdlink, ClueBot NG, WikiTatik, Movses-bot, O.Koslowski, Masssly, Joel B. Lewis, MerllwBot, Helpful Pixie Bot, BG19bot, Theotherscripto12, Virus2801, Brad7777, Spasoev, Sfarney, Vedsuthar, Cyberbot II, None but shining hours, Khazar2, Houbn, Dtotoo, Dexbot, Deltahedron, Cwobeel, Jochen Burghardt, BrooksMaxwell, Greenjello77, Xaqron, MarshalWalter, Kirstenlovesyouu, The Annoyed Logician, Dunditschia, Editor of the wiki swag, Aarsh A Chotalia, StephenDunker, Greatingsworld, Tony the but hole, KasparBot, Chery Ann Arguelles, Shahryar1976, Βοιοῦης, Gari chetty, Lr0^^k, X1X2X3, Baking Soda, GreenC bot, Longisquama, L3X1, Nihlus, JOEMANDUDEPERSONCOOLNESSOFAWSOMENESS, AvalerionV, KolbertBot, Investigation11111 and Anonymous: 203

- **Mathematical statistics** *Source:* https://en.wikipedia.org/wiki/Mathematical_statistics?oldid=801017550 *Contributors:* Deljr, Cvore, Den fjättrade ankan~enwiki, Charles Matthews, Dysprosia, Power~enwiki, Creidieki, HedgeHog, Rich Farmbrough, Dallashan~enwiki, Joolz, Avenue, Igny, Acerperi, NeoUrfahraner, Joe Decker, Algebraist, Neilbeach, Number 57, Bota47, Sardanaphalus, JJL, SmackBot, Boris Barowski, MalafayaBot, Dreadstar, Richard001, G716, Bjankuloski06en-enwiki, Andrew Davidson, CBM, Panda17, Lehalle, Chrislk02, Thijs!bot, Soulviver, JAnDbot, Ugajin, R'n'B, J.delanoy, KylieTastic, VolkovBot, JohnBlackburne, Oshwah, AlleborgoBot, Melcombe, ClueBot, Turbojet, Qwfp, Addbot, Luckas-bot, Yobot, AnomieBOT, Jim1138, Materialscientist, Obersachsebot, Xqbot, Capricorn42, GrouchoBot, AstaBOTh15, DrilBot, Kiefer.Wolfowitz, PlyrStar93, H.ehsaan, EmausBot, Dcirovic, ZéroBot, Nightfury, ChuispastonBot, ClueBot NG, Helpful Pixie Bot, Marcocapelle, Brad7777, Illia Connell, Enterprisey, Dexbot, Brirush, Seppi333, KasparBot, Marvellous Spider-Man, Schistocyte, Prinsipe Ybarro and Anonymous: 41

- **Theoretical computer science** *Source:* https://en.wikipedia.org/wiki/Theoretical_computer_science?oldid=801884463 *Contributors:* Stan Shebs, David.Monniaux, Tea2min, Giftlite, Mahanga, Ruud Koot, BD2412, Arbor, Protez, SLi, Intgr, Lmatt, RussBot, Arado, Dv82matt,

- **Outline of physical science** *Source:* https://en.wikipedia.org/wiki/Outline_of_physical_science?oldid=802225531 *Contributors:* Zundark, Jzcool, Ghakko, DavidLevinson, Michael Hardy, Ixfd64, Kosebamse, Ronz, Stefan-S, Poor Yorick, Big iron, IceKarma, Fvincent, Phoebe, Zandperl, Lowellian, Merovingian, Academic Challenger, Pengo, Chowbok, Andycjp, Antandrus, Beland, DragonflySixtyseven, Icairns, Talrias, D6, DanielCD, Random contributor, Discospinster, Vsmith, Bishonen, Paul August, RoyBoy, Bobo192, Circeus, AnyFile, Maureen, 9SGjOS-fyHJaQVsEmy9NS, Mdd, Jumbuck, Joolz, Andrewpmk, Paleorthid, Ynhockey, Velella, SteinbDJ, Gene Nygaard, Alai, Oleg Alexandrov, Woohookitty, Linas, Hard Raspy Sci, Joke137, Liface, Skylion, Phlebas, Graham87, BD2412, Jclemens, Melesse, Josh Parris, Vegaswikian, Yellowmellow45, Gurch, Jrtayloriv, X42bn6, Chaos, Wimt, NawlinWiki, Harro, Stevenwmccrary58, DeadEyeArrow, IceCreamAntisocial, Leptictidium, Www.wikinerds.org, 21655, Pb30, Katieh5584, FrobozzElectric, Pentasyllabic, DVD R W, Yvwv, SmackBot, David Kernow, Jagged 85, Gabi bart, WookieInHeat, CMD Beaker, Binarypower, Onebravemonkey, Edgar181, Powo, Stephan202-enwiki, Gilliam, ERcheck, MalafayaBot, Papa November, SchfiftyThree, Go for it!, Hallenrm, Can't sleep, clown will eat me, TheGerm, Zalmoxe, Savidan, TechPurism, Richard001, RandomP, Just plain Bill, Serein (renamed because of SUL), Evenios, Needlenose, Cyberstrike2000x, Special-T, RichardF, Caiaffa, TheFarix, Levineps, Iridescent, Courcelles, Generalcp702, Postmodern Beatnik, CmdrObot, Dojikami, Yaris678, Cydebot, Christian75, Alaibot, Savitr, Epbr123, Mbell, CharlotteWebb, Dawnseeker2000, I already forgot, Cyclonenim, AntiVandalBot, Pariomasial, Volcanoguy, Leuko, The Transhumanist, True Genius, Staib, Bongwarrior, VoABot II, Professor marginalia, SineWave, Rich257, Kjhskj75, Loonymonkey, RisingStick, Gwern, MartinBot, Sm8900, Mrdoc, El0i, Shellwood, J.delanoy, Maurice Carbonaro, Jalaldn, NerdyNSK, Space-Age Meat, Cpiral, Katalaveno, DarkFalls, McSly, Mrceleb2007, The Transhumanist (AWB), NewEnglandYankee, SJP, Redrocket, Feer, Funandtrvl, Sam Blacketer, ABF, Indubitably, AlnoktaBOT, Oshwah, Technopat, Hqb, Lradrama, DoktorDec, Cremepuff222, Maxim, Mouse is back, Falcon8765, Twooars, Palaeovia, Matthe20, Twopenguins, WereSpielChequers, Malcolmxl5, WTucker, Caltas, Yintan, Bentogoa, Quest for Truth, Tiptoety, Cbone9056, Ssting, Oxymoron83, Faradayplank, Avnjay, KoshVorlon, Spartan-James, Yhkhoo, JL-Bot, Roraem, Martarius, ClueBot, Avenged Eightfold, DFRussia, Tanglewood4, Mild Bill Hiccup, DanielDeibler, Xenon54, Alpha Ralpha Boulevard, NuclearWarfare, Promethean, Jo Weber, ChrisHodgesUK, La Pianista, Calor, Thingg, Qwfp, SoxBot III, Saeed.Veradi, Viccarothers, Mifter, Toomanylies, ZooFari, Ncaaballa12, Destinylee, King Pickle, GoldenMedian, Elron WolfBane, Jojhutton, Captain-tucker, Crazysane, Karmin90, Fgnievinski, GD 6041, MrOllie, Glane23, Sseedaf, Issyl0, Tide rolls, Verbal, Lightbot, Fryed-peach, Swarm, Legobot, Yobot, The Grumpy Hacker, THEN WHO WAS PHONE?, Tempodivalse, AnomieBOT, Jim1138, Piano non troppo, Materialscientist, Xqbot, Cureden, Capricorn42, MZK77QRH9, Ataleh, ChristopherKingChemist, Ckit13, Doulos Christos, Shadowjams, A.amitkumar, Thehelpfulbot, Sky Attacker, Anterior1, Jumanji95, Finalius, Jamesooders, Jayachopra, Winterst, Pinethicket, Jonesey95, RedBot, Serols, Gamewizard71, كاشرف عقيل, Sweet xx, Jonkerz, January, DARTH SIDIOUS 2, Whisky drinker, The Utahraptor, Hajatvrc, Josephcunningham, RA0808, Wikipelli, Dcirovic, Hhhippo, Captain Screebo, Wayne Slam, RockMagnetist, Peter Karlsen, Kleopatra, Petrb, Xanchester, ClueBot NG, Peter James, Incompetence, CaroleHenson, Throw3345, Titodutta, Calabe1992, Wbm1058, BG19bot, Hallows AG, Ployer1, Mark Arsten, Johnisnaked, Cpgdaarfob, Smartweirdo, Polarbearsstare, TBrandley, Shaun, Cimorcus, LHcheM, CarrieVS, Enterprisey, Mogism, Saehry, Isarra (HG), Frosty, OakRunner, Nikolas Tales, Gyan-nehru, Jporter52515, CsDix, EzA+lSeb Nnakari, DavidLeighEllis, Pokedora, A-Z Raju, Finnusertop, Vinny Lam, Susan.grayeff, Suelru, Sharkeisha233232323, Trollman700, Ujmmbնղ, Amortias, Loraof, NekoKatsun, DiscantX, Dietic, CaptainPiggles, AlexanderThe3rd, CV9993, Squiver, GreenMeansGo, S.grayeff, Webslinger3423, Katzrockso, Boomer Vial, Articlecreate1234, Bear-rings, 72, Sassysquirrell, Edthat2, Magic links bot, Jyoti kumari, Tornado chaser and Anonymous: 442

- **List of life sciences** *Source:* https://en.wikipedia.org/wiki/List_of_life_sciences?oldid=802971067 *Contributors:* Karen Johnson, Someone else, Gabbe, Edcolins, Stevietheman, Bender235, Billlion, Anthony Appleyard, VoluntarySlave, BDD, Dennis Bratland, WadeSimMiser, Rjwilmsi, ErikHaugen, Chobot, DVdm, Bgwhite, Wavelength, Malcolma, Daniel Mietchen, Yeryry, Pegship, Sandstein, Rrburke, NickPenguin, Lambiam, IronGargoyle, Hgrobe, JForget, NickW557, ShelfSkewed, QuiteUnusual, The Transhumanist, Fabrictramp, Adrian J. Hunter, Edward321, Philologia Sæculāres, Katharineamy, Chiswick Chap, Cmichael, Medicineman84, Funandtrvl, Guillaume2303, Csapdani, Quest for Truth, Flyer22 Reborn, JL-Bot, Jjkutch, MenoBot, ClueBot, GorillaWarfare, Fadesga, Niceguyedc, Excirial, Thingg, Jengirl1988, Roxy the dog, Jytdog, Addbot, Fgnievinski, Tide rolls, Luckas Blade, Luckas-bot, Yobot, Amirobot, AnomieBOT, Jim1138, JackieBot, Materialscientist, Citation bot, E0steven, Peter Buch, Dan6hell66, Pinethicket, I dream of horses, MBirkholz, Tom.Reding, Gamewizard71, Writeswift, Merlinsorca, Tbhotch, DARTH SIDIOUS 2, Seanhealey, EmausBot, Look2See1, Racerx11, GoingBatty, KIbrain, Solarra, Moswento, Dcirovic, K6ka, ZéroBot, DJ Tricky86, Midas02, Vanished user sh304hjsj3hsl43, Bamyers99, Thine Antique Pen, Tomásdearg92, RockMagnetist, 28bot, Kleopatra, Tequilastoner, ClueBot NG, NobuTamura, Ypnypn, MelbourneStar, Biovisionlyon, Demon 83, Delusion23, Theopolisme, Strike Eagle, Titodutta, Tuke admin, BG19bot, Northamerica1000, Graham11, Zyxwv99, Snow Rise, Atomician, Vanischenu, Loriendrew, Horai 551, BattyBot, EuroCarGT, JYBot, Dexbot, Evad37, MisterShiney, Telfordbuck, Theo's Little Bot, CsDix, I am One of Many, Jamesmcmahon0, Iztwoz, DavidLeighEllis, Wikiuser13, Dalutgens, Chrisandres, Ginsuloft, Curalelan, Rasheedavail, Vinny Lam, Adarshpp300, Angkp nuss, Chaya5260, Monkbot, DSCrowned, Loraof, Rambo12212, Dietic, Squiver, KasparBot, Diego441, CLCStudent, InternetArchiveBot, Bear-rings, Jagatroy, Bender the Bot, Linguist91, Kinberley, Dockabo, ATPhosphate, AlchemTarun, Anthunterwiki, Volunteer1234, Forensichic and Anonymous: 136

- **Social science** *Source:* https://en.wikipedia.org/wiki/Social_science?oldid=803366986 *Contributors:* WojPob, The Anome, Ed Poor, Larry Sanger, Grouse, Rgamble, William Avery, SimonP, Ryguasu, Earth, Lexor, Ixfd64, TakuyaMurata, Kingturtle, Nikai, Andres, Rob Hooft, BRG, Charles Matthews, Reddi, Pedant17, Tpbradbury, Hyacinth, Studymore, J D, Shizhao, Rbellin, Flockmeal, Gakmo, JorgeGG, PuzzletChung, Bearcat, Robbot, Tomchiukc, Tobias, Jxg, Nilmerg, Gidonb, Sunray, Hadal, Quadalpha, Cyrius, Filemon, Alan Liefting, Stirling Newberry, Ancheta Wis, Aphaia, Everyking, Yekrats, Elmindreda, Wildt-enwiki, Stevietheman, Andycjp, Jdevine, Beland, Piotrus, Gsociology, Secfan, Kevin B12, Pethan, Scott Burley, Karl-Henner, Tharenthel-enwiki, Asbestos, Mike Rosoft, D6, CALR, Discospinster, Rich Farmbrough, Uni-Ace, Florian Blaschke, Bender235, Andrejj, El C, Mwanner, Marcok, Art LaPella, Bookofjude, Jlin, Bobo192, Brendansa, Viriditas, Tmh, Maureen, Nlight, Jojit fb, MPerel, Nsaa, Mdd, Jumbuck, Alansohn, Gary, Rodw, Logologist, Wikidea, Neilmckillop, Velella, Grenavitar, Vol-

26.10. TEXT AND IMAGE SOURCES, CONTRIBUTORS, AND LICENSES 277

untarySlave, TriNotch, Y0u, Megan1967, Velho, Woohookitty, RHaworth, Rattus, Kzollman, Mcphee-enwiki, Jeff3000, Tabletop, Wikiklrsc, Smmurphy, BrenDJ, Iniobong, Joe Roe, Graham87, KaisaL, FreplySpang, Island, Icey, Rjwilmsi, Mayumashu, Erebus555, Dpark, Helvetius, Eyu100, XLerate, Ground Zero, Margosbot-enwiki, GünniX, Jameshfisher, TheSpook, RexNL, Gurch, YashaBK, Chobot, DVdm, Roboto de Ajvol, The Rambling Man, RussBot, ROYGBIV, Fabartus, Barfoos, Bhny, Mark Ironie, Anomalocaris, Frybread, Jgrantduff, Rjensen, Jpbowen, Dolsson5, Moe Epsilon, RL0919, Dbfirs, Aaron Schulz, M3taphysical, Action potential, Maunus, Dv82matt, Mamawrites, CQ, Zzuuzz, Pb30, KGasso, Josh3580, Sean Whitton, Feedmymind, JoanneB, Katieh5584, NeilN, DVD R W, Gbalaji82, Luk, SmackBot, RedHouse18, Ma8thew, Ariedartin, Bomac, Jagged 85, Jfgrcar, NickShaforostoff, Binarypower, Hardyplants, Gaff, Yamaguchi??, Gilliam, Hmains, Skizzik, Ppntori, Andy M. Wang, Frédérick Lacasse, The monkeyhate, Chris the speller, Koryakov Yuri, Te24409nsp, Jprg1966, Thumperward, Fplay, Moshe Constantine Hassan Al-Silverburg, Jerome Charles Potts, CSWarren, J. Spencer, Go for it!, Baronnet, DHN-bot-enwiki, Colonies Chris, Arges, Darth Panda, Battlecry, Aldaron, Cybercobra, EPM, Hoof Hearted, Cordless Larry, RandomP, Nexus Seven, Salt Yeung, Ice tres, Starghost, Clicketyclack, Cast, Z-d, Arnoutf, OcarinaOfTime, J 1982, Ishmaelblues, Perfectblue97, JorisvS, Commons@tiac.net, Astuishin, Vir, RichardF, Hu12, BranStark, Iridescent, K, Bogle, LeyteWolfer, Tophtucker, JHP, Igoldste, Fsotrain09, Shoreranger, Marysunshine, Tawkerbot2, CmdrObot, Dycedarg, Makeemlighter, Page Up, DaveDixon, Yarnalgo, NickW557, Thomasmeeks, ButterApple, Neelix, Sanspeur, Penbat, Peripitus, UncleBubba, Gogo Dodo, Wiknik, Garik, Omicronpersei8, Mattisse, Barticus88, Nhelm83, TheFearow, JustAGal, PolarisSLBM, Escarbot, Mentifisto, Cyclonenim, Hires an editor, AntiVandalBot, Guy Macon, Seaphoto, Little Rachael, Edokter, Osubuckeyeguy, Mack2, Danger, Gökhan, Crissidancer88, Husond, Hijklmno, MER-C, The Transhumanist, Kevinharbin, Douglas R. White, Yahel Guhan, MaxPont, Reswik, Magioladitis, Gsaup, VoABot II, Dekimasu, (Didie), Lkpotts, Catgut, WhatamIdoing, Davies69, Cooper-42, Vssun, Foregone conclusion, Gwern, S3000, MartinBot, Kostisl, R'n'B, Laleena, Thirdright, Shellwood, J.delanoy, Pharaoh of the Wizards, Adavidb, Uncle Dick, AThousandYoung, Nigholith, Arcette, NerdyNSK, Jkaplan, Mannschaftskapitän, Smeira, DarwinPeacock, JayJasper, Sunidesus, Loohcsnuf, The Transhumanist (AWB), NewEnglandYankee, SJP, Treisijs, The Fat Guy, DASonnenfeld, M. Frederick, Izno, Funandtrvl, VolkovBot, Profdrmendoza, Dom Kaos, WOSlinker, TXiKiBoT, Cosmic Latte, Jacob Lundberg, Tomsega, Miranda, Guillaume2303, Gerrish, Gen. Quon, Nickipedia 008, Stuttgart1950, Justinfr, George Sagi, Mikehoffman, Synthebot, Kiliwa, Asamind, Sesshomaru, Insanity Incarnate, Pjoef, Logan, Moax18, SieBot, StuartGaia, France3470, Keilana, Bentogoa, Happysailor, Flyer22 Reborn, Qst, HighburyVanguard, Yerpo, Oxymoron83, Lightmouse, Fratrep, Svick, Catrope, Segregold, Szalagloria, Miss loves, Wetwarexpert, Mr. Stradivarius, Superbeecat, Denisarona, Escape Orbit, Jaipurite, ClueBot, SummerWithMorons, GorillaWarfare, RisingSunWiki, Gorillasapiens, EastCoast1111, Fioravante Patrone en, The Thing That Should Not Be, ImperfectlyInformed, Yiannis vet, Mild Bill Hiccup, Niemeyerstein en, BlondGirl, CharlieRCD, Masterpiece2000, DragonBot, Awickert, Excirial, Alexbot, Jusdafax, Hezarfenn, Rkh10, Tnxman307, Redthoreau, Doktor Mephisto, Bilbaosr, Aleksd, Taranet, Pularoid, Patricusrex, MelonBot, DumZiBoT, 3pointswish, Yodaki, Little Mountain 5, WikHead, SilvonenBot, Zaloom, Quinntaylor, Addbot, Fgnievinski, Riadismet, Mootros, SpellingBot, Leszek Jańczuk, Mattgizzy14, Cst17, Chamal N, CarsracBot, Glane23, Favonian, Blaylockjam10, Numbo3-bot, Tide rolls, Zorrobot, Simon J Kissane, Elm, Megaman en m, Jessika Folkerts, Legobot, Kurtis, Luckas-bot, Yobot, TaBOT-zerem, Abasass, Carnoy, Andresswift, Moptan2007, Sumail, AnomieBOT, Rjanag, Pgm8693, Galoubet, JackieBot, Ularevalo98, Ulric1313, Materialscientist, E2eamon, Eumolpo, SeventhHell, LilHelpa, Obersachsebot, Xqbot, JJ cool D, Jesse627, Phil hutchinson, Anna Frodesiak, Miguel in Portugal, Inferno, Lord of Penguins, Susan Chan, J04n, GrouchoBot, Omnipaedista, GhalyBot, Verbum Veritas, Fleetfist, BignBad, Joaquin008, Qarana, Zojiji, FrescoBot, FalconL, Zeyi, Zero Thrust, HamburgerRadio, Citation bot 1, Killian441, Redrose64, Midiom, Boxplot, Otto S. Knottnerus, Pinethicket, HRoestBot, Ibn khaldun78, DTMGO, Jonesey95, MastiBot, LanecoC, Doktore, Nafile, Gamewizard71, FoxBot, TobeBot, Darigan, Vrenator, LilyKitty, Duoduoduo, Arjunjaidka, Onel5969, Mean as custard, TjBot, Ripchip Bot, Floydman66, DASHBot, AtTheNecropolis, EmausBot, John of Reading, Josephcunningham, Hirsutism, Ajraddatz, K6ka, Acategory, Milindkrishna1, Efierros, Oswaldoalvizarb, H3llBot, Gz33, Wayne Slam, Swentworth25, Erianna, MercWithMouth, Phronetic, Michtan12, LS C HIST, RockMagnetist, ClamDip, Vox Pluvia, ResidentAnthropologist, Jackofhats, ClueBot NG, Chen969, CocuBot, This lousy T-shirt, Frietjes, Yahudikiwi, Vrangbaek, Widr, Paterfranz, Mouramoor, Saxoamacad, Heakins, GuyHimGuy, Helpful Pixie Bot, Gob Lofa, BG19bot, Tengwang777, Aloharumi, Drjmh, Graham11, MagicCinemas, Frze, Davidiad, Mark Arsten, Robert Dal, DCMaynard, Altair, Harizotoh9, Pendulum Dowsing, Dayshade, Princess843, Amalo98, Timbrooker, Purnava1827, Taief.shahed, Kleinbaum, Cyberbot II, Ushau97, GoShow, Littlesoup, Faryalnasir22, Codeh, Loupiotte, EagerToddler39, Dexbot, Hmainsbot1, Webclient101, Makecat-bot, UseTheCommandLine, Wecoexist, Lugia2453, Frosty, Jamesx12345, RotlinkBot, GabeIglesia, AsthreeID, Abhijeet Sinha Sinha, C5st4wr6ch, SocSciPhD, CsDix, CGBoas, Tentinator, EvergreenFir, Backendgaming, New worl, CouvGeek, Finnusertop, Wolfgang.molnar, Bryanf222, Vinny Lam, Averruncus, Joan Ricart-Huguet, Param Mudgal, Linuxrox, Plunkersiniapes, Keshab Ray, JaconaFrere, Crossswords, Kingxdahz, Proudcommunist111111, Mutluri Abraham, KeanonRitchie, RC711, Mxschumacher, KH-1, Zabshk, IahBessy, Squiver, Supdiop, Groenewilde, Johnlgreen, BU Rob13, Semak mwu, CAPTAIN RAJU, Manname, Jacklin417, InternetArchiveBot, Santoshone, Wpmantra, GreenC bot, Dr.KOLLMANN J. György, Jossavanleeuwen, Bear-rings, Marshy maburner, Cherrypicker5454, Empiricism667, Magic links bot, Prinsipe Ybarro, Gtravel, KolbertBot, Pranjal jain, Matt7899 and Anonymous: 692

- **Applied science** Source: https://en.wikipedia.org/wiki/Applied_science?oldid=794468140 Contributors: (, Ellywa, Llull, Andres, Mxn, Robbot, Alan Liefting, Mintleaf-enwiki, Tom harrison, Wellparp, Discospinster, Paul August, Remuel, Kjkolb, Jumbuck, Danski14, Gary, Liao, Paleorthid, Blaxthos, Phlebas, Leslie Mateus, Srleffler, Chobot, Metropolitan90, Vmenkov, YurikBot, Thomas E. Goodwin, G.G., Tachyon01, Xaxafrad, AdBo, Mais oui!, Snaxe920, SmackBot, Unschool, Melchoir, Gilliam, Tito4000, Fplay, Addshore, Cybercobra, Khukri, EPM, Dr. Gabriel Gojon, SashatoBot, Oskilian, Bjankuloski06en-enwiki, Caroline.marshall, Childzy, RichardF, Satish.murthy-enwiki, K, RekishiEJ, Dp462090, Heqs, JPilborough, Cydebot, Thijs!bot, Epbr123, Headbomb, Aasimar, Escarbot, AntiVandalBot, Courtjester555, Osubuckeyeguy, Teilhardo, Kaobear, The Transhumanist, Bubba hotep, Lenticel, DGG, J.delanoy, Pharaoh of the Wizards, Maurice Carbonaro, Jorfer, Treisijs, Rjclaudio, The Fat Guy, Coolmanoj, Oshwah, BotKung, Kilmer-san, Yk Yk Yk, Gsb212, Falcon8765, SieBot, Matthew Yeager, Flyer22 Reborn, Oiws, ImageRemovalBot, Leranedo, ClueBot, MBD123, Dawdler, Snigbrook, Healthwise, The Thing That Should Not Be, Excirial, Jusdafax, IForTheMoney, XLinkBot, Addbot, CubBC, J05HYYY, Quercus solaris, Maltesedove, Zorrobot, Luckas-bot, Yobot, TaBOT-zerem, The Grumpy Hacker, გოგო, MacTire02, JackieBot, Minnecologies, ??, Xqbot, GrouchoBot, Shirik, Shadowjams, Erik9bot, Pepper, Lukich1014, Bellsniff, Pinethicket, Inbamkumar86, فرانك يوقى, SchreyP, Comet Tuttle, Hyarmendacil, EmausBot, WikitanvirBot, ZéroBot, Toshi93, Pingu.dbl96, Rangoon11, RockMagnetist, ClueBot NG, Northamerica1000, AvocatoBot, Joydeep, Jionpedia, Lugia2453, Timothy Perseus Wordsworthe, Brian.doyle88, Zenibus, Mrm7171, Sam Sailor, Nizam655, Nainsal, Kylo Ren, Squiver, KasparBot, Milkstout39, SlimKH45, Samdog1243, Fmadd, Wmdly and Anonymous: 125

- **Interdisciplinarity** Source: https://en.wikipedia.org/wiki/Interdisciplinarity?oldid=802843724 Contributors: Tobias Hoevekamp, Mav, The Anome, Slrubenstein, Andre Engels, Ted Longstaffe, XJaM, SimonP, Heron, Edward, Lexor, Rossami, Cherkash, Morwen, Buridan, Robbot, ShaunMacPherson, Jfdwolff, Jeremykemp, M.R.Forrester, ??, Noisy, Rich Farmbrough, El C, Rgdboer, Alison9, Gary, Arthena, Paleorthid, SidP, Dsimcha, Woohookitty, Liface, Stefanomione, Mandarax, Rjwilmsi, Vivekdse, Smithfarm, Tiedau, Chobot, Vorpal Suds, Bgwhite, YurikBot, Wavelength, X42bn6, RussBot, Fabartus, Gaius Cornelius, Steviejay, Madcoverboy, Leutha, Deskana, Berlin Stark, Supten, MaxVeers, Guillom, HereToHelp, SmackBot, Pfaff9, Ppntori, Jprg1966, Alieseraj, Jon Awbrey, Adamarthurryan, BikeStrong, SliceNYC, Bendzh, Dstokols, Matatigre36, Levineps, RekishiEJ, DavidOaks, Lsiskin, Covalent, Fufyk, JamesX, CmdrObot, Picaroon, Ejph, Cydebot, Dancter, After Midnight, Letranova, Al Lemos, Marek69, CharlotteWebb, Jj137, Gregalton, Alphachimpbot, Wayiran, Gloriaoriggi, Glenntwo, Skomorokh, Inks.LWC, −1g, Wasell, Snowded, Fabrictramp, RogerSun, Betsyrosalen, The Nixinator, Marieprecious, Thyroidpsychic, LvInglafoftneatsteak, Ejm634, Tddwigg, Fmandog85, Tokidoki27, Meencantayoga, Devastator1906, Jim.henderson, Erkan Yilmaz, The One I Love, Nemo bis, Mrg3105, Hadgraft, Kenneth M Burke, Jcs2006, Funandtrvl, Burlywood, Tonytypoon, Djcremer, Don4of4, Room429, Jcorry, SieBot, Taskoh, Pahndeepah, Correogsk, Aibdescalzo, Denisarona, Wambugwe, ClueBot, SummerWithMorons, Vendavel, Bradka, Gobeshock Gobochondro Gyanotirtho, BirgerH, Dansville Foster, Thesunkenroad, Altacc, Marshamarshamarsha, Ttm1974, Bluemosquito, Askahrc, Behavioralethics, DumZiBoT, Libcub, Fiskbil, Good Olfactory, Addbot, Fgnievinski, Victor-435, Download, Lightbot, Abduallah mohammed, Margin1522, Luckas-bot, Lauriehaycock, Yobot, Ptbotgourou, I.A.Contino, Brougham96, Pganas, Bbb23, 1exec1, ArthurBot, LilHelpa, Xqbot, DSisyphBot, Gumruch, Suvojtc, GreekAlexander, Sahehco, Reinhard Hartmann, FrescoBot, Interdisciplinarity, Wiki3ditorial, D'ohBot, Steve Quinn, Pinethicket, RedBot, Wagersmith, Bgpaulus, Anticent, RobertHuaXia, Gamewizard71, Atarroyo, PhilStrauss, Curious1949, Jfmantis, Pinkbeast, EmausBot, Golfandme, Moaiivow, Tornado Maker, GrindtXX, Staceyctobin, RockMagnetist, ClueBot NG, Elongatedmoose, Killerscene, Ouso1999, Wbm1058, Marcocapelle, Jmcquaid79, Wikih101, DMSchneider, Khazar2, JYBot, Cerabot~enwiki, The Vintage Feminist, Settdigger, Me, Myself, and I are Here, Studyyear, Star767, Skourki, Superegz, AwesomeSky, HMSLavender, Loraof, KasparBot, Ira Leviton, CAPTAIN RAJU, 1848jag, GlobalStrategy, Brachney, Chrissymad, Bear-rings, Bender the Bot, Carlowen, Magic links bot, KolbertBot and Anonymous: 144

- **Philosophy of science** *Source:* https://en.wikipedia.org/wiki/Philosophy_of_science?oldid=803389343 *Contributors:* Mav, The Anome, Tim Chambers, --April, Ed Poor, RK, Anthere, Ajdecon, B4hand, R Lowry, ChrisSteinbach, Michael Hardy, Fred Bauder, Isomorphic, BoNoMoJo (old), 168..., Snoyes, JWSchmidt, Poor Yorick, Andres, Evercat, Sethmahoney, Mxn, Hike395, Charles Matthews, Timwi, Scmarney, Reddi, Ww, Jm34harvey, Dysprosia, Markhurd, Talkingtoaj, OverZealousFan, Darwindecks, Omegatron, Buridan, Topbanana, Banno, ThereIsNoSteve, Lumos3, Robbot, MrJones, Murray Langton, Jaredwf, Goethean, Rursus, Rasmus Faber, Hadal, Matthew Stannard, Giftlite, Polsmeth, Kim Bruning, Everyking, Duncharris, Christofurio, JRR Trollkien, Neilc, Andycjp, Popefauvexxiii, LiDaobing, Piotrus, Karol Langner, APH, JimWae, Jokestress, Karl-Henner, Sam Hocevar, Starx, Marcos, Jmeppley, Tsemii, Caton~enwiki, Fermion, Robin klein, Eduardoporcher, ELApro, D6, Mormegil, Rich Farmbrough, Vsmith, Bender235, ESkog, Kharhaz, Mjk2357, Mwanner, Wareh, CDN99, Bobo192, Icut4you, Johnkarp, Polocrunch, Nk, Pearle, Mdd, Alansohn, Sextus~enwiki, WhiteC, Burn, Gene Nygaard, Kazvorpal, Falcorian, Mahanga, Patrice Létourneau, Mel Etitis, Jpers36, Barrylb, Kzollman, Jeff3000, Wileycount, Eleassar777, Jok2000, Tabletop, Wikiklrsc, Ivar Y, GregorB, Macaddct1984, Plrk, Aidje, BD2412, Rjwilmsi, Zbxgscqf, KYPark, Jweiss11, HolyApocalypse, Mike Peel, Vegaswikian, Cassowary, Margosbot~enwiki, Twipley, Gark, Truman Burbank, Chreliot, Nick81, Ahwaz, DVdm, Bgwhite, Roboto de Ajvol, The Rambling Man, Hal4, Wavesmikey, Gaius Cornelius, KSchutte, CarlHewitt, NawlinWiki, A314268, Leutha, Grafen, Welsh, Yahya Abdal-Aziz, ETTan, Ragesoss, Philosofool, Shotgunlee, WAS 4.250, Enormousdude, Bondegezou, Brianlucas, CWenger, Fram, Palthrow, MullerHolk, Infinity0, Snalwibma, Otheus, JJL, SmackBot, Rtc, Jim62sch, Johnrcrellin, Jagged 85, Srnec, Brothers, Kmarinas86, David Ludwig, Chris the speller, JMSwtlk, Wicherink, MartinPoulter, Bazonka, Go for it!, Sbharris, WikiPedant, Skoglund, Милан Јелисавчић, Avsn, Snowmanradio, Avb, Jajhill, Gavin Moodie, Normxxx, Elimisteve, BullRangifer, Jon Awbrey, Just plain Bill, Metamagician3000, Byelf2007, Drewarrowood, ArglebargleIV, Rory096, Nareek, Rigadoun, Lapaz, Wtwilson3, Dialecticas, Ajbird, Tal.yaron, Grumpyyoungman01, Mr Stephen, Ryulong, RichardF, Kripkenstein, DabMachine, OnBeyondZebrax, HisSpaceResearch, K, Xinyu, RekishiEJ, Danarothrock, J Milburn, CRGreathouse, MicahDCochran, Myasuda, Gregbard, Maestrojohnstone, TheQuickBrownFox, M a s, Teratornis, Letranova, Brahmajnani, Pphysics, Bunzil, Bpv, Klausness, Beeezy, WinBot, BrownApple, RDT2, Smartse, Danger, Wayiran, Steelpillow, JAnDbot, Ristonet, Stephanhartmannde, Skomorokh, Matthew Fennell, Ikanreed, MegX, Roidroid, MaxPont, Arno Matthias, JamesBWatson, Tito-, Chrisdel, Cic, Lucaas, Timothy J Scriven, Snowded, KConWiki, Ben Ram, Gomm, Vincent douzal, Exiledone, Jacobko, Philosophy Junkie, JaGa, Nowletsgo, Otvaltak, Brodemi, Anarchia, SuperMarioMan, Dionysiaca, Lilac Soul, Mreeves51, HEL, Tikiwont, Maurice Carbonaro, Nigholith, TheSeven, Metrax, SharkD, Chiswick Chap, Antony-22, Trilobitealive, DadaNeem, Jwiley80, Delmlstan, Michelferrari, DASonnenfeld, Artblakey, Frguerre, DDSaeger, Joeoettinger, Johnfos, Jimmaths, TXiKiBoT, Oshwah, Calwiki, Knock-kneed, Tomsega, Jazzwick, The Tetrast, DennyColt, Cerebellum, Don4of4, Shanata, Aphilo, Eldredo, Pderr, Earlynaval, ELeng, Shaoweifang, Newbyguesses, SieBot, Dawn Bard, Thickey3, Nigel E. Harris, Lightmouse, Vanished user kijsdion3i4jf, Sunrise, Svick, Myrvin, Tautologist, Martarius, ClueBot, SummerWithMorons, Mpdimitroff, UGD, Drmies, Rotational, Ljasie, SamuelTheGhost, Masterpiece2000, Leetviper, Estirabot, BirgerH, Brews ohare, Lightsaver~enwiki, Vegetator, Indopug, EdChem, Heironymous Rowe, Pfhorrest, Queensgirl, Addbot, DOI bot, Elmondo21st, Solatido, MrOllie, Tassedethe, Thi, DK4, Legobot, Luckas-bot, Yobot, Andresswift, Trinitrix, AnomieBOT, Piano non troppo, Trabucogold, NickK, Flewis, Materialscientist, Citation bot, Xqbot, TheAMmollusc, Tasudrty, GaroGarabedyan, Postulant, Gilo1969, The Land Surveyor, DavidCBeck, Hi878, GrouchoBot, Omnipaedista, Eugene-elgato, Manawiki, FreeKnowledgeCreator, Hugetim, FrescoBot, Paine Ellsworth, Rotideypoc41352, Machine Elf 1735, Citation bot 1, I dream of horses, Jonesey95, Shahidur Rahman Sikder, Jandalhandler, TRBP, Trappist the monk, WolBalston, Wotnow, Jordgette, Hickorybark, GregKaye, Zvn, Nickanc, Calciumpower, DARTH SIDIOUS 2, Jesuszamorabonilla, TjBot, Onancastrovejano, Uanfala, Semmendinger, 1337junior, Pradeu, EmausBot, And we drown, John of Reading, Look2See1, GoingBatty, Dcirovic, PBS-AWB, Josve05a, Emdelrio, Stovl, Simweir, H3llBot, Usb10, Phronetic, RockMagnetist, Cassowary Rider, Terra Novus, ClueBot NG, Mdlevin, Hamard Evitiatini, Lahedoniste, Snotbot, Braincricket, O.Koslowski, Helpful Pixie Bot, HMSSolent, BG19bot, Kevbonham, Valentindedu, Zyxwv99, Winfredtheforth, Mthoodhood, Philosopherofscience, Dobrich, Polmandc, KropotkinsLibrary, Mr.amitg, PremierAndrew, BattyBot, Cyberbot II, Jeremyhowick, JYBot, DavidIwinkler, EagerToddler39, Sminthopsis84, Zeitgeistpage, TheTahoeNatrLuvnYaho, Mdpacer, Healing toolbox, Jochen Burghardt, Alexis1812w, Mark viking, Kilternom, Gladtobeherenow, Lemnaminor, Ruby Murray, Marswuzhere, I am One of Many, Melonkelon, Lee Bunce, Star767, CassandraBlair, Zenibus, Citrusbowler, Liz, LCcritic, JacobWeiser, Stringertheory, Monkbot, Thoth.Esmeralda, Gergnoswad, Moreeditit, Sohrab cu, Ghost Lourde, Inorout, KasparBot, Feminist, Brandon B Lunga, DatGuy, Brisbane.psychologist, Pantextual, Asastev, Bender the Bot, PrimeBOT, Vbygyggyvgvgyb, AcademeEditorial, ScienticGuy, KolbertBot and Anonymous: 374

- **History of science** *Source:* https://en.wikipedia.org/wiki/History_of_science?oldid=802217963 *Contributors:* Sodium, The Anome, Slrubenstein, Ed Poor, Alex.tan, Danny, SimonP, DavidLevinson, Zoe, Heron, ChrisSteinbach, Michael Hardy, Llywrch, Lexor, Ixfd64, SebastianHelm, Ihcoyc, Aarchiba, Poor Yorick, Charles Matthews, Reddi, Ike9898, Markhurd, Kaal, SEWilco, Thue, Jackson~enwiki, Wetman, Qertis, Owen, Huangdi, Phil Boswell, Chuunen Baka, Sander123, Fredrik, ABVR, Nurg, Modulatum, Arkuat, Gandalf61, Sverdrup, Rursus, Auric, Alan

Liefting, Ancheta Wis, Giftlite, DocWatson42, Jyril, Levin, Fastfission, Joe Kress, Matthead, OldakQuill, Gadfium, Formeruser-81, Quadell, Beland, Piotrus, Woofles, Huwr, Neutrality, Rosentredere, Robin klein, Deglr6328, Ashami, Noisy, Discospinster, Rich Farmbrough, NrDg, Vsmith, Florian Blaschke, ArnoldReinhold, Dave souza, Roo72, Dbachmann, Stereotek, Bender235, Kbh3rd, S.K., Bdk, Brian0918, El C, Rgdboer, Chalst, Worldtraveller, Phoenix Hacker, Sietse Snel, RoyBoy, Sole Soul, Circeus, NetBot, Longhair, Smalljim, Maureen, 9SGjOSfy-HJaQVsEmy9NS, Nk, Thialfi, Pharos, Hooperbloob, Mdd, Alansohn, 119, Mac Davis, Fawcett5, CunningLinguist, Bart133, Metron4, Snowolf, Cugel~enwiki, Suruena, Evil Monkey, Dominic, Jguk, Heida Maria, Itsmine, Alai, Angr, Woohookitty, Mindmatrix, Natcase, Brunnock, Deeahbz, Polyparadigm, Wijnand, Ruud Koot, WadeSimMiser, Eleassar777, WikiklrSc, Psneog, Eilthireach, Ggonnell, Allen3, PeregrineAY, LexCorp, Jan van Male, BD2412, David Levy, Acestorides~enwiki, Melesse, Nickradford, Porcher, Drbogdan, Rjwilmsi, Koavf, KYPark, Mike s, DonSiano, BoomHitch, Maurog, RobertG, Andy85719, Mitsukai, Akhenaten0, Hatch68, Bgwhite, Gwernol, The Rambling Man, Wavelength, Daverocks, RussBot, Mark Ironie, Gaius Cornelius, Ksyrie, Shaddack, ANaughty, LiniShu, Welsh, Rjensen, Howcheng, Daanschr, Ragesoss, Aaron Brenneman, Wysoka60, Jpbowen, Aleichem, RonCram, Tomisti, Wknight94, AjaxSmack, Pawyilee, Emijrp, Andrew Lancaster, Speisert, CWenger, Anclation~enwiki, Che829, Allens, NickelShoe, Sardanaphalus, Veinor, SmackBot, FocalPoint, Derek Andrews, Davepape, Jagged 85, Jfurr1981, Edgar181, Gilliam, Hmains, Squiddy, David Ludwig, Bluebot, Persian Poet Gal, Grimhelm, Papa November, Tianxiaozhang~enwiki, Sadads, CSWarren, Colonies Chris, Darth Panda, Leinad-Z, Jefffire, Michael.Pohoreski, Yidisheryid, Jajhill, RedHillian, Jon Awbrey, Mwtoews, Sammy1339, Byelf2007, John, JorisvS, Hans van Deukeren, Peterlewis, IronGargoyle, Ckatz, Samfreed, Waggers, SandyGeorgia, Battem, AdultSwim, Ryulong, RichardF, P199, Novangelis, K, TwistOfCain, JoeBot, Twas Now, Trialsanderrors, Tawkerbot2, Chris55, CmdrObot, Blue-Haired Lawyer, ShelfSkewed, Neelix, Gregbard, Shanew2, Logicus, Icek~enwiki, Kanags, M a s, Dchristle, Evolve17, DBaba, Kozuch, SteveMcCluskey, Gimmetrow, Epbr123, Barticus88, InSpace~enwiki, Mereda, Oliver202, Headbomb, Pjvpjv, Missvain, Bobblehead, RFerreira, Nick Number, Eddius, Liebgard~enwiki, Muski27, Jj137, MECU, Kadros~enwiki, Kaobear, MER-C, The Transhumanist, OhanaUnited, BlueRbt, RainbowCrane, Tstrobaugh, Dr mindbender, Gsrgsr, Demophon, Acroterion, Magioladitis, Creationlaw, VoABot II, Dekimasu, Agassi~enwiki, Sodabottle, Sfu, DerHexer, Waninge, Gun Powder Ma, Designquest10, MartinBot, Kangal~enwiki, Mtevfrog, Jim.henderson, David J Wilson, R'n'B, CommonsDelinker, AlphaEta, J.delanoy, Bongomatic, Snowfalcon cu, Krishnachandranvn, Chiswick Chap, GhostPirate, Antony-22, Rjclaudio, DASonnenfeld, EcoRover, Aesopos, Sherip23, Jazzwick, Aymatth2, Littlealien182, Dendodge, Wikiway, Dmcq, Monty845, HiDrNick, PericlesofAthens, Arjun024, Romuald Wróblewski, Ttony21, Nihil novi, WereSpielChequers, Smsarmad, Renatops, GlassCobra, Wilson44691, Fratrep, Sunrise, Correogsk, Gamall Wednesday Ida, 3rdAlcove, Khirurg, WikipedianMarlith, Tomasz Prochownik, ClueBot, Phoenix-wiki, Jncc0, The Thing That Should Not Be, Kafka Liz, Rjd0060, Argentina 678, LizardJr8, SamuelTheGhost, Masterpiece2000, Excirial, Tiniti, BirgerH, Rhododendrites, Garing, Cenarium, Arjayay, Gnip, Atalkcostsky, Egmontaz, Urgehip, Feldspaar, XLinkBot, Olvegg, Yodaki, Qgil-WMF, NellieBly, Sfaoldguy, Felix Folio Secundus, Addbot, Vero.Verite, Fyrael, Scince man, Kapaleev, Cst17, Download, CarsracBot, Tassedethe, Uuda, Mnnaw, Carpe Carpio, Justpassin, Bootyy, LarryJeff, Bfigura's puppy, ???755, Legobot, Rradulak, Yobot, Washburnmav, Houutata, Placejata, Jimjilin, SwisterTwister, Smmalut, Wheenguta, Salah Eddine, AnomieBOT, Daniele Pugliesi, Madesfuga, Quispiam, Materialscientist, Citation bot, Jtamad, Eumolpo, DynamoDegsy, Shogartu, Drosdaf, Quebec99, LilHelpa, Marshallsumter, Capricorn42, Drilnoth, GenQuest, Jmundo, Sophivorus, BookWormHR, Craemell, Cuauti, Earlypsychosis, Yeafvnl, Gabsvillalobos, Verbum Veritas, MaryBowser, FrescoBot, Paine Ellsworth, Tobby72, Michael93555, ComputScientist, Steve Quinn, Machine Elf 1735, Lordharrypotter, DivineAlpha, Citation bot 1, Dunstanne, DrilBot, Winterst, Pinethicket, Fat&Happy, Danwhite2010, PlyrStar93, Tim1357, Cha0s6983, Mercy11, Trappist the monk, Diblidabliduu, ItsZippy, Lotje, Bluefist, ZhBot, Big shiny quarter, RjwilmsiBot, NerdyScienceDude, Tesseract2, DASHBot, Mr. Thrasymachus, John of Reading, Immunize, Super48paul, Northernsoutherner, Faolin42, Syncategoremata, GoingBatty, CD1872, Dcirovic, Aaliasache, PBS-AWB, Hashemi1971, Sf5xeplus, Aeonx, Zloyvolsheb, Bushmillsmccallan, L Kensington, Donner60, Chewings72, Mcc1789, Rmashhadi, 28bot, ClueBot NG, Corduroydog, Miamelha, Matthiaspaul, Satellizer, Hazhk, Rezabot, Brickmack, NCAR Archives, Widr, Reify-tech, Sameenahmedkhan, MerllwBot, Helpful Pixie Bot, Lolm8, Tholme, Wbm1058, Gob Lofa, Bibcode Bot, Azhermajidsiddiqui, NewsAndEventsGuy, Quarkgluonsoup, MusikAnimal, LouisAlain, Davidiad, Solomon7968, Iczero, Harizotoh9, Deltaseeker, Aisteco, Anbu121, Scotthendrix1970, Toploftical, Cyberbot II, Shallowhai, Khazar2, Faberglas, Sminthopsis84, Numbermaniac, Blackredstart, IVORK, Fedelis4198, SUCKIT!! 42, Gladtobeherenow, Alfy32, Decentman12, Sol1, Vyc11994, Aubreybardo, Sravsandbobby, Manul, Winged Blades of Godric, Suelru, Lauramuseo, Monkbot, Kattits69, Gmparsons33, Chaudeau, 115ash, JonathanHopeThisIsUnique, 2Mars4S2Billion, Bransonroskelley, MichelleSmith8, Batsgasps, TheFalseEditor, Quivico, Serten II, Johnretrica, User000name, Pierceunique, Rleakey, KasparBot, Brachney, Blazearon21, Qzd, InternetArchiveBot, Sakialrap, Chackerian, Nodayrt, Afdhal123, GreenC bot, Gulumeemee, BruisedWF, Bender the Bot, AndrewK2005, Elfslayer22, Magic links bot, Alexhummels and Anonymous: 378

- **Outline of science** Source: https://en.wikipedia.org/wiki/Outline_of_science?oldid=803076506 Contributors: Markhurd, Timrollpickering, Jurema Oliveira, Robin Hood~enwiki, Dbachmann, Giraffedata, Mdd, Alansohn, BryanD, Eukesh, Bsadowski1, Woohookitty, LOL, Mandarax, BD2412, Qwertyus, Rjwilmsi, Quiddity, Mred64, Bgwhite, Wavelength, Bhny, Aeusoes1, Tvtonightokc, Ragesoss, Tony1, Jeh, Closedmouth, David Biddulph, Auroranorth, SmackBot, David Kernow, Jpvinall, Gilliam, Chris the speller, Whpq, Hu12, Cydebot, Second Quantization, 2bornot2b, The Transhumanist, D1doherty, Eldumpo, Parveson, R'n'B, Nigholith, The Transhumanist (AWB), Cmichael, SieBot, RESEARCHFAN, Sunrise, Kumioko (renamed), Correogsk, Yoda of Borg, Mr. Stradivarius, JL-Bot, ClueBot, MBD123, VQuakr, Robert Skyhawk, Sun Creator, SchreiberBike, Ottawa4ever, Htmlcoderexe, Saeed.Veradi, Addbot, Fyrael, OlEnglish, Jarble, Alfie66, Yobot, Fraggle81, DemocraticLuntz, Minnecologies, Materialscientist, Helpfulbot, Tamunagegi, FrescoBot, Spartan S58, Pinethicket, I dream of horses, Gamewizard71, Mean as custard, Alph Bot, Brian345, BertSeghers, WildBot, Dewritech, Staszek Lem, Lawstubes, Scientific29, RockMagnetist, Llightex, ClueBot NG, Widr, CasualVisitor, Helpful Pixie Bot, EuropeBeSavedByJesus, John Michael Rasing, BG19bot, PhnomPencil, 143vivienne, Marikafragen, FitzJD, Numb25, Rhinopias, Jag140, Elcid1212, Mogism, Wrdstck, Samuseal, Camyoung54, Yadsalohcin, Vinny Lam, Irkavarice, Daddysfirstgirl0824, Loraof, CraftyChemist, Yusifdheaa, Ajion, Cyrej, Mike goodwill, Jeunine, Ammarpad, CAPTAIN RAJU, Podoguru, Anktplwl91, POLONIUM and Anonymous: 64

- **Objectivity (philosophy)** Source: https://en.wikipedia.org/wiki/Objectivity_(philosophy)?oldid=789012941 Contributors: The Anome, Andre Engels, ChrisSteinbach, Elian, Smelialichu, Michael Hardy, Fred Bauder, Dominus, Liftarn, Ducker, NEWONE, Rob Hooft, Charles Matthews, Trontonian, Wik, Raul654, Rbellin, David.Monniaux, Banno, Robbot, DocWatson42, Fastfission, Wikiwikifast, 20040302, Quadell, Beland, Vina, Jareha, Jh51681, Random account 47, Eep², Lucidish, JoeSmack, Carlon, Lycurgus, Vervin, NetBot, Mhuben, Makawity, Pearle, Oncogene, Amerindianarts, Crimson117, Someoneinmyheadbutit'snotme, Velho, Woohookitty, TheNightFly, NotSuper, Bbatsell, Inventm, Deltabeignet, BD2412, Kbdank71, Dpv, Salix alba, MZMcBride, FlaBot, Who, Gurch, Alberto Molina~enwiki, Roboto de Ajvol, Borgx, Splash, Chris Capoccia, Emmanuelm, NawlinWiki, Grafen, Locksdonkey, Tomisti, Mike Dillon, Langdell~enwiki, Raijinili, Jwissick, Chromakode, Allens, Boss1000, Sardanaphalus, Chiok, SmackBot, Rtc, Josephprymak, Nethency, Uxejn, Gilliam, NewName, David

Ludwig, JASKN, Fuzzform, Octahedron80, Zime2005, Jefffire, LoveMonkey, Noblige, Clicketyclack, Byelf2007, Akendall, Lapaz, Osbus, A.Z., Feraudyh, Grumpyyoungman01, SandyGeorgia, Martian.knight, LaMenta3, K, Melander, George100, J Milburn, Peter1c, Wolfdog, Postmodern Beatnik, Aherunar, Neelix, Gregbard, Cydebot, Atomaton, Peterdjones, Miguel de Servet, FrederikSchack, Mystar, SigmaAlgebra, Sigma-Algebra, AntiVandalBot, Stev17, Modernist, JAnDbot, Skomorokh, I.I, Hurmata, Dekimasu, JNW, Midgrid, Snowded, Ethan a dawe, JaGa, JohnCatlin, Kjmarino, N4nojohn, Pharaoh of the Wizards, Maurice Carbonaro, DarkFalls, JayJasper, VolkovBot, Oshwah, Monkey Bounce, SCwatcher, Bashboy91787, Sanfranman59, ^demonBot2, Honew, Newbyguesses, KyZan, SieBot, Fixer1234, Scarian, Gabethenerd, ConcernedScientist, Mimihitam, LaidOff, Onopearls, Jblev2, Ratiuglink, Emptymountains, Mx. Granger, Martarius, ClueBot, SummerWithMorons, ICAPTCHA, Humbert.h.humbert, Mild Bill Hiccup, SuperHamster, Shooksides, Vanisheduser12345, Rhododendrites, SchreiberBike, 1ForTheMoney, Krohn211, Fledgeaaron, Against the current, XLinkBot, Alexius08, Kbdankbot, Addbot, MarioColbert, OlEnglish, Zorrobot, عمر‎, Luckas-bot, Yobot, TaBOT-zerem, Azcolvin429, AnomieBOT, Wæng, Xqbot, J04n, Jorijnsmit, Omnipaedista, RibotBOT, Stmannew, Sariom, Metaphysicalnaturalist, Hugetim, LegendFSL, Jeffrey Boyd, Pteridonut, Gamewizard71, Wikidiculous, Lotje, Tbhotch, Reach Out to the Truth, Faolin42, Tommy2010, AvicBot, PBS-AWB, Robertsan, Libertaar, Donner60, Deepski, ClueBot NG, Widr, Langchri, Tomasz Raburski, BattyBot, DarafshBot, Electricmuffin11, Faizan, Uberaccount, Aniketsodhiya, Abdulla348, Chuck Burns the 4th, SamWilson989, Ferrousfelangaius, Bethany.Lucas, Shyshyinitan, PrimeBOT, Volunteer1234, Philosophogos and Anonymous: 157

- **Logical reasoning** Source: https://en.wikipedia.org/wiki/Logical_reasoning?oldid=758163010 Contributors: Wjbeaty, Zuidervled, Eliazar, Jaberwocky6669, Flammifer, Ryanmcdaniel, Arthena, Velella, Forderud, Linas, SMcCandlish, Salamurai, ArglebargleIV, Banedon, Gregbard, Widefox, Minhtung91, Arno Matthias, Waninge, Jeff G., The Thing That Should Not Be, Jusdafax, Addbot, Fryed-peach, Wcmead3, TaBOTzerem, Materialscientist, E235, LilHelpa, Machine Elf 1735, I dream of horses, Gamewizard71, Etincelles, Lotje, Lemih91, ClueBot NG, Jack Greenmaven, Masssly, Widr, BG19bot, Achowat, Wannabemodel, Mohamed-Ahmed-FG, YiFeiBot, Sensisparis and Anonymous: 36

- **Scientific method** Source: https://en.wikipedia.org/wiki/Scientific_method?oldid=803325318 Contributors: AxelBoldt, Chenyu, Derek Ross, Lee Daniel Crocker, Robert Merkel, Zundark, The Anome, Slrubenstein, Manning Bartlett, Malcolm Farmer, Ed Poor, RK, Eclecticology, Josh Grosse, Youssefsan, Fredbauder, William Avery, DavidLevinson, Anthere, Ajdecon, Heron, Blue-enwiki, Dwheeler, Yaginuma, R Lowry, ChrisSteinbach, Olivier, Stevertigo, Edward, Lir, Mtmsmile, Chas zzz brown, Boud, Michael Hardy, Pit-enwiki, Lexor, Grizzly, Kku, BoNoMoJo (old), Liftarn, MartinHarper, Ixfd64, Moses ben Nachman, Fwappler, Karada, Skysmith, Alfio, Kosebamse, 168..., Ahoerstemeier, Ronz, William M. Connolley, Snoyes, CatherineMunro, Angela, Nichtich-enwiki, JWSchmidt, Alvaro, Mark Foskey, Cyan, AugPi, Poor Yorick, Rotem Dan, Cimon Avaro, TonyClarke, Lancevortex, WouterVH, GRAHAMUK, Hemmer, Hawthorn, RickK, Reddi, Ww, Gutza, Wik, Zoicon5, Quux, Lord Kenneth, DJ Clayworth, Markhurd, Patrick0Moran, Marshman, Furrykef, Hyacinth, Saltine, Tempshill, SEWilco, Populus, Bevo, SH-enwiki, Wiwaxia, Prisonblues, Bloodshedder, Pir, Jusjih, Gurry, Banno, Jon Roland, King brosby, Lumos3, Huangdi, Dimadick, Robbot, Plehn, Murray Langton, Pigsonthewing, ChrisG, Moriori, Fredrik, PBS, Zandperl, NPOV-enwiki, Stephan Schulz, Seglea, Modulatum, Arkuat, Sam Spade, Lowellian, Tualha, Sverdrup, Pcr, Sunray, Mr-Natural-Health, Hadal, Michael Snow, Aetheling, Diberri, Azwaldo, Carnildo, David Gerard, Enochlau, Wjbeaty, Ancheta Wis, Thebriguy, Exploding Boy, Giftlite, DocWatson42, Christopher Parham, Barbara Shack, Kim Bruning, Netoholic, Tom harrison, Hagedis, Michael Devore, FeloniousMonk, Bensaccount, BalthCat, Cpk, Yekrats, 20040302, Sundar, SWAdair, Wmahan, Isidore, Geni, Sonjaaa, Pcarbonn, Antandrus, Beland, Scott MacLean, The Trolls of Navarone, Piotrus, Scottperry, Kaldari, Nick-in-South-Africa, Karol Langner, Gmlk, Kesac, Icairns, GeoGreg, Histrion, Gscshoyru, Creidieki, Neutrality, Burschik, Benzh-enwiki, Grunt, Safety Cap, Zro, Rfl, Mercurius-enwiki, RTCearly, Discospinster, KillerChihuahua, Laoma, John FitzGerald, Inkypaws, Vsmith, Damien Prystay, Florian Blaschke, Dave souza, Notinasnaid, AlexKepler, Exabyte, Carptrash, SpookyMulder, Bender235, ESkog, S.K., Melamed, Eric Forste, El C, Chalst, Edward Z. Yang, RoyBoy, Triona, Etimbo, Orlady, Guettarda, Bobo192, Adraeus, John Vandenberg, Viriditas, .:Ajvol:., Skywalker, Nk, Vanished user 19794758563875, Sam Korn, Thialfi, Nsaa, Mdd, 578, Siim, Poweroid, Gary, Guy Harris, AllTom, Ricky81682, Derumi, Lightdarkness, Mac Davis, SMesser, Titanium Dragon, Bart133, Cortonin, Cugel-enwiki, Clockwork-Soul, Knowledge Seeker, *Kat*, Nof20, RainbowOfLight, LFaraone, H2g2bob, BDD, Sleigh, Gene Nygaard, Capecodeph, Yurivict, Omnist, Noz92, Bobrayner, Roylee, Mel Etitis, OwenX, Mindmatrix, Syamil, Carlos Porto, LOL, KeithStump, Shadeofblue, Kurzon, JFG, Pol098, WadeSimMiser, MONGO, Pdn-enwiki, Ivar Y, Stefanomione, PeregrineAY, Cataclysm, Paxsimius, GSlicer, Askewmind, Graham87, BD2412, Qwertyus, Kbdank71, Acestorides-enwiki, Dpr, Mendaliv, Achaeus, Acrotatus, Rjwilmsi, KYPark, Vary, ErikHaugen, Vegaswikian, DonSiano, Crazynas, Kalogeropoulos, Ems57fcva, Bubba73, Afterwriting, TBHecht, TheIncredibleEdibleOompaLoompa, Williamborg, FlavrSavr, Maurog, Yamamoto Ichiro, FuelWagon, Agasicles, Wragge, FlaBot, AED, Felixdakat, Mathiastck, Wknight8111, CarolGray, Pathoschild, RexNL, Gurch, TheDJ, Agathocles of Bactria, Simonsimpson, Spencerk, Accurate Nuanced Clear, DVdm, Monkofthetrueschool, Bgwhite, YurikBot, Wavelength, TexasAndroid, Koveras, Arado, Mark Ironie, The.orpheus, Markus Schmaus, SpuriousQ, Hydrargyrum, Okedem, Ksyrie, Eleassar, Salsb, Wimt, CarlHewitt, Jrbouldin, NawlinWiki, A314268, DragonHawk, Wiki alf, Grafen, Dumoren, Welsh, Holon, Joelr31, Mccready, Ragesoss, Brandon, Jpbowen, PM Poon, Nate1481, Aaron Schulz, RonCram, Shotgunlee, Action potential, DeadEyeArrow, Haemo, Sarkar112, Rsugden, Mugunth Kumar, WAS 4.250, FF2010, Enormousdude, Phgao, Chase me ladies, I'm the Cavalry, Ronasi, Dspradau, GraemeL, Red Jay, Trumpetman2, LeonardoRob0t, Spliffy, ArielGold, Markbenecke, Benandorsqueaks, Eric Norby, GrinBot-enwiki, Airconswitch, Samuel Blanning, Evolver, DVD R W, Finell, Soir, Luk, Snalwibma, RichG, Yvwv, Sardanaphalus, Pharrison, Amalthea, SmackBot, Theasus, RedHouse18, Rtc, Reedy, KnowledgeOfSelf, Hydrogen Iodide, JoshDuffMan, Pgk, C.Fred, Jim62sch, Jagged 85, WookieInHeat, Delldot, Kslays, Powo, Commander Keane bot, Xaosflux, Hmains, BrotherGeorge, David Ludwig, Wigren, Teemu Ruskeepää, Chris the speller, Jibbajabba, Persian Poet Gal, Bduke, Miquonranger03, Papa November, SchfiftyThree, Terraguy, Portnadler, Dragice, Mladifilozof, Bov, TidyCat, Rrelf, Can't sleep, clown will eat me, Faaaa, Jefffire, HarrisX, HoodedMan, Brimba, OrphanBot, Onorem, Nixeagle, Thisisbossi, Addshore, Edivorce, SundarBot, Jerrch, Shrine of Fire, Bowlhover, Nakon, Steve Pucci, Dreadstar, Dacoutts, Nrcprm2026, Brainyiscool, Derek R Bullamore, BullRangifer, Jon Awbrey, EdGl, Wisco, Mitchumch, Vina-iwbot-enwiki, Curly Turkey, DKEdwards, Pilotguy, Will Beback, JLogan, Byelf2007, SashatoBot, Mukadderat, Wtwilson3, Muthaofdamatrix, JoshuaZ, Minna Sora no Shita, Rundquist, IronGargoyle, Dwxyzq, Ckatz, BillFlis, Slakr, Dr Smith, Beetstra, Xiaphias, Macellarius, Meco, Ryulong, Pseudoanonymous, Zapvet, Nabeth, Koweja, KJS77, Hu12, Quaeler, JYi, Roland Deschain, Iridescent, K, Paul venter, RekishiEJ, Chris55, Nightson, Rabbitcarrot, Patrickwooldridge, JForget, SleekWeasel, Wafulz, Sir Vicious, BeenAroundAWhile, Picaroon, Dr.Bastedo, CWY2190, Jsd, FlyingToaster, Pgr94, Dgw, Cydebot, Peterdjones, Hollow are the Ori, JFreeman, Frosty0814snowman, Corpx, Islander, Revdrace, Tawkerbot4, DumbBOT, ADude, NaLalina, Ssilvers, NC cousin, Geometricmean, Garik, SteveMcCluskey, Mattisse, Thijs!bot, Epbr123, Daa89563, Dr Aaron, Wikid77, D4g0thur, Qwyrxian, Jobber, Ucanlookitup, Mojo Hand, Headbomb, John254, Bobblehead, Second Quantization, Matthew Joseph Harrington, Davidmack, Dfrg.msc, AgentPeppermint, Muaddeeb, Sean William, SvenAERTS, Northumbrian, David D., AntiVandalBot, Yonatan, Luna Santin, Opelio, Bigtimepeace, SPetersen, VI7361, TimVickers, Mdotley, Smartse, Math Teacher, Spencer, Scruple, Mhagerman, JAnDbot, Narssarssuaq, Deflective, Husond, Zachblume,

Hans Mayer, MER-C, Brainbuster, The Transhumanist, Arch dude, Fabometric, Honette, BenB4, Hut 8.5, Kerotan, Savoylettuce, Yahel Guhan, Fisherm77, MaxPont, Magioladitis, VoABot II, Fusionmix, JNW, Swpb, SineWave, CTF83!, Singularity, Pixel ;-), Recurring dreams, Avicennasis, Snowded, Indon, Animum, Depressedrobot, -Bobby, Adrian J. Hunter, Ashadeofgrey, Allstarecho, Spellmaster, Faro0485, THobern, DerHexer, Grunge6910, Baristarim, Oicumayberight, Will2green, Gjd001, JohnDCampbell, Hdt83, MartinBot, Mermaid from the Baltic Sea, Uvainio, Padillah, Mschel, R'n'B, Dionysiaca, N4nojohn, Wlodzimierz, J.delanoy, Filll, Rgoodermote, EscapingLife, Mike.lifeguard, JVersteeg, SharkD, Andareed, Katalaveno, McSly, Mikael Häggström, Pyrospirit, AntiSpamBot, Chiswick Chap, Lbeaumont, Ohms law, MetsFan76, Usp, Oopsla, Merzul, Axle12693, DASonnenfeld, BernardZ, Dkreisst, The Great Redirector, Bionictulip, UnicornTapestry, VolkovBot, Joeoettinger, ABF, DagnyB, DSRH, The Duke of Waltham, VasilievVV, Gaianauta, Jimmaths, Yunje76, Barneca, Philip Trueman, SamMichaels, Cosmic Latte, Flyte35, The Original Wildbear, Sherip23, RobertDP, Hqb, Dchall1, Ridernyc, Alphaios~enwiki, Fredrick day, Qxz, Ocolon, Uurtamo, Ontoraul, The Tetrast, Corvus cornix, Philogo, AllGloryToTheHypnotoad, LeaveSleaves, Rjm at sleepers, PDFbot, Whatiguana, Eubulides, 2112 rush, Yk Yk Yk, Hanjabba, DJFishlips, Synthebot, Laokoön, Klenole, Oliepedia, C0N6R355, Isis07, Grepol, SieBot, StAnselm, Alessgrimal, Tiddly Tom, Triwbe, Redhookesb, LeadSongDog, JohnManuel, Keilana, Djayjp, Toddst1, JSpung, Strife911, Lobas~enwiki, Oxymoron83, Lightmouse, Tombomp, RyanParis, Sunrise, Correogsk, Katzen03, C'est moi, Reneeholle, Mojoworker, Maralia, Firefly322, Escape Orbit, Myrvin, Pickledh, WickerGuy, ImageRemovalBot, Khirurg, Loren.wilton, Immutable1, ClueBot, Jbening, Fox, Rjd0060, Jagun, ILikeVocab, Ndenison, Knepflerle, Francine3, CooPs89, Wispanow, Meekywiki, Der Golem, Boing! said Zebedee, CounterVandalismBot, Niceguyedc, Chase101, Christian Skeptic, Excirial, Canis Lupus, Mdebellis, Staffansvensson, Lartoven, ZuluPapa5, Tyler, Bracton, B-man79210, SchreiberBike, MilesAgain, Shinobi0888, Vegetator, Aitias, Ranjithsutari, Ubardak, Crowsnest, ClanCC, Darkicebot, EENola, Against the current, XLinkBot, Jocf101, Jayangaw, Maijinsan, Vianello, The Rationalist, Tayste, Addbot, Cxz111, AVand, Jafeluv, Jhend1234, CL, Fgnievinski, Metagraph, Ronhjones, Vishnava, NjardarBot, Ka Faraq Gatri, Mentisock, Urbanette, MapleHero, Doctor Correct, جمال, Gail, Qwertyytrewqqwerty, GarethH1, Krukouski, MissAlyx, Legobot, Luckas-bot, Yobot, 6u56u, Senator Palpatine, Legobot II, Bdog9121, Andresswift, ArchonMagnus, MrBurns, AnomieBOT, Rockypedia, Balamanti2, Piano non troppo, LlywelynII, Zahnay183029, Scan't you see when we all lalalal, Materialscientist, Citation bot, Notandrw, LilHelpa, Hve2hold, Carturo222, Xqbot, TheAMnollusc, ManningBartlett, Addihockey10, Stefanson, BurntSynapse, Locos epraix, Srich32977, Dr Oldekop, J04n, GrouchoBot, RibotBOT, Chris.urs-o, Methcub, 김치, Constructive editor, Hugetim, FrescoBot, Tobby72, Thayts, Mbcannell, Endofskull, Argumzio, Machine Elf 1735, Soufray, Armigo~enwiki, Cannolis, Anaphysik, Citation bot 1, Чаховіч Уладзіслаў, AstaBOTh15, Boxplot, Pinethicket, Tom.Reding, Bejinhan, RedBot, Jandalhandler, Tim1357, Trappist the monk, Douglasbell, Livingrm, Vrenator, Katerenka, Dandrestor, Jesse V., DARTH SIDIOUS 2, The Utahraptor, RjwilmsiBot, Ripchip Bot, Foxymamma, WildBot, Tesseract2, DASHBot, EmausBot, Obamafan70, Gfoley4, Dewritech, Primefac, Syncategoremata, Alexkvaskov, Tommy2010, Mz7, ZéroBot, PBS-AWB, Checkingfax, Knight1993, Empty Buffer, Dotoree, Sgerbic, AvicAWB, Jimmy Maxwell Jaffa, Erianna, Khaydock, Thouny, L Kensington, Peter M. Brown, Phronetic, Elmorang, Emperyan, RockMagnetist, JanetteDoe, Teapeat, Llightex, DJDunsie, Shivanshu3, CocuBot, Snotbot, Biophil.o, Widr, Australopithecus2, Oxford73, Helpful Pixie Bot, Candleabracadabra, Bibcode Bot, Andolan1, BG19bot, Quarkgluonsoup, Pacerier, Solomon7968, CarloMartinelli, Ramos1990, Benzband, CitationCleanerBot, Harizotoh9, MrBill3, Pertin1x, Rolandwilliamson, Rhinopias, BattyBot, Bart49, Darylgolden, Ninmacer20, ChrisGualtieri, IjonTichyIjonTichy, Timelezz, Dexbot, Mogism, Cheerioswithmilk, Razibot, NathanWubs, Ruby Murray, Biogeographist, Star767, New worl, JackBrad419, Sol1, Surfscoter, AntiPOVmagnet, Jackmcbarn, Aubreybardo, Lizia7, LucSaffre, Barjimoa, Suelru, Csutric, IStoleThePies, MelaniePS, Teaksmitty, SkateTier, Filedelinkerbot, Historian7, ArchitectOfIdeas, James343e, Loraof, Batsgasps, Charlotte Aryanne, Mohammed al-Bukhari, Anonimmuz, LadyLeodia, Isambard Kingdom, User000name, Jeunine, AppliedStatistics, Jerodlycett, Asterixf2, Roe.ese, Cleopatran Apocalypse, BU Rob13, Barbara (WVS), InternetArchiveBot, GreenC bot, Fmadd, Bear-rings, Jmcgnh, BrandiKnapp21, Sparkyscience, PrimeBOT, Cookiemonster005, PaleoNeonate, Magic links bot, X1\, KolbertBot and Anonymous: 831

- **Outline of scientific method** *Source:* https://en.wikipedia.org/wiki/Outline_of_scientific_method?oldid=779327280 *Contributors:* Edward, Michael Hardy, Banno, Fredrik, JesseW, Alan Liefting, Ancheta Wis, Andries, Zigger, Brequinda, Jrdioko, Wmahan, Antandrus, D6, Aranel, Shenme, Hipocrite, Mel Etitis, Quiddity, Nihiltres, RussBot, DanMS, Doubleg, Tevildo, David Kernow, Jagged 85, Commander Keane bot, Leinad-Z, Dreadstar, FlyHigh, DabMachine, JeffW, CmdrObot, Cydebot, Enoch the red, Legotech, Bobblehead, The Transhumanist, RainbowCrane, VoABot II, Steven J. Anderson, Neparis, Sunrise, Klauys, MrKIA11, Panyd, NuclearWarfare, Materialscientist, Decstop, Loraof, Jeunine, ThePlatypusofDoom and Anonymous: 22

- **Empirical research** *Source:* https://en.wikipedia.org/wiki/Empirical_research?oldid=801430433 *Contributors:* The Cunctator, Jfitzg, Trontonian, Robbot, Wikibot, Lupo, Wikilibrarian, Rj, Costyn, Karol Langner, Humblefool, Vsmith, Brainy J, Ceyockey, Graham87, Drbogdan, Lockley, Pathoschild, Bhny, Vivaldi, SmackBot, Gilliam, Jon Awbrey, Goodnightmush, Aleenf1, K, CmdrObot, Fyrberd, Gregbard, Cs california, Mattisse, Thijs!bot, Andyjsmith, Bengreen5, Shirt58, Stalik, JHartley, Skomorokh, Gsaup, J.delanoy, Oopsla, VolkovBot, Zidonuke, Grepol, CutOffTies, Twinsday, Gaschroeder, Ost316, WikHead, Necropirate, Addbot, Quercus solaris, Lightbot, Yobot, AnomieBOT, Wall3d, Materialscientist, Wrelwser43, LilHelpa, TesseUndDaan, Srich32977, Abce2, Omnipaedista, Mhotep, Kuphrer, Thehelpfulbot, Pinethicket, Serols, Beao, Олексій Гейтенко, Michel nivard, 7mike5000, DASHBot, Dcirovic, Hhhippo, ZéroBot, GeorgeBarnick, Noggo, Puffin, ClueBot NG, Dinie g, Plantdrew, Sleepsinhammock, Pratyya Ghosh, Snbx, Fox2k11, Gladtobeherenow, Biogeographist, Shrikarsan, YiFeiBot, Skr15081997, Mxschumacher, Loraof, Barneecadd, KasparBot, DatGuy, Ceonathomas, ThePlatypusofDoom, TitaniumOne, NoToleranceForIntolerance, Rascally rarebit, Lagerthon, Magic links bot and Anonymous: 89

- **Deductive-nomological model** *Source:* https://en.wikipedia.org/wiki/Deductive-nomological_model?oldid=798074880 *Contributors:* Hyacinth, Graeme Bartlett, Leonig Mig, Karol Langner, Bender235, GregorB, Bgwhite, Welsh, Tomisti, Dast, CCR, SmackBot, Jaymay, Colonies Chris, Jon Awbrey, Neelix, Gregbard, Thijs!bot, Escarbot, Frog Splash, KConWiki, Jbessie, Tokyo Watcher~enwiki, Synthebot, Harmonicemundi, StAnselm, WereSpielChequers, Soul77, SchreiberBike, Addbot, Yobot, The Banner, Ajahnjohn, Omnipaedista, FrescoBot, LittleWink, John of Reading, Eichhoernchen, JimsMaher, Helpful Pixie Bot, Plantdrew, DPL bot, Dexbot, BreakfastJr, Irami.oseifrimpong, Junaid usa2002, Occurring, Bender the Bot, Apollo The Logician, KolbertBot and Anonymous: 33

- **Scientific modelling** *Source:* https://en.wikipedia.org/wiki/Scientific_modelling?oldid=803211322 *Contributors:* Ronz, Andres, Ike9898, Patrick0Moran, Jon Roland, Mayooranathan, Centrx, Tom harrison, Bender235, El C, Dalf, Bobo192, Evolauxia, Mdd, Atlant, Velella, Marc A. Dubois, Firsfron, Woohookitty, Linas, Prashanthns, Nobbie, BD2412, Josh Parris, Sjö, Spencerk, YurikBot, MadeYouReadThis, Pseudomonas, Anomalocaris, Grafen, Yahya Abdal-Aziz, ScottyWZ, Erik Sandberg, SmackBot, DCDuring, Cazort, Benjaminevans82, Bluebot, Robth, Mwtoews, FlyHigh, Tktktk, Bjankuloski06en~enwiki, Stwalkerster, Iridescent, Aeternus, CmdrObot, Floridi~enwiki, Pgr94, Gregbard, Cydebot, Al Lemos, Mojo Hand, Headbomb, West Brom 4ever, Mailseth, Luna Santin, JAnDbot, Stephanhartmannde, Sukratu Barve, MER-C, VoABot II, EagleFan, Squidonius, MartinBot, Anne97432, CommonsDelinker, Erkan Yilmaz, Eliz81, TheSeven, MONODA, Gusfre, Funandtrvl,

VolkovBot, AlnoktaBOT, Begewe, Malinaccier, Kilmer-san, Monty845, Pdfpdf, Noiseball, Equilibrioception, Gerakibot, Katonal, Oculi, Sanya3, Sunrise, DancingPhilosopher, Sgagnon, SlackerMom, ClueBot, Bmotoc, Niceguyedc, Excirial, Brews ohare, Hans Adler, SchreiberBike, Chacen, Dellexxx, Davemody, Gmeltser, SilvonenBot, Addbot, DougsTech, Aktsu, Teles, Yobot, Ptbotgourou, KamikazeBot, AnomieBOT, DemocraticLuntz, Galoubet, Hommeles, Materialscientist, Srich32977, RjbotBOT, MLauba, Spiralforward, Some standardized rigour, FrescoBot, Kdn1982, Wgbh66, Tom.Reding, JokerXtreme, Emble64, The Stick Man, EmausBot, John Cline, LÊ TÄN LỘC, Architectchao, Cortes IDS5717C, 罪罪, Wcfios, Rocketrod1960, ClueBot NG, Onanoff, Chillllls, MerllwBot, Helpful Pixie Bot, PhnomPencil, Wingroras, Pmuskee, Neils51, Frosty, Gladtobeherenow, Johnbandler, EvergreenFir, Pamphilia, ModalPeak, BillWhiten, Cmattison387, Aubreybardo, Raad Z Homod, Noyster, WikiTikiTaki 53, Loraof, Hayman30, SolidPhase, Isambard Kingdom, GeneralizationsAreBad, Adam9007, VerdenalH, Canis lupulus, Reshmiiyer, Robot psychiatrist, Bear-rings, Acopyeditor, NOTME, SauciestPasta and Anonymous: 158

- **Hypothetico-deductive model** *Source:* https://en.wikipedia.org/wiki/Hypothetico-deductive_model?oldid=798983351 *Contributors:* DavidLevinson, BoNoMoJo (old), Karada, Bueller 007, EdH, Hike395, Markhurd, Banno, Frazzydee, Clngre, Blainster, Ancheta Wis, Rj, Rynelm, Piotrus, Guppyfinsoup, Bender235, Viriditas, TheDJ, YurikBot, Atfyfe, Tomisti, Kronocide, SmackBot, Rtc, Srnec, MartinPoulter, Chlewbot, Jon Awbrey, SashatoBot, CmdrObot, Thomasmeeks, Pgr94, Gregbard, JPalonus, Nick Number, Magioladitis, Cic, Jaggerblade, Deor, TXiKiBoT, SieBot, Antipoeten, Sunrise, Mccaskey, Addbot, Santoemma, GargoyleBot, Yobot, Xqbot, Thehelpfulbot, Machine Elf 1735, EmausBot, WikitanvirBot, Faolin42, BG19bot, ChrisGualtieri, YFdyh-bot, Me, Myself, and I are Here, CAPTAIN RAJU, Apollo The Logician, Sparkyscience, Gitchygoomy, Magic links bot, Berlioz825 and Anonymous: 19

- **Branches of science** *Source:* https://en.wikipedia.org/wiki/Branches_of_science?oldid=801634437 *Contributors:* Ed Poor, Markhurd, Alan Liefting, Mike Rosoft, Discospinster, Florian Blaschke, Bender235, Smalljim, Mdd, Liao, Amorymeltzer, Graham87, Quiddity, GünniX, Jared Preston, Aethralis, Bgwhite, Pburka, Welsh, Brian Crawford, Ef.brazil, Sardanaphalus, Gilliam, Rrburke, רדרד, Chaleyer61, Storkk, The Transhumanist, Greensburger, Dane, Shellwood, Trusilver, Cpiral, NewEnglandYankee, DadaNeem, KylieTastic, Squids and Chips, Philip Trueman, Oshwah, Meters, Gbawden, WereSpielChequers, Malcolmxl5, Yintan, Flyer22 Reborn, Sanya3, Denisarona, JL-Bot, Gaia Octavia Agrippa, Excirial, The Red, SchreiberBike, Stickee, WikHead, Addbot, Jncraton, OlEnglish, Yobot, Eric-Wester, AnomieBOT, Jim1138, Materialscientist, Xqbot, Darthvatrayen, Amaury, Dan6hell66, FrescoBot, DivineAlpha, Pinethicket, I dream of horses, Jujutacular, Gamewizard71, FoxBot, SchreyP, Cowlibob, EngineerFromVega, EmausBot, Gfoley4, Gcastellanos, Super48paul, RA0808, Tekeek, Solarra, Wikipelli, Deirovic, K6ka, Bamyers99, Thine Antique Pen, Brandmeister, Sahimrobot, Puffin, RockMagnetist, 28bot, ClueBot NG, MelbourneStar, Asukite, Widr, Helpful Pixie Bot, HMSSolent, Titodutta, DBigXray, BG19bot, Jwchong, Rose Marie 123, Dayshade, Zujua, David.moreno72, Darylgolden, -riley, Pratyya Ghosh, ChrisGualtieri, Cares789, TwoTwoHello, SFK2, Telfordbuck, 069952497a, Eagleash, Faizan, Melonkelon, Biomedicinal, Tentinator, Everymorning, Flat Out, Babitaarora, Ramesh indian, Haminoon, Classicwiki, JoshuaChen, Ginsuloft, Quenhitran, Manul, Daemyth, DilanDJ, Monkbot, BrightonC, HMSLavender, Mark D. Marquez, Thereppy, Thewickedkid, JonathanHopeThisIsUnique, Spongebobsquarepantsx, Goutham999, RationalBlasphemist, Eurodyne, Stillmorepeople, Gladamas, JQTriple7, Kethrus, Atvica, Dietic, CV9933, Minions8398, Jchmrt, Adam9007, InternetArchiveBot, Zawl, GreenC bot, Van Kleiss, Marvellous Spider-Man, Bender the Bot, Linguist91, Ecliptica, MadEmperorYuri, DrStrauss, Ghugbdhdg, Magic links bot, Asd1921hj, Marvel20041022, Argyle Warrior, KolbertBot, Marsman101 and Anonymous: 273

- **Exact sciences** *Source:* https://en.wikipedia.org/wiki/Exact_sciences?oldid=802546699 *Contributors:* Michael Hardy, Warofdreams, Bearcat, Robbot, Ancheta Wis, Jorend, Edcolins, Beefalo, Lumidek, Florian Blaschke, El C, Mdd, Forderud, Shreevatsa, Knowledge-is-power, Uleph, Phlebas, Rjwilmsi, Bubba73, YurikBot, Alynna Kasmira, Mr. Delayer, SmackBot, Jagged 85, Verne Equinox, Can't sleep, clown will eat me, Anthon.Eff, Radagast83, Ne0Freedom, Bjankuloski06en-enwiki, 16@r, Jimmy Pitt, K, Sam Staton, SteveMcCluskey, Thijs!bot, Second Quantization, I already forgot, Luna Santin, Bestuardo, Rlsheehan, VolkovBot, RiverStyx23, Botev, SieBot, YonaBot, Jdaloner, Sanya3, Sunrise, Capitalismojo, Fadesga, El bot de la dieta, MystBot, Addbot, Zorrobot, Luckas-bot, Xqbot, DSisyphBot, Davshul, Omnipaedista, I dream of horses, HRoestBot, Fabio.kon, MastiBot, Dinamik-bot, Ripchip Bot, ZéroBot, RockMagnetist, StarryGrandma, GabeIglesia, Trackteur, GeoffreyT2000, JonathanHopeThisIsUnique, Sizeofint, KasparBot and Anonymous: 36

- **History of scientific method** *Source:* https://en.wikipedia.org/wiki/History_of_scientific_method?oldid=802409294 *Contributors:* ChrisSteinbach, Edward, William M. Connolley, Charles Matthews, Ww, Lumos3, Psychonaut, Alan Liefting, Ancheta Wis, Kaldari, Karol Langner, Rich Farmbrough, Dave souza, Paul August, Bender235, Brian0918, Pearle, Wtmitchell, Kenyon, Tabletop, MarcoTolo, BD2412, Rjwilmsi, Nightscream, Koavf, Nihiltres, Crazycomputers, Nivix, Jameshfisher, RussBot, NawlinWiki, Ragesoss, PTSE, Kungfuadam, Sardanaphalus, SmackBot, Jagged 85, Hmains, Chris the speller, Colonies Chris, Zsinj, Leinad-Z, BullRangifer, Jon Awbrey, Clicketyclack, JzG, John, JHunterJ, K, CmdrObot, Geremia, Pgr94, Amack, FilipeS, Tkynerd, Doug Weller, SteveMcCluskey, Epbr123, Barticus88, Nick Number, Mdotley, DuncanHill, MER-C, Matthew Fennell, Charlesreid1, KConWiki, Fliegen, JaGa, Schmloof, STBot, Alsee, R'n'B, Thirdright, Fred.e, Chiswick Chap, NewEnglandYankee, Michelferrari, Jonas Mur-enwiki, VasilievVV, Jimmaths, Aymatth2, Sintaku, Ontoraul, Rjm at sleepers, Robert1947, Graymornings, Nagy, Thony C., Dawn Bard, Orthorhombic, Carlw4514, JerroldPease-Atlanta, Sunrise, Khirurg, YSSYguy, Sfan00 IMG, ClueBot, Der Golem, J8079s, Rockfang, Excirial, CohesionBot, LaosLos, Sun Creator, Johnuniq, Heironymous Rowe, Guydauncey, Maijinsan, MystBot, Addbot, Fgnievinski, Glass Sword, David0811, Yobot, 2D, Eric-Wester, AnomieBOT, Piano non troppo, Citation bot, LilHelpa, Poetaris, FrescoBot, PRC 07, Citation bot 1, Rbh00, Pinethicket, Kiefer.Wolfowitz, Jandalhandler, Bdmclean, Reconsider the static, Trappist the monk, Lam Kin Keung, Decstop, Mr.98, JV Smithy, Keegscee, DASHBot, Syncategoremata, Boleroinferno, Auró, ZéroBot, Stovl, Dotoree, Ὁ οἶστρος, Bamyers99, Mcc1789, ClueBot NG, Bulldog73, Widr, Saadia02, Helpful Pixie Bot, Bibcode Bot, BG19bot, MusikAnimal, Solomon7968, Ramos1990, CitationCleanerBot, Arubio1, Pertin1x, David.moreno72, Mollskman, Panini12345, Malinadams, FoCuSandLeArN, Cwobeel, Hmainsbot1, Mogism, Jochen Burghardt, I am One of Many, Spivorg, Filedelinkerbot, VanishedUser sdu9aya9fs654654, OracleNola, SWAGMONEY666, FTOFAccount, Simplexity22, Brodimus24, Magic links bot, CO837, KolbertBot and Anonymous: 111

26.10.2 Images

- **File:1543,Visalius'{}OpticChiasma.jpg** *Source:* https://upload.wikimedia.org/wikipedia/commons/5/5c/1543%2CVisalius%27OpticChiasma.jpg *License:* Public domain *Contributors:* Originally from en.wikipedia; description page is (was) here *Original artist:* User Ancheta Wis on en.wikipedia

- **File:6n-graf.svg** *Source:* https://upload.wikimedia.org/wikipedia/commons/5/5b/6n-graf.svg *License:* Public domain *Contributors:* Image: 6n-graf.png simlar input data *Original artist:* User:AzaToth

26.10. TEXT AND IMAGE SOURCES, CONTRIBUTORS, AND LICENSES

- **File:ADN_animation.gif** *Source:* https://upload.wikimedia.org/wikipedia/commons/8/81/ADN_animation.gif *License:* Public domain *Contributors:* Own work *Original artist:* brian0918™
- **File:Abacus_6.png** *Source:* https://upload.wikimedia.org/wikipedia/commons/a/af/Abacus_6.png *License:* Public domain *Contributors:*
- Article for "abacus", 9th edition Encyclopedia Britannica, volume 1 (1875); scanned and uploaded by Malcolm Farmer *Original artist:* Encyclopædia Britannica
- **File:AdamSmith.jpg** *Source:* https://upload.wikimedia.org/wikipedia/commons/0/0a/AdamSmith.jpg *License:* Public domain *Contributors:* http://www.library.hbs.edu/hc/collections/kress/kress_img/adam_smith2.htm *Original artist:* Etching created by Cadell and Davies (1811), John Horsburgh (1828) or R.C. Bell (1872).
- **File:Albert_Einstein_Head.jpg** *Source:* https://upload.wikimedia.org/wikipedia/commons/d/d3/Albert_Einstein_Head.jpg *License:* Public domain *Contributors:* This image is available from the United States Library of Congress's Prints and Photographs division under the digital ID cph.3b46036.
  This tag does not indicate the copyright status of the attached work. A normal copyright tag is still required. See Commons:Licensing for more information. *Original artist:* Photograph by Orren Jack Turner, Princeton, N.J.
- **File:AlfedPalmersmokestacks.jpg** *Source:* https://upload.wikimedia.org/wikipedia/commons/a/aa/AlfedPalmersmokestacks.jpg *License:* Public domain *Contributors:* Library of Congress CALL NUMBER LC-USW36-376, reproduction number LC-DIG-fsac-1a35072 *Original artist:* Alfred T. Palmer
- **File:Alhazen,_the_Persian.gif** *Source:* https://upload.wikimedia.org/wikipedia/commons/6/63/Alhazen%2C_the_Persian.gif *License:* Public domain *Contributors:* www.levity.com/alchemy/islam09.html *Original artist:* Unknown<a href='https://www.wikidata.org/wiki/Q4233718' title='wikidata:Q4233718'><img alt='wikidata:Q4233718' src='https://upload.wikimedia.org/wikipedia/commons/thumb/f/ff/Wikidata-logo.svg/20px-Wikidata-logo.svg.png' width='20' height='11' srcset='https://upload.wikimedia.org/wikipedia/commons/thumb/f/ff/Wikidata-logo.svg/30px-Wikidata-logo.svg.png 1.5x, https://upload.wikimedia.org/wikipedia/commons/thumb/f/ff/Wikidata-logo.svg/40px-Wikidata-logo.svg.png 2x' data-file-width='1050' data-file-height='590' /></a>
- **File:Amartya_Sen_NIH.jpg** *Source:* https://upload.wikimedia.org/wikipedia/commons/e/e0/Amartya_Sen_NIH.jpg *License:* Public domain *Contributors:* http://dir.niehs.nih.gov/ethics/past.htm
  Original artist: NIH (according to picture caption)
- **File:Ambox_important.svg** *Source:* https://upload.wikimedia.org/wikipedia/commons/b/b4/Ambox_important.svg *License:* Public domain *Contributors:* Own work based on: Ambox scales.svg *Original artist:* Dsmurat, penubag
- **File:Antikythera_mechanism.svg** *Source:* https://upload.wikimedia.org/wikipedia/commons/e/e8/Antikythera_mechanism.svg *License:* Public domain *Contributors:* File:Meccanismo_di_Antikytera.jpg *Original artist:* Lead Holder
- **File:Arbitrary-gametree-solved.svg** *Source:* https://upload.wikimedia.org/wikipedia/commons/d/d7/Arbitrary-gametree-solved.svg *License:* Public domain *Contributors:*
- Arbitrary-gametree-solved.png *Original artist:*
- derivative work: Qef (talk)
- **File:Archimedes_pi.svg** *Source:* https://upload.wikimedia.org/wikipedia/commons/c/c9/Archimedes_pi.svg *License:* Public domain *Contributors:* Own work based on: Archimedes pi.png: <a href='//commons.wikimedia.org/wiki/File:Archimedes_pi.png' class='image'><img alt='Archimedes pi.png' src='https://upload.wikimedia.org/wikipedia/commons/thumb/0/03/Archimedes_pi.png/32px-Archimedes_pi.png' width='32' height='11' srcset='https://upload.wikimedia.org/wikipedia/commons/thumb/0/03/Archimedes_pi.png/48px-Archimedes_pi.png 1.5x, https://upload.wikimedia.org/wikipedia/commons/thumb/0/03/Archimedes_pi.png/64px-Archimedes_pi.png 2x' data-file-width='707' data-file-height='238' /></a> by Fredrik *Original artist:* Fredrik
- **File:Aristoteles_Louvre.jpg** *Source:* https://upload.wikimedia.org/wikipedia/commons/a/a4/Aristoteles_Louvre.jpg *License:* CC BY-SA 2.5 *Contributors:* Eric Gaba (User:Sting), July 2005. *Original artist:* After Lysippos
- **File:Aristotle_Altemps_Inv8575.jpg** *Source:* https://upload.wikimedia.org/wikipedia/commons/a/ac/Aristotle_Altemps_Inv8575.jpg *License:* Public domain *Contributors:* Jastrow (2006) *Original artist:* After Lysippos
- **File:Atmosphere_composition_diagram-en.svg** *Source:* https://upload.wikimedia.org/wikipedia/commons/a/a3/Atmosphere_composition_diagram-en.svg *License:* Public domain *Contributors:* Strategic Plan for the U.S. Climate Change Science Program *Original artist:* Phillipe Rekacewicz
- **File:BS-12-Begriffsschrift_Quantifier1-svg.svg** *Source:* https://upload.wikimedia.org/wikipedia/commons/c/ce/BS-12-Begriffsschrift_Quantifier1-svg.svg *License:* CC0 *Contributors:* Own work *Original artist:* Majo statt Senf
- **File:BernoullisLawDerivationDiagram.svg** *Source:* https://upload.wikimedia.org/wikipedia/commons/2/20/BernoullisLawDerivationDiagram.svg *License:* CC-BY-SA-3.0 *Contributors:* Image:BernoullisLawDerivationDiagram.png *Original artist:* MannyMax (original)
- **File:Blochsphere.svg** *Source:* https://upload.wikimedia.org/wikipedia/commons/f/f3/Blochsphere.svg *License:* CC-BY-SA-3.0 *Contributors:* Transferred from en.wikipedia to Commons. *Original artist:* MuncherOfSpleens at English Wikipedia
- **File:Braid-modular-group-cover.svg** *Source:* https://upload.wikimedia.org/wikipedia/commons/d/da/Braid-modular-group-cover.svg *License:* Public domain *Contributors:* Own work, created as per: en:meta:Help:Displaying a formula#Commutative diagrams; source code below. *Original artist:* Nils R. Barth
- **File:Brain,_G_Reisch.png** *Source:* https://upload.wikimedia.org/wikipedia/commons/c/c8/Brain%2C_G_Reisch.png *License:* Public domain *Contributors:* ? *Original artist:* ?

- **File:CH4-structure.svg** *Source:* https://upload.wikimedia.org/wikipedia/commons/0/0f/CH4-structure.svg *License:* GPLv3 *Contributors:* File:Ch4-structure.png *Original artist:* Own work
- **File:CMS_Higgs-event.jpg** *Source:* https://upload.wikimedia.org/wikipedia/commons/1/1c/CMS_Higgs-event.jpg *License:* CC BY-SA 3.0 *Contributors:* http://cdsweb.cern.ch/record/628469 *Original artist:* Lucas Taylor / CERN
- **File:Caesar3.svg** *Source:* https://upload.wikimedia.org/wikipedia/commons/2/2b/Caesar3.svg *License:* Public domain *Contributors:* Own work *Original artist:* Cepheus
- **File:Candle-light-animated.gif** *Source:* https://upload.wikimedia.org/wikipedia/commons/2/2a/Candle-light-animated.gif *License:* CC BY-SA 3.0 *Contributors:* Own work *Original artist:* Andrikkos
- **File:Carl_Friedrich_Gauss.jpg** *Source:* https://upload.wikimedia.org/wikipedia/commons/9/9b/Carl_Friedrich_Gauss.jpg *License:* Public domain *Contributors:* Gauß-Gesellschaft Göttingen e.V. (Foto: A. Wittmann). *Original artist:* After Christian Albrecht Jensen
- **File:Cerebrum_lobes.svg** *Source:* https://upload.wikimedia.org/wikipedia/commons/d/d8/Cerebrum_lobes.svg *License:* CC BY-SA 3.0 *Contributors:* http://training.seer.cancer.gov/module_anatomy/unit5_3_nerve_org1_cns.html *Original artist:* vectorized by Jkwchui
- **File:Charles_Darwin_seated_crop.jpg** *Source:* https://upload.wikimedia.org/wikipedia/commons/2/2e/Charles_Darwin_seated_crop.jpg *License:* Public domain *Contributors:*
- Charles_Darwin_seated.jpg *Original artist:* Charles_Darwin_seated.jpg: Henry Maull (1829–1914) and John Fox (1832–1907) (Maull & Fox) [#cite_note-FDarwin-2 [2]]
- **File:Chicken_farmer_in_Ghana_(5926941911).jpg** *Source:* https://upload.wikimedia.org/wikipedia/commons/c/c2/Chicken_farmer_in_Ghana_%285926941911%29.jpg *License:* Public domain *Contributors:* Chicken farmer in Ghana *Original artist:* USAID Africa Bureau
- **File:Clinton&1998NobelLaureates.jpg** *Source:* https://upload.wikimedia.org/wikipedia/commons/9/9d/Clinton%261998NobelLaureates.jpg *License:* Public domain *Contributors:* http://clinton3.nara.gov/WH/EOP/OSTP/html/19981125.html *Original artist:* The White House
- **File:Commons-logo.svg** *Source:* https://upload.wikimedia.org/wikipedia/en/4/4a/Commons-logo.svg *License:* PD *Contributors:* ? *Original artist:* ?
- **File:Commutative_diagram_for_morphism.svg** *Source:* https://upload.wikimedia.org/wikipedia/commons/e/ef/Commutative_diagram_for_morphism.svg *License:* Public domain *Contributors:* Own work, based on en:Image:MorphismComposition-01.png *Original artist:* User:Cepheus
- **File:Composite_trapezoidal_rule_illustration_small.svg** *Source:* https://upload.wikimedia.org/wikipedia/commons/d/dd/Composite_trapezoidal_rule_illustration_small.svg *License:* Attribution *Contributors:*
- Composite_trapezoidal_rule_illustration_small.png *Original artist:*
- derivative work: Pbroks13 (talk)
- **File:Conformal_grid_after_Möbius_transformation.svg** *Source:* https://upload.wikimedia.org/wikipedia/commons/3/3f/Conformal_grid_after_M%C3%B6bius_transformation.svg *License:* CC BY-SA 2.5 *Contributors:* By Lokal_Profil *Original artist:* Lokal_Profil
- **File:Corncobs.jpg** *Source:* https://upload.wikimedia.org/wikipedia/commons/7/7d/Corncobs.jpg *License:* CC BY-SA 2.0 *Contributors:* Own work *Original artist:* User:Asbestos
- **File:DFAexample.svg** *Source:* https://upload.wikimedia.org/wikipedia/commons/9/9d/DFAexample.svg *License:* Public domain *Contributors:* Own work *Original artist:* Cepheus
- **File:DIMendeleevCab.jpg** *Source:* https://upload.wikimedia.org/wikipedia/commons/c/c8/DIMendeleevCab.jpg *License:* Public domain *Contributors:* кабинет академика Михаила Михайловича Шульца - фото любезно передано мне в собственность вдовой М.М.Шульца Ниной Дмитриевной Шульц. *Original artist:* ?
- **File:DNA_icon_(25x25).png** *Source:* https://upload.wikimedia.org/wikipedia/commons/9/93/DNA_icon_%2825x25%29.png *License:* Public domain *Contributors:* Own work *Original artist:* en:User:Markus Schmaus
- **File:DNA_replication_split.svg** *Source:* https://upload.wikimedia.org/wikipedia/commons/7/70/DNA_replication_split.svg *License:* CC-BY-SA-3.0 *Contributors:* Own work *Original artist:* Madprime
- **File:DrustveneNauke.png** *Source:* https://upload.wikimedia.org/wikipedia/commons/c/c0/DrustveneNauke.png *License:* CC-BY-SA-3.0 *Contributors:* Transferred from bs.wikipedia to Commons. *Original artist:* The original uploader was Mhare at Bosnian Wikipedia
- **File:EQbrain_optical_stim_en.jpg** *Source:* https://upload.wikimedia.org/wikipedia/commons/a/ad/EQbrain_optical_stim_en.jpg *License:* CC BY-SA 3.0 *Contributors:* Own work *Original artist:* ManosHacker
- **File:EastHanSeismograph.JPG** *Source:* https://upload.wikimedia.org/wikipedia/commons/a/ae/EastHanSeismograph.JPG *License:* CC-BY-SA-3.0 *Contributors:* en:File:EastHanSeismograph.JPG *Original artist:* en:user: Kowloonese
- **File:Edit-clear.svg** *Source:* https://upload.wikimedia.org/wikipedia/en/f/f2/Edit-clear.svg *License:* Public domain *Contributors:* The Tango! Desktop Project. *Original artist:*
  The people from the Tango! project. And according to the meta-data in the file, specifically: "Andreas Nilsson, and Jakub Steiner (although minimally)."
- **File:Edwin_Smith_Papyrus_v2.jpg** *Source:* https://upload.wikimedia.org/wikipedia/commons/b/b4/Edwin_Smith_Papyrus_v2.jpg *License:* Public domain *Contributors:* Edited version of Image:EdSmPaPlateVIandVIIPrintsx.jpg *Original artist:* Jeff Dahl
- **File:Einstein_cross.jpg** *Source:* https://upload.wikimedia.org/wikipedia/commons/c/c8/Einstein_cross.jpg *License:* Public domain *Contributors:* http://hubblesite.org/newscenter/archive/releases/1990/20/image/a/ *Original artist:* NASA, ESA, and STScI

26.10. TEXT AND IMAGE SOURCES, CONTRIBUTORS, AND LICENSES

- **File:Elliptic_curve_simple.png** *Source:* https://upload.wikimedia.org/wikipedia/commons/7/76/Elliptic_curve_simple.png *License:* CC-BY-SA-3.0 *Contributors:* Upload from English Wikipedia. The original description is/was here. *Original artist:* Created by Sean κ. + 23:33, 27 May 2005 (UTC)
- **File:Elliptic_curve_simple.svg** *Source:* https://upload.wikimedia.org/wikipedia/commons/d/da/Elliptic_curve_simple.svg *License:* CC-BY-SA-3.0 *Contributors:*
- Elliptic_curve_simple.png *Original artist:*
- derivative work: Pbroks13 (talk)
- **File:Emile_Durkheim.jpg** *Source:* https://upload.wikimedia.org/wikipedia/commons/2/24/Emile_Durkheim.jpg *License:* Public domain *Contributors:* http://www.marxists.org/glossary/people/d/pics/durkheim.jpg *Original artist:* Unknown<a href='https://www.wikidata.org/wiki/Q4233718' title='wikidata:Q4233718'><img alt='wikidata:Q4233718' src='https://upload.wikimedia.org/wikipedia/commons/thumb/f/ff/Wikidata-logo.svg/20px-Wikidata-logo.svg.png' width='20' height='11' srcset='https://upload.wikimedia.org/wikipedia/commons/thumb/f/ff/Wikidata-logo.svg/30px-Wikidata-logo.svg.png 1.5x, https://upload.wikimedia.org/wikipedia/commons/thumb/f/ff/Wikidata-logo.svg/40px-Wikidata-logo.svg.png 2x' data-file-width='1050' data-file-height='590' /></a>
- **File:Empirical_Cycle.svg** *Source:* https://upload.wikimedia.org/wikipedia/commons/5/53/Empirical_Cycle.svg *License:* CC BY 3.0 *Contributors:*
- Empirical_Cycle.png *Original artist:* Empirical_Cycle.png: TesseUndDaan
- **File:Encarsia_formosa,_an_endoparasitic_wasp,_is_used_for_whitefly_control.jpg** *Source:* https://upload.wikimedia.org/wikipedia/commons/a/a7/Encarsia_formosa%2C_an_endoparasitic_wasp%2C_is_used_for_whitefly_control.jpg *License:* CC BY-SA 3.0 *Contributors:* Own work *Original artist:* Dekayem
- **File:Epicycle_and_deferent.svg** *Source:* https://upload.wikimedia.org/wikipedia/commons/f/fb/Epicycle_and_deferent.svg *License:* Public domain *Contributors:* Based upon data from Thomas S. Kuhn, *La rivoluzione copernicana* (*The Copernican Revolution*), Einaudi, Torino, 2000, p. 78. Also based upon File:Epicycle et deferent.png by Julo *Original artist:* Own work MLWatts
- **File:Euclid.jpg** *Source:* https://upload.wikimedia.org/wikipedia/commons/2/21/Euclid.jpg *License:* Public domain *Contributors:* ? *Original artist:* ?
- **File:Ferdinand_de_Saussure.jpg** *Source:* https://upload.wikimedia.org/wikipedia/commons/8/8f/Ferdinand_de_Saussure.jpg *License:* Public domain *Contributors:* ? *Original artist:* ?
- **File:Feuer123.JPG** *Source:* https://upload.wikimedia.org/wikipedia/commons/9/98/Feuer123.JPG *License:* CC BY-SA 3.0 *Contributors:* Own work *Original artist:* Achates
- **File:Fibonacci.jpg** *Source:* https://upload.wikimedia.org/wikipedia/commons/a/a2/Fibonacci.jpg *License:* Public domain *Contributors:* Scan from "Mathematical Circus" by Martin Gardner, published 1981 *Original artist:* unknown 19th-century artist
- **File:Flowchart_MBR.png** *Source:* https://upload.wikimedia.org/wikipedia/commons/0/00/Flowchart_MBR.png *License:* CC BY-SA 4.0 *Contributors:* Own work *Original artist:* Reshmiiyer
- **File:Folder_Hexagonal_Icon.svg** *Source:* https://upload.wikimedia.org/wikipedia/en/4/48/Folder_Hexagonal_Icon.svg *License:* Cc-by-sa-3.0 *Contributors:* ? *Original artist:* ?
- **File:Francis_Bacon_statue._Gray'{}s_Inn.jpg** *Source:* https://upload.wikimedia.org/wikipedia/commons/e/e6/Francis_Bacon_statue%2C_Gray%27s_Inn.jpg *License:* CC BY-SA 2.0 *Contributors:* http://www.geograph.org.uk/photo/1266685 *Original artist:* Mike Quinn
- **File:Friedrich_Hegel_mit_Studenten_Lithographie_F_Kugler.jpg** *Source:* https://upload.wikimedia.org/wikipedia/commons/e/e0/Friedrich_Hegel_mit_Studenten_Lithographie_F_Kugler.jpg *License:* Public domain *Contributors:* Das Wissen des 20.Jahrhunderts, Bildungslexikon, Rheda, 1931 *Original artist:* Franz Kugler
- **File:GDP_PPP_Per_Capita_IMF_2008.svg** *Source:* https://upload.wikimedia.org/wikipedia/commons/d/d4/GDP_PPP_Per_Capita_IMF_2008.svg *License:* CC BY 3.0 *Contributors:* Sbw01f's work, but converted to an SVG file instead. Data from International Monetary Fund World Economic Outlook Database April 2009 *Original artist:* Powerkeys
- **File:Galileo.arp.300pix.jpg** *Source:* https://upload.wikimedia.org/wikipedia/commons/c/cc/Galileo.arp.300pix.jpg *License:* Public domain *Contributors:*

  http://www.nmm.ac.uk/mag/pages/mnuExplore/ViewLargeImage.cfm?ID=BHC2700

  Original artist: Justus Sustermans
- **File:GalileosInclinedPlane.jpg** *Source:* https://upload.wikimedia.org/wikipedia/commons/c/c0/GalileosInclinedPlane.jpg *License:* Attribution *Contributors:* http://www.scitechantiques.com *Original artist:* Jim and Rhoda Morris
- **File:GodfreyKneller-IsaacNewton-1689.jpg** *Source:* https://upload.wikimedia.org/wikipedia/commons/3/39/GodfreyKneller-IsaacNewton-1689.jpg *License:* Public domain *Contributors:* http://www.newton.cam.ac.uk/art/portrait.html *Original artist:* After Godfrey Kneller
- **File:Gottfried_Wilhelm_von_Leibniz.jpg** *Source:* https://upload.wikimedia.org/wikipedia/commons/6/6a/Gottfried_Wilhelm_von_Leibniz.jpg *License:* Public domain *Contributors:* /gbrown/philosophers/leibniz/BritannicaPages/Leibniz/LeibnizGif.html *Original artist:* Christoph Bernhard Francke
- **File:Gravitation_space_source.png** *Source:* https://upload.wikimedia.org/wikipedia/commons/2/26/Gravitation_space_source.png *License:* CC-BY-SA-3.0 *Contributors:* ? *Original artist:* ?
- **File:Gravitational_lens-full.jpg** *Source:* https://upload.wikimedia.org/wikipedia/commons/0/02/Gravitational_lens-full.jpg *License:* Public domain *Contributors:* ? *Original artist:* ?

- **File:Group_diagdram_D6.svg** *Source:* https://upload.wikimedia.org/wikipedia/commons/0/0e/Group_diagdram_D6.svg *License:* Public domain *Contributors:* Own work *Original artist:* User:Cepheus
- **File:Hutton_James_portrait_Raeburn.jpg** *Source:* https://upload.wikimedia.org/wikipedia/commons/0/0e/Hutton_James_portrait_Raeburn.jpg *License:* Public domain *Contributors:* Scottish National Portrait Gallery *Original artist:* Henry Raeburn
- **File:Hyperbolic_triangle.svg** *Source:* https://upload.wikimedia.org/wikipedia/commons/8/89/Hyperbolic_triangle.svg *License:* Public domain *Contributors:* ? *Original artist:* ?
- **File:Illustration_to_Euclid'{}s_proof_of_the_Pythagorean_theorem.svg** *Source:* https://upload.wikimedia.org/wikipedia/commons/2/26/Illustration_to_Euclid%27s_proof_of_the_Pythagorean_theorem.svg *License:* WTFPL *Contributors:* ? *Original artist:* ?
- **File:Integral_as_region_under_curve.svg** *Source:* https://upload.wikimedia.org/wikipedia/commons/f/f2/Integral_as_region_under_curve.svg *License:* CC-BY-SA-3.0 *Contributors:* Own work, based on JPG version *Original artist:* 4C
- **File:Islamic_MedText_c1500.jpg** *Source:* https://upload.wikimedia.org/wikipedia/commons/6/6d/Islamic_MedText_c1500.jpg *License:* Public domain *Contributors:* Unknown *Original artist:* User Allen3 on en.wikipedia
- **File:Issoria_lathonia.jpg** *Source:* https://upload.wikimedia.org/wikipedia/commons/2/2d/Issoria_lathonia.jpg *License:* CC-BY-SA-3.0 *Contributors:* ? *Original artist:* ?
- **File:James_Clerk_Maxwell_profile.jpg** *Source:* https://upload.wikimedia.org/wikipedia/commons/9/9e/James_Clerk_Maxwell_profile.jpg *License:* Public domain *Contributors:* Portrait of James Clerk Maxwell (1831-1879), Physicist *Original artist:* Unidentified photographer, Smithsonian Institution from United States
- **File:Jean_Louis_Théodore_Géricault_001.jpg** *Source:* https://upload.wikimedia.org/wikipedia/commons/f/f3/Jean_Louis_Th%C3%A9odore_G%C3%A9ricault_001.jpg *License:* Public domain *Contributors:* The Yorck Project: *10.000 Meisterwerke der Malerei.* DVD-ROM. 2002. ISBN 3936122202. Distributed by DIRECTMEDIA Publishing GmbH. *Original artist:* Théodore Géricault
- **File:JeremiahHorrocks.jpg** *Source:* https://upload.wikimedia.org/wikipedia/commons/c/cf/JeremiahHorrocks.jpg *License:* Public domain *Contributors:* http://www.economist.com/science/displayStory.cfm?story_id=2705523 *Original artist:* **William Richard Lavender** (1877-1915)
- **File:Johannes_Kepler_1610.jpg** *Source:* https://upload.wikimedia.org/wikipedia/commons/d/d4/Johannes_Kepler_1610.jpg *License:* Public domain *Contributors:* ? *Original artist:* Unknown<a href='https://www.wikidata.org/wiki/Q4233718' title='wikidata:Q4233718'><img alt='wikidata:Q4233718' src='https://upload.wikimedia.org/wikipedia/commons/thumb/f/ff/Wikidata-logo.svg/20px-Wikidata-logo.svg.png' width='20' height='11' srcset='https://upload.wikimedia.org/wikipedia/commons/thumb/f/ff/Wikidata-logo.svg/30px-Wikidata-logo.svg.png 1.5x, https://upload.wikimedia.org/wikipedia/commons/thumb/f/ff/Wikidata-logo.svg/40px-Wikidata-logo.svg.png 2x' data-file-width='1050' data-file-height='590' /></a>
- **File:Johannesmagistris-square.jpg** *Source:* https://upload.wikimedia.org/wikipedia/commons/c/ca/Johannesmagistris-square.jpg *License:* Public domain *Contributors:* Own work *Original artist:* Peter Damian
- **File:Justus_Sustermans_-_Portrait_of_Galileo_Galilei,_1636.jpg** *Source:* https://upload.wikimedia.org/wikipedia/commons/d/d4/Justus_Sustermans_-_Portrait_of_Galileo_Galilei%2C_1636.jpg *License:* Public domain *Contributors:* http://www.nmm.ac.uk/mag/pages/mnuExplore/PaintingDetail.cfm?ID=BHC2700 *Original artist:* Justus Sustermans
- **File:Kapitolinischer_Pythagoras_adjusted.jpg** *Source:* https://upload.wikimedia.org/wikipedia/commons/1/1a/Kapitolinischer_Pythagoras_adjusted.jpg *License:* CC-BY-SA-3.0 *Contributors:* First upload to Wikipedia: de.wikipedia; description page is/was here. *Original artist:* The original uploader was Galilea at German Wikipedia
- **File:Karl_Popper.jpg** *Source:* https://upload.wikimedia.org/wikipedia/commons/4/43/Karl_Popper.jpg *License:* No restrictions *Contributors:* http://www.flickr.com/photos/lselibrary/3833724834/in/set-72157623156680255/ *Original artist:* LSE library
- **File:Kepler-solar-system-2.gif** *Source:* https://upload.wikimedia.org/wikipedia/commons/1/1d/Kepler-solar-system-2.gif *License:* Public domain *Contributors:* ? *Original artist:* ?
- **File:Lattice_of_the_divisibility_of_60.svg** *Source:* https://upload.wikimedia.org/wikipedia/commons/5/51/Lattice_of_the_divisibility_of_60.svg *License:* CC-BY-SA-3.0 *Contributors:* No machine-readable source provided. Own work assumed (based on copyright claims). *Original artist:* No machine-readable author provided. Ed g2s assumed (based on copyright claims).
- **File:Laurentius_de_Voltolina_001.jpg** *Source:* https://upload.wikimedia.org/wikipedia/commons/f/fc/Laurentius_de_Voltolina_001.jpg *License:* Public domain *Contributors:* The Yorck Project: *10.000 Meisterwerke der Malerei.* DVD-ROM. 2002. ISBN 3936122202. Distributed by DIRECTMEDIA Publishing GmbH. *Original artist:* Laurentius de Voltolina
- **File:Leonhard_Euler_2.jpg** *Source:* https://upload.wikimedia.org/wikipedia/commons/6/60/Leonhard_Euler_2.jpg *License:* Public domain *Contributors:*
    - 2011-12-22 (upload, according to EXIF data)

    Original artist: Jakob Emanuel Handmann
- **File:Leonhart_Fuchs_Historia_Stirpium.jpg** *Source:* https://upload.wikimedia.org/wikipedia/commons/0/0d/Leonhart_Fuchs_Historia_Stirpium.jpg *License:* Public domain *Contributors:* http://www.rarebookroom.org/Control/leodeh/index.html *Original artist:* Michel Isingrin
- **File:Libr0310.jpg** *Source:* https://upload.wikimedia.org/wikipedia/commons/2/24/Libr0310.jpg *License:* Public domain *Contributors:* ? *Original artist:* ?
- **File:Limitcycle.svg** *Source:* https://upload.wikimedia.org/wikipedia/commons/9/91/Limitcycle.svg *License:* CC BY-SA 3.0 *Contributors:* Own work *Original artist:* Gargan
- **File:Linear_regression.svg** *Source:* https://upload.wikimedia.org/wikipedia/commons/3/3a/Linear_regression.svg *License:* Public domain *Contributors:* Own work *Original artist:* Sewaqu

26.10. TEXT AND IMAGE SOURCES, CONTRIBUTORS, AND LICENSES

- **File:Lock-green.svg** *Source:* https://upload.wikimedia.org/wikipedia/commons/6/65/Lock-green.svg *License:* CC0 *Contributors:* en:File:Free-to-read_lock_75.svg *Original artist:* User:Trappist the monk
- **File:Logic_portal.svg** *Source:* https://upload.wikimedia.org/wikipedia/commons/7/7c/Logic_portal.svg *License:* CC BY-SA 3.0 *Contributors:* Own work *Original artist:* <a href='//commons.wikimedia.org/wiki/File:Watchduck.svg' class='image'><img alt='Watchduck.svg' src='https://upload.wikimedia.org/wikipedia/commons/thumb/d/d8/Watchduck.svg/40px-Watchduck.svg.png' width='40' height='46' srcset='https://upload.wikimedia.org/wikipedia/commons/thumb/d/d8/Watchduck.svg/60px-Watchduck.svg.png 1.5x, https://upload.wikimedia.org/wikipedia/commons/thumb/d/d8/Watchduck.svg/80px-Watchduck.svg.png 2x' data-file-width='703' data-file-height='806' /></a> Watchduck (a.k.a. Tilman Piesk)
- **File:Lorenz_attractor.svg** *Source:* https://upload.wikimedia.org/wikipedia/commons/f/f4/Lorenz_attractor.svg *License:* CC BY 2.5 *Contributors:* Own work *Original artist:* Dschwen
- **File:MagCompas3.jpg** *Source:* https://upload.wikimedia.org/wikipedia/commons/e/e5/MagCompas3.jpg *License:* CC BY 2.5 *Contributors:* No machine-readable source provided. Own work assumed (based on copyright claims). *Original artist:* No machine-readable author provided. Clipper assumed (based on copyright claims).
- **File:Mandel_zoom_07_satellite.jpg** *Source:* https://upload.wikimedia.org/wikipedia/commons/b/b3/Mandel_zoom_07_satellite.jpg *License:* CC-BY-SA-3.0 *Contributors:* ? *Original artist:* ?
- **File:Map_of_Medieval_Universities.jpg** *Source:* https://upload.wikimedia.org/wikipedia/commons/6/68/Map_of_Medieval_Universities.jpg *License:* Public domain *Contributors:* William R. Shepherd: *Historical Atlas*, New York, Henry Holt and Company, 1923, in the Public Domain (also to be found in the 1911 edition, p. 100) *Original artist:* ?
- **File:Maquina.png** *Source:* https://upload.wikimedia.org/wikipedia/commons/3/3d/Maquina.png *License:* Public domain *Contributors:* en.wikipedia *Original artist:* Schadel (http://turing.izt.uam.mx)
- **File:Marie_Curie_c1920.jpg** *Source:* https://upload.wikimedia.org/wikipedia/commons/7/7e/Marie_Curie_c1920.jpg *License:* Public domain *Contributors:* Christie's, [1] *Original artist:* Unknown<a href='https://www.wikidata.org/wiki/Q4233718' title='wikidata:Q4233718'><img alt='wikidata:Q4233718' src='https://upload.wikimedia.org/wikipedia/commons/thumb/f/ff/Wikidata-logo.svg/20px-Wikidata-logo.svg.png' width='20' height='11' srcset='https://upload.wikimedia.org/wikipedia/commons/thumb/f/ff/Wikidata-logo.svg/30px-Wikidata-logo.svg.png 1.5x, https://upload.wikimedia.org/wikipedia/commons/thumb/f/ff/Wikidata-logo.svg/40px-Wikidata-logo.svg.png 2x' data-file-width='1050' data-file-height='590' /></a>
- **File:Market-Chichicastenango.jpg** *Source:* https://upload.wikimedia.org/wikipedia/commons/6/62/Market-Chichicastenango.jpg *License:* CC-BY-SA-3.0 *Contributors:* ? *Original artist:* ?
- **File:Market_Data_Index_NYA_on_20050726_202628_UTC.png** *Source:* https://upload.wikimedia.org/wikipedia/commons/4/46/Market_Data_Index_NYA_on_20050726_202628_UTC.png *License:* Public domain *Contributors:* ? *Original artist:* ?
- **File:MathModel.svg** *Source:* https://upload.wikimedia.org/wikipedia/commons/f/f2/MathModel.svg *License:* Public domain *Contributors:* Own work, inspired by Bertuglia & Vaio 2005 *Original artist:* Tomaschwutz
- **File:Maximum_boxed.png** *Source:* https://upload.wikimedia.org/wikipedia/commons/1/1a/Maximum_boxed.png *License:* Public domain *Contributors:* Created with the help of GraphCalc *Original artist:* Freiddy
- **File:Maya.svg** *Source:* https://upload.wikimedia.org/wikipedia/commons/1/1b/Maya.svg *License:* CC-BY-SA-3.0 *Contributors:* Image:Maya.png *Original artist:* Bryan Derksen
- **File:Measure_illustration.png** *Source:* https://upload.wikimedia.org/wikipedia/commons/a/a6/Measure_illustration.png *License:* Public domain *Contributors:* self-made with en:Inkscape *Original artist:* Oleg Alexandrov
- **File:Mergefrom.svg** *Source:* https://upload.wikimedia.org/wikipedia/commons/0/0f/Mergefrom.svg *License:* Public domain *Contributors:* ? *Original artist:* ?
- **File:Modeling_and_Simulation_Integrated_Use.jpg** *Source:* https://upload.wikimedia.org/wikipedia/commons/c/c0/Modeling_and_Simulation_Integrated_Use.jpg *License:* Public domain *Contributors:* ? *Original artist:* ?
- **File:NASA-Apollo8-Dec24-Earthrise.jpg** *Source:* https://upload.wikimedia.org/wikipedia/commons/a/a8/NASA-Apollo8-Dec24-Earthrise.jpg *License:* Public domain *Contributors:* http://www.hq.nasa.gov/office/pao/History/alsj/a410/AS8-14-2383HR.jpg *Original artist:* NASA / Bill Anders
- **File:Navier_Stokes_Laminar.svg** *Source:* https://upload.wikimedia.org/wikipedia/commons/7/73/Navier_Stokes_Laminar.svg *License:* CC BY-SA 4.0 *Contributors:* Own work

Brief description of the numerical method

The following code leverages some numerical methods to simulate the solution of the 2-dimensional Navier-Stokes equation.
We choose the simplified incompressible flow Navier-Stokes Equation as follows:

<img src='https://wikimedia.org/api/rest_v1/media/math/render/svg/1b352a66970b542690aff9810ff1514eca0952bd' class='mwe-math-fallback-image-inline' aria-hidden='true' style='vertical-align: −2.505ex; width:27.565ex; height:6.176ex;' alt='{\displaystyle \rho \left({\frac {\partial \mathbf {v} }{\partial t <i' />Original artist: S5+\mathbf {v} \cdot \nabla \mathbf {v} \right)=\mu \nabla ^{2}\mathbf {v} .}'>

The iterations here are based on the velocity change rate, which is given by

<img src='https://wikimedia.org/api/rest_v1/media/math/render/svg/84351a8157ffbc3af56ed19583c97d062bfd428d' class='mwe-math-fallback-image-inline' aria-hidden='true' style='vertical-align: −2.338ex; width:23.36ex; height:5.843ex;' alt='{\displaystyle {\frac {\partial \mathbf {v} }{\partial t <h2' />Content license $3={\frac {\mu }{\rho }}\nabla ^{2}\mathbf {v} -\mathbf {v} \cdot \nabla \mathbf {v} .}'>

Or in X coordinates:

```
<img src="https://wikimedia.org/api/rest_v1/media/math/render/svg/d009b926bc255277cd55f75f9773d3721861f3ac" class="mwe-math-fallback-image-inline" aria-hidden="true" style="vertical-align: -2.505ex; width:45.497ex; height:6.343ex;" alt="{\displaystyle {\frac {\partial v_{x}}{\partial t}}={\frac {\mu }{\rho }}({\frac {\partial ^{2}v_{x}}{\partial x^{2}}}+{\frac {\partial ^{2}v_{x}}{\partial y^{2}}})-v_{x}{\frac {\partial v_{x}}{\partial x}}-v_{y}{\frac {\partial v_{x}}{\partial y}}.}">
```

The above equation gives the code. The case of Y is similar.IIkamusumeFan}}

- **File:Newton'{}s_reflecting_telescope.jpg** *Source:* https://upload.wikimedia.org/wikipedia/commons/a/aa/Newton%27s_reflecting_telescope.jpg *License:* Public domain *Contributors:* Retrieved May 28, 2014 from <a data-x-rel='nofollow' class='external text' href='http://books.google.com/books?id=4hY6AAAAcAAJ,,&,,pg=PA46'>David Brewster (1855) *Memoirs of the Life, Writings, and Descoveries of Sir Isaac Newton, Vol. 1*, Thomas Constable and Co., Edinburgh, p. 46, fig. 7</a> on Google Books *Original artist:* David Brewster

- **File:Nuvola_apps_edu_mathematics_blue-p.svg** *Source:* https://upload.wikimedia.org/wikipedia/commons/3/3e/Nuvola_apps_edu_mathematics_blue-p.svg *License:* GPL *Contributors:* Derivative work from Image:Nuvola apps edu mathematics.png and Image:Nuvola apps edu mathematics-p.svg *Original artist:* David Vignoni (original icon); Flamurai (SVG conversion); bayo (color)

- **File:Nuvola_apps_kalzium.svg** *Source:* https://upload.wikimedia.org/wikipedia/commons/8/8b/Nuvola_apps_kalzium.svg *License:* LGPL *Contributors:* Own work *Original artist:* David Vignoni, SVG version by Bobarino

- **File:Office-book.svg** *Source:* https://upload.wikimedia.org/wikipedia/commons/a/a8/Office-book.svg *License:* Public domain *Contributors:* This and myself. *Original artist:* Chris Down/Tango project

- **File:Old_Bailey_Microcosm_edited.jpg** *Source:* https://upload.wikimedia.org/wikipedia/commons/f/f3/Old_Bailey_Microcosm_edited.jpg *License:* Public domain *Contributors:* Ackermann, Rudolph; Pyne, William Henry; Combe, William (1904) [1808] "Old Bailey" in *The Microcosm of London; or, London in Miniature*, Volume 2, London: Methuen and Company Retrieved on 9 January 2009. *Original artist:* Thomas Rowlandson and Augustus Pugin

- **File:Oldfaithful3.png** *Source:* https://upload.wikimedia.org/wikipedia/commons/0/0f/Oldfaithful3.png *License:* Public domain *Contributors:* ? *Original artist:* ?

- **File:Open_Access_logo_PLoS_transparent.svg** *Source:* https://upload.wikimedia.org/wikipedia/commons/7/77/Open_Access_logo_PLoS_transparent.svg *License:* CC0 *Contributors:* http://www.plos.org/ *Original artist:* art designer at PLoS, modified by Wikipedia users Nina, Beao, and JakobVoss

- **File:Oxyrhynchus_papyrus_with_Euclid'{}s_Elements.jpg** *Source:* https://upload.wikimedia.org/wikipedia/commons/8/8d/P._Oxy._I._29.jpg *License:* Public domain *Contributors:* http://www.math.ubc.ca/~{}cass/Euclid/papyrus/tha.jpg *Original artist:* Euclid

- **File:P_philosophy.png** *Source:* https://upload.wikimedia.org/wikipedia/commons/b/bb/P_philosophy.png *License:* CC-BY-SA-3.0 *Contributors:* ? *Original artist:* ?

- **File:Painting_of_Volta_by_Bertini_(photo).jpeg** *Source:* https://upload.wikimedia.org/wikipedia/commons/0/0e/Painting_of_Volta_by_Bertini_%28photo%29.jpeg *License:* Public domain *Contributors:* ? *Original artist:* ?

- **File:Papyrus_text:_fragment_of_Hippocratic_oath._Wellcome_L0034090.jpg** *Source:* https://upload.wikimedia.org/wikipedia/commons/4/4c/Papyrus_text%3B_fragment_of_Hippocratic_oath._Wellcome_L0034090.jpg *License:* CC BY 4.0 *Contributors:* http://wellcomeimages.org/indexplus/obf_images/0d/0e/e9d3d979e732af7f22a1b41d1110.jpg
  Original artist: ?

- **File:Partial_ordering_of_the_sciences_Balaban_Klein_Scientometrics2006_615-637.svg** *Source:* https://upload.wikimedia.org/wikipedia/commons/0/04/Partial_ordering_of_the_sciences_Balaban_Klein_Scientometrics2006_615-637.svg *License:* CC-BY-SA-3.0 *Contributors:* ? *Original artist:* ?

- **File:Paul_Feyerabend_Berkeley.jpg** *Source:* https://upload.wikimedia.org/wikipedia/commons/e/e2/Paul_Feyerabend_Berkeley.jpg *License:* Attribution *Contributors:* The uploader on Wikimedia Commons received this from the author/copyright holder. *Original artist:* Grazia Borrini-Feyerabend

- **File:People_icon.svg** *Source:* https://upload.wikimedia.org/wikipedia/commons/3/37/People_icon.svg *License:* CC0 *Contributors:* OpenClipart *Original artist:* OpenClipart

- **File:Perihelion_precession.jpg** *Source:* https://upload.wikimedia.org/wikipedia/commons/8/83/Perihelion_precession.jpg *License:* CC-BY-SA-3.0 *Contributors:* Transferred from en.wikipedia to Commons. *Original artist:* Markus Schmaus at English Wikipedia

- **File:Persian_Khwarazmi.jpg** *Source:* https://upload.wikimedia.org/wikipedia/commons/b/b7/Persian_Khwarazmi.jpg *License:* Public domain *Contributors:* Based on the USSR stamp of him, shaded by me, see File:Abu Abdullah Muhammad bin Musa al-Khwarizmi.jpg for a version of the stamp, although perhaps not the one used here. *Original artist:* Unknown<a href='https://www.wikidata.org/wiki/Q4233718' title='wikidata:Q4233718'><img alt='wikidata:Q4233718' src='https://upload.wikimedia.org/wikipedia/commons/thumb/f/ff/Wikidata-logo.svg/20px-Wikidata-logo.svg.png' width='20' height='11' srcset='https://upload.wikimedia.org/wikipedia/commons/thumb/f/ff/Wikidata-logo.svg/30px-Wikidata-logo.svg.png 1.5x, https://upload.wikimedia.org/wikipedia/commons/thumb/f/ff/Wikidata-logo.svg/40px-Wikidata-logo.svg.png 2x' data-file-width='1050' data-file-height='590' /></a>

- **File:Pharmacologyprism.jpg** *Source:* https://upload.wikimedia.org/wikipedia/commons/8/8f/Pharmacologyprism.jpg *License:* Public domain *Contributors:* ? *Original artist:* ?

- **File:Physical_world.jpg** *Source:* https://upload.wikimedia.org/wikipedia/commons/9/9c/Physical_world.jpg *License:* Public domain *Contributors:* CIA World Factbook [1]
  Original artist: Central Intelligence Agency

- **File:Plato'{}s_Academy_mosaic_from_Pompeii.jpg** *Source:* https://upload.wikimedia.org/wikipedia/commons/4/48/Plato%27s_Academy_mosaic_from_Pompeii.jpg *License:* Public domain *Contributors:* http://www.departments.bucknell.edu/History/Carnegie/plato/academy.html *Original artist:* Unknown<a href='https://www.wikidata.org/wiki/Q4233718' title='wikidata:Q4233718'><img alt='wikidata:Q4233718' src='https://upload.wikimedia.org/wikipedia/commons/thumb/f/ff/Wikidata-logo.svg/20px-Wikidata-logo.svg.png' width='20' height='11' srcset='https://upload.wikimedia.org/wikipedia/commons/thumb/f/ff/Wikidata-logo.svg/30px-Wikidata-logo.svg.png 1.5x, https://upload.wikimedia.org/wikipedia/commons/thumb/f/ff/Wikidata-logo.svg/40px-Wikidata-logo.svg.png 2x' data-file-width='1050' data-file-height='590' /></a>
- **File:Portal-puzzle.svg** *Source:* https://upload.wikimedia.org/wikipedia/en/f/fd/Portal-puzzle.svg *License:* Public domain *Contributors:* ? *Original artist:* ?
- **File:QtubIronPillar.JPG** *Source:* https://upload.wikimedia.org/wikipedia/commons/3/3f/QtubIronPillar.JPG *License:* Public domain *Contributors:* Original photograph *Original artist:* Photograph taken by Mark A. Wilson (Department of Geology, The College of Wooster). [1]
- **File:Question_book-new.svg** *Source:* https://upload.wikimedia.org/wikipedia/en/9/99/Question_book-new.svg *License:* Cc-by-sa-3.0 *Contributors:*
Created from scratch in Adobe Illustrator. Based on Image:Question book.png created by User:Equazcion *Original artist:*
Tkgd2007
- **File:Question_dropshade.png** *Source:* https://upload.wikimedia.org/wikipedia/commons/d/dd/Question_dropshade.png *License:* Public domain *Contributors:* Image created by JRM *Original artist:* JRM
- **File:Register_transfer_level_-_example_toggler.svg** *Source:* https://upload.wikimedia.org/wikipedia/commons/0/01/Register_transfer_level_-_example_toggler.svg *License:* CC-BY-SA-3.0 *Contributors:* Transferred from en.wikipedia to Commons. *Original artist:* Alinja at English Wikipedia
- **File:Roger-bacon-statue.jpg** *Source:* https://upload.wikimedia.org/wikipedia/commons/8/8a/Roger-bacon-statue.jpg *License:* CC-BY-SA-3.0 *Contributors:* ? *Original artist:* ?
- **File:Rough_diamond.jpg** *Source:* https://upload.wikimedia.org/wikipedia/commons/d/d7/Rough_diamond.jpg *License:* Public domain *Contributors:* Original source: USGS "Minerals in Your World" website. Direct image link: [1] *Original artist:* Unknown USGS employee
- **File:Rubik'{}s_cube.svg** *Source:* https://upload.wikimedia.org/wikipedia/commons/a/a6/Rubik%27s_cube.svg *License:* CC-BY-SA-3.0 *Contributors:* Based on Image:Rubiks cube.jpg *Original artist:* This image was created by me, Booyabazooka
- **File:Science-symbol-2.svg** *Source:* https://upload.wikimedia.org/wikipedia/commons/7/75/Science-symbol-2.svg *License:* CC BY 3.0 *Contributors:* en:Image:Science-symbol2.png *Original artist:* en:User:AllyUnion, User:Stannered
- **File:Sea_island_survey.jpg** *Source:* https://upload.wikimedia.org/wikipedia/commons/5/5c/Sea_island_survey.jpg *License:* Public domain *Contributors:* 1726 Tu Shu Ji Cheng 圖書集成 *Original artist:* Liu Hui
- **File:Signal_transduction_pathways.svg** *Source:* https://upload.wikimedia.org/wikipedia/commons/b/b0/Signal_transduction_pathways.svg *License:* CC BY-SA 3.0 *Contributors:* This file was derived from: Signal transduction v1.png
Original artist: cybertory
- **File:Simple_feedback_control_loop2.svg** *Source:* https://upload.wikimedia.org/wikipedia/commons/9/90/Simple_feedback_control_loop2.svg *License:* CC BY-SA 3.0 *Contributors:* This file was derived from Simple feedback control loop2.png: <a href='//commons.wikimedia.org/wiki/File:Simple_feedback_control_loop2.png' class='image'><img alt='Simple feedback control loop2.png' src='https://upload.wikimedia.org/wikipedia/commons/thumb/4/45/Simple_feedback_control_loop2.png/50px-Simple_feedback_control_loop2.png' width='50' height='17' srcset='https://upload.wikimedia.org/wikipedia/commons/thumb/4/45/Simple_feedback_control_loop2.png/75px-Simple_feedback_control_loop2.png 1.5x, https://upload.wikimedia.org/wikipedia/commons/thumb/4/45/Simple_feedback_control_loop2.png/100px-Simple_feedback_control_loop2.png 2x' data-file-width='439' data-file-height='150' /></a>
Original artist: Simple_feedback_control_loop2.png: Corona
- **File:SimplexRangeSearching.svg** *Source:* https://upload.wikimedia.org/wikipedia/commons/e/e3/SimplexRangeSearching.svg *License:* Public domain *Contributors:* This file was derived from: SimplexRangeSearching.png
Original artist: Gfonsecabr at English Wikipedia
- **File:Sinusvåg_400px.png** *Source:* https://upload.wikimedia.org/wikipedia/commons/8/8c/Sinusv%C3%A5g_400px.png *License:* Public domain *Contributors:* Own work *Original artist:* User Solkoll on sv.wikipedia
- **File:Socrates.png** *Source:* https://upload.wikimedia.org/wikipedia/commons/c/cd/Socrates.png *License:* Public domain *Contributors:* Transferred from en.wikipedia to Commons. *Original artist:* The original uploader was Magnus Manske at English Wikipedia Later versions were uploaded by Optimager at en.wikipedia.
- **File:Su_Song_Star_Map_1.JPG** *Source:* https://upload.wikimedia.org/wikipedia/commons/d/df/Su_Song_Star_Map_1.JPG *License:* Public domain *Contributors:* Joseph Needham, *Science and Civilization in China: Volume 3, Mathematics and the Sciences of the Heavens and the Earth* (page 277) *Original artist:* PericlesofAthens
- **File:SumerianClayTablet,palm-sized422BCE.jpg** *Source:* https://upload.wikimedia.org/wikipedia/commons/1/16/SumerianClayTablet%2Cpalm-sized422BCE.jpg *License:* CC BY 2.0 *Contributors:* Transferred from en.wikipedia to Commons. *Original artist:* The original uploader was Ancheta Wis at English Wikipedia
- **File:Supply-demand-P.png** *Source:* https://upload.wikimedia.org/wikipedia/commons/4/43/Supply-demand-P.png *License:* CC-BY-SA-3.0 *Contributors:* Transferred from en.wikipedia to Commons. *Original artist:* CSTAR Later versions were uploaded by Guanaco, Everlong at en.wikipedia.
- **File:Symbol_book_class2.svg** *Source:* https://upload.wikimedia.org/wikipedia/commons/8/89/Symbol_book_class2.svg *License:* CC BY-SA 2.5 *Contributors:* Mad by Lokal_Profil by combining: *Original artist:* Lokal_Profil

- **File:Symbol_list_class.svg** *Source:* https://upload.wikimedia.org/wikipedia/en/d/db/Symbol_list_class.svg *License:* Public domain *Contributors:* ? *Original artist:* ?
- **File:Symbol_question.svg** *Source:* https://upload.wikimedia.org/wikipedia/en/e/e0/Symbol_question.svg *License:* Public domain *Contributors:* ? *Original artist:* ?
- **File:TSP_Deutschland_3.png** *Source:* https://upload.wikimedia.org/wikipedia/commons/c/c4/TSP_Deutschland_3.png *License:* Public domain *Contributors:* https://www.cia.gov/cia/publications/factbook/maps/gm-map.gif *Original artist:* The original uploader was Kapitän Nemo at German Wikipedia
- **File:Terasawa5.jpg** *Source:* https://upload.wikimedia.org/wikipedia/commons/a/a1/Terasawa5.jpg *License:* CC BY-SA 3.0 *Contributors:* jpg *Original artist:* yuzi
- **File:Text-x-generic.svg** *Source:* https://upload.wikimedia.org/wikipedia/commons/2/20/Text-x-generic.svg *License:* Public domain *Contributors:* The Tango! Desktop Project *Original artist:* The people from the Tango! project
- **File:Text_document_with_red_question_mark.svg** *Source:* https://upload.wikimedia.org/wikipedia/commons/a/a4/Text_document_with_red_question_mark.svg *License:* Public domain *Contributors:* Created by bdesham with Inkscape; based upon Text-x-generic.svg from the Tango project. *Original artist:* Benjamin D. Esham (bdesham)
- **File:The_Horse_in_Motion_high_res.jpg** *Source:* https://upload.wikimedia.org/wikipedia/commons/d/d2/The_Horse_in_Motion_high_res.jpg *License:* Public domain *Contributors:* Provided directly by Library of Congress Prints and Photographs Division *Original artist:* Eadweard Muybridge
- **File:The_Scientific_Method_as_an_Ongoing_Process.svg** *Source:* https://upload.wikimedia.org/wikipedia/commons/5/5c/The_Scientific_Method_as_an_Ongoing_Process.svg *License:* CC BY-SA 4.0 *Contributors:* Own work *Original artist:* ArchonMagnus
- **File:The_Scientific_Universe.png** *Source:* https://upload.wikimedia.org/wikipedia/commons/7/75/The_Scientific_Universe.png *License:* CC BY-SA 3.0 *Contributors:* Own work *Original artist:* Efbrazil
- **File:Thesaurus_opticus_Titelblatt.jpg** *Source:* https://upload.wikimedia.org/wikipedia/commons/c/c6/Thesaurus_opticus_Titelblatt.jpg *License:* Public domain *Contributors:* ? *Original artist:* ?
- **File:Three_models_of_theory_change.png** *Source:* https://upload.wikimedia.org/wikipedia/en/6/62/Three_models_of_theory_change.png *License:* Cc-by-sa-3.0 *Contributors:* ? *Original artist:* ?
- **File:Torus.png** *Source:* https://upload.wikimedia.org/wikipedia/commons/1/17/Torus.png *License:* Public domain *Contributors:* This image was created with POV-Ray *Original artist:* LucasVB
- **File:Triangle_illustration.svg** *Source:* https://upload.wikimedia.org/wikipedia/commons/4/45/Triangle_illustration.svg *License:* Public domain *Contributors:* self-made with en:Inkscape *Original artist:* Oleg Alexandrov
- **File:Trinity_Test_Fireball_25ms.jpg** *Source:* https://upload.wikimedia.org/wikipedia/commons/7/77/Trinity_Test_Fireball_25ms.jpg *License:* Public domain *Contributors:* http://www.atomicarchive.com/Photos/Trinity/image8.shtml *Original artist:* Courtesy of US Govt. Defense Threat Reduction Agency
- **File:Two_red_dice_01.svg** *Source:* https://upload.wikimedia.org/wikipedia/commons/3/36/Two_red_dice_01.svg *License:* CC0 *Contributors:* Open Clip Art Library *Original artist:* Stephen Silver
- **File:Tycho_instrument_sextant_mounting_19.jpg** *Source:* https://upload.wikimedia.org/wikipedia/commons/4/40/Tycho_instrument_sextant_mounting_19.jpg *License:* Public domain *Contributors:* ? *Original artist:* ?
- **File:Ulugh_Beg'{}s_Astronomic_Observatory.jpg** *Source:* https://upload.wikimedia.org/wikipedia/commons/7/7c/Ulugh_Beg%27s_Astronomic_Observatory.jpg *License:* GFDL *Contributors:* Own work *Original artist:* Igor Pinigin
- **File:Vector_field.svg** *Source:* https://upload.wikimedia.org/wikipedia/commons/c/c2/Vector_field.svg *License:* Public domain *Contributors:* Own work *Original artist:* Fibonacci.
- **File:Venn_A_intersect_B.svg** *Source:* https://upload.wikimedia.org/wikipedia/commons/6/6d/Venn_A_intersect_B.svg *License:* Public domain *Contributors:* Own work *Original artist:* Cepheus
- **File:Wang_tiles.svg** *Source:* https://upload.wikimedia.org/wikipedia/commons/f/f3/Wang_tiles.svg *License:* Public domain *Contributors:* Own work *Original artist:* Anomie
- **File:Wegener.jpg** *Source:* https://upload.wikimedia.org/wikipedia/commons/8/88/Wegener.jpg *License:* CC BY 2.0 *Contributors:* [1] *Original artist:* Ancheta Wis [2]
- **File:Wikibooks-logo-en-noslogan.svg** *Source:* https://upload.wikimedia.org/wikipedia/commons/d/df/Wikibooks-logo-en-noslogan.svg *License:* CC BY-SA 3.0 *Contributors:* Own work *Original artist:* User:Bastique, User:Ramac et al.
- **File:Wikibooks-logo.svg** *Source:* https://upload.wikimedia.org/wikipedia/commons/f/fa/Wikibooks-logo.svg *License:* CC BY-SA 3.0 *Contributors:* Own work *Original artist:* User:Bastique, User:Ramac et al.
- **File:Wikidata-logo.svg** *Source:* https://upload.wikimedia.org/wikipedia/commons/f/ff/Wikidata-logo.svg *License:* Public domain *Contributors:* Own work *Original artist:* User:Planemad
- **File:Wikinews-logo.svg** *Source:* https://upload.wikimedia.org/wikipedia/commons/2/24/Wikinews-logo.svg *License:* CC BY-SA 3.0 *Contributors:* This is a cropped version of Image:Wikinews-logo-en.png. *Original artist:* Vectorized by Simon 01:05, 2 August 2006 (UTC) Updated by Time3000 17 April 2007 to use official Wikinews colours and appear correctly on dark backgrounds. Originally uploaded by Simon.
- **File:Wikiquote-logo.svg** *Source:* https://upload.wikimedia.org/wikipedia/commons/f/fa/Wikiquote-logo.svg *License:* Public domain *Contributors:* Own work *Original artist:* Rei-artur

- **File:Wikisource-logo.svg** *Source:* https://upload.wikimedia.org/wikipedia/commons/4/4c/Wikisource-logo.svg *License:* CC BY-SA 3.0 *Contributors:* Rei-artur *Original artist:* Nicholas Moreau
- **File:Wikiversity-logo-Snorky.svg** *Source:* https://upload.wikimedia.org/wikipedia/commons/1/1b/Wikiversity-logo-en.svg *License:* CC BY-SA 3.0 *Contributors:* Own work *Original artist:* Snorky
- **File:Wikiversity-logo.svg** *Source:* https://upload.wikimedia.org/wikipedia/commons/9/91/Wikiversity-logo.svg *License:* CC BY-SA 3.0 *Contributors:* Snorky (optimized and cleaned up by verdy_p) *Original artist:* Snorky (optimized and cleaned up by verdy_p)
- **File:Wikivoyage-Logo-v3-icon.svg** *Source:* https://upload.wikimedia.org/wikipedia/commons/d/dd/Wikivoyage-Logo-v3-icon.svg *License:* CC BY-SA 3.0 *Contributors:* Own work *Original artist:* AleXXw
- **File:Wiktionary-logo-v2.svg** *Source:* https://upload.wikimedia.org/wikipedia/en/0/06/Wiktionary-logo-v2.svg *License:* CC-BY-SA-3.0 *Contributors:* ? *Original artist:* ?
- **File:Wundt-research-group.jpg** *Source:* https://upload.wikimedia.org/wikipedia/commons/a/a3/Wundt-research-group.jpg *License:* Public domain *Contributors:* the English language Wikipedia (log) *Original artist:* uploaded to Wikipedia by Kenosis

- Creative Commons Attribution-Share Alike 3.0